United Kingdom National Accounts Concepts, Sources and Methods

Editor: E A Doggett

1998 *Edition*

London:
The Stationery Office

Foreword

This volume is one of a pack of six books published to mark the extensive changes to the UK national accounts in September 1998.

The most significant of the changes are the adoption by the UK of the new European System of Accounts (ESA 95) and the transition of balance of payments data to the standards of the fifth edition of the IMF Balance of Payments Manual (BPM 5). The pack of six books is designed to provide all the information users will need to adjust to the new concepts and revised presentation.

The six books are:

- United Kingdom National Accounts - The Blue Book
- United Kingdom Balance of Payments - The Pink Book
- United Kingdom Input-Output Suppy and Use Balances, 1992-96

- United Kingdom National Accounts Concepts, Sources and Methods
- Introducing the European System of Accounts 1995 in the United Kingdom
- United Kingdom Sector Classification for the National Accounts

A supporting CD-ROM is also available:

- United Kingdom National Accounts, 1998

The books and CD-ROM are available through the Stationery Office (tel: 0171 873 9090).

The Office for National Statistics welcomes comments from users on any aspect of these publications via *na.info@ons.gov.uk* or through the specific contact point given below for this volume.

Stephen Penneck
Office for National Statistics
Room D3/04
1 Drummond Gate
London SW1V 2QQ
Fax: 0171 533 5937

Preface

Objectives of this book

The United Kingdom economic accounts are from this year published on the basis of the European System of Accounts 1995 (ESA95) which itself is based on the System of National Accounts 1993 (SNA93) sponsored by all major international organisations and being adopted worldwide. The European System is consistent with SNA93 but more specific and prescriptive in certain parts. The United Kingdom is producing ESA95-based accounts for use by the European Union from April 1999 and has adopted the system for domestic purposes with effect from the *Blue Book* published in September 1998. This will make international comparisons of United Kingdom and other countries' data much more straightforward.

United Kingdom National Accounts Concepts, Sources and Methods is intended to provide users with a guide to the structure and contents of the new system of accounts and the methods used to derive the figures. It identifies and describes the important changes in coverage and presentation in the new *Blue Book* compared with previous editions.

Format of the book

Concepts, Sources and Methods is in three parts:

*   Chapters 1 to 10 are an introduction to the concepts of the ESA. They are written in general terms and may be useful not only in interpreting the United Kingdom statistics but also in relation to other countries and for determining how to deal with new activities which might emerge in future.

*   Chapters 11 to 24 describe the United Kingdom's application of those principles with regard to such matters as the treatment of valuables, of illegal activity and of financial intermediation services indirectly measured. For some items the methodology has not yet been fully developed and the detail given here will be subject to change.

*   Chapters 25 and 26 contain information which will be useful as background to the accounts. Chapter 25 provides a tabulation of the main statistical sources used while Chapter 26 gives a chronology of events which may have had an impact on the accounts. Besides assisting with interpretation these may be useful in their own right for reference purposes.

The book concludes with a glossary, bibliography and index.

Credits

The radical change which has taken place in the national accounts is a massive exercise in itself which has placed enormous demands on the ONS's staff of national accountants. This edition of *Concepts, Sources and Methods*, in common with the Initial Draft released in 1997, has therefore been written and edited mainly by people from outside the national accounts group. The chapters were written by Ken Mansell, Martin Hargreaves, Adèle Barklem, Denise Whitting, Henry Neuburger, Helen Shanks and Ted Doggett and the project team consisted of Henry Neuburger, Graham Jenkinson, Margaret Dolling, David Blunt, Martin Hargreaves and Callum Foster. However, the ultimate source of detailed knowledge about the accounts is necessarily the national accounts compilers. We are grateful to all those who, despite all the pressures on them, found time to provide information, read and correct drafts.

Ted Doggett

Contents

Methods used in the UK accounts

Sources

Chapter 1

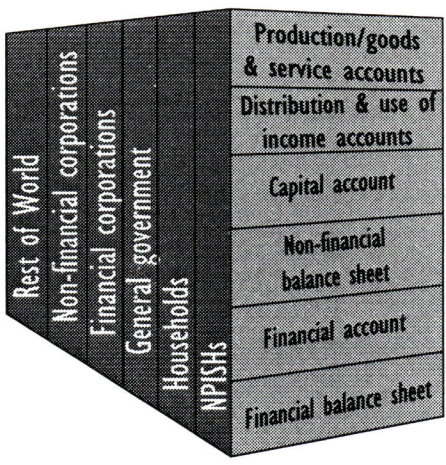

The labels within the image read: Rest of World, Non-financial corporations, Financial corporations, General government, Households, NPISHs; Production/goods & service accounts, Distribution & use of income accounts, Capital account, Non-financial balance sheet, Financial account, Financial balance sheet

The Economy and its Measurement

Chapter 1 The economy and its measurement

1.1   Economic accounts for the United Kingdom and its regions are designed to tell us about production, how effective it is and what it is for. Accounts can help to measure progress and compare different parts of the United Kingdom and the World. So what is progress, and why do we want to know about it?

1.2   The present system of measurement is about fifty years old but economic measurement is far older. There is data about production and prices over at least five hundred years prior to the development of the first economic accounts in the United States and the United Kingdom. The innovation of the founders of economic accounts was to provide a framework in which a range of measurements could be brought together to ensure systematic coverage. When Rita Maurice began her classic exposition of the economic accounts she felt no need for justification or explanation. She began with the accounting identities which form the core insights of economic accounting

1.3   *'The national income is a measure of the money value of goods and services becoming available to the nation from economic activity. It can be regarded in three ways: as a sum of incomes derived from economic activity, which can be broadly divided, for example, between income from employment and incomes from profits; as a sum of expenditure, being that between consumption and adding to wealth (or investment); or as a sum of the products of various industries of the nation.'*
    (National Accounts Statistics: Sources and Methods, 2nd Edition, HMSO, 1968)

1.4   Since she wrote, economies and the way they are measured have changed a great deal. Nonetheless, the fundamental insights remain the same. What has perhaps changed is the status of the economic accounts and the way in which they are seen. The occasion for the present edition, *Concepts, Sources and Methods*, is, as explained in the introduction, the adoption of new international standards by the European Union and other international agencies. Between the 1968 and the 1993 editions of the United Nations *System of National Accounts* there has developed a substantial literature seeking to extend economic accounts in various ways. It has no longer been seen as sufficient to assess the money value of goods and services becoming available to the nation from economic activity. It has been necessary to explain how these relate to people's understanding of the economy and the concerns of policy makers. The provisions made in the 1993 *SNA* for such elaborations and extension are introduced in the sections of this chapter on social accounting matrices and satellite accounts. This chapter sets out how the new system describes the traditional core of economic activity.

Why would you read economic accounts?

1.5   Very few people do. But nearly anyone who takes any interest in economic affairs uses them. Economic accounts provide the integrating framework into which nearly all indicators of economic performance can be fitted. Most famously, gross domestic product (GDP) – supplanting national income as used by Rita Maurice – provides the single measure which encapsulates what we want to know about how much the economy is and can deliver. As is inevitable in a portmanteau measure it represents a range of aspects loosely, but its pre-eminence as a measure of the economy has probably grown as more and more countries have contrived to produce it. But the role of economic accounts is more than the generation of single summary measures; it is the basis of assurance that different indicators of performance tell, as far as possible, a coherent story. Most people do not need to be familiar with such bases, but they are nonetheless dependent on the structural coherence that the accounts provide. In the same way this is not a book that many people will read from cover to cover, just as most people will not read cookery books from cover to cover. Most users of cookery books use them as works of reference, to be opened when they want to perform a particular culinary act. But, just as some people look for greater understanding of food through reference to cookery books without having an immediate meal in mind, so it is hoped this book will be useful to those who wish to gain a better feel for the economy and its measurement.

What would you expect to find there?

1.6   The UN *System of National Accounts (SNA)*, and more specifically the *European System of Accounts (ESA)* which forms the basis for the United Kingdom economic accounts, consist of a coherent, consistent and integrated set of accounts and balance sheets based on a set of internationally agreed concepts, definitions, classifications and accounting rules. The accounts are compiled for a succession of time periods and cover economic activities, the economy's productive assets and the wealth of its inhabitants. They include an external account that displays the links between an economy and the Rest of the World.

1.7   For the purpose of the economic accounts countries are divided into basic institutional units (for example households or companies). These are grouped into sectors. A sector is a group within which all the units tend to act (from the viewpoint of economic analysis) in a similar way (for example households, and government). Accounts are prepared for the whole economy and for these sectors. The System is built around a sequence of interconnected flow accounts together with balance sheets that record the values of the stocks of assets and liabilities held by institutional units or sectors at the beginning and end of the period. Each flow account relates to a particular kind of activity such as production, or the generation, distribution, redistribution or use of income. Each account is balanced by introducing a balancing item, the difference between the total resources and uses recorded on the two sides of the account. The balancing item from one account is carried forward as the first item in the following account, thereby making the sequence of accounts an articulated whole. The balancing items, like value added, disposable income and saving, are of particular interest as summary indicators. The carry forward of balances also links the flow accounts and the balance sheets, as all the changes affecting assets or liabilities held by institutional units or sectors are systematically recorded in one or another of the flow accounts. The closing balance sheet reflects the opening balance sheet and the transactions or other flows recorded in the sequence of accounts.

1.8   Most of the accounts described in this chapter are described with no reference to their price base. As everyone reading this chapter will know, general inflation means that measuring flows of money is not equivalent to flows of the quantity of goods. To get an idea of the scale and economic significance of any transactions, and in particular to make comparisons over time and place, adjustments need to be made for different price levels. The method of doing this, using index numbers of prices and output and creating spending figures in the prices of a base year, is described in Chapter 2. These constant-price figures, particularly GDP and its expenditure and output components, represent some of the most important and widely discussed indicators of economic activity. Current-price GDP and gross national product (GNP) also play a significant role. The former provides a useful scalar; the criteria for convergence in the European Union (EU), for example, use it as the denominator against which government debt and deficit are compared. It is also used in appraising pressure, such as the velocity of circulation of money (the ratio of money stock to GDP). GNP is used as one of the tax bases for various international organisations, most importantly the EU.

The sequence of accounts

1.9   **Current accounts** record the production of goods and services, the generation of incomes by production, the subsequent distribution and redistribution of incomes among institutional units and the use of incomes for purposes of consumption or saving.

1.10   The production account records the output of the economy and the materials used up (called intermediate consumption). Its balancing item is gross value added. Subtracting capital goods used up (depreciation) we arrive at net value added. Gross value added is a measure of the contribution to GDP made by an individual producer, industry or sector, and primary incomes are paid out of it; so it is carried forward into the distribution of income accounts.

1.11   Distribution and use of income accounts show how incomes are:

- Generated by production.

- Distributed to institutional units with claims on the *value added* created by production. This shows income from employment (wages and salaries, employer contributions to pension schemes), operating surplus (profit) and mixed income (mostly self-employment income).

- Redistributed among institutional units, mainly by government units through social security contributions and benefits and taxes. This shows how taxes on income and transfers made by governments and overseas modify the original distribution.

- Eventually used by households, government units or non-profit institutions serving households (NPISH) for purposes of final consumption or saving. This shows how some activities of collective consumption like health spending then accrue to people in households.

1.12   The balancing item emerging from the complete set of the distribution and use of income accounts is saving. The accounts show how the final consumers – individuals, non-profit institutions and governments – use products to satisfy individual and collective needs and wants of households and the community. The balancing item, saving, is carried forward into the capital account, the first in the System's sequence of accumulation accounts.

1.13   **Accumulation accounts** like current accounts are flow accounts that record the acquisition and disposal of financial and non-financial assets and liabilities by institutions mostly through transactions. The capital account records acquisitions and disposals of non-financial assets as a result of transactions with other units or internal bookkeeping transactions linked to production (changes in inventories and consumption of fixed capital). The financial account records acquisitions and disposals of financial assets and liabilities, also through transactions.

1.14   A third account, the *other changes in assets account*, consists of two sub-accounts. The first, the *other changes in volume of assets account*, records changes in the amounts of the assets and liabilities held by institutional units or sectors as a result of factors other than transactions; for example, destruction of fixed assets by natural disasters. The second, the *revaluation account*, records those changes in the values of assets and liabilities that result from changes in their prices.

1.15   The link between the accumulation accounts and the income accounts is provided by the fact that saving is used to acquire financial or non-financial assets of one kind or another. When saving is negative the excess of consumption over disposable income must be financed by disposing of assets or incurring liabilities. The financial account shows the way in which funds are channelled from one group of units to another, especially through financial intermediaries.

1.16   The balance sheets show the values of the stocks of assets and liabilities held by institutional units or sectors at the beginning and end of an accounting period. The values of the assets and liabilities held at any moment in time vary automatically whenever any transactions, price changes or other changes affecting the volume of assets or liabilities held take place. These are all recorded in one or another of the accumulation accounts so that the difference between the values in the opening and closing balance sheets is entirely accounted for within the System, provided of course that the assets and liabilities recorded in the balance sheets are valued consistently with the transactions and other changes.

1.17   In addition to the flow accounts and balance sheets described earlier the central framework of the System also contains detailed *supply and use* tables in the form of matrices that record how supplies of different kinds of goods and services originate from domestic industries and imports and how those supplies are allocated between various intermediate or final uses, including exports. These tables involve the compilation of a set of integrated production and generation of income accounts for industries; that is, groups of establishments (for example plants) as distinct from institutional units (for example corporations). The *supply and use* tables provide an accounting framework within which the commodity flow method of compiling economic accounts, in which the total supplies and uses of individual types of goods and services have to be balanced with each other, can be systematically exploited. Only a very summary set of tables for industrial sectors is published with the economic accounts. The more detailed tables which lie at the core of the annual balancing processes are published separately. The *supply and use* tables also provide the basic information for the derivation of detailed *input-output* tables that can be used for purposes of economic analysis and projections.

Activities and transactions

1.18   In economic accounts activity is measured by transactions. While there are accounts which directly measure work or the flow of materials, such accounts can only compare a limited range of activities or particular aspects of them. For example, the accounts do not record the eating of food or the burning of fuel. Instead they record the purchase of food or fuel on the assumption that most people use them in the same way. Most transactions involve the expenditure of money but some, like consumption of owner-occupied housing, impute such a flow.

1.19   The use of transactions data has important advantages. A transaction that takes place between two different institutional units has to be recorded for both parties to the transaction and therefore generally appears twice in a system of macroeconomic accounts. This enables important linkages to be established. For example, output is obtained by summing the amounts sold, bartered or transferred to other units *plus* the amounts entered into *less* the amounts withdrawn from inventories. In effect the value of output can be obtained by recording the various uses of that output by means of data on transactions. In this way flows of goods and services can be traced through the economic system from their producers to their eventual users. Some transactions are only internal book-keeping transactions that are needed when a single unit engages in two activities, such as the production and consumption of the same good or service, but the great majority of transactions take place between different units in markets.

The institutional sectors of the economy

1.20   The System identifies two kinds of institution, consuming units – mostly households – and production units, mainly corporations and non-profit institutions (NPIs) or government units. Units can own goods and assets, incurring liabilities and engaging in economic activities and transactions with other units in their own right. All units within the country are put in one of five sectors. Also, the Rest of the World is treated as a sector in respect of its dealings with the United Kingdom.

1.21   **Financial corporations** are those engaged primarily in financial activities such as banking and insurance.

1.22   **Non-financial corporations** are those which exist to produce goods and non-financial services. They are, in the United Kingdom, mainly public limited companies, private companies and partnerships. They are mostly owned privately but there are some public corporations.

1.23   **General government** comprises central government and local authorities. They produce public services, some of which are transferred to households, as well as making cash transfers. They also invest in public assets. Their activities are funded by levying taxes and selling financial instruments.

1.24   **Non profit institutions serving households (NPISH).** These include productive units such as charities and universities. In the former system they were part of the personal sector.

1.25   The **household** sector contains all the resident people of the United Kingdom as receivers of income and consumers of products. It includes those in institutions such as prisons and hospitals as well as conventional family units. It also contains one-person businesses where household and businesses accounts cannot be separated.

1.26   Units not in the country are allocated to the **Rest of the World** sector. No specific account is prepared for them but for many purposes their account is equivalent to the balance of payments.

1.27   Some *sub*-sectors are also identified and some accounts prepared for these. Accounts as described in the previous section are prepared for each sector and the complete set of accounts at the level of the five main sectors is shown in the annual *Blue Book*.

1.28   Some accounts in some sectors take on particular significance. In the corporate sector accounts, perhaps surprisingly, there are not many economic accounts magnitudes which correspond to widely-used indicators. The current accounts for the corporate sectors differ significantly from those with which readers of company accounts would be familiar, mainly because companies account net interest received as profit while the economic accounts in general treat interest flows as redistribution of primary income; that is, property income, not profit. Economic accounts distinguish between profit and holding gains (stock appreciation) so that profit is calculated on a replacement cost basis. Not all companies do this.

1.29   Government accounts are more familiar. While the balance on current account is of no great significance the balance on the accumulation account is the financial deficit. The liabilities side of the balance sheet shows the Government's debt. These two are used as measures of convergence for European Monetary Union. The public sector net cash requirement (PSNCR) can be defined in terms of financial transactions but is not an ESA concept.

1.30   In previous versions of economic accounts the personal sector, at least as far as current and accumulation accounts were concerned, contained both of what are in the ESA the household and NPISH sectors. The level of personal disposable income, saving and the saving ratio as traditionally viewed are now represented by the balance on the current accounts (after secondary distribution) of both of these sectors added together. The new system will also make possible the measurement of saving ratios, more specifically for the household sector.

1.31   The balances on the Rest of the World accounts are of particular significance. The balance of exports and imports of goods is known as the balance of trade in goods, the balance on current account is known as the current balance of payments.

Categories of spending

1.32   Economic accounts identify several types of expenditure. The main items are current consumption, capital spending and financial transactions. Consumption uses up goods and services. There are two different kinds of consumption: intermediate consumption consists of inputs of production, final consumption consists of goods and services used by individual households or the community to satisfy their current individual or collective needs or wants. Capital spending, or gross capital formation, divides into gross fixed capital formation and inventories. Acquisition is restricted to institutional units in their capacity as producers, being defined as the value of their acquisitions *less* disposals of fixed assets. Fixed assets are produced assets (both tangible like computers and intangible like software) that are used repeatedly or continuously in production over several years. Inventories consist of materials awaiting use, productive processing or sale. The net change in these is an element of expenditure. In order to bring expenditure into line with production two other elements of expenditure need to be mentioned, trade and adjustment to basic prices. Clearly it is possible for a country to spend more than it produces by buying from abroad. Therefore exports are recorded positively and imports negatively in expenditure. All expenditures are recorded at purchasers' prices, that is inclusive of taxes on products like VAT and exclusive of subsidies on products. (These terms are defined later in the book.) Production and income, however, are recorded at basic prices which exclude these taxes and subsidies. These net product taxes are added back to balance the GDP accounts at purchasers' prices (unlike the former accounts which balanced at factor cost).

1.33   The distinction between consumption and capital formation is fundamental for economic analysis and policy-making. The balance between consumption and investment is a key indicator of the future orientation of an economy. Nevertheless, the borderline between consumption and gross fixed capital formation is not always easy to determine in practice. Certain activities contain some elements that appear to be consumption and at the same time others that appear to be capital formation. In order to try to ensure that the *SNA* was implemented in a uniform way decisions were taken about the ways in which certain difficult, even controversial, items are to be classified. The conventions adopted are described in Chapter 2 (2.27–2.33).

What are economic accounts useful for?

Monitoring the behaviour of the economy

1.34   Economic accounts make it possible to monitor the movements of major economic flows such as production, household consumption, government consumption, capital formation, exports, imports, wages, profits, taxes, lending, borrowing, the budget surplus or deficit, the share of income which is saved, or invested, by individual sectors of the economy or the economy as a whole, the trade balance, etc. Like most countries the United Kingdom system of accounts provides quarterly estimates while many short-term indicators are monthly.

1.35   Economic accounts are also used to investigate the causal mechanisms at work within an economy. Such analysis usually takes the form of the estimation of the parameters of functional relationships between different economic variables by applying econometric methods to time series of economic accounts data. The types of macroeconomic model used for such investigations may vary according to the school of economic thought of the investigator as well as the objectives of the analysis, but the System is sufficiently flexible to accommodate the requirements of different economic theories or models provided only that they accept the basic concepts of production, consumption, income, etc, on which the System is based.

Macroeconomic policy

1.36   Economic policy in the United Kingdom is formulated on the basis of an assessment of the recent behaviour and current state of the economy and a forecast of likely future developments. Forecasts are typically made using econometric models of the type just described. Over the medium- or long-term economic policy has to be formulated in the context of a broad economic strategy which may need to be quantified in the form of a plan. A new framework for macroeconomic policy was announced by the Chancellor of the Exchequer on June 12th 1997. This is overviewed by a new policy committee.

International comparisons

1.37   The *SNA* is the system used for reporting to international or supranational organisations economic accounts data that conform to standard, internationally-accepted concepts, definitions and classifications. The resulting data are widely used for international comparisons of the volumes of major aggregates, such as GDP or GDP per head, and also for comparisons of structural statistics, such as ratios of investment, taxes or government expenditures to GDP. Such comparisons are used by economists, journalists and other analysts to evaluate the performance of one economy against that of other similar economies. They can influence popular and political judgements about the relative success of economic programmes in the same way as developments over time within a single country. Databases consisting of sets of economic accounts for groups of countries can also be used for econometric analyses in which time-series and cross-section data are pooled to provide a broader range of observations for the estimation of functional relationships.

1.38   Levels of GDP or, alternatively, gross national income (GNI) per head in different countries are also used by international organisations to determine eligibility for loans, aid or other funds or to determine the terms or conditions on which such loans, aid or funds are made available. When the objective is to compare the volumes of goods or services produced or consumed per head, data in national currencies must be converted into a common currency by means of purchasing power parities and not exchange rates. (See *Economic Trends*, July 1994.) In general neither market nor fixed exchange rates reflect the relative internal purchasing powers of different currencies. Exchange rate converted data cannot be interpreted as measures of the relative volumes of goods and services concerned.

Links with business accounting and economic theory

1.39   The accounting rules and procedures used in the System are based on those long used in business accounting. The traditional double-entry book-keeping principle, whereby a transaction gives rise to a pair of matching debit and credit entries within the accounts of each of the two parties to the transaction, is a basic axiom of economic accounting. For example, recording the sale of output requires not only an entry in the production account of the seller but also an entry of equal value, often described as the counterpart, in the seller's financial account to record the cash, or short-term financial credit, received in exchange for the output sold. As two entries are also needed for the buyer the transaction must give rise to four simultaneous entries of equal value in a system of macroeconomic accounts covering both the seller and the buyer. In general a transaction between two different institutional units always requires four equal, simultaneous entries in the accounts of the System – i.e. quadruple entry accounting – even if the transaction is a transfer and not an exchange and even if no money changes hands. These multiple entries enable the economic interactions between different institutional units and sectors to be recorded and analysed.

1.40   The design and structure of the System draw heavily on economic theory and principles as well as business accounting practices. Basic concepts such as production, consumption and capital formation are meant to be rooted in economic theory. A good example is the economic concept of opportunity cost.

1.41   Business accounts often record costs on an historic basis, partly to ensure that they are completely objective. Historic cost accounting requires goods or assets used in production to be valued by the expenditures actually incurred to acquire those goods or assets, however far back in the past those expenditures took place. In the SNA, however, the concept of opportunity cost as defined in economics is employed. In other words, the cost of using, or using up, some existing asset or good in one particular process of production is measured by the amount of the benefits that could have been secured by using the asset or good in alternative ways. Opportunity cost is calculated with reference to the opportunities foregone at the time the asset or resource is used, as distinct from the costs incurred at some time in the past to acquire the asset. As an approximation to opportunity cost accounting the economic accounts seek to ensure that assets and goods used in production are valued at their actual or estimated current market prices at the time the production takes place. However there is no uniquely correct way of achieving this. The accounting profession issued five sets of proposals during the decade when inflation was high and variable. Following the report of the Inflation Accounting (Sandilands) Committee in 1975 one set of proposals reached the status of a Statement of Standard Accounting Practice (SSAP16) in 1980. However, this was mandatory for only a short time and interest in accounting for inflation has since declined with lower rates of inflation.

1.42   When there is persistent inflation, even moderate inflation, the use of historic costs tends to underestimate the opportunity costs of production in an economic sense so that historic cost profit may be much greater than the operating surplus as defined in the System. Profits at historic costs can give very misleading signals as to the profitability of the production processes.

1.43   Measuring consumption of fixed capital at current costs is equivalent to measuring the operating surplus from production after deducting the costs of maintaining intact the stock of fixed assets used in production – that is, after deducting the costs of replacing assets used up in production. Even when the fixed assets used up are not actually replaced the amount of consumption of fixed capital charged as a cost of production should be sufficient to enable the assets to be replaced if desired.

1.44   Similarly, the concept of disposable income used is based on the underlying idea that it represents the maximum amount available to a household for purposes of consumption after maintaining its net worth intact, i.e. its assets *minus* its liabilities valued at current prices. However, the System **excludes** from the calculation of income:

- any assets received or disposed of as a result of capital transfers that merely redistribute wealth between different units,

- any assets received or disposed of as a result of 'other volume changes' as described,

- holding gains or losses on assets or liabilities due to changes in their prices.

At a macro level the aggregate income of a group of units is not changed by redistributing wealth within the group. Other volume changes and holding gains or losses are recorded in the accumulation accounts of the units concerned and not in their income accounts.

Micro-macro links

1.45   Accounts and balances could, in principle, be compiled at any level of aggregation, even that of an individual institutional unit. It might therefore appear desirable for the macroeconomic accounts for sectors or the total economy to be obtained directly by aggregating corresponding data for individual units. There would be considerable analytical advantages in having micro-databases that were fully compatible with the corresponding macroeconomic accounts for sectors or the total economy. Data in the form of aggregates, or averages, often conceal a great deal of useful information about changes occurring within the populations to which they relate.

1.46   In practice, however, macroeconomic accounts can seldom be built up by simply aggregating the relevant micro-data. Even when individual institutional units keep accounts or records the concepts that are needed or appropriate at a micro level may not be suitable at a macro level. Individual units may be obliged to use concepts designed for other purposes, such as for taxation. The accounting conventions and valuation methods used at a micro level typically differ from those required by the System. For example, as already noted, the widespread use of historic cost accounting means that the accounts of individual enterprises may differ significantly from those used in the System. In such situations it is impracticable to try to adjust the individual accounts of thousands of enterprises before aggregating them. It may be much easier to adjust the data after they have been aggregated to some extent. Similarly imputed rent on owner-occupied housing does not feature in the accounts of most households. So aggregation from microdata is rarely simple or straightforward.

The aggregates of the system as indicators of economic activity and welfare

1.47   The *SNA* consists of a set of accounts and tables designed for a variety of analytical and policy purposes. Nevertheless, certain key aggregates of the System, such as GDP and GDP per head of population, have acquired an identity of their own and are widely used by analysts, politicians, the press, the business community and the public at large as summary, global indicators of economic activity and welfare. Movements of such aggregates, and their associated price and volume measures, are used to evaluate the overall performance of the economy and hence to judge the relative success or failure of economic policies pursued by governments.

Capacity utilisation

1.48   GDP is a measure of production. The level of production is important because it largely determines how much a country can afford to use and it also affects the level of employment. One of the earliest documents in United Kingdom economic accounts was Lord Keynes's *How to Pay for the War*. It was designed to assess the resources available to the UK to pursue the Second World War. Economic accounts play a similar role in the analysis of UK economic policy now. Movements in GDP relative to its long-term trend are taken as an indication of pressure on resources and thus on inflation.

Welfare

1.49   Consumption of goods and services, both individually and collectively, is one of the most important factors influencing the welfare of a community, but it is only one of several factors. Transactions and production are not the same thing. Products can be supplied without being sold and some are kept by the producers for their own use. Such production needs to be represented by an imputed transaction. The System does not record all outputs, however, because domestic and personal services produced and consumed by members of the same household are omitted. Subject to this one major exception, GDP is intended to be a comprehensive measure of the total gross value added in production by all resident institutional units.

1.50   In a market economy the prices used to value different goods and services should reflect not only their relative costs of production but also the relative benefits to be derived from using them for production or consumption. This establishes the link between changes in aggregate production and consumption and changes in welfare. However, changes in the volume of consumption, for example, are not the same as changes in welfare. The distinction between the quantity of some good or service and the welfare derived from consuming it is clear enough at the level of an individual good or service. For example, the quantity of sugar consumed by households is measured in physical units. It is not the same as the welfare it produces.

1.51   Welfare depends on many other factors besides the amounts of goods and services consumed. Apart from natural events such as epidemics, droughts or floods, welfare also depends on political factors such as freedom and security. Obviously, as a measure of production GDP is not intended to embrace non-economic events such as political revolutions, wars, natural disasters or epidemics.

1.52   Consider the effects of an exceptionally severe winter combined with an influenza epidemic. Other things being equal the production and consumption of a number of goods and services may be expected to rise in response to extra demands created by the cold and the epidemic; the production and consumption of fuels, clothing and medical services will tend to increase. As compared with the previous year people may consider themselves to be worse off overall because of the exceptionally bad weather and the epidemic, notwithstanding the fact that production and consumption may have increased in response to the additional demand for heating and health services. Total welfare could fall even though GDP increased in volume terms.

1.53   This kind of situation does not mean that welfare cannot be expected to increase as GDP increases, other things being equal. Given the occurrence of the cold and the epidemic the community presumably finds itself much better off with the extra production and consumption of heating and health services than without them. There may even be a general tendency for production to rise to remedy the harmful effects of events that reduce people's welfare in a broad sense. For example, production may be expected to increase in order to repair the damage caused by wars or such natural disasters as earthquakes, hurricanes and floods. Given that the disaster has occurred the extra production presumably increases welfare, but obviously not up to the level it would have been without the disaster.

1.54   Similar considerations arise with respect to so-called 'regrettable necessities' in general. When production and consumption increase in order to compensate for the loss of welfare created by damage or 'bads' that did not previously exist, the community may be no better off than if the damage had not occurred. However, this should not be allowed to obscure the fact that without the extra production and consumption the community would actually be worse off still. The extra production and consumption, in itself, actually increases welfare. Goods and services are consumed by households to satisfy their needs and wants. Some of these needs or wants may be created or increased by factors or events over which households have little or no control and which they may resent – bad weather, natural disasters, pollution, etc. – but this in no way diminishes the fact that they do derive benefits from consuming the goods and services in question. Quite ordinary consumer goods such as food and drink could be characterised as 'regrettable necessities' which merely satisfy the recurrent basic needs of hunger and thirst without leaving the individuals any better off than before the onset of the hunger and thirst. Pushed to its logical conclusion, scarcely any consumption improves welfare in this line of argument. Later in this chapter we shall discuss how the use of satellite accounts can improve the role of economic accounts in measuring welfare.

Introducing social accounting matrices

1.55   A social accounting matrix (SAM) is a presentation of accounts in matrix terms that incorporates whatever degree of detail is of special interest. By arranging tables which show payers on the columns and payees in rows transactions are shown not twice, as in double entry, but only once. This allows a more compact presentation but one which lends itself less readily to time series. To date, builders of SAMs have exploited the available flexibility to highlight special interests and concerns more than compilers of regular economic accounts, displaying the interconnections, disaggregating the household sector, showing the link between income generation and consumption, etc. In the United Kingdom the sources of information used for such disaggregations are often not those used for compiling the accounts or, if they are, are modified before use. Consequently the development of a SAM for the United Kingdom involves a great deal of reconciliation work.

Introducing satellite accounts

1.56   In some cases working with the central framework, even in a flexible way, is not sufficient. Even when conceptually consistent the central framework could be overburdened with details. Moreover, some requirements may conflict with the central conceptual framework and its architecture. In the United Kingdom we make a distinction between internal and external satellites. Internal satellites involve a focus on particular sub-aggregates and may lead to the modification of the core accounting concepts. An external satellite will typically involve activity beyond the production or asset boundary. It may use measures in natural units like tonnage rather than money measures.

1.57   Tourism is a good example of an internal satellite. Various aspects of producing and consuming activities connected with tourism may appear in detailed classifications of activities, products and purposes. However, specific tourism transactions and purposes appear separately only in a few cases. In order to describe and measure tourism in a economic accounts framework it is necessary to make a choice between two approaches: either subdivide many elements in the accounts of the central framework to get the required figures for tourism and pay the price of overburdening and unbalancing the various components of the accounts *or* elaborate a specific framework for tourism. The latter approach, the only feasible one actually, also allows adaptation of the various classifications and measurement of additional aggregates, such as national expenditure on tourism, which may cover intermediate as well as final consumption.

1.58   Research and development is another possible topic for a mainly internal satellite. It would rearrange transactions within the production boundary so as to give a presentation which treats R&D as capital formation rather than intermediate expenditure. If this were done in the core accounts it would increase GDP and capital stock would be larger and very hard to measure. Preparing a satellite enables the features of R&D to be addressed without disturbing the overall continuity and comparability of the accounts.

1.59   Environmental accounts present an example of an external satellite where the pollution generated by economic activity within the production boundary itself falls outside it. The same is true of the depletion of natural raw materials. Pollutants are typically measured in physical units and no attempt is made to monetise them or produce a 'green' GDP.

Geographical coverage

1.60   The economic accounts are prepared for the United Kingdom. Although the *SNA* makes provision for regional accounts such accounts as are available for the United Kingdom show only a limited range of the accounts prepared at the United Kingdom level. This range is mainly limited to the production and distribution of income accounts. Some elements of accumulation accounts are shown. The 'regions' are Scotland, Wales, Northern Ireland and the standard regions known as government office regions. More limited still are the figures published for counties. (*See* December 1995, March 1996 and May 1996 editions of *Economic Trends*.) There is no adjustment for price differences over time or between regions. While it is possible to identify most production and some items of expenditure by region, much of government spending and international trade is not allocable to regions.

Chapter 2

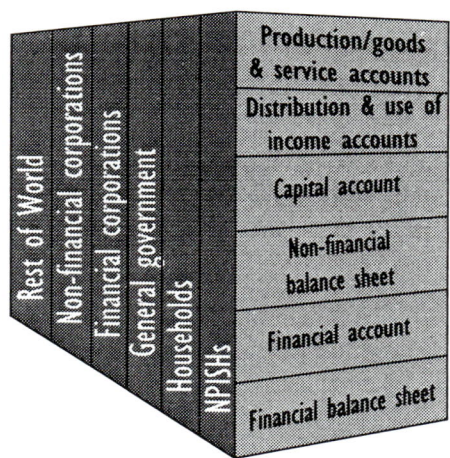

Basic Concepts

Chapter 2 Basic concepts

2.1   Any system of accounts requires rules so that people using them can understand what compilers have done. The aim of this chapter is to introduce some of the more important of these. The first section looks at what is included and what is excluded – the production boundary. It includes a discussion of the illegal and hidden economies and of the output of some special sectors like finance. The second section looks at accounting rules like double entry, time of recording, valuation and consolidation, and market and non-market transactions. The final section introduces price and volume changes.

What is production?

2.2   Economic production is carried out using inputs of labour, capital, and goods and services to produce outputs of goods or services. There must be an institutional unit that assumes responsibility for the process and owns any goods produced as outputs or is entitled to be paid for them. A wholly natural process without any human involvement or direction is not production in an economic sense. For example, the unmanaged growth of fish is not production, whereas fish farming *is* production.

2.3   While production of goods can be identified without difficulty it is not always so easy to distinguish the production of services from other activities that may be both important and beneficial. Activities that are not productive in an economic sense include basic human activities such as eating, drinking, sleeping, taking exercise, etc, that it is impossible for one person to do for another. Paying someone else to take exercise is no way to keep fit. On the other hand activities such as washing, preparing meals, caring for children, the sick or aged are all activities that can be provided by others and, therefore, fall within the general production boundary. Many households employ paid domestic staff to carry out these activities for them.

2.4   The production boundary in the economic accounts is more restricted than the general production boundary. Production accounts are not compiled within the economic accounts for work within the home unless someone is paid to do it. So the core economic accounts include all economic activity by units other than households *plus* production of goods and services by self-employed members of the household, the production of paid household staff and the use of owner-occupied housing. Although the *SNA* makes provision for own-account production of goods (production by the household to be consumed within the household) it is designed for economies where such activity is significant, not the UK. Household production of services may be accounted for within satellite accounts.

2.5   The production of housing services for their own final consumption by owner-occupiers has always been included within the production boundary in economic accounts, although it constitutes an exception to the general exclusion of own-account service production. The ratio of owner- occupied to rented dwellings can vary significantly between countries and even over short periods of time within a single country, so that both international and intertemporal comparisons of the production and consumption of housing services could be distorted if no imputation were made for the value of own-account housing services. The imputed value of the income generated by such production is taxed in some countries and used to be taxed in Britain. While 'do-it-yourself' activity should in principle be included in production, in practice it is regarded as too hard to measure and is excluded by international convention.

Illegal production

2.6   One of the most important changes introduced by the ESA 95 is the requirement that illegal production be included in measured activity. Previous international standards had been silent on the subject. The result was not their exclusion but a lack of clarity as to whether or not they were included. Their inclusion will not necessarily imply their explicit measurement in all cases but a reassurance that allowance has been made for them. Setting an explicit standard ensures that production comparisons between different countries or between time periods in the same country are less distorted by differences in the legal status of certain activities. Thus the United Kingdom economic accounts probably recorded a rise in the 1950s in economic activity when off-course betting on horses was made legal, though there was informal evidence that much of this apparent increase in activity had been going on illegally before. The illegality of an activity clearly adds greatly to the difficulty of measuring it. Consequently, while there will be an obligation to include illegal activities in accounts ultimately, the accounts currently exclude such activity.

2.7   There are two kinds of illegal production:

(a)   the production of goods like drugs or services like prostitution whose sale, distribution or possession is forbidden by law;

(b)   production activities which are usually legal but which become illegal when carried out by unauthorized producers, e.g. unlicensed casinos.

2.8   Both kinds of production are included within the production boundary provided their outputs consist of goods or services for which there is an effective market demand. Production must involve willing parties. Criminal activity in which there are unwilling victims such as theft or extortion are not productive. Thus the infliction of pollution whether legal or not is excluded. It is accounted for in environmental accounts (*see* paragraph 1.62). The units which purchase such outputs may not be involved in any kind of illegal activity other than the illegal transactions themselves. Transactions in which illegal goods or services are bought and sold need to be recorded not simply to obtain comprehensive measures of production and consumption but also to prevent errors appearing elsewhere in the accounts if the funds exchanged in illegal transactions are presumed to be used for other purposes. The incomes generated by illegal production may be disposed of quite legally while, conversely, expenditures on illegal goods and services may be made out of funds obtained quite legally.

Concealed production and the underground economy

2.9   Certain activities may be both productive in an economic sense and also, in themselves, quite legal (provided certain standards or regulations are complied with) but deliberately and possibly illegally concealed from public authorities for the following kinds of reason:

(a)   To avoid the payment of income, value added or other taxes;

(b)   To avoid the payment of social security contributions;

(c)   To avoid having to meet certain legal standards such as minimum wages, maximum hours, safety or health standards, etc.;

(d)   To avoid complying with certain administrative procedures.

2.10   Such activities fall within the production boundary. Producers engaged in this type of production may be described as belonging to the 'underground economy'. The underground economy is more common in certain industries – for example, construction. There may be no clear borderline between the underground economy and illegal production. For example, production which does not comply with certain safety, health or other standards could be described as illegal. Similarly, the evasion of taxes is itself usually a criminal offence. However, it is not now necessary to try to fix the precise borderline between underground and illegal production as both are included within the production boundary in any case.

2.11   Because certain kinds of producer try to conceal their activities from public authorities it does not follow that they are not included in economic accounts in practice. The UK economic accounts have for many years contained a range of adjustments to allow for under-recording of activity arising from concealment of various kinds from the surveys and official sources used. It is important to realise that what is hidden from the tax authorities, for example, is often not hidden from the national accountant. It is also important to remember that income concealed from tax authorities may involve transfer or property income and so under-recording would not reduce GDP. Tax gatherers have only one source of information while national accountants have many; neither does revelation carry the same costs. In the UK with its system of GDP measurement based on three different concepts it is possible to estimate the hidden economy by comparing imbalances, e.g. the overt spending money earned from concealed activity.

Special measurement problems

2.12   The measurement of output is mostly straightforward for both goods and services. However, there are some components which present special features. These are described in the following sections.

Managing natural processes, long-term and multistage production

2.13   Most agricultural production occurs naturally. Agriculture is the most obvious example of the fact that the management of natural processes is itself production. Growing crops on a farm or cultivating forests is a productive activity, which forms part of GDP. Unmanaged growth, like that of plants or trees in the wild, falls outside the production boundary. The main measurement difficulty arises from the fact that growth, whether of crops or livestock, takes time, typically longer than a single accounting period. The process of agricultural production may thus be seen as the creation of work in progress, and on completion its conversion to stocks (called 'inventories' in the new system) of finished goods and then sale. The value of output in each stage is measured by the costs incurred in production. It can be quite hard to measure how much work has gone into each stage of production in a multistage process. This is described further in Chapter 15. A similar problem arises in the manufacture or construction of large capital items like buildings or ships. Here the work is measured using similar principles to allocate the work to stages, except where there is explicit commitment to stage payments in a contract in which case these may be used.

Transport, storage, leasing

2.14   Transportation, storage and distribution do not change the physical appearance or nature of goods but change their time or place. Their *value added* is therefore the difference in value of the good when it started and when it finished being kept or moved. In the case of transport the value is nonetheless measured by the payment for the service. In the case of storage and distribution the payment for the service may be inseparable from the change in price of the good stored. The value added in distribution is given as the actual receipts from sales *less* the purchase of goods for resale *less* recurrent losses due to wastage, theft, etc *plus* net change in distributors' inventories.

2.15   The leasing of machinery takes two forms: operating leasing and finance leasing. In the former case the lessor rents out equipment etc for a short period, typically far less than the life of the machine, maintains it and keeps a sufficient stock to be able to supply customers at short notice. The lessors' add value making equipment available to those who then do not need to buy it outright. Finance leasing on the other hand involves the transfer of a piece of equipment to the lessee in all but name. It is in effect a financial transaction. This distinction is recognised in the accounts by treating operating leasing as production and finance leasing as a financial operation. The *value added* attributed is no more than the cost of making the financial arrangement – far less than the 'rental charge' because the payment is regarded as property income in respect of a loan.

Financial intermediation services indirectly measured (FISIM)

2.16   The output of financial institutions has always presented difficulties of measurement. Each successive revision of the system of economic accounts has come up with a different proposal for measurement. Neither national accountants nor economists seem entirely clear what it is that banks and financial intermediaries produce. *SNA93* is no exception. It has proposed a measure called *financial intermediation services indirectly measured* (FISIM). Under this proposal the *value added* of financial institutions is seen as intermediation — that is bringing together borrowers and lenders. Neither borrower nor lenders themselves are deemed to add value.

1968 SNA treatment

The convention proposed in the 1968 version of the *SNA* recorded the whole of the output as the intermediate consumption of a nominal industry. This convention makes total GDP for the economy as a whole invariant to the size of the estimated output of the financial sector.

1993 SNA proposal

A major innovation of the 1993 *SNA* arises from the recognition that the way in which financial institutions charge for their services often lies in the way in which they set their borrowing or lending of interest. In this way the principles of measurement are not very different from the principles of measurement of distributors of goods. The implication of this observation is that lenders as well as borrowers derive benefit from the operation of financial intermediaries. The depositors in banks and building societies receive interest rates below the market rates because they value the financial security and services provided by the institution. That difference as well as the additional interest charged to borrowers is measured as part of financial sector output. *FISIM* is defined as the receipts of property income by financial institutions *less* the earnings they receive on their own assets. Most purchases of *FISIM* are intermediate consumption and so netted off in the consolidated measure of GDP (as in the 1968 system). These include purchases by enterprises, and by households in respect of business activity and in respect of mortgage interest (since owner-occupation is within the production boundary).

However, purchases of *FISIM* that enter final consumption will increase GDP. Like other products *FISIM* can be traded internationally. When the output is allocated among different users one possible way of proceeding is to base the allocation on the difference between the actual rates of interest payable and receivable and a 'reference' rate of interest. The reference rate to be used is to represent the pure cost of borrowing funds – that is, a rate from which the risk premium has been eliminated to the greatest extent possible and which does not include any intermediation services. At the time of writing the process for allocation had not been finally decided.

Pending international agreement on the methods for implementing the 1993 proposals the United Kingdom practice is based on the 1968 methodology. (*See* Annex to Chapter 20.)

Insurance

2.19 Insurance companies and pension funds perform a dual function; they offer cover against certain risks and they provide a vehicle for holding savings. Here again the charges for two services are combined. In this case the output is the premium or contributions charged *less* claims or pensions paid out *less* the changes in reserves (known as technical reserves) required to make sure that future claims can be covered.

Intellectual property

2.20 Production of artistic originals is capitalised. The production of books, recordings, films, software, tapes, disks, etc, is a two-stage process of which the first stage is the production of the original and the second stage the production and use of copies of the original. The output of the first stage is the original itself over which legal or *de facto* ownership can be established by copyright, patent or secrecy. This is recorded as capital formation. The value of the original depends on the actual or expected receipts from the sale or use of copies at the second stage, which have to cover the costs of the original as well as costs incurred at the second stage. Consumption of fixed capital is recorded in respect of the use of the asset in the same way as for any other fixed asset used in production.

Accounting rules

2.21 The *SNA* utilises the term *resources* for the side of the current accounts where transactions which add to the amount of economic value of a unit or a sector appear. For example, wages and salaries are a resource for the unit or sector receiving them. Resources are by convention put on the right side, or at the top of tables arranged vertically. The left side (or bottom section) of the accounts, which relates to transactions that reduce the amount of economic value of a unit or sector, is termed *uses*. To continue the example, wages and salaries are a use for the unit or sector that must pay them. Balance sheets are presented with liabilities and net worth (the difference between assets and liabilities) on the right side and assets on the left. Comparing two successive balance sheets one gets changes in liabilities and net worth and changes in assets.

2.22 The accumulation accounts and balance sheets being fully integrated the right side of the accumulation accounts is called *changes in liabilities and net worth* and their left side is called *changes in assets*. In the case of transactions in financial instruments the changes in liabilities are often referred to as (net) *incurring of liabilities* and the changes in assets as (net) *acquisition of financial assets*.

Double entry/quadruple entry

2.23 For a unit or sector economic accounting is based on the principle of double entry, as in business accounting. Each transaction must be recorded twice, once as a *resource* (or a change in liabilities) and once as a *use* (or a change in assets). The total of transactions recorded as resources or changes in liabilities and the total of transactions recorded as uses or changes in assets must be equal, thus permitting a check of the consistency of the accounts. Economic flows that are not transactions have their counterpart directly as changes in net worth, by construction.

A household's purchase on credit of a consumer good will appear as a use under final consumption expenditure and as the incurring of a liability under loans, for example. If this good is paid for in cash, however, the picture is less simple: the counterpart of a use under final consumption is now a negative acquisition of assets, under currency and deposits for instance. Other transactions are even more complicated. Output of goods is recorded as a resource in the account of a producer, its counterpart among uses is recorded as a positive change in inventories. When the output is sold there is a negative change in inventories – that is, a negative acquisition of assets – balanced by a positive acquisition of assets, for instance under currency and deposits.

2.24   In many instances the difficulty of seeing how the double entry principle applies is due to the fact that the categories of transaction in the System are compacted. In principle economic accounts – with all units and all sectors – are based on a principle of quadruple entry, because most transactions involve two institutional units. Each transaction of this type must be recorded twice by the two transactors involved. For example, a social benefit in cash paid by a government unit to a household is recorded in the accounts of government as a use under the relevant type of transfers and a negative acquisition of assets under currency and deposits; in the accounts of the household sector it is recorded as a resource under transfers and an acquisition of assets under currency and deposits.

Intermediate consumption

2.25   One of the most difficult concepts in economic accounting is intermediate consumption. Typically any product which reaches its final destination as, say, personal consumption has gone through various stages from its beginnings as raw material, work, etc. These stages will include the production of goods in intermediate stages. Thus a car will at various stages in its production exist as iron ore and rubber, as steel ingots, pistons and electronic components and as a new car sitting in a factory car park. At each stage value will be added to take it on to the next stage. The output of any institutional unit in the chain will be the value of the product sold on to the next stage. Its value added will consist of the output *less* the value of the goods bought in from the previous stage. If we were to add the output of each stage then the total would change every time there was a change in the arrangement of institutional units. A car assembler taking over one of its suppliers would reduce total output without changing any processes. That is why output is not a very useful concept and we put the emphasis on *value added* which will give the same total whatever the institutional arrangement of productive units. Thus for any given unit or set of units value added is output *less* intermediate consumption. The output of any unit can be divided into sales to final use (including inventories) and intermediate consumption of some other unit.

2.26   Certain goods and services used by enterprises do not enter directly into the process of production itself but are consumed by employees working on that process. In such cases it is necessary to decide whether the goods and services are intermediate consumption or, alternatively, remuneration in kind to employees. In general, when the goods or services are used by employees in their own time and at their own discretion for the direct satisfaction of their needs or wants they constitute remuneration in kind. However, when employees are obliged to use the goods or services in order to enable them to carry out their work they constitute intermediate consumption. This is discussed more fully in Chapter 4 (paragraph 4.30)

The boundary between intermediate consumption and capital expenditure

2.27   Intermediate consumption measures the value of goods and services that are transformed or entirely used up in the course of production during the accounting period. It does not cover the costs of using fixed assets owned by the enterprise nor expenditures on the acquisition of fixed assets. The boundary between these kinds of expenditures and intermediate consumption can be complex. The issues are discussed in more detail in Chapters 3 and 4.

2.28 Expenditures on durable producer goods which are small, inexpensive and used to perform relatively simple operations may be treated as intermediate consumption when such expenditures are made regularly and are very small compared with expenditures on machinery and equipment. Examples of such goods are hand tools such as saws, spades, knives, axes, hammers, screwdrivers, spanners and so on. However, in countries where such tools account for a significant part of the stock of producers' durable goods they may be treated as fixed assets.

2.29 The distinction between maintenance and repairs and gross fixed capital formation is not clear-cut. The ordinary, regular maintenance and repair of a fixed asset used in production constitutes intermediate consumption. Ordinary maintenance and repair, including the replacement of defective parts, are typical ancillary activities but such services may also be provided by a separate establishment within the same enterprise or purchased from other enterprises.

2.30 The practical problem is to distinguish ordinary maintenance and repairs from major renovations, reconstructions or enlargements which go considerably beyond what is required simply to keep the fixed assets in good working order. Major renovations, reconstructions, or enlargements of existing fixed assets may enhance their efficiency or capacity or prolong their expected working lives. They must be treated as gross fixed capital formation as they add to the stock of fixed assets in existence. The issue is discussed in more detail in Chapter 4 (paragraphs 4.32–4.34).

'Investment for the future'

2.31 Businessmen often refer to 'investment for the future' by undertaking programmes of research and development or similar activities which do not necessarily result in the acquisition or improvement of fixed capital assets. Research and development are undertaken with the objective of improving efficiency or productivity or deriving other future benefits so that they are inherently investment (in a general sense) rather than consumption-type activities. Other activities, such as staff training, market research or environmental protection, may have similar characteristics. However, in order to classify activities as investment in a system of economic accounts it is necessary to have clear criteria for delineating them from other activities, to be able to identify and classify the assets produced, to be able to value such assets in an economically meaningful way and to know the rate at which they depreciate over time. In practice it is difficult to meet all these requirements. By convention, therefore, all the outputs produced internally or sold to another unit by research and development, staff training, market research and similar activities are treated as being consumed as intermediate inputs even though some of them may bring future benefits.

Mineral exploration

2.32 Expenditures on mineral exploration are not treated as intermediate consumption. Whether successful or not they are needed to acquire new reserves and are, therefore, all classified as gross fixed capital formation.

Defence expenditure

2.33 Accounting for defence production has always presented problems in economic accounts. Chapter 5 describes the treatment in detail. Briefly, ships, aircraft, vehicles and other equipment acquired by military establishments, and the construction of buildings, roads, airfields, docks, etc, are in general treated like their civilian counterparts as capital, but military weapons such as rockets, missiles and their warheads and the warships, aircraft and other vehicles on which the weapons are mounted are regarded as consumption goods.

The boundary between intermediate consumption and value added

2.34   The boundary between intermediate consumption and value added is not a rigid one fixed purely by the technology of production. It is also influenced by the way in which the production is organized and distributed between different establishments or enterprises.

2.35   The types of services produced by ancillary activities can either be produced for own use within the same establishment or obtained from outside, i.e. from specialist market enterprises. If an establishment obtains the services from outside instead of from ancillary activities its value added is reduced and intermediate consumption increased, even though its principal activity remains completely unchanged. As ancillary activities themselves have intermediate inputs, however, the increase in intermediate consumption is likely to be less than the value of the additional services purchased. Nevertheless, the distribution of value added between establishments and enterprises is bound to be influenced by the extent to which the services of ancillary activities are produced in-house or obtained from outside. Similarly, observed input-output ratios may vary significantly for the same reason, even between equally efficient establishments utilizing the same technology for their principal activity.

2.36   The decision to rent rather than purchase buildings, machinery or equipment, can also have a major impact on the ratio of intermediate consumption to value added and the distribution of value added between producers. Rentals paid on buildings or on machinery or equipment under an operating lease constitute purchases of services that are recorded as intermediate consumption. However, if an enterprise owns its buildings, machinery and equipment, most of the costs associated with their use are not recorded under intermediate consumption. The capital consumption on the fixed assets forms part of gross value added while interest costs, both actual and implicit, have to be met out of the net operating surplus. Only the costs of the materials needed for maintenance and repairs appear under intermediate consumption.

Market and non market

2.37   Recorded output may be analysed into three fundamental categories: market output, output produced for own final use and *other non-market* output. *Market output* is output that is sold at prices that are economically significant or otherwise disposed of on the market, or intended for sale or disposal on the market. Prices are said to be economically significant when they have a significant influence on the amounts the producers are willing to supply and on the amounts purchasers wish to buy.

2.38   *Output for own final use* consists of goods or services that are retained for their own final use by the owners of the enterprises in which they are produced. As corporations have no final consumption output for own final consumption is produced only by unincorporated enterprises: for example, agricultural goods produced and consumed by members of the same household. The output of domestic and personal services produced for own consumption within households is not included in the UK accounts, however, for reasons already given, although housing services produced for own consumption by owner-occupiers and services produced on own account by employing paid domestic staff are included under this heading.

2.39   Goods or services used for own gross fixed capital formation can be produced by any kind of enterprise, whether corporate or unincorporated. They include, for example, the special machine tools produced for their own use by engineering enterprises and dwellings, or extensions to dwellings, produced by households. A wide range of construction activities may be undertaken for the purpose of own gross fixed capital formation in rural areas in some countries, including communal construction activities undertaken by groups of households.

2.40   *Other non-market output* consists of goods and individual or collective services produced by non- profit institutions serving households (NPISH sector) or government that are supplied free, or at prices that are not economically significant, to other institutional units or the community as a whole. These services are produced without a market price either because it is impossible to charge for them or because the government chooses as a matter of policy that it will not do so. The National Health Service is a good example of the latter while military defence is an example of the former. Producers are allocated to market or non-market categories according to their main output. Many enterprises may produce both market and non-market output. Services produced by the NPISH sector include social housing and education.

Time of recording

2.41   The times at which sales are to be recorded are when the receivables and payables are created: that is, when the ownership of the goods passes from the producer to the purchaser or when the services are provided to the purchaser. When payments are made in advance or in arrears the value of sales should not include any interest or other charges incurred by the producer or purchaser. Such charges are recorded as separate transactions. Goods or services provided to employees as compensation in kind, or used for other payments in kind, should be recorded when the ownership of the goods is transferred or the services are provided.

2.42   The principles governing the recording of changes in inventories and work-in-progress are the same for both market and non-market output. Output should be recorded at the time it is produced and valued at the same price whether it is immediately sold or otherwise used or entered into inventories for sale or use later. No output is recorded when goods produced previously are withdrawn from inventories and sold or otherwise used.

2.43   When there is inflation significant price increases may occur while goods are held in inventory. Holding gains accruing on goods held in inventory after they have been produced must not be included in the value of output. The methodology ensures their exclusion by valuing goods withdrawn from inventory at the prices prevailing at the time they are withdrawn ('current costs') and not at the prices at which they entered, or their 'historic costs'. This method of valuation can lead to much lower figures for both output and profits in times of inflation than those obtained by business accounting methods based on historic costs.

2.44   When goods are first produced they may be held in store for a time in the expectation that they may be sold, exchanged or used more advantageously in the future. In these circumstances storage can be regarded as an extension of the production process over time. The storage services become incorporated in the goods, thereby increasing their value while being held in store. Thus, in principle, the values of additions to inventories should include not only the values of the goods at the time they are stored but also the value of the additional output produced while the goods are held in store.

2.45   When the process of production takes a long time to complete the output must be recognized as being produced continuously as work-in-progress. As the process of production continues intermediate inputs are continually being consumed so that it is necessary to record some corresponding output to avoid obtaining meaningless figures for value added by recording the inputs and outputs as if they took place at different times, or even in different accounting periods. Work-in-progress is essentially incomplete output that is not yet marketable. Such output is recorded whenever the process of production is not completed within a single accounting period so that work-in-progress is carried forward from one period to the next. In this case the current value of the work-in-progress completed up to the end of the first period is recorded in the closing balance sheet that also serves as the opening balance sheet for the next period.

2.46   Additions to work-in-progress take place continuously as work proceeds. Within any given accounting period, such as a year or a quarter, it is therefore necessary to record the cumulative amount of work-in-progress produced within that period. Until a sale is ultimately recorded the addition to work-in-progress is the only component of output recorded each period. When the production process is terminated the whole of the work-in-progress accumulated up to that point is effectively transformed into an inventory of finished product ready for delivery or sale. When a sale takes place the value of the sale must be cancelled out by a withdrawal from inventory of equal value so that only the additions to work-in-progress recorded while production was taking place remain as measures of output. In this way the output is distributed over the entire period of production.

2.47   Domestic or personal services produced by members of households for each other are, by convention, treated as falling outside the production boundary of the System. There are, however, two specific categories of services produced for own final consumption whose output must be valued and recorded. Government units or non-profit institutions serving households may engage in non-market production because of market failure or as a matter of deliberate economic or social policy. Such output is recorded at the time it is produced, which is also the time of delivery in the case of non-market services.

2.48   The intermediate consumption of a good or service is recorded at the time when the good or service enters the process of production, as distinct from the time it was acquired by the producer. The two times coincide for inputs of services but not necessarily for goods, which may be acquired some time in advance of their use in production.

2.49   In practice establishments do not usually record the actual use of goods in production directly. Instead they keep records of purchases of materials and supplies intended to be used as inputs and also of any changes in the amounts of such goods held in inventory. An estimate of intermediate consumption during a given accounting period can then be derived by subtracting the value of changes in inventories of materials and supplies from the value of purchases made.

Valuation

2.50   The System utilizes three kinds of output prices, namely basic prices, producers' and purchasers' prices:

(a)   The *basic price* is the amount receivable by the producer from the purchaser for a unit of a good or service produced as output *minus* any tax payable, and *plus* any subsidy receivable on that unit as a consequence of its production or sale. It excludes any transport charges invoiced separately by the producer.

(b)   The *producer's price* is the amount receivable by the producer from the purchaser for a unit of a good or service produced as output *minus* any VAT or similar deductible tax invoiced to the purchaser. It excludes any transport charges invoiced separately by the producer.

(c)   The *purchasers' price* is the amount payable by purchasers to take delivery at the time and place of their choice. It includes non-deductible VAT, any other tax (*less* subsidy) on the product, and separately invoiced transport costs.

2.51   In summary then:

Producers' price = Amount receivable - Value added tax

Basic price = Amount receivable - Value added tax   -   Other taxes on products
less subsidies on products
(both excluding separately-invoiced transport charges)

and Purchasers' price = Amount receivable - Deductible VAT
(including separately-invoiced transport charges).

2.52   Neither the producers' nor the basic price includes any amounts receivable in respect of value added tax (VAT) invoiced on the output sold. The difference between the two is that to obtain the basic price any other tax payable per unit of output is deducted from the producers' price while any subsidy receivable per unit of output is added. Both producers' and basic prices are actual transaction prices which can be directly observed and recorded. Basic prices are often reported in statistical inquiries and some official 'producer price' indices actually refer to basic prices rather than to producers' prices as defined here.

2.53   When the purchaser trades directly with the seller the purchasers' price exceeds the basic price by the value of non-deductible VAT, any other tax (*less* subsidy) on the product and separate transport charges. Where intermediate transactions are involved the difference will also include trade margins.

2.54   When, as in the UK accounts, output is recorded at basic prices, any tax on the product actually payable on the output is treated as if it were paid by the purchaser directly to the government instead of being paid to the producer. Conversely, any subsidy on the product is treated as if it were received directly by the purchaser and not the producer. The basic price measures the amount retained by the producer and is, therefore, the price most relevant for the producer's decision-taking. It is becoming increasingly common in many countries for producers to itemise taxes separately on their invoices so that purchasers are informed about how much they are paying to the producer and how much as taxes to the government.

2.55   Value added tax (VAT) is a percentage tax on products which is collected by enterprises. 'Invoiced VAT' is shown separately on the seller's invoice but the full amount of this is not paid over to the government as producers are allowed to withold the amount ('deductible VAT') that they themselves have paid in VAT on goods and services purchased for their own internal consumption, gross fixed capital formation or resale. VAT paid by households for purposes of final consumption or fixed capital formation in dwellings is not deductible.

2.56   VAT itself covers most goods and services and any other taxes of a similar nature but with narrower scope are treated in the same way in the economic accounts. The concept of 'market price' becomes somewhat blurred with VAT because the price from the buyer's point of view can be different from that from the seller's. The producers' price as defined above is particularly difficult to interpret as it excludes some but not all taxes on products, and for this reason the basic price is generally preferred for valuing outputs.

2.57   The economic accounts convention is that outputs and imports are valued excluding invoiced VAT while purchases of goods and services are recorded including non-deductible VAT. The tax is regarded as falling not on sellers but on purchasers, and then only on those not able to deduct it. In effect it is paid in respect of final consumption, mainly by households though small amounts may be borne by businesses for purchases where VAT is not deductible.

2.58   A disadvantage is that different prices must be recorded for the two parties to the same transaction when the VAT is not deductible. The price recorded for the producer does not include invoiced VAT whereas the price recorded for the purchaser does include the invoiced VAT whenever it is not deductible. Thus, on aggregate, the total value of the expenditures recorded for purchasers must exceed the total value of the corresponding sales receipts recorded for producers by the total amount raised in non-deductible VAT.

2.59   Goods and services produced for sale on the market at economically significant prices may be valued either at basic prices or at producer's prices. The UK maintains both systems, although the *SNA93* expresses a preference for purchasers' prices. This is used wherever there is no market transaction e.g. production for own use or remuneration in kind. If no price is available from market transactions such non-market output should be valued by the total production costs incurred, including consumption of fixed capital and any taxes (*less* subsidies) on production other than taxes or subsidies on products. The non-market output produced by government units and non-profit institutions and supplied free, or at prices that are not economically significant, to other institutional units or the community as a whole is valued by total production costs, including consumption of fixed capital and taxes (*less* subsidies) on production other than taxes or subsidies on products.

2.60   Expenditures by enterprises on goods or services intended to be used for intermediate consumption should be valued at purchasers' prices. Intermediate inputs obtained from other establishments belonging to the same enterprise should be valued at the same prices as were used to value them as outputs of those establishments *plus* any additional transport charges not included in the output values.

Price and volume measures

2.61   Changes over time in the values of flows of goods and services can be directly factored into two components reflecting changes in the prices of the goods and services concerned and changes in their volumes. Similarly, changes in the values of inventories can usually also be decomposed into their own equivalent of price (often known as stock appreciation) and volume components.

> Many flows in the System, such as cash transfers, do not have price and quantity dimensions of their own and cannot, therefore, be decomposed in their way. Such flows cannot be measured at constant prices but can nevertheless be measured 'in real terms' by deflating their values by price indices in order to measure their real purchasing power over some selected basket of goods and services that serves as numeraire.
>
> However, there may be more than one way to do this. There may be no obvious choice of numeraire in which to measure the purchasing power and it may well be appropriate to choose different numeraires for the units paying and receiving the same transfers when these are different kinds of units – e.g. government units and households. Moreover, a flow such as wages and salaries can be treated in two quite different ways. For purposes of analysing production and productivity in which wages and salaries constitute costs of production, it may be necessary to measure inputs of labour at constant input prices – i.e. at constant wage and salary rates – whereas when wages and salaries are recorded as receivables in the primary distribution of income account they may need to be measured in terms of their purchasing power over some basket of final household consumption goods and services.

2.62 Value (v) at the level of a single, homogeneous good or service is equal to the price per unit of quantity (p) multiplied by the number of quantity units (q); that is,

$$v = pq$$

In contrast to price, value is independent of the choice of quantity unit. Certain important properties of quantities, prices and values may be briefly noted:

(a) *Quantities* are additive only for a single homogeneous product. Quantities of different products are not commensurate and not additive even when measured in the same kinds of physical unit. For example, it is not economically meaningful to add 10 tons of coal to 20 tons of sugar, even though their combined weight of 30 tons may provide relevant information for other purposes, such as loading ships or vehicles. Less obviously, the addition of 10 cars of one type to 20 cars of another type may also not be economically meaningful.

(b) The *price* of a good or service is defined as the value of one unit of that good or service. It varies directly with the size of the unit of quantity selected and can therefore be made to vary arbitrarily in many cases by choosing to measure in tons, for example, instead of in kilos. Prices, like quantities, are not additive across different goods or services. A simple average of the prices of different goods or services has no economic significance and cannot be used to measure price changes over time.

(c) *Values* are expressed in terms of a common unit of currency and are commensurate and additive across different products. As already noted, values are invariant to the choice of quantity unit.

2.63 The aggregation of the values of different goods and services is justified by the theory that, in a market system, the relative prices of different goods and services should reflect both their relative costs of production and their relative use to purchasers, whether the latter intend to use them for production or consumption. Such theories are rarely wholly valid in most circumstances; market imperfections such as externalities and oligopoly cause prices to diverge from marginal cost or usefulness. Nonetheless the adoption of basic prices and market prices for measurement provides a consistent basis for measurement.

Volumes

2.64 A volume index is an average of the proportionate changes in the quantities of a specified set of goods or services between two periods of time. The quantities compared must be homogeneous while the changes for different goods and services must be weighted by their economic importance as measured by their values in one or other, or both, periods. The concept of a volume index may be illustrated by a simple example.

Imagine an industry producing two kinds of cheese, a basic cooking cheese valued at £5 a kilo and an eating cheese at £10 a kilo. In the first period 50 tonnes of each are produced. An increase in the total value of production of 10% can arise

by increasing the production of both cheeses to 55 tonnes each
or
by increasing the production of cooking cheese to 65 tonnes
or
by increasing the production of eating cheese to 57.5 tonnes
or
by increasing the production of eating cheese to 67.5 tonnes and reducing the production of cooking cheese to 30 tonnes

2.65   In the last case the tonnage of cheese is actually reduced but the value increases because the eating cheese is so much more expensive than the cooking cheese. While the aim is always to construct price and volume indices according to the above principles there is a long tradition in areas such as international trade of measuring commodities in physical units or numbers. The resulting measures are called *unit value indices* and *quantities* rather than prices and volumes. Such measures are in the process of being replaced by direct measurement of international trade prices to estimate volumes, but there remain in place some unit value index numbers and quantities.

Types of intertemporal index numbers of price and volume

Laspeyres and Paasche indices

2.66   The two most commonly used indices are the Laspeyres and Paasche indices. Both may be defined as weighted averages of price or quantity relatives, the weights being the values of the individual goods or services in one or other of the two periods being compared.

2.67   Let $v_{ij} = p_{ij} q_{ij}$ the value of the ith product in period j

The Laspeyres price index (L_p) is defined as a weighted arithmetic average of the price relatives using the values of the earlier period 0 as weights:

$$L_{pi} = \frac{\sum v_{io}(p_{ij}/p_{io})}{\sum v_{io}}$$

(1)

where the summation takes place over different goods and services. The Laspeyres volume index (L_q) is a similar weighted average of the quantity relatives, that is:

$$L_{qi} = \frac{\sum v_{io}(q_{ij}/q_{io})}{\sum v_{io}}$$

(2)

2.68   The period that provides the weights for an index is described as the 'base' period. It usually (but not always) coincides with the reference period to which the comparisons relate. As the summation always takes place over the same set of goods and services it is possible to dispense with the subscript i in expressions such as (1) and (2). As v_j is equal to $p_j q_j$ by definition it is also possible to substitute for v_o in (1) and (2) to obtain:

$$L_p = \frac{\sum p_j q_o}{\sum p_o q_o}$$

(3)

$$L_q = \frac{\sum p_o q_j}{\sum p_o q_o}$$

(4)

Expressions (1) and (3) are algebraically identical with each other, as are (2) and (4).

2.69   Paasche price and volume indices are defined reciprocally to Laspeyres indices by using the values of the later period j as weights and a harmonic average of the relatives instead of an arithmetic average. A Paasche index $(P_p$ or $P_q)$ is defined as follows:

$$P_p = \frac{\sum v_j}{\sum v_j (p_o/p_j)} = \frac{\sum p_j q_j}{\sum p_o q_j)}$$

(5)

$$P_q = \frac{\sum v_j}{\sum v_j (q_o/q_j)} = \frac{\sum p_j q_j}{\sum p_j q_o)}$$

(6)

When a time series of Paasche indices is compiled, the weights therefore vary from one period to the next.

2.70 The Paasche index may also be interpreted as the reciprocal of a 'backward looking' Laspeyres: that is, the reciprocal of a 'Laspeyres' index for period o that uses period j as the base period. Because of this reciprocity between Laspeyres and Paasche indices there are important symmetries between them. In particular, the product of a Laspeyres price (volume) index and the corresponding Paasche volume (price) index is identical with the proportionate change in the total value of the flow of goods or services in question; that is:

$$L_p \times P_q = \frac{\Sigma \, p_j \, q_j}{\Sigma \, p_o q_o} = \frac{v_j}{v_o} \tag{7}$$

$$L_q \times P_p = \frac{\Sigma \, p_j \, q_j}{\Sigma \, p_o q_o} = \frac{v_j}{v_o} \tag{8}$$

Fisher indices

2.71 A development from Laspeyres and Paasche indices is Fisher's 'ideal' index, calculated as the geometric mean of the correspondng Laspeyres and Paasche indices; that is:

$$F_p = \sqrt{L_p . P_p}$$

$$F_q = \sqrt{L_q . P_q}$$

The relationship between Laspeyres and Paasche indices

2.73 The *System of National Accounts* states:

'16.20 Before considering other possible formulas, it is necessary to establish the behaviour of Laspeyres and Paasche indices vis-a-vis each other. In general, a Laspeyres index tends to register a larger increase over time than a Paasche index:

that is, in general

both $L_p > P_p$ and $L_q > P_q$

It can be shown that this relationship holds whenever the price and quantity relatives (weighted by values) are negatively correlated. Such negative correlation is to be expected for price takers who react to changes in relative prices by substituting goods and services that have become relatively less expensive for those that have become relatively more expensive. In the vast majority of situations covered by index numbers, the price and quantity relatives turn out to be negatively correlated so that Laspeyres indices tend systematically to record greater increases than Paasche with the gap between them tending to widen with the passage of time.'

2.74 This piece of doctrine is not as widely valid as this passage implies. There are important examples of prices and volumes being positively correlated. The largest relative price change in the history of the United Kingdom accounts – the various gyrations of the price of oil in 1973, 1979 and 1985 – were all positively associated with the production of oil. So this view, while widely believed, should be treated with caution. In periods of relatively stable oil prices, however, the conventional assumption is likely to hold good.

Chain indices

2.75   It is convenient to start by considering the example of a time series of Laspeyres volume indices on a fixed base period. In the course of time the pattern of relative prices in the base period tends to become progressively less relevant to the economic situations of later periods, to the point at which it becomes unacceptable to continue using them to measure volume measures from one period to the next. It may then be necessary to update the base period and to link the old series to the series on the new base period. This 'rebasing' is a regular feature of the United Kingdom accounts. Chain indices are produced by changing the base year periodically so that the volume change between each period and the next is based on up-to-date weights and the longer-term changes is obtained by linking the individual changes together.

2.76   For a single index taken in isolation linking is a simple arithmetic operation. However, within an accounting framework it is not possible to preserve the accounting relationships between an aggregate and its components while at the same time linking the aggregate and its components separately. Over the *link years* which occur inbetween the five-yearly rebasing years there is a 'join' which means that constant price components do not sum to the constant price total before the latest link year. The fundamental problem is that volume measures in 'constant prices' of the base year are in truth no more than index numbers scaled up to a conventional cash figure using base year values. Volume measures can only truly be used to measure the change over time in a given indicator.

2.77   Both *SNA93* and *ESA95* recommend the use of annually chain-linked Fisher indices as the preferred measure of movements in GDP volume. However, since Fisher indices may not be acceptable because they present computational difficulties and are not additively consistent for any period, annually chained Laspeyres volume indices are recognised as acceptable alternatives.

2.78   At present the United Kingdom rebases every five years, chaining rebased indices for recent periods to those for earlier periods *via* a link year. Constant-price data in the 1998 *Blue Book* have been rebased onto the year 1995. The previous base year was 1990 and the link year 1994. So constant price magnitudes for 1993 and before do not add up. Annual chain linking is required by the *ESA*, and the UK is considering how to adopt it in the near future.

2.79   If the objective is to measure the actual movements of prices and volumes from period to period then indices should be compiled only between consecutive time periods. Changes in prices and volumes between periods that are separated in time are then obtained by cumulating the short-term movements, i.e. by linking the indices between consecutive periods together to form 'chain indices'. Such chain indices have a number of practical as well as theoretical advantages. For example, it is possible to obtain a much better match between products in consecutive time periods than between periods that are far apart, given that products are continually disappearing from markets to be replaced by new products or new qualities. Chain Laspeyres and Paasche indices tend on the whole to be closer together than their unchained equivalents. On the other hand, chain indices can give rise to problems where prices fluctuate. While unchained indices return to their original level if all prices and quantities are the same as in the base, the same is not true for chained indices. Because intermediate changes use different weights it is possible to have a series of changes in a chain index which do not cancel out.

Conclusion

2.80   This chapter has introduced some of the accounting rules by which the United Kingdom economic accounts are assembled. In the following chapters the practical application of these rules will be described in more detail.

Chapter 3

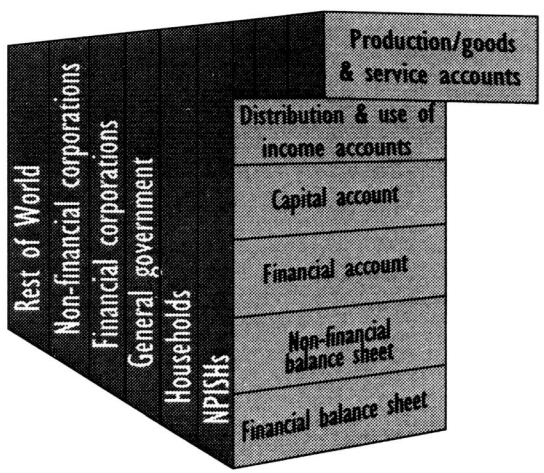

| Rest of World | Non-financial corporations | Financial corporations | General government | Households | NPISHs |

Production/goods & service accounts

Distribution & use of income accounts

Capital account

Financial account

Non-financial balance sheet

Financial balance sheet

The Goods and Services account

Chapter 3 The goods and services account

Summary

3.1   This chapter is concerned more with classifications than with concepts. It describes in broad terms:

- the different types of good and service entering into economic transactions;
- the types of transaction recorded in the economic accounts.

In doing so, it introduces a number of aggregates which will be discussed more fully in subsequent chapters, namely:

- output;
- imports and exports;
- taxes and subsidies on production and trade;
- intermediate and final consumption;
- inventories, fixed capital and valuables.

In each case the emphasis at this stage is on identification and definition. Boundaries are established without as yet being fully justified.

Some technical terms are introduced, such as:

- the 'production boundary' for economic accounts (which will be explained more fully in Chapter 4);
- 'actual consumption' as opposed to 'consumption expenditure' (which will be explored in Chapter 5).

The chapter also provides some necessary background on:

- the specification of institutional and productive units whose activities are to be studied;
- some general principles and conventions of recording and valuation;
- the underlying view of economic processes themselves which is implicit in economic accounts.

This and the following six chapters each focus on one type of account in the new regime. They have some common features, in each case outlining the purpose of the account and its characteristic features, describing the individual elements incorporated with their coverage, definition and inter-relationships and also examining the linkages with other parts of the integrated system. However, in this first chapter the emphasis is mainly on classifications, general conventions and background information, whereas in subsequent ones it is on the specific concepts applying to one aspect of economic activity. In none of these chapters is there any reference to specific sources and methods except where these help to explain the thinking behind the structure. The treatment is essentially in terms of the economy as a whole, the issue of sectorisation being addressed in Chapter 10.

Broad structure of the account

3.2   The goods and services account is the most fundamental in the whole system, in recognition of which it is referred to as 'account zero' in some literature. It balances total resources in the form of goods and services against the various uses of those resources. This is a 'transaction account', showing how much product has been traded and with whom – a summary of the detailed information to be found in the accounts of particular sectors and industries – and it can be compiled either for the economy as a whole or for individual categories of product. At the global level the total of resources is by definition equal to that of uses, without any 'balancing item', but this identity does not apply to individual kinds of transaction.

3.3   In the broadest terms resources come from output and imports while the uses are consumption, investment, changes in inventories, and exports, but each of these needs to be subdivided:

- Output can be 'market' or 'non-market' output, the latter including goods and services for the producer's own use. Taxes on products (*less* subsidies) are included on the same side of the account.
- Imports and exports can be of goods or services.
- Consumption can be final or intermediate. Final expenditure can be looked at on the basis of which sector pays for the goods and services (consumption expenditure) or which sector consumes them (actual final consumption'). Final consumption of general government can be individual or collective. This is discussed in more detail in Chapter 5.
- Investment (fixed capital formation) can be in either tangible or intangible assets, and take place through acquisition of new or existing assets or through disposal of existing ones.
- Stockbuilding (changes in inventories) needs to be seen alongside additions to the value of non-produced assets' (such as land, and mineral deposits) and acquisitions (*less* disposals) of valuables.

The full account, using the nomenclature of ESA1995, is shown in Table 3.1. In the presentation of this account resources are shown on the left and uses on the right, reflecting the fact that product flows are the direct counterpart of monetary flows.

Table 3.1: Components of the goods and services account

Resources	Uses

Output (P1)
    Market output (P11)
    Output for own use (P12)
    Other non-market output (P13)
Taxes on products (D21)
Subsidies on products (D31)
Imports of goods and services (P7)
	Intermediate consumption (P2)
	Final consumption expenditure (P3)/Actual final consumption (P4)
	    Individual consumption expenditure (P31)/Actual individual consumption (P41)
	    Collective consumption expenditure (P32)/Actual collective consumption (P42)
	Gross fixed capital formation (P51)
	Changes in inventories (P52)
	Acquisitions *less* disposals of valuables (P53)
	Exports of goods and services (P6)

Note: Codes in parentheses are as used in SNA1993 and ESA1995

© National Accounts Concepts, Sources and Methods

Definition of goods and services

3.4   In national accounting terms goods are physical objects used either to satisfy directly the needs and wants of households and the community or to produce other goods and services. By virtue of this they are in demand, and must be such that ownership can be transferred from one institutional unit to another through market transactions. Production and exchange are quite separate activities for goods. Some goods are never exchanged while others change hands many times.

3.5   Services, in contrast, are not owned and cannot be traded separately from their production. They are heterogeneous outputs produced to order and by the time their production is complete they must have been provided with a customer. In order for a market to develop the activity must be capable of being carried out by one unit for the benefit of another, even if such services are sometimes produced for a unit's own consumption.

3.6   Service activities can take various forms:

•   Working directly on consumers' own goods (e.g. cleaning and repairing)
•   Changing people's physical or mental condition (e.g. by providing transport, medical treatment or education)
•   Changing the economic state of a person or institution (e.g. by providing insurance).

The changes brought about may be temporary or permanent but either way are normally embodied in the consumers or their possessions. The producers do not have ownership of the services and cannot hold them in inventories or trade them. A single process of production may provide services for a group of people or institutions simultaneously (e.g. on a train or in a school class) and some services are provided collectively to the community as a whole (e.g. defence).

3.7   Some industries which are generally classified as service industries produce outputs which have many of the characteristics of goods, namely those dealing with information, advice and entertainment. These products possess the essential characteristic of services that they are produced by one unit and supplied to another. However, ownership rights *can* sometimes be established over them and the outputs often take the form of physical objects (e.g. paper or disks) which can be stored and traded. Such products could be treated as either goods or services.

3.8   It is on the basis of these characteristics of goods and services that it is possible to define the concepts of production and the production boundary which were introduced in Chapter 2 above and are fully described in Chapter 4. The following section discusses the nature, organisation and classification of production in more detail as a prelude to coverage of the first component of the goods and services account – output.

The production boundary

3.9   Economic production is defined as any activity carried on by an institutional unit which uses inputs of materials, labour and capital to produce output of goods and services. The need for an institutional unit to control and be responsible for the activity rules out purely natural processes. It is relatively easy to apply this definition in the case of goods – for example it includes fish farming and excludes the unmanaged growth of fish stocks in international waters – but it is sometimes more difficult for services. It includes the production of financial services, even when these are paid for by interest flows rather than charges. Income from the renting of buildings is included, but rent from the ownership of land is not (the reason for the distinction is explained in Chapter 5).

3.10   Particular problems arise over the production of domestic and personal services by households for their own final consumption. These include the cleaning, maintenance and minor repair of both houses and household durables, the preparation of meals, care of children or the infirm and transportation of household members. Such activities involve a large amount of labour and contribute greatly to economic welfare but they have traditionally been excluded from measured production in economic accounts for various reasons:

- The services in question are not generally traded so it is hard to choose appropriate market prices with which to value them;
- Even if meaningful incomes and expenditures could be imputed it would be misleading for purposes of economic analysis and policy-making to regard these as equivalent to cash amounts.

3.11   For the UK the main exception to the general exclusion of households' own-account production is that owner-occupiers' housing services for their own final consumption have always been included within the production boundary, based on imputed values (which in some countries are regarded as sufficiently firmly-based to be taxed). On reason for this inclusion is that the ratio of owner-occupied to rented housing varies considerably over time and between countries so comparisons of consumption would be distorted if no allowance were made for the owner-occupied element. Where there is domestic own-account production of goods, that is included in production where possible.

3.12   As already noted, minor repairs and maintenance by households of both dwellings and durables are excluded from production, any materials purchased being treated as final consumption expenditure. However, materials for more substantial repairs (of a sort which would not be carried out by tenants) are regarded as intermediate expenditure incidental to the production of housing while major renovations or extensions are treated separately as fixed capital formation.

3.13   The fact that some production is illegal does not put it outside the production boundary. Provided the activity is a genuine production process generating goods or services for which there is an effective market demand it should be included even if carried out by unauthorised people or wholly forbidden by law. Possible examples are prostitution, smuggling, the manufacture or distribution of banned drugs, medical services offered by unlicensed practitioners and trading on unofficial markets in foreign exchange or price-controlled goods. Such activities need to be recorded not only to obtain comprehensive measures of production and consumption but also to prevent errors appearing in other parts of the accounts though misallocation, e.g. where income from illegal production is disposed or legitimately or where expenditure on illegal goods is made out of honestly-obtained funds. However, in the first instance, illegal activity is not being estimated (*see* Chapter 2)

3.14   Illegal production needs to be distinguished from the generation of harmful externalities, e.g. through discharge of pollutants (for which no values are imputed) and from the theft of legally-produced output (which, as noted in Table 3.9, is treated as intermediate consumption).

3.15   Finally, the production boundary encompasses activities which may well be legal in themselves but are being deliberately concealed from the authorities so as to avoid paying tax or social security contributions, meeting statutory standards or complying with official procedures such as the completion of statistical returns. In some countries this 'underground economy' or 'black economy' may account for much of the total output of certain industries where small firms predominate (e.g. some service industries and the building trade). It is therefore important that estimates of production attempt to include such activity (e.g. by using the commodity flow method) even if the quality of estimates of such activity is necessarily poor.

Table 3.2: Inclusions in and exclusions from production

Inclusions	Exclusions
Individual or collective goods or services supplied (or intended for supply) to other units. Goods retained by their producers for their own final consumption or gross fixed capital formation, the latter including fixed assets such as construction, software development and mineral exploration. Own-account production of housing services by owner-occupiers. Domestic and personal services produced by employing paid domestic staff. Ownership and management of buildings and other assets. Voluntary activities that result in goods (e.g. construction of buildings).	Domestic and personal services produced and consumed within the same household, apart from those of paid domestic staff and owner-occupied dwellings as listed under *Inclusions*. Cases in point are shown below. Cleaning, decorating and maintenance of the dwelling, to the extent that these activities are commonly undertaken by tenants. Cleaning, servicing and repair of household durables. Preparation and serving of meals. Care, training and instruction of children. Care of sick, infirm or old people. Transportation of members of the household or their goods.

Note: Activities are not excluded on the grounds that they are illegal or not registered with public authorities

3.16 To recapitulate and summarise, the production boundary for economic account includes in principle:

- Production of goods and services supplied (or intended for supply) to units other than their producers, together with those used up in the production process;
- Own-account production of goods retained by their producers for their own final consumption or gross capital formation;
- Own-account production of housing services produced by owner-occupiers along with domestic and personal services produced by the employment of paid staff.

The next section describes some aspects of the measurement of production defined in this way.

Enterprises

3.17 Institutional units which are involved in production are called 'enterprises'. These are usually corporations but may be non-profit institutions or unincorporated enterprises, the latter being those *parts* of households or government units which are concerned with production. Large enterprises can be engaged in a wide range of activities and, for purposes of economic analysis, need to be partitioned into units which are reasonably homogeneous with respect to output, cost structure and technology. These are called 'kind-of-activity units' (KAUs), and correspond to the class (i.e. 4-digit) level of the European classification of economic activities (NACE). They are further disaggregated on the basis of geographical situation into 'local KAUs' (sometimes called 'establishments') which are the smallest elements identified for economic accounting purposes. The enterprise's information system needs to be capable of recording certain statistical data for each local KAU: as a minimum this comprises the value of production, intermediate consumption, compensation of employees, operating surplus, employment and gross fixed capital formation.

3.18　A distinction is made between principal, secondary and ancillary activities of local KAUs. The principal activity is that with the highest value added, determined by reference to NACE, secondary activities are those with lesser value added and ancillary activities are support functions whose output is not intended for use outside the enterprise, e.g. purchasing, sales, marketing, accounting, data processing, transportation, storage, maintenance, cleaning and security services (but not own-account capital formation). Enterprises may have a choice between undertaking ancillary activities themselves and buying them in from specialist service providers. If undertaken internally they are treated as integral parts of the principal or secondary activities with which they are associated: there is no separate recording of either their outputs or the inputs consumed (materials, labour, etc.). Primary and secondary activities may produce some output which is used internally but this must be such that it *could* be delivered to other units.

Components of the account

Output (P.1)

3.19　Output consists of the products created during a given accounting period. It includes goods and services supplied by one local KAU to another within the same enterprise (these not being netted out on aggregation) and goods remaining in inventories at the end of the period in which they are produced. Goods produced and consumed within the same local KAU in the same accounting period are ignored, not being recorded either as output or as intermediate consumption.

3.20　A distinction is made between market (P.11) and non-market (P.12 and P.13) output. The former consists mainly of goods sold at 'economically significant' prices defined in the ESA as those which allow more than half of production costs to be covered by sales) but it also includes products which are bartered, used for payments in kind (including employees' compensation), supplied to another local KAU within the same enterprise or added to inventories *intended* for the market. By convention it also includes *all* output of households which is sold to other institutional units.

3.21　A second category of output is that for the producer's own final use (P.12, by definition, non-market output). 'Final' in the economic accounts context implies that the use is not intermediate; that is it is not consumed in any further productive process. Output for use for final consumption can only apply to the household sector (e.g. agricultural produce retained by farmers, housing services for owner-occupiers and household services produced by employing staff) but own-use for gross fixed capital formation can be in any sector (from machine tools produced by engineering companies to home extensions built by households for themselves).

3.22　The third and last category of output ('other non-market', P.13) is supplied to other units either free or at prices that are not economically significant.

3.23　Corresponding to the distinction between 'market', 'own final use' and 'other non-market' output is a distinction between producers in these three groups, each unit belonging to whichever category constitutes the greater part of its output.

3.24 Bound up with the distinction between market and non-market producers is that between public and private. Public producers are those controlled by the government, in the sense that the government determines the corporate policy or programme, e.g. by appointing directors or owning more than half the shares. Such producers are classified as corporations if more than 50 per cent of their production costs are covered by sales, and otherwise to the general government sector.

3.25 All other producers are classified as private, and they are found in all sectors *except* general government. Some are unincorporated enterprises owned by households and, as already noted, these are always either market producers or producers for their own final use, classified to the household sector, *except* in the case of quasi-corporate enterprises which are treated as market producers like corporations. For other private producers a further distinction is made between non-profit institutions (NPIs) and the rest.

3.26 NPIs are entities whose status does not permit them to be a source of financial gain to the units that establish, control or finance them. Any surpluses they generate cannot be appropriated. In determining their 'market' status a criterion is applied which is similar to that for public producers. If more than 50 per cent of production costs are covered by sales then NPIs are regarded as market producers and classified as corporations; otherwise they are 'other non-market' and classified either to the general government sector (if controlled by the government) or as 'non-profit institutions serving households' (NPISH). A further case is that of private NPIs serving businesses, which are usually financed by subscriptions from the businesses concerned. These subscriptions are treated not as transfers but as payments for services rendered, i.e. sales, so the institutions are market producers and classified as corporations, as are all other private producers not already mentioned.

3.27 The three-way split can be summarised as in Table 3.3.

3.28 For purposes of the 50 per cent criterion, sales exclude taxes on products but include any payments from government or EU institutions which are granted to all producers of the same type in proportion to the volume or value of their output, but excluding payments to cover overall deficits. Production costs are the sum of intermediate consumption, compensation of employees, consumption of fixed capital and taxes on production, including subsidies for production but excluding costs of own-account capital formation. The 50 per cent criterion itself is intended to be applied over a run of years: units should not be reclassified on account of minor fluctuations in sales from year to year.

Table 3.3: Allocation of enterprises as between 'market' (M), 'own final use' (OFU) and 'other non-market' (ONM).

	Sales as percentage of production cost	
	50%+	Other
Public producers	M	ONM
Quasi-corporate enterprises owned by households	M	M
Other unincorporated enterprises owned by households	M/OFU	M/OFU
Private non-profit institutions	M	ONM
All other private producers	M	M

3.29 For some NPIs it would be misleading to include in sales all payments linked to output, e.g. payments from government to schools on the basis of their numbers of pupils. Where a school is mainly financed and controlled by the government it is regarded as an 'other non-market' producer classified to the general government sector.

3.30 Once an enterprise has been classified to 'market', 'own final use' or 'other non-market', the distinction is applied to local KAUs within it. If the enterprise is a market producer then so is its principal local KAU but the secondary local KAUs can be either market producers or producers for own final use though not, by convention, 'other non-market' producers. If the enterprise as a whole is an 'other non-market' producer then the secondary local KAUs can be either market or 'other non-market' producers, depending upon the 50 per cent criterion. The classification of producers is discussed further in Chapter 10 (10.23–10.24).

3.31 Having classified production units the next stage is to classify their output. By convention local KAUs which are market producers or producers for their own final use can only supply output in those categories: not 'other non-market output'. For local KAUs which are 'other non-market' producers the situation is more complex. They can supply in all three categories (any output for their own final use being own-account capital formation). In principle the 50 per cent criterion for identifying market output should be applied to individual products, e.g. where hospitals charge economic prices for some of their services, where museums sell reproductions to visitors in a shop or meteorological offices sell weather forecasts. In practice it may be difficult to distinguish revenues and costs in this detail, in which case all secondary activities may be treated as producing a single type of market output. Furthermore it may be difficult for non-market producers to distinguish between market output at economically significant prices and non-market output at non-significant prices (e.g. in the case of museums between sales from the shop and tickets for admission). In such cases the two can be combined into whichever type of output generates more revenue.

3.32 The distinction between 'market', 'own final use' and 'other non-market' has been described at some length because it determines the way the output is valued for purposes of the goods and services account. The principles involved in this are discussed in the next section.

Time of recording and valuation of output

3.33 Output is recorded and valued when it is generated by the production process, the valuation generally being at basic prices.

3.34 Output for own final use is valued at the basic prices of similar products sold on the market, except for own-account construction which normally has to be valued by costs of production.

3.35 Additions to work-in-progress are valued in proportion to the basic prices of the finished products (including any for own final use or going into inventories) but preliminary estimates may be based on costs incurred plus a mark-up.

3.36 For buildings acquired in an incomplete state a value is estimated based either on costs incurred plus a mark-up or on stage payments. In the case of own-account construction spanning several accounting periods the value of output in each period is estimated by applying the appropriate fraction of total production costs to the basic price of the finished structure or, if the latter cannot be estimated, by taking costs of production themselves. If labour is provided free then an estimate of the cost of equivalent paid labour should be included.

Table 3.4: Special cases of output valuation

Activity	Treatment
Agricultural production	Output is regarded as being produced continuously over the entire period of production: not just when crops are harvested or animals slaughtered. Growing crops, animals being reared for food, etc. are treated as inventories of work-in-progress and transformed into inventories of finished stocks when the process is complete.
Construction extending over several periods according to a contract agreed in advance	Output produced in each period is treated as being sold at the end of that period, becoming the purchaser's fixed capital formation rather than the constructor's work-in-progress. The value of the output may be approximated by stage payments (if any). In the absence of a contract incomplete output *is* treated as work-in-progress.
Wholesale and retail trade	Output is measured by the trade margins on goods purchased for resale, this being the difference between the price realised when the good is sold and the cost to the distributor of replacing it at that time. By convention holding gains and losses are excluded from trade margins but may be impossible to separate out in practice.
Hotels and restaurants	Output includes the value of food and drink consumed.
Transport services (for goods or people)	Output is the amounts receivable but transportation for own use within a local KAU is considered an ancillary activity and not separately recorded.
Storage services	Output is the additional value of work-in-progress which accrues over time, including any physical change (e.g. from maturing of wine).
Travel agencies	Output is the fees and commission charged, not the full expenditure (as the agency is only an intermediary).
Tour operators	Output is the full expenditure made by travellers to the operator (who creates a new product the various components of which, eg travel and accommodation, are not separately priced as far as the consumer is concerned).
Financial intermediation	If fees or commissions are explicitly charged (e.g. for currency exchange, arranging housing finance or providing advice) then these are used to value the output. Otherwise output is the property income received by the service providers *minus* their total interest payments. Left out of the calculation are income from investment of their own funds or of insurance technical reserves, and holding gains or losses, though trade margins on dealing in foreign exchange and securities should be included in output.

Table 3.4 *cont...*

Moneylenders advancing only their own funds	These are not regarded as providing services.
Insurance	Output is actual premiums *plus* premium supplements (equal to income from investment of technical reserves) *minus* claims due *minus* change in reserves (actuarial and with-profits), holding gains and losses being ignored.
Pension funding	Output is actual contributions *plus* supplementary contributions (equal to income from investment of technical reserves) *minus* benefits due *minus* change in actuarial reserves.
Services of owner-occupied dwellings	These are valued at the rental a tenant could expect to pay for the same accommodation. Similar imputations are made for domestic garages but not those used for parking near workplaces. Rental values of owner-occupied dwellings abroad (e.g. holiday homes) are recorded as imports of services, and any net surplus as primary income received from the rest of the world (with analogous entries for dwellings in the UK owned by non-residents).
Property services for non-residential buildings	Output is the value of the rentals due.
Leasing services for renting out equipment etc	Output is the value of the rentals due.
Financial leasing	This is essentially making a loan to finance the acquisition of fixed assets, the rentals consisting mainly of repayments and interest with only a small element of service.
Research and development	Wherever possible a separate local KAU should be designated for these activities; otherwise they should be recorded as a secondary activity. The valuation of output depends upon the nature of the activity, as follows. (a) *Specialised commercial laboratories*: Output is revenue from sales, contracts, commissions, fees, etc. (b) *R&D for use within the enterprise*: Output is in principle valued at the basic prices that would be paid if the work were sub-contracted commercially, but in practice is usually based on total production costs. (c) *R&D by government units, universities, non-profit institutes, etc*: This is usually 'other non-market' production and valued at costs of production. Revenues are recorded as being from (secondary) market output.

Table 3.4 *cont...*

Development of computer software	Not to be classified as R&D. Software is a produced intangible asset, and is not patented.
Education and training	The accounting treatment is like that of research and development.
Public administration, defence and compulsory social security	These are always provided as 'other non-market' services and valued accordingly.
Education and health services	Here institutions may charge only nominal fees for some courses or treatments but commercial tariffs for others, and the same type of service may be provided both by government and by private institutions between which there are large differences in price and quality of service. In such cases it is necessary to draw precise borderlines between market and 'other non-market' outputs.
Production of books, recordings, films, software, tapes, disks, etc	This is a two-stage process and is measured accordingly. (a) Output from the production of originals (intangible fixed assets) is measured by the price paid if sold or, if not sold, by the basic price of similar originals, the production costs or the discounted value of expected future receipts from reproduction. (b) If the owners of such assets license others to make reproductions then the output of the owners' service is the fees, commissions, royalties, etc which are received, but the sale of the assets themselves is regarded as negative fixed asset formation.
Domestic staff employed by private households	By convention the output of such staff is valued by the compensation paid them, including payments in kind.

3.37 For 'other non-market' producers (local KAUs) *total* output is valued at the costs of production, including intermediate consumption, compensation of employees, consumption of fixed capital and taxes (*less* subsidies) on production but excluding imputed rents of owned non-residential buildings and, by convention, interest payments (even where these are substantial, e.g. for housing associations). However, valuation of *component* outputs depends upon circumstances, as follows:

- If there is *no* (secondary) market output then the 'other non-market' output is valued at the costs of production;

- If there *is* (secondary) market output then this is valued by receipts from sales (though in principle it should be by basic prices) and 'other non-market' output is then valued by subtraction from the total, i.e. as a residual.

These general rules for the valuation of output are subject to certain clarifications and exceptions which are listed in Table 3.4.

Imports and exports of goods and services (P.7 and P.6)

3.38   The second major block of resources in the goods and services account, after output, is imports. Its counterpart on the other side of the account – exports – is essentially a mirror image, governed by the same concepts and symmetrical definitions, so the two are dealt with together in this section.

3.39   Imports of goods and services consist of transactions from non-residents of the United Kingdom to residents (see definition of residence in Chapter 24), and exports *vice versa*. The transactions need not be sales, as barter, gifts and grants are also included, but three types of transfer are excluded:

- Deliveries to UK residents by UK affiliates of foreign enterprises and, conversely, deliveries to non-residents by foreign affiliates of UK enterprises (i.e. enterprises owned or controlled by UK residents);
- Primary income flows from and to the rest of the world, e.g. compensation of employees and revenues from direct investment (the latter possibly including an indistinguishable component in respect of the provision of services);
- Purchase or sale of financial assets and non-produced assets, e.g. land and patents.

3.40   Imports and exports of goods occur when there are changes of ownership between UK residents and non-residents but this general principle is modified in some cases, namely:

- *Financial leasing*: Change of ownership is regarded as taking place when the lessee takes possession of the goods.
- *Deliveries between affiliated enterprises*: Change of ownership is regarded as taking place at the time of delivery.
- *Goods for major repair or processing*: These are recorded in both imports and exports though there is no change of ownership.
- *Merchanting*: No import or export is recorded when UK merchants or commodity dealers buy from and sell to non-residents within the same accounting period (and similarly for non-resident merchants dealing with UK residents).

3.41   Imported and exported goods do not necessarily move across frontiers. For example:

- Goods produced by foreign units operating in international waters can be imported directly into the UK (e.g. fish and oil).
- Movable equipment can be bought and thus imported by a UK resident from a non-resident without physically moving.
- Imported goods may be lost or destroyed after changing ownership but before leaving the country of origin.

Table 3.5 lists types of good which are included in or excluded from imports and exports, some of the exclusions involving moves across national frontiers.

Table 3.5: Inclusions in and exclusions from imports and exports of goods

Inclusions	Exclusions
Non-monetary gold (i.e. gold not used for purposes of monetary policy)	Goods in transit through a country.
Precious metals and stones (including silver bullion and diamonds)	Goods shipped to or from embassies, military bases and other enclaves inside another country's borders.
Trade with foreign embassies etc. within the country.	Construction and other movable equipment leaving a country temporarily for work abroad without any change of ownership.
Banknotes and coins not in circulation and unissued securities (valued as goods, not at face value)	Goods sent abroad for *minor* processing, maintenance, service or repair.
Electricity, gas and water	Other goods leaving a country temporarily, being returned within (generally) a year in their original state and without change of ownership, e.g. goods sent abroad for exhitition or entertainment purposes, goods under operating leases (including leases for several years) and goods returned because expected sales did not materialise.
Livestock driven across frontiers	
Parcel post	
Government exports including goods financed by grants and loans	
Goods transferred to or from buffer stocks	Goods on consignment lost or destroyed after crossing a frontier but before change of ownership occurs.
Goods transferred to or from non-resident affiliates	
Smuggled goods	
Other unrecorded shipments, e.g. gifts and low-value items	
Goods sent abroad for processing involving substantial physical change	
Investment goods sent abroad for substantial repairs	

3.42   Though in principle trade should be recorded at the time when ownership is transferred in practice recording takes place when the parties to the transaction enter it in their books or accounts: not necessarily the contract date, transfer date or payment date. Goods should be valued 'free on board' (f.o.b.) at the frontier of the *exporting* country, this value consisting of:

- The value of the goods at basic prices

    plus Transport costs up to the frontier, including those of loading for onward transportation
    plus any taxes (*less* subsidies) on the goods, including for intra-EU deliveries VAT and other taxes paid in the exporting country.

Though this convention applies to the goods and services account a different one is used for imports in overseas trade statistics. This is the 'cost-insurance-freight' (c.i.f.) price – that of a good delivered at the frontier of the *importing* country before payment of any duties or taxes on imports.

3.43   Proxies for the f.o.b. value may be necessary in certain circumstances, e.g.:

- Bartered goods should be valued at the basic prices that would have applied to a cash transaction.
- Transactions between affiliated enterprises should be priced at actual transfer values unless market prices are significantly different and can be estimated.
- Goods transferred under financial leases are valued at the purchasers' price paid by the lessor, not by the rental payments.
- Customs data on imports use c.i.f. valuation and need to be adjusted to f.o.b., but only at the most aggregate level.
- Survey-based and *ad hoc* data on imports and exports usually record only purchasers' prices, not f.o.b. values.

3.44   Imports of services consist of all services rendered by non-residents to UK residents and exports of services consist of all services rendered by UK residents to non-residents. A major element of trade in services is the transportation of imported and exported goods where the actual situation differs from that in which UK carriers handle exports up to the UK border but not beyond and handle imports from the border of the country of origin but not hitherto. The types of imports and exports of services to which this gives rise are listed in Table 3.6 together with a number of other borderline cases. It should be particularly noted that transportation of UK-exported goods beyond the UK frontier by non-resident carriers does not produce any import of services even if this transportation is actually paid for by the exporter under 'export-c.i.f.-contracts'.

3.45   All direct purchases by UK residents abroad, whether of goods or services, are by convention treated as imports of services but their treatment in the accounts differs according to whether the expenditure is by business travellers (classified as intermediate consumption) or on personal trips (classified to household final consumption expenditure).

3.46   Imports and exports of services are recorded at the time they are rendered, which is normally the time at which they are produced. Imports are valued at purchasers' prices and exports at basic prices.

Table 3.6: Inclusions in imports and exports of services

Imports	Exports
Transport services	**Transport services**
Transport of UK-exported goods to the UK frontier when provided by non-resident carriers (to offset the amount included in the f.o.b. value of the goods)	Transport of UK-exported goods beyond the UK frontier when provided by UK carriers.
Transport of UK-imported goods from the frontier of the country of origin when provided as a separate service by non-UK carriers.	Transport of UK-imported goods up to the frontier of the country of origin when provided by UK carriers.
Transport of UK residents' goods by non-resident carriers which does not involve import or export of goods (e.g. transport of goods in transit or outside the UK)	Transport of non-residents' goods by UK carriers which does not involve import or export of goods.
International or national passenger transportation for UK residents by non-resident carriers.	International or national passenger transportation for non-residents by UK carriers.
Other services	**Other services**
Minor processing and repairs carried out for UK residents by non-residents	Minor processing and repairs carried out by UK residents for non-residents.
Construction work in UK by non-residents which lasts less than a year and whose output is not gross fixed capital formation	Construction work abroad by UK residents which lasts less than a year and whose output is not gross fixed capital formation.
Installation of equipment in UK by non-residents where the project by its nature is of limited duration	Installation of equipment abroad by UK residents where the project by its nature is of limited duration.
Explicit commission and fees for financial services provided for UK residents by non-residents	Explicit commission and fees for financial services provided by UK residents for non-residents.
Service charges for insurance services provided for UK residents by non-residents	Service charges for insurance services provided by UK residents for non-residents.
All expenditure abroad by UK-resident tourists and business travellers	All expenditure in the UK by non-resident tourists and business travellers.
All expenditure by UK residents on health and education services provided by non-residents, whether in the UK or abroad	All expenditure by non-residents on health and education services provided by UK residents, whether in the UK or abroad.
Services of holiday homes abroad owner-occupied by UK residents	Services of holiday homes in the UK owner-occupied by non-residents.
Royalties and licence fees paid by UK residents to non-residents	Royalties and licence fees paid to UK residents by non-residents.

Taxes on products (D.21)

3.47   Associated with the two principal resources on the left hand side of the goods and services account – output and imports – are two supplementary items which are dealt with in this and the following section, namely taxes and subsidies on products.

3.48   A tax on production or imports is defined as a compulsory unrequited payment in cash or in kind levied by the government or EU institutions in respect either of production and imports themselves or the other resources used in production (e.g. labour, land and buildings). Taxes on production are of two types: taxes on products, and other taxes on production. They are described in more detail in Chapter 5.

3.49   Taxes on products are those that are levied per unit of quantity or on an *ad valorem* basis. Three types can be distinguished, and are defined in more detail in Chapters 5 and 21:

- Value added taxes (VAT);
- Taxes on imports;
- Other taxes on products.

Subsidies on products (D.31)

3.50   Subsidies are defined as current unrequited payments made by the government or EU institutions to producers with the objective of influencing their levels of production, their prices or the remuneration of factors of production. As with taxes, the payment can be related either to products or production. The former may be per unit of quantity, *ad valorem* or based on the difference between a specified target price and the market price. By convention, subsidies on products only apply to market output or output for own final use, not to 'other non-market' output. Subsidies are also defined in more detail in Chapters 5 and 21.

3.51   Goods and services that are the output of the production process can be used in four ways:

- They can be consumed as inputs to further processing (intermediate consumption);

- They can be consumed in direct satisfaction of collective needs and wants (final consumption);
- They can be used as capital, to facilitate the continued production of other goods and services (gross fixed capital formation);
- They may find themselves in inventories during the process.

The next four sections describe these categories in more detail.

Intermediate consumption (P.2)

3.52   The first item in the 'uses' section of the goods and services account is intermediate consumption, defined as the value of products, other than fixed assets, which are transformed or used up as inputs to a process of further production. The products concerned should be recorded at the time they enter the production process, the valuation being at purchasers' prices for similar goods and services at that time. In practice, however, producers do not usually record the actual use of goods but rather their purchases of goods intended for use as inputs and the changes in inventories of such goods. Intermediate consumption therefore has to be estimated by subtracting the latter from the former. Table 3.9 lists the borderline cases included in and excluded from the definition.

Table 3.7: Coverage of intermediate consumption

Inclusions	Exclusions
Goods and services used as inputs to ancillary activities (which are not distinguished from those consumed by the main activities of the local KAU)	Investment-type expenditure which *is* treated as capital formation, e.g. on valuables, mineral exploration, major improvements to fixed assets and software purchased or produced on own account.
Goods and services received from another local KAU within the same enterprise (provided they comply with the general definition of intermediate consumption)	Wages and salaries in kind.
Costs of using rented fixed assets (i.e operational leasing)	Social security contributions.
Subscriptions, contributions and dues paid to non-profit business associations	Use of government services by market or own-account producers.
Investment-type expenditures which are not treated as gross capital formation, e.g. small tools (costing up to 500 ecu at 1995 prices) and military weapons and delivery systems (though not light weapons and armoured vehicles acquired by police and security forces)	Goods and services produced and consumed within the same accounting period and local KAU.
Ordinary maintenance and repair of fixed assets used in production	Payments for government licences and fees.
R&D, staff training, market research and similar services bought in from an outside agency or provided by a separate local KAU within the enterprise	
Payments for the use of intangible non-produced assets, e.g. patented assets and trade marks (but excluding the purchase of such property rights)	
Necessary expenditure by employees reimbursed by the employer (e.g. for safety wear)	
Expenditure by employers which benefits both them (i.e. their production) and their employees, e.e. reimbursements of travelling or removal expenses and provision of workplace amenities	
Insurance service charges paid by local KAUs (i.e. premiums *less* claims) excluding life	

Final consumption (P.3)

3.53   Final consumption expenditure is that on goods and services used for the direct satisfaction of individual or collective needs and wants. The expenditure must be by UK institutional units but may take place either in the UK or abroad.

3.54   By definition, the bulk of expenditure on goods and services must be by households. Table 3.10 shows various inclusions in and exclusions from household final consumption expenditure. However, not all final consumption expenditure is by the household sector. Non-profit institutions serving households (NPISH) and government each make final consumption expenditure. This expenditure may be in the form of goods and services they produce themselves, or on goods and services from market producers which are then passed on (without any transformation) to households. Corporations, on the other hand, do not have final consumption expenditures. Their purchases of consumption-type goods are either for further processing (i.e. intermediate consumption) or passed on to employees as compensation in kind.

Table 3.8: Coverage of household final consumption expenditure

Inclusions	Exclusions
Services of owner-occupied dwellings	Social transfers in kind, e.g. expenditures initially incurred by households but subsequently reimbursed by social security.
Income in kind, e.g. that received by employees or produced as output of unincorporated enterprises owned by households and retained for conumption	Items of intermediate consumption or gross capital formation, e.g. business expenditure of households owning unincorporated enterprises; owner-occupiers' expenditure on types of maintenance and repair not typically carried out by tenants; purchase of dwellings; and expenditure on valuables.
Materials for small repairs to and interior decoration of dwellings of a type typically carried out by tenants as well as owners and materials for repair and maintenance of consumer durables (including vehicles)	Purchase of land and other acquisitions of non-produced assets.
Consumer durables that continue to perform their function over several accounting periods	
Financial services directly charged for	Licence payments by households which are regarded as taxation, e.g. to own vehicles, hunt, shoot or fish.
Insurance and pension fund services (implicit service charges only)	Subscriptions, contributions and dues paid by households to non profit institutions, e.g. trade unions, professional societies, consumer organisations, churches and clubs.
Payments by households for licences, permits, etc. which are regarded as buying services	
Purchases at prices which are not economically significant, e.g. National Health Service charges	Voluntary transfers in cash or in kind from households to charities etc.

3.55 The goods and services account distinguishes between individual and collective consumption, which have different characteristics as follows:

- *Individual consumption* of goods and services is only recognised when it is possible to record each acquisition and its timing, the household concerned must have agreed to the provision and made it possible (e.g. by attending a school or clinic) and the good or service itself must be such that acquisition by one person or small group precludes its acquisition by others.
- *Collective consumption* of services requires that they can be delivered simultaneously to every member of the community or sections of the community. The use of such services is normally 'passive' (not requiring the explicit agreement or active participation of everyone concerned) and there is no rivalry in acquisition – provision of a collective service to one individual does not reduce the amount available to others.

By definition all final consumption by households is individual and by convention so is all that of the NPISH sector.

3.56 The characteristic of individual consumption is that it relates to the goods and services which at some periods or in some economies households might purchase for themselves. By aggregating the individual consumption expenditure of households, the NPISH and general government sectors we arrive at a concept, the actual final consumption of households (P.4) which is more comparable internationally and over time.

3.57 The concepts and their implementation in the consequent *Redistribution of income in kind account* are described in more detail in Chapters 5 and 21.

3.58 As regards the timing and valuation of consumption the general rules apply – that recording takes place when ownership of a good changes or delivery of a service is completed, and that expenditure is recorded at purchasers' prices – but with the following qualifications:

- Expenditure on a good acquired under a hire purchase or similar credit agreement is recorded when the good is delivered, even if legally there is no change of ownership at that time;
- Own-account consumption is recorded when the output concerned is produced, and valued at basic prices;
- Goods and services supplied as compensation in kind of employees when produced by the employer are valued at basic prices;
- Final consumption expenditures by the government and NPISH sectors on products produced by themselves or supplied *via* market producers are recorded at the time of delivery.

Gross fixed capital formation (P.51)

3.59 The terms 'gross' and 'net' are used in various different ways in economic literature but in this manual they mean before and after consumption of fixed capital. This and the following sections discuss gross capital formation, which has three elements: formation of fixed capital (P.51), changes in inventories, previously called stocks (P.52) and acquisitions *less* disposals of valuables (P.53).

3.60 Fixed capital can be either tangible or intangible. The former includes not only buildings and equipment but also cultivated assets such as trees and livestock while intangible acquisition covers such items as the costs of mineral exploration, computer software (where used in production for more than one year) and original works of literature or art. Entries in the account can be either positive (acquisitions) or negative (disposals) and Table 3.9 lists the main types of each.

Table 3.9: Positive and negative gross fixed capital formation

Positive acquisition	Negative acquisition (i.e. disposals)
Purchases of fixed assets	Sales of fixed assets
Retention for own use of fixed assets produced (including those not yet completed or mature)	
Bartered assets acquired	Bartered assets surrendered.
Capital transfers in kind received	Capital transfers in kind paid or surrendered.
Acquisitions under financial leases	
Major improvements to fixed assets, going well beyond ordinary maintenance and repairs and classified with acquisition of new assets of the same kind; includes work on historical monuments.	
Growth of renewable natural resources, notably trees cultivated year after year and farm animals kept for use in production (e.g. breeding stock, dairy cattle, sheep reared for wool and draught animals)	

Note: Purchases, assets and capital receipts can be of either new or existing assets. Disposals do not include consumption of fixed capital (e.g. through anticipatable accidental damage) or exceptional loss of assets through natural disasters.

3.61   Acquisition of fixed assets includes acquisition by purchase, by barter, by financial leasing or by capital transfer in cash or kind. Details of the types of asset covered are given in Chapter 6 (6.6–6.15). The distinction between gross fixed capital formation and intermediate consumption is described in Chapter 4 (4.33–4.38).

3.62   In the matter of timing gross fixed capital formation is recorded at the time when ownership of the assets is transferred to the unit which will use them in production. Where there is a financial lease this time is assumed to be when the user takes possession and for own-account capital formation it is when the asset is produced. Valuation is at producers' prices (including costs of ownership transfer) subject to the following special cases:

- Own account production is valued at the basic prices of similar assets or, if these are not available, at costs of production;
- Mineral exploration is valued at cost;
- Computer software is valued at purchasers' prices when purchased in the market and, when developed in-house, at estimated basic prices or, failing that, costs of production;
- Entertainment and original works of literature or art are valued at the price paid by the purchaser (if any) *or* the basic price of similar items *or* the production cost *or* the discounted value of future receipts from use.

Disposals by sale are valued at basic prices *less* any ownership transfer costs incurred by the seller.

Changes in inventories (P.52)

3.63   Inventories (formerly called stocks) are of four types:

- Materials and supplies stored with the intention of using them in production as intermediate inputs (including precious materials);
- Work-in-progress, including growing crops, maturing trees and livestock, uncompleted structures (except production on own account or under a contract of sale agreed in advance), other uncompleted assets (e.g. ships) and partially-completed research, film productions and computer programs;
- Finished goods, i.e. output which the producer does not intend to process further (including that for use as intermediate input to some other production process);
- Goods for resale, i.e. that were acquired with a view to reselling in their existing state.

Changes in inventories are measured by the value of entries *less* withdrawals and recurrent losses. The valuation is made at the time when the entry or withdrawal takes place and, consistently with other flows, is at basic prices for producers' inventories or, in the case of work-in-progress, proportionate to the basic prices of finished products. Goods for resale which are added to wholesalers' or retailers' inventories are valued at purchasers' prices while those withdrawn are valued at the purchasers' prices at which they could be replaced at that time.

3.64   Recurrent losses from inventories can occur as a result of physical deterioration, accidental damage or pilfering. When these affect materials and supplies they are treated as if the stores had actually been used up in production (i.e. as intermediate consumption), for work-in-progress they are treated as an offset to the additions accruing from production within the sane period, while for finished goods and goods for resale they are treated as if withdrawn in the usual way at current prices.

3.65   It is recognised that in practice the conceptually correct valuation as described above is often difficult to apply and that approximate methods need to be used. In particular, when changes in the volume of inventories are regular or the price is fairly constant it is reasonable to apply the *average* price for the period (basic price for producers, purchasers' price for traders). Where both volume and price change substantially over short periods it is necessary to use more sophisticated approximation methods, taking account of the phasing of (for example) seasonal fluctuations. If hard information is only available for the beginning and end of each accounting period (e.g. because there are many individual products) then it may be possible to estimate intervening changes in volume from turnover rates. It should be noted that very marked seasonal changes in nominal prices (e.g. for fresh fruit and vegetables) may reflect differences in quality, which should be treated as *volume* effects.

Acquisition and disposal of valuables

3.66   Valuables are non-financial goods that are not used primarily for consumption, do not normally deteriorate over time and are acquired and held primarily as stores of value. They include jewellery, precious stones, art objects, antiques and collectors' items. In internationally-accepted economic accounting practice acquisitions and disposals of these are separately distinguished for:

- Financial intermediaries (including central banks) dealing in non-monetary precious metals;
- Enterprises which do not *trade* in these products (and for whom they therefore do not count as intermediate consumption or fixed capital formation);
- Households (for whom they do not count as final consumption expenditure).

The European System of National Accounts goes further than this and also separates them out for:

- Jewellers and art dealers (though in principle their transactions in such items should count as changes in inventories);
- Museums (though in principle their transactions should count as fixed capital formation).

These additional conventions avoid the need for frequent reclassification between the three types of capital formation as goods change hands within the art and antiquities market. *Production* of 'valuables' is valued at basic prices and all other acquisitions at purchasers' prices (including agents' fees or commissions and dealers' margins). Disposals are valued at the prices received by sellers *less* fees and commissions.

Chapter 4

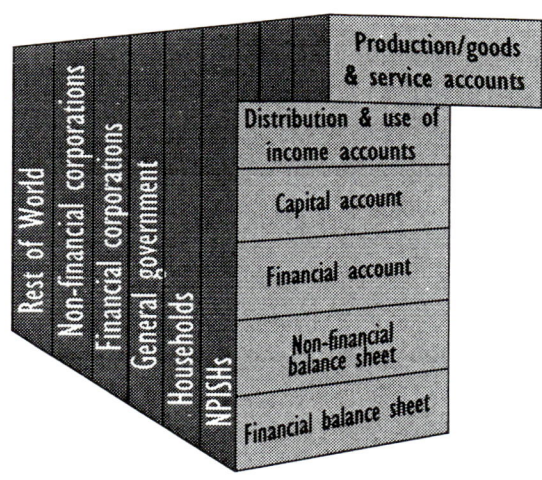

Rest of World
Non-financial corporations
Financial corporations
General government
Households
NPISHs

Production/goods & service accounts

Distribution & use of income accounts

Capital account

Financial account

Non-financial balance sheet

Financial balance sheet

The Production Account

Chapter 4 The production account

Summary

4.1 This chapter introduces the concept of valued added, and the associated concepts of:

- intermediate consumption;
- consumption of fixed capital.

with emphasis on the fundamental accounting identities involved.

It gives and explains the definitions of:

- market and non-market output;
- inventories and fixed assets;
- basic prices, producers' prices and purchasers' prices;
- alternative measures of value added;
- GDP.

In the course of this it touches on, *inter alia*:

- the effects of inflation on inventories;
- goods and services which are difficult to classify, e.g. military equipment, R&D and collective services;
- the nature and impact of VAT.

There is discussion of the practical problems and possibilities of recording and valuing output, inventories, work in progress and fixed assets.

Particular mention is made of ways in which economic accounts conventions differ from those generally followed in the preparation of business accounts.

4.2 The production account (referred to in the ESA as 'Account I') is at the heart of economic accounts as it records the activity of producing goods and services. Its balancing term is value added, which measures the contribution to GDP made by individual producers, industries or sectors. Gross value added is measured as the value of output *less* intermediate consumption. It is the source from which primary incomes are generated and it is carried forward into the distribution and use of income accounts which will be described in Chapter 5.

4.3 The production account has only one item on the 'resource' side, namely output, though this can be valued in different ways according to the choice made between basic prices and producers' prices and the treatment of taxes. In the 'uses' section the account differentiates between intermediate consumption (i.e. the goods and services directly used up in producing output) and consumption of fixed capital (resulting from progressive wear and tear). The latter makes the difference between gross and net value added. Thus the fundamental identities of the account are:

> Output
> *less*
> intermediate consumption
> *less*
> (taxes *less* subsidies on products)

> = Gross value added

> = Net value added
> *plus*
> consumption of fixed capital

For the economy as a whole gross/net value added corresponds to gross/net domestic product. Institutional sectors have their own production accounts but in these output and intermediate consumption are not broken down by product.

4.4 Exceptional treatment is accorded to financial intermediation services indirectly measured (FISIM) which is described in more detail in Chapter 2 and Annex 1 to Chapter 20. FISIM are financial services charged for by a differential interest rate rather than by explicit charges. The effect is that these services are recorded in the goods and services account and the production account as being sold by financial intermediaries to a notional sector and industry, of which they are intermediate consumption.

4.5 Several of the items in the production account were introduced in Chapter 3 in the context of the goods and services account, namely output, intermediate consumption and taxes/subsidies on products. This chapter does not repeat the definitions, classifications and valuation conventions described there but explores further the concept of the production boundary and the principles governing the measurement of output from the point of view of understanding the process of production *per se*. It goes on to look at the new element in this particular account – consumption of fixed capital – and concludes with an examination of value added and associated aggregates which feed into the rest of the economic accounts.

Output, sales and inventories

4.6 The measurement of output was discussed in Chapter 3 mainly from the standpoint of definition and classification, with a view to fitting it into the overall framework of transactions in goods and services. This section goes more deeply into the principles involved so as to clarify not only what is done but why it is done that way, with particular reference to inventories.

4.7 Output is a concept which applies to producer units rather than the processes of production. The output of an establishment (or, in the terminology of the ESA, a kind-of-activity unit, KAU) consists only of those goods and services produced within it, and the output of an enterprise is the sum of the outputs of its component establishments. Some goods and services are used up in the production process within the same establishment and accounting period: they do not leave the establishment and are therefore not counted as output but as intermediate consumption. Others are not consumed but, at the end of the accounting period, are not yet in a state to be marketed: they are treated as work-in-progress.

4.8   When finally produced as output goods and services may be disposed of in any of seven ways:

As market output:
- Sold at economically significant prices;
- Bartered in exchange for other goods, services or assets;
- Used as payments in kind (e.g. to employees);
- Added to the producer's inventories;
- Supplied to another establishment within the same enterprise for use in production;

As non-market output:
- Retained for the producer's own final consumption, e.g. gross fixed capital formation;
- Supplied free or at prices which are not economically significant.

As sales, barters, etc. may come out of inventories rather than directly from production it is necessary to subtract any such withdrawals from the additions to inventories, giving rise to the important accounting identity:

Value of output	=	sales and other uses of output	+	change in inventories

Market output (P.11)

4.9   Market output is subject to the following conventions. Sales are recorded when ownership of goods passes, or a service is provided, to the purchaser, and valued at the basic prices at which they are sold or, failing that, at producers' prices, not including any interest or other supplementary charges (which are recorded as separate transactions). Bartered goods and those used as payments in kind are valued at the basic prices that *would* have been received if they had been sold. Goods and services passed between establishments within an enterprise are counted as outputs of the producer establishment, valued at current basic prices, and become either intermediate consumption or gross fixed capital formation for the receiving establishment, valued there at the same prices *plus* any additional transportation costs paid to third parties. The use of artificial transfer prices (used for internal accounting purposes within enterprises) is avoided wherever possible.

Measuring changes in inventories

4.10   The need to value output at the basic prices of the time at which it is produced, and sales or other disposals at prices (*see* paragraph 4.18) ruling at the time of disposal has implications for the valuation of changes in inventories. If the identity

Output = sales + change in inventories

is to be fulfilled, then additions to inventories should be valued at the time it is produced (at the ruling basic prices); and withdrawals from inventories are valued at the prices at which they are sold or otherwise disposed of. Thus the positive amount of sales etc. from inventories is cancelled out by the negative amount of the withdrawals from the inventories. This means of valuation, which has been called the 'perpetual inventory method' (PIM), is not always easy to apply in practice and can lead to counter-intuitive results.

4.11   The problems arise in periods of inflation when price increase can occur while goods are being held in inventory. Business accounting methods using historic costs (always valuing stocks at the same prices as when they were entered) have the effect of counting such 'holding gains' as additional output, thus producing correspondingly inflated estimates of operating surplus. The PIM avoids this, but at the cost that all entries to and withdrawals from inventories need to be recorded continuously as they occur, which may be difficult even with modern methods of management and control. These difficulties have given rise to the widespread use of the approximate methods of valuation described in Chapter 3, taking the difference in quantities over the accounting period and applying the *average* price prevailing, but these are theoretically inappropriate and can be highly inaccurate in circumstances where prices or inventories show short-period fluctuations, most notably perhaps in the case of agricultural production.

4.12   As noted in Chapter 3, goods held in inventory may be subject to recurrent losses and these need to be subtracted from entries-less-withdrawals to get the total change in inventories. In the opposite direction, and notwithstanding the exclusion of 'holding gains', the change in inventories may include not only the value of the goods as originally stored but also additional 'output' produced through their being stored. This recognises that in some circumstances the storage can become an important process of production in its own right, whereby goods are 'transported' from one point of time to another, become qualitatively better and thus command a higher price quite apart from inflationary effects. The maturing of alcoholic drinks over a period of years is an example.

4.13   Also included in inventories is work-in-progress. At the end of each accounting period, producers have some output which is incomplete. Appropriate valuation of this is crucial for the production account because the balancing item – value added – would be meaningless if production processes continuing through more than one period were recorded as consuming intermediate inputs but not producing any output. The longer the gestation period the more important it is to record the current value of work-in-progress at the end of each period, and carry it forward as the opening balance for the next period. Cases in point include the production of films, some agricultural commodities and large construction projects.

4.14   Though work-in-progress is treated as an element of producers' inventories the borderline with gross fixed capital formation is not always clear. The latter cannot be recorded until ownership of the assets is transferred from their producers to their users. This does not normally take place until completion, so all output produced up to that point must be treated as work-in-progress even for the largest structures. However, if a contract of sale has been concluded in advance the transfer may be deemed (legally as well as conceptually) to happen in stages and capital formation be accrued on an incremental basis, though it should be noted that stage payments are sometimes made in advance or in arrears of actual work. The definition of work-in-progress can thus depend upon the form of contractual arrangements rather than the nature and timing of the production process itself.

4.15   Additions to and withdrawals from work-in-progress are accounted for in the same way as inventories of finished goods, being recorded at the appropriate time and (basic) prices, but in view of their special characteristics some further explanation of timing and valuation is required. Additions to work-in-progress are recorded continuously until completion, at which point the accumulated total is transferred *en bloc* to an inventory of finished goods. Thereafter when a sale takes place its value is exactly cancelled out by withdrawals from that inventory. The only record of output then remaining in the accounts is the successive additions to work-in-progress, appropriately distributed over the whole period of production.

4.16   In the absence of inflation the addition to work-in-progress in each period can be obtained by taking the value (at basic prices) of the final output and applying to it the proportion of total production costs incurred during the period in question. If the price which will eventually be realised is not known in advance then the additions can be estimated from production costs plus a mark-up, such estimates being revised when the actual sale takes place. A further complication arises when the anticipated sale price is continually increasing as a result of general inflation, and in this case the correct procedure is to use whatever is the expected price at the time the estimate is made. This is the only way to match inputs to outputs in each period in order to obtain economically meaningful measures of added value. It is an application of the general rule that additions to inventories must be valued at the prices prevailing when they take place, even if these need to be estimated.

Non-market output (P.12 and P.13)

4.17   This section deals with output for producers' own final use and other non-market output.

Output for producers' own final use (P.12)

4.18   Some output is retained by producers for their own final use, which can be either consumption or gross fixed capital formation. However, corporations have no final consumption and, as noted above, households' output of domestic and personal services for their own consumption is not measured in the accounts. The consumption element is therefore quite restricted, but in principle it does include owner-occupiers' housing services, services provided by employing domestic staff and goods produced and consumed by the same household.

4.19   Heads of owner-occupier households are treated as owners of unincorporated enterprises producing housing services, whose output is valued by reference to the rental that a tenant would pay for the same accommodation, i.e. the price of a comparable service sold in a well-organised market, in line with the general rule for own-account production. Rather similarly, domestic staff are treated as employees of enterprises owned and managed by the head of the household, whose compensation (including that in kind, e.g. food and accommodation) is by convention deemed to represent the value of the services produced, implicitly ignoring any intermediate costs or using up of fixed capital. For both owner-occupiers' costs and domestic services the same value as that recorded for output is entered in the accounts as final consumption expenditure.

4.20   Own-account output in the form of gross fixed capital formation can be undertaken by any kind of enterprise holding fixed assets for use in future production, such as engineering companies manufacturing special machine tools for their own use and households building their own house extensions. Total output for own use also includes changes in the relevant inventories, though where these take the form of work-in-progress on structures they are classified as acquisitions of assets rather than inventories as such.

4.21   Output of goods and services for own use should in principle be recorded at the time their production takes place, and valued at the basic prices at which they could be sold. This valuation is only possible if goods and services of the same type are actually bought and sold in sufficient quantities for reliable market prices to emerge. If they are not then a 'second best' procedure is to value by cost of production (i.e. the sum of intermediate consumption, compensation of employees, consumption of fixed capital and taxes *less* subsidies on production). In the case of construction the second best procedure is the usual one, and even this may be difficult to apply to household or communal production where most of the input is unpaid labour. The valuation here should be on the basis of wage rates paid for similar work in local labour markets.

Other non-market output (P.13)

4.22   Another type of non-market output is that of government and non-profit institutions serving households (NPISH sector) which is supplied free or at prices that are not economically significant. This includes collective services such as defence, or external relations for which it is technically impossible to charge individuals, and it also applies to goods and services for which government and the NPISH sector *could* levy a charge but choose not to as a matter of policy, e.g. education and health services.

4.23   A price which is not economically significant is one which is fixed as a matter of policy, has little or no influence on the quantities supplied or demanded but is charged in order to raise some revenue or moderate the excess demand which would arise if the item were absolutely free. It may not reflect either production cost or consumer preferences and so does not provide a suitable basis for valuing goods and services. In principle the output sold at such prices, or supplied free, could be valued by reference to similar output produced for sale but in practice this is seldom feasible. There are no markets for collective services such as public administration and defence and even for those offered to individual households (e.g. education and health) there are serious difficulties. Such services may indeed be produced on a market basis and sold alongside the non-market versions but there are usually major differences of type and quality between the two. For these reasons, and to ensure that all government and NPISH non-market services are treated consistently with one another, they are all valued according to costs of production, the net operating surplus being taken as zero. The output is recorded at the time it is produced (or delivered in the case of services).

4.24   Establishments mainly engaged in supplying goods and services free or at prices that are not economically significant are non-market producers, but not the only ones. Also included in this category are any establishments which market little or none of their production, such as subsistence farmers or units engaged in gross fixed capital formation on behalf of the enterprises of which they form part. Non-market output therefore includes, in addition to the components already described, the value of goods and services supplied by one establishment to another belonging to the same non-market producer to be used as intermediate input. Also included, as usual, is the related change in inventories.

4.25   Finally, what is primarily a non-market establishment may nevertheless engage in some market production, e.g. non-commercial museums selling reproductions as a secondary activity. Even so the total output should be valued at production cost. Subtracting from this the value of *market* output (given by sales) then yields the value of the non-market output as a residual, which subsumes the value of any sales at prices which are not economically significant.

Intermediate consumption (P2)

4.26   The concept of intermediate consumption was introduced very briefly in Chapter 3 together with some inclusions and exclusions chosen to distinguish this element of the accounts from other uses of goods and services. This chapter describes the theory of intermediate consumption as a part of the production process with particular reference to its recording and valuation and the rationale for the boundaries between it and three closely-related aggregates, namely compensation of employees, gross fixed capital formation and value added.

4.27   It is in the nature of the process of production that it consumes certain goods and services, either by using them up completely (e.g. electricity) or transforming them into something else (e.g. from grain into flour). The value of these inputs is called intermediate consumption. It includes inputs to ancillary activities (e.g. marketing, accounting, transportation, maintenance and security), which are not separately identified from inputs to principal or secondary activities. Also included are rentals paid for the operational leasing of fixed assets from other industrial units, and fees, commissions, royalties, etc., payable under licensing arrangements. Excluded are expenditure on 'valuables' (e.g. precious metals and stones, jewellery and works of art), unless used in production, as these are not used up and do not deteriorate physically. The gradual consumption of fixed assets in the course of production is also excluded.

4.28   Intermediate consumption is recorded when the good or service in question enters the production process, and is valued at the purchasers' prices then prevailing, i.e. what the producer would have to pay to replace it, including transportation costs, trade margins and tax (*less* subsidy) on the product, as well as the basic price.

4.29   In practice establishments do not usually record directly the use of goods in production but rather their purchases of materials and supplies *intended* for use as inputs and changes in inventories of these. Subtracting the latter from the former gives intermediate consumption. Withdrawals from inventories reduce the (positive) change in them and thus increase intermediate consumption. Entries to and withdrawals or losses from inventories are all valued at purchasers' prices i.e. the same as purchases of materials and supplies.

4.30 Goods or services passed from one establishment to another (even within the same enterprise) and then fed back into production are counted as output of the first establishment and intermediate consumption of the second (e.g. flour passed from a grain-milling plant to a bakery) but if both processes take place within the same establishment then nothing is recorded.

4.31 Certain goods and services are consumed not by the production process itself but by employees working on it. It is immaterial to the employer whether these are treated as intermediate consumption or as employee compensation (in kind) since the choice does not affect the net operating surplus, but it *does* affect value added and hence GDP. In general the goods and services are regarded as intermediate consumption where employees need to use them in order to do the job, as with tools, special clothing, accommodation at the workplace, travel on business and necessary changing, washing or medical facilities. (These cases are subject to exceptions, for example, tools or clothing which employees *choose* to use when off duty or accommodation which can also be used by families. The cost of these exceptions would be included in the compensation of employees. However, it is immaterial whether the employer pays the bill directly or the employee makes the purchase and is subsequently reimbursed.)

4.32 However, goods and services provided to employees for use in their own time and at their own discretion for the satisfaction of needs and wants are treated as remuneration in kind, and include such items as regular meals, tied houses, company cars and travel to work.

4.33 The boundary between intermediate consumption and gross fixed capital formation requires more explanation, particular problems being posed by tools, maintenance and repair services, research and development, mineral exploration and military equipment. These are dealt with in turn in the following paragraphs.

4.34 Tools which are large, expensive, elaborate or account for a significant proportion of the producer's expenditure on durable goods are treated as fixed assets but items worth less than 500 ecu in 1995 prices are treated as intermediate consumption.

4.35 The problem with maintenance and repairs is to differentiate between ordinary regular activity required to keep fixed assets in good working order and, on the other hand, major renovations, reconstructions or enlargements which increase efficiency, capacity or life expectancy, effectively adding to the stock of fixed assets and therefore needing to be treated as fixed capital formation. The criteria shown in Table 4.1 come into play here.

4.36 In some areas, where activities are concerned with future efficiency rather than current consumption the distinction between consumption and investment is particularly problematical. As explained in Chapter 2, by international convention, all output of research and development, staff training, market research, environmental protection and similar activities is treated as intermediate consumption. Where an enterprise carries out research and development on a significant scale it is desirable to identify a separate establishment to cover it so that inputs and outputs can be distinguished, the latter usually being valued by costs of production because relevant price data are unobtainable. Typically the output of the research establishment will be distributed between some or all of the other establishments making up the enterprise and included in their intermediate consumption. Equally, where an enterprise contracts an outside agency to undertake R&D on its behalf this purchase of services is treated as intermediate consumption.

Table 4.1: Categorisation of expenditure on maintenance and repairs

Ordinary maintenance and repairs	Major improvements
Expenditure *must* be made if assets are to continue functioning properly. Process does not change the asset but just preserves or restores it, e.g. by replacing defective parts.	Discretionary investment decision is made, not dictated by the state of the asset and often well before it becomes unserviceable. Performance is significantly enhanced by enlargement, restructuring or refitting.

4.37 Mineral exploration, on the other hand, is always classed as gross fixed capital formation, even if unsuccessful, because it is undertaken for the purpose of acquiring new reserves.

The former treatment of general government consumption

The treatment in the UK Blue Books up to 1997 has been somewhat ambiguous. Both old and new versions of the SNA indicate that where, say, a government department purchases supplies for consumption, the purchase of these supplies, and the wages and salaries of the staff should be recorded in a production account. The output would be the service provided by the department - for example defence or environmental protection. Final consumption would be the purchase of this output on behalf of the community. Thus purchase of supplies is a purchase of intermediate output.

This is the way the transactions were recorded in Table 2.1 of the 1997 Blue Book (input-output tables). However in Sections 7–9 (central, local, and general government) no separate production account was produced, so several of the tables show analyses of final consumption which look through to the production account components of wages and salaries, procurement, and the imputed charge for capital consumption.

4.38 Awkward conceptual issues arise in the case of military equipment. Traditionally all goods required for purposes of national defence have been treated by the SNA as intermediate consumption rather than capital formation, and this is clearly appropriate in the case of weapons such as missiles and their warheads which are not durable, cannot be used repeatedly and, if used at all, result in destruction rather than production. By extension it is reasonable to regard as intermediate consumption any systems whose sole purpose is the delivery of such weapons (e.g. warships, fighter aircraft and tanks). Nevertheless, the provision of defence requires long-term use of durable goods which are similar to, and used in much the same way as, some of those with non–military purposes (e.g. docks, airfields and bridges) and which may indeed be shared between military and civilian roles. The 1993 SNA requires expenditure on such structures to be treated as gross fixed capital formation. The same treatment applies to equipment (e.g. office machinery and vehicles used for transport) and, exceptionally, to light weapons used by non-military establishments engaged in internal security or policing, even though similar goods bought by military establishments would be regarded as intermediate. The essential distinction that is made is according to the purpose for which goods are acquired: whether they are ordinary durables that could be used by either military or civilian producers or, on the other hand, specialised items designed only to destroy.

4.39 The intermediate consumption of enterprises excludes collective services supplied to them by government because such services benefit many different enterprises and households and it is impossible to allocate them appropriately. By convention, collective services even include specific non-market output provided to individual market producers (e.g. free veterinary services for farmers). Included in intermediate consumption is a different form of collective provision, namely the services rendered by non-profit making business associations, which are used up by the membership of market producers and valued by the subscriptions they pay.

4.40 The treatment of the intermediate consumption of financial intermediation services is described in Annex 1 to Chapter 20.

4.41 Social transfers in kind which are produced by market producers and provided without further processing to households are treated as final consumption expenditure by the government establishments or NPISHs that supply them. They form part of the actual consumption of households. (*See* Chapter 5 for further discussion.)

4.42 Finally under intermediate consumption it should be noted that the borderline with value added is not rigidly defined by the technology of production but is also influenced by the way production is organised – in particular the extent to which ancillary services are bought in from outside the establishment as opposed to being produced in-house. At disaggregated levels the former may reduce value added, increases intermediate consumption and changes input-output ratios, even though technology and efficiency may be completely unaffected.

Consumption of fixed capital (K1)

4.43 Consumption of fixed capital is the decline during an accounting period in the current value of the assets used by producers, as a result of physical deterioration, normal obsolescence and accidental damage. It is not identical to depreciation as recorded in business accounts or as allowed for taxation purposes. Rather it is defined in a way that is consistent with the rest of the production account and is relevant for economic analysis, which means that it reflects underlying resource costs and relative demands at the time production takes place, calculated using the prices prevailing at that time rather than when they were originally acquired.

4.44 Consumption of capital cannot be gauged simply from the rentals payable by the users of assets to the owners for the right to use them in production for a specified period of time, because those rentals also include the interest cost of the owners' capital tied up in the assets and any other costs they incur, even if owners and users are one and the same. If the assets are actually rented under operating leases or similar arrangements then the whole of the rental is recorded in intermediate consumption as the purchase of a service provided by the lessors. (The lessor, as owner of the asset, would record capital consumption.)

4.45 Capital consumption covers both tangible and intangible assets, the former including those constructed to improve or traverse land (e.g. drainage systems and roads). Some of these might appear to have infinite lives if properly maintained but it must be remembered that assets may lose value not only because they deteriorate physically but also because they are no longer needed or have become obsolete. Obsolescence is allowed for to the extent that it takes place at a normal rate but not where it results from unexpected technological developments. Similarly losses due to accidental damage are included in capital consumption when they occur with predictable regularity and a representative average rate can be calculated, but not when they result from war or major natural disasters which occur very infrequently. Losses falling outside these criteria are recorded in a different account: that for other changes in the value of assets.

4.46 Capital consumption does not cover valuables, or the depletion and degradation of non-produced assets such as land and mineral deposits.

4.47   In principle the value of a fixed asset is determined by the discounted present value of the stream of future rentals expected over its remaining life, valued at current prices or rentals, and consumption of fixed capital is the decrease in this value over an accounting period. This is a forward-looking measure, depending on the benefit which producers expect to derive *in future*, unlike depreciation in business accounts which usually represents an allocation of *past* expenditures over subsequent accounting periods. At the very least these latter would need to be adjusted from historic to current prices, and depreciation allowances for tax purposes are best ignored altogether as they have often been arbitrarily manipulated for commercial reasons. A more satisfactory solution is to estimate consumption of fixed capital independently in conjunction with estimates of the capital stock. This can be done by the perpetual inventory method using, *inter alia*, data on gross fixed capital formation in the past and information or assumptions about the rate at which its efficiency will decline over time.

4.48   The mechanics of the perpetual inventory method are described in Chapter 16 but the logical sequence of steps in the process is broadly as follows:

- The 'gross capital stock' for the current period is obtained by estimating how much of past gross fixed capital formation has survived, and then revaluing it at current prices (or the prices of some base year if a constant price time series is required).
- The rate at which efficiency declines over time is estimated for each type of asset according to a pattern which may be linear *or* show constant geometric decline *or* a constant level of efficiency until the assets suddenly disintegrate (as in the case of a light bulb) *or* some combination of these. The relative efficiencies determine the rentals users will pay, and hence the current value of the asset.
- This decline in the efficiency of assets is not the same as consumption of capital, as is most clearly seen from the light bulb example where the former takes place all at once but the latter is continuous. Capital consumption tends to accelerate as assets get older but it becomes constant over time ('straight line depreciation') in the special and simple case where rentals fall at a linear rate up to a cut-off point and then drop instantaneously to zero. A more 'accelerated' pattern of capital consumption, which is more prudent and perhaps more realistic to work with, is the 'double-declining balance' method whereby the depreciation is calculated as a constant fraction $(2/n)$ of the written-down value of the assets at the start of the period, where n is the expected service life. This is in fact a good approximation to the case of constant geometric decline.
- Once the pattern and rate of depreciation for a particular asset has been determined the capital consumption is calculated by simply applying this to the current purchase price of a new asset of the same type (which can be obtained by uprating the original price by an appropriate price index).
- Subtracting the cumulative capital consumption up to a given point of time from the purchasers' price of a new asset gives the 'written-down value' and the sum of written-down values is described as the net capital stock and goes into the balance sheet.

4.49   Because consumption of fixed capital is an imputed value its economic significance is different from that of parts of the accounts based on market transactions. Though the concepts governing its measurement are clear, their implementation requires assumptions about future asset lives at a degree of detail which can make practical estimation very difficult. For this reason balancing items such as value added and saving are often recorded 'gross', i.e. before capital consumption. However, for analytical purposes it is usually more appropriate to derive a net figure and doing so can make a very large difference to the results. Capital consumption in the United Kingdom accounts for 10 per cent or more of total GDP.

Definition and measurement of prices

4.50   This section describes various different methods of valuation for inputs and outputs in the production account. The differences arise from alternative treatments of taxes and subsidies on products, value added taxes and transport charges.

4.51   Chapter 2 (2.50–2.58) described the price bases defined in ESA95:

Producers' price = Amount receivable – Value added tax

Basic price = Amount receivable – Value added tax – Other taxes on products
less subsidies
on products

(both excluding separately-invoiced transport charges).

Purchasers' = Amount – Deductible
price receivable VAT

(including separately-invoiced transport charges).

4.52   As stated at the beginning of this chapter value added is the balancing item of the production account. It measures the value created by production, either before deducting consumption of fixed capital (gross value added) or after (net). Neither involves any duplication, as the value of intermediate consumption is deducted from both. As a balancing item valued added is not an independent entity but a function of all of the other entries in the production account. It cannot be measured as the sum of a specific set of transactions each having a unit price and a quantity. However, the various inputs and outputs which affect value added *do* have prices and quantities and by applying different price vectors to the given quantity vector it is possible to generate alternative measures of value added. The names conventionally given to these do not fully describe their definition, as follows:

- *'Value added at basic prices'*. Here output is valued at basic prices and intermediate consumption at purchasers' prices. The measure is particularly relevant for producers as it uses the prices actually paid and received.

- *'Value added at producers' prices'*. With this alternative output is valued at producers' prices and intermediate consumption at purchasers' prices. In the absence of VAT this is an economically meaningful 'market price' measure, since intermediate consumption valued at purchasers' prices is the same as that valued – in conformity with output – at producers' prices (the only difference being whether distribution and transport services are regarded as being traded separately). However, with VAT the two prices are different. Value added at producers' prices differs from that at basic prices by the amount of taxes (*less* subsidies) on products.

- *'Value added at factor cost'*. This is obtained from either of the measures already described by subtracting any taxes (*less* subsidies) payable out of value added as defined, e.g. taxes on payroll, vehicles or buildings. However, by their nature such taxes cannot be eliminated from prices so this is not strictly a measure of value added but rather of income. It represents the amount remaining for distribution and equals the total value of factor incomes generated by production.

4.53 Conceptually the sum of gross values added for all United Kingdom producers (derived from the production accounts by the 'production approach') should equal GDP as given, quite independently, by the 'expenditure approach'. However, the latter is defined as

| GDP at market prices (from expenditure approach) | = | Final expenditure at purchasers' prices | – | Imports f.o.b. |

which includes (in final expenditure) the value of non-deductible VAT and excludes (from imports) the value of import taxes (*less* subsidies). However, in the United Kingdom, gross value added is measured at basic prices, which exclude VAT and other taxes *less* subsidies on products. It follows that, to preserve the identity of GDP, both non-deductible VAT and taxes (*less* subsidies) on products need to be added to the sum of gross values added.

4.54 However, where as in ESA79 value added is measured at producers prices, it *excludes* non-deductible VAT and measures imports at purchasers' prices which *include* import taxes (*less* subsidies). In this case the GDP identity requires that only non-deductible VAT and import taxes (*less* subsidies) need to be added to the sum of gross values added.

4.55 GDP defined in this way measures the value created by the productive activity of United Kingdom residents, which is not the same as production taking place within the economic territory of the United Kingdom. Some production by United Kingdom producers takes place overseas (e.g. through the installation, repair and servicing of equipment or the provision of services directly to clients). Equally some activity within the UK is carried out by institutional units from abroad. The distinction between resident and non-resident units is therefore an important one for the definition and coverage of GDP, and is becoming ever more so as developments in transport and communications make national borders progressively less significant for trade.

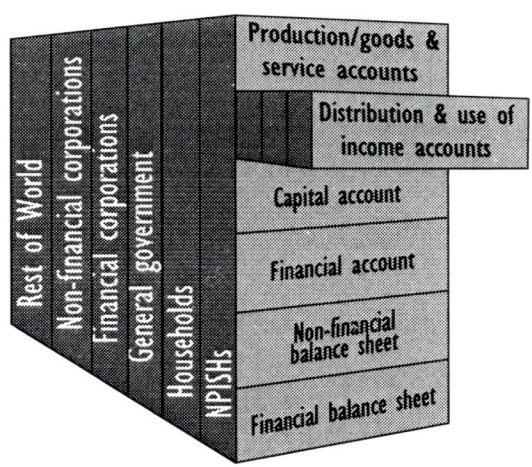

Distribution and use of income accounts

Chapter 5 Distribution and use of income accounts

5.1   This sequence of accounts traces flows of income in four distinct stages, summarised in Table 5.1

Table 5.1: Distribution and use of income accounts

Stage	Accounts	Result yielded

1	Primary distribution of income
	– Generation of income (from production process)
	
	– Allocation of primary income (between labour, capital and taxes)
	    – Entrepreneurial income
	    – Other primary income
		Operating surplus (or mixed income)
		
		Balance of primary incomes (or national income for UK economy)

2	Secondary distribution of income (via transfers)	Disposable income
3	Redistribution of income in kind (via social transfers)	Adjusted disposable income
4	Use of income – Disposable income – Adjusted disposable income	Saving

These are dealt with in turn below.

Primary distribution of income accounts

5.2   Primary incomes are those that accrue to institutional units as a consequence of their involvement in production or ownership of productive assets. They include property incomes (from lending or renting assets) and taxes on production and imports, but exclude taxes on income or wealth, social contributions or benefits and other current transfers. The primary distribution of income is the way these are distributed among institutional units and sectors. It starts with the generation of income account, compiled for United Kingdom enterprises, which shows the sectors in which primary incomes *originate*. For example, the compensation of employees recorded for the household sector is that payable by unincorporated enterprises owned by households, not that receivable by the household sector.

5.3.   The upper part of the generation of income account shows only one 'resource' carried forward from the production account. This is value added, in the case of the sectors of the United Kingdom and gross domestic product in the case of the whole UK economy. The lower part shows as 'uses' the compensation of employees and taxes (*less* subsidies) on production, including any payable to non-residents (e.g. the EU). The taxes include those on labour and assets used in production but not those on income accrued from production. Taxes on production shown in the sector accounts are those which are an integral part of the price and therefore regarded as being paid by the producer (for example, in the United Kingdom, business rates). For the sectors, they do not include taxes such as VAT or other taxes on products linked to the quantity of output produced. Invoiced VAT is never treated as part of the price receivable by the producer and is therefore omitted from this account for producing sectors as it is not a charge against value added. However, for the whole economy taxes on products, including VAT, are, by convention, included in GDP in the production account and they are therefore also recorded as a charge against GDP in the uses side of the generation of income account.

5.4   The balancing item of the generation of income account measures the surplus or deficit accruing from production before taking account of any interest or rent either payable or receivable. This is called the operating surplus, except in the case of unincorporated enterprises owned by households where the balance implicitly incorporates some remuneration for work done by the owner or other household members, inseparable from the return to the owner as entrepreneur, and is therefore called 'mixed income'. See box below.

The balancing item of the generation of income account corresponds *broadly* to the following concepts in the former SNA:

  gross trading profits of companies
  gross trading surplus of public corporations and general government
  income from self-employment
  rental on buildings (including ground rents on leasehold property, but *not* rent on land alone)

before or after deduction of capital consumption, as appropriate. See the section on rents and rentals later in this chapter.

5.5   No net operating surplus is generated by non-market producers owned by government or non-profit institutions, whose output is valued by production costs rather than market prices. For other producers the operating surplus excludes holding gains on inventories or fixed assets (by virtue of the way the components of value added are measured), unlike profit as reported in business accounts. The operating surplus for the whole economy includes not only the corresponding total for the standard institutional sectors but also the value of financial intermediation services other than those directly measured and accounted for as financial services. For the time being, these 'financial intermediation services indirectly measured (FISIM)' are measured as the property income receipts less interest payments of financial intermediaries other than insurance corporations and pension funds. They are part of financial intermediaries' output but cannot yet be attributed to the actual sectors receiving the services. Until agreement has been reached on the methodology to be used for attribution, they will continue to be treated as intermediate consumption by a notional sector, which therefore has negative value added and a negative operating surplus as a result. That notional sector is included in the UK totals along with the other, identified, sectors and the sum of the identified sectors' operating surplus is thereby reduced.

5.6   The second part of the primary distribution of income concerns allocation, focusing on institutional units or sectors in their capacity as income recipients rather than income-generators. Two kinds of income are shown as resources in the account, the first being the primary incomes already described while the second consists of property incomes in the form of interest and dividends from financial assets and rents from land or sub-soil assets. The 'uses' are the property incomes payable, to creditors, shareholders, landowners, etc. Payments/receipts other than rents can be to/ from non-residents of the United Kingdom as well as residents. The nature of the balance of primary income varies considerably between sectors, according to the types of income received: for example, apart from non-residents, taxes go only to government and compensation of employees only to households. At the level of the whole economy the balancing item is described as 'national income' while the aggregate difference between primary incomes receivable from and payable to non-residents is 'net income from abroad'.

5.7   Numerically, net income from abroad is equal to the difference between gross national income (GNI) and gross domestic product (GDP) as previously defined. GNI was traditionally called 'gross national product' though based on income rather than output, but this terminology is no longer used in the economic accounts. Both GNI and GDP refer to the same set of institutional units, namely UK residents. (The term 'gross national product' will remain in use for the time being with reference to data produced under the old ESA 1979 definition. That measure will continue to be used for the immediate future as the basis for assessing EU Member States' 'fourth resource' contributions to the EU budget.)

The allocation of primary income account is sometimes partitioned into two sub-accounts in order to identify an additional balancing item, namely the entrepreneurial income of institutional units and sectors. This is defined as:

| Entre-preneurial income | = | Operating surplus | + | Property income receivable | – | Interest and rents payable. |

This is clearly an income concept, similar to that of profit and loss in business accounting except that the latter often incorporates holding gains. Entrepreneurial income may be difficult to identify for unincorporated enterprises because of the problem of separating the assets and liabilities of the enterprise itself from those of the owners personally. Such a separation is possible in the case of quasi-corporations but often not where an unincorporated business is owned by a household.

Whenever an entrepreneurial income account is compiled it is accompanied by an 'other primary income' account in which the first 'resource' is entrepreneurial income, followed by compensation of employees received by households, taxes receivable by government (*less* subsidies) and property incomes receivable (other than those included in entrepreneurial income). These are set against the 'use' represented by property income payable (other than enterprises' rents) and the resulting balancing item is equal to that of a consolidated allocation of primary income account.

This subdivision is not implemented in the United Kingdom.

5.8   The following tables summarise the contents of the primary distribution of income accounts.

Table 5.2: Generation of income account

Uses	Resources
Compensation of employees Taxes on production and on imports *less* Subsidies	Value added
Operating surplus	

Table 5.3: Allocation of primary income account

Uses	Resources
	Operating surplus
Property Income	Compensation of employees Taxes on production and on imports Subsidies Property income
	Balance of primary incomes

Compensation of employees (D.1)

5.9   Compensation appears in both the generation of income account (as a 'use' for the sectors which pay it) and in the allocation of primary income account (as a 'resource' for the households and rest of the world sectors, which receive it). This section describes its coverage and definition.

5.10   Compensation of employees is defined as the total remuneration payable by enterprises in cash or in kind, and comprises not only wages and salaries but also the value of social contributions payable by the employer (including imputed contributions for unfunded benefits), but not taxes paid by the employer. It is recorded on an accrual basis; that is in respect of entitlement arising out of work done during the accounting period whether paid in advance, simultaneously, or in arrears. It does not cover unpaid work (including that done by household members within their own households) or the earnings of the self-employed.

5.11   To be classified as 'occupied' a person must be engaged in an activity that is within the production boundary. Of these, 'employees' are those who have an agreement (formal or informal) with an enterprise to work in return for remuneration, normally based on time spent or work done. The 'self-employed', on the other hand, are people who own unincorporated enterprises in which they work, these being neither separate legal entities nor separate institutional units. Such people receive mixed incomes rather than compensation.

5.12   For the purpose of classifying incomes in the economic accounts, any occupied person producing entirely for their own final consumption or capital formation, whether individually or collectively, is treated as self-employed, as are unpaid family members. That is to say, their remuneration takes the form of mixed income. However, where a single shareholder or small group not only own a corporation but also work for it and are paid remuneration (apart from dividends) they are treated as employees. Students are generally regarded as consumers (of educational services) rather than employees unless they have a formal commitment to provide labour, (for example as apprentices, articled clerks or research assistants) in which case they are treated as employees even if they receive no remuneration at all in cash.

5.13   Self-employed persons can be either employers (i.e. those with paid employees) or own-account workers. A special category of the latter is outworkers, who have a prior arrangement or contract to work for a particular enterprise but whose place of work is not within any of its establishments, generally being at their home. They therefore meet at least some of their own production costs. Outworkers have some characteristics of employees and some of the self-employed, and their classification is determined by the basis on which they are remunerated: those paid for the amount of work done (i.e. inputs) are employees whereas those paid according to the value of their outputs are self-employed, as are those who themselves employ others to do the same kind of work. A supplementary criterion, as already suggested, is that employees have implicit or explicit contracts whereas own-account workers do not. The distinction has important implications for the economic accounts as employees are paid out of the enterprise's value added while payments to own-account workers are purchases of intermediate goods and services. The income of the self-employed is 'mixed income' (B.3) (*see* below).

5.14   Compensation of employees is divided into:

- wages and salaries
- employers' contributions

Wages and salaries (D.11)

5.15   Wages and salaries comprise a number of elements:

- Regular earnings, including those for piecework, overtime, working away from home, etc;
- Supplementary allowances in respect of housing, travel to work, etc;
- Holiday or lay-off pay for employees away from work for short periods;
- *Ad hoc* bonuses and other exceptional payments linked to the overall performance of the enterprise;
- Commissions and gratuities received by employees, which are treated as payments for services rendered by the enterprise and included in its output and value added;
- Social contributions, income taxes, etc, payable by the employee, even if withheld by the employer for payment directly to the authorities.

Wages and salaries do not include reimbursement of expenses incurred by employees in taking up new jobs or equipping themselves with tools, clothing, etc, needed wholly or mainly for their work. These are treated as intermediate consumption of employers. Any necessary expenditure which is not reimbursed is deducted from wages and salaries and added to intermediate consumption – not regarded as household final consumption.

Employers' contributions (D.12)

5.16   These comprise contributions in cash (D.121) by the employer to social security schemes or to private pension funds, insurance or medical schemes; and imputed contributions (D.122) in respect of unfunded benefits. Employers' contributions exclude those payable by the employee, even if they are withheld by the employer and paid directly to the scheme.

5.17   Unfunded social benefits are sometimes paid by employers in the form of (for example) education allowances for employees' dependents, payments for sickness or maternity leave and severance pay. These are not strictly a form of remuneration because they are not related to the amount of work done but provided selectively to individual employees meeting certain criteria. However, the contingent liability incurred by the employer is treated as a form of employee remuneration. Ideally its value should be based on an actuarial calculation of what it would cost to fund *future* claims but in practice an estimate may have to be derived from information on *current* actual payments. From an accounting point of view imputed social contributions for unfunded benefits are recorded in the generation of income account as a payment from employers to existing employees, though in the secondary distribution of income account (see below) employees are recorded as paying it back as if to a separate social insurance scheme.

5.18   Wages and salaries in kind are goods and services provided by an employer which are not necessary for work and can be used by employees or members of their households in their own time and at their own discretion for the satisfaction of needs and wants. Income in this form may be less welcome than cash but it still needs to be valued consistently with other goods and services, using purchasers' prices when the items have been purchased by the employer and producers' prices when they have been produced by the employer. Typical examples of wages and salaries in kind are given in Table 5.4. Some of these (e.g. transport to work, car parking and crèches) have some of the characteristics of intermediate consumption but they are treated as compensation because they are not related to the production process or working conditions and many employees have to pay for such things out of their own incomes as final consumption.

Treatment of taxes on production and imports (D2)

5.19   Taxes in general are compulsory unrequited payments made by institutional units to government, or institutions of the European Union, in cash or in kind. Taxes on production and imports include: *taxes on products* which are payable on goods and services when they are produced or disposed of; *taxes on imports*, payable when goods enter the UK economic territory or services are delivered to resident units by non-resident; and *other taxes on production*, which comprise taxes on the ownership and use of assets employed in production (including land), or those paid in order to be allowed to produce, and on labour employed or compensation paid (excluding social security contributions). Other taxes on production are paid irrespective of the level of production. Together they correspond to 'indirect taxes' as traditionally understood, which might be passed on to other institutional units by increasing prices. They are recorded in the generation of income account under 'uses' and in the allocation of primary income account under 'resources'.

> Outside the European Union the SNA allows the alternative of measuring output at producers' prices, which include the taxes on products other than VAT. In that case these taxes on products need to be charged to the generation of income account.

5.20 For the purpose of the generation of income account individual units do not record import taxes, non-deductible VAT or other taxes on products, these having already been deducted from the value of output at basic prices in the production account. *Other taxes on production* (taxes on assets and labour) always need to be recorded as they cannot be deducted from producers' prices but are payable out of value added. However, receipts by general government and the European Union of all types of taxes on production are included as resources in the allocation of primary incomes account.

5.21 A distinction needs to be drawn between taxes and fees paid for licences or certificates issued by the government. Where the issuing of these involves a service to the applicant the payment is treated as a purchase of services (unless the amount demanded is out of all proportion to the work involved) but where authorisation is granted automatically without any check on the suitability of the applicant, nor the provision of any services, it is regarded as a form of taxation. Some licence fees do not fit unambiguously into either category. In these circumstances it may be possible to split the licence fee into a payment for a service and tax.

5.22 Taxes are recorded so that tax receipts are allocated to the time when the tax liability was created. In the United Kingdom this is estimated by applying a lag to the cash receipts. Taxes are recorded net of refunds. In principle, interest on overdue tax should be recorded separately from the tax itself but this is usually not practicable and it is recorded as extra tax.

5.23 A sub-category of taxes on production is taxes on products (D.21), i.e. those levied per unit of good or service or *ad valorem*. They usually become payable when products are sold or imported. The main categories are:

- *Value added type taxes* (D211) as described in Chapter 4 above. These are payable on imports as well as domestic production. This type of tax is ultimately charged in full to final purchasers but is collected from enterprises in stages during the production process. Producers at each stage pay only the *difference* between the VAT on their sales and that on their purchases for their own intermediate consumption or gross fixed capital formation. In the accounts output and imports are recorded exclusive of VAT while purchases are recorded inclusive of that part of VAT which is non-deductible. The tax is therefore seen to fall mainly on final users, essentially for household consumption, though some is paid by enterprises (e.g. those exempted from VAT). (*See* Annex 2 to Chapter 14).
- *Other taxes on imports* (D.212). These become payable when goods enter the UK economic territory or services are delivered to residents by non-residents. They do not include taxes which subsequently become payable as goods, whether imported or home-produced, pass through the distribution chain (e.g. excise duties). Some taxes on imports (e.g. customs duties) are payable only on goods or services of particular types. Also included in this category (in principle) are implicit taxes resulting from the operation of multiple exchange rates.
- *Export taxes* (D.213) which become payable when goods leave the economic territory of the United Kingdom or services are delivered to non-residents. They include (in principle) taxes arising from multiple exchange rates.
- *Other taxes on products* (D.214), payable on the production, sale, transfer, leasing, delivery or use of goods and services. These include any general sales or turnover taxes, excise duties and taxes on specific services including the transactions of financial intermediaries.

Examples of taxes on products are shown in the following table.

Table 5.4: Taxes on products other than VAT and import duties

Tax	Coverage
Excise duties and consumption taxes	Excluding those levied on imports when the goods are imported. Including taxes on specific products such as alcohol, tobacco and road fund licences.
Stamp duties and taxes on capital transactions	Levied on legal documents and purchases or sales of assets (including foreign exchange)
Taxes on gambling	Excluding taxes on winnings
Taxes on insurance premiums	Insurance premium tax
Monetary compensatory amounts collected on exports	These are levied on agricultural products and paid to EU institutions.

5.24   Finally under 'other taxes on production' (D.29) it is possible to distinguish certain taxes levied on production but not on individual products. These include:

- Taxes on *payroll or workforce* (excluding taxes and compulsory social security contributions paid by employees) – for example, Selective Employment Tax which was levied in UK in the 1960s ;
- Recurrent taxes on the use or ownership of *land, buildings and other structures* – for example, business rates ;
- *Licences* to carry out a particular business or profession (unless these involve the provision of services to the applicant at reasonable cost, in which case they are part of intermediate consumption). This includes television franchise payment and radio spectrum charges levied by government ;
- Taxes on the *use of vehicles or other equipment* needed for production, normally levied at a flat rate irrespective of usage;
- Taxes on *pollution* (excluding charges for the collection and disposal of waste by public authorities – part of intermediate consumption).

Such taxes are payable irrespective of the profitability of production. Taxes on profits and other incomes received by businesses are excluded from this category.

Subsidies (D.3)

5.25   Subsidies are current unrequited payments that government units or the European Union make to resident producers or importers based on the type of production and in some cases linked to the volume of production. They are for many purposes equivalent to negative taxes. They do not include current transfers to households, and non-profit institutions serving households, or grants to enterprises to finance capital formation or compensate for capital losses.

5.26   The breakdown of subsidies parallels that for taxes on production. Subsidies on products (D.31) may be amounts per unit of quantity; *ad valorem* rates or calculated to make up the difference between the market price and a specified target price; and amounts paid to ensure a specific good or service continues in production at levels of output and price that the market would not supply without subsidy. Import/export subsidies become payable when goods enter/leave the United Kingdom economic territory. Import subsidies exclude any subsidies that become payable once goods have entered the country, while export subsidies exclude repayment at the customs frontier of taxes previously paid and the waiving of taxes that would be due if goods were not being exported.

5.27 Other subsidies on products include any to United Kingdom producers in respect of output used domestically, losses *deliberately* incurred by government trading organisations and regular transfers to public corporations and quasi-corporations to compensate for persistent losses on their market output. However, a transfer to a public corporation to cover losses built up over several years is classified as a capital transfer.

5.28 Other subsidies on production (D.39) as such include those on payroll or workforce instituted for economic or social reasons (including the promotion of training, such as the 'welfare to work' programme) and subsidies for additional processing to reduce pollution. Some examples of such subsidies on production are shown in Table 5.5.

Table 5.5: Coverage of subsidies on production

Inclusions	Exclusions
Subsidies on payroll or workforce, either in total or according to employment of particular groups, e.g. the handicapped or long-term unemployed	Current transfers from government to households in their capacity as consumers
Subsidies on the cost of training schemes organised or financed by enterprises	Current transfers between different parts of the general government sector in their capacities as producers of non-market goods and services
Subsidies to reduce environmental pollution, covering some or all of the cost of additional processing	Investment grants
	Extraordinary payments into social insurance funds designed to increase their actuarial reserves
Interest relief granted to encourage capital formation (treated in the accounts as subsidy to producers even if paid directly by the government to credit institutions	Transfers from general government to non-financial corporations and quasi-corporations to cover accumulated or exceptional losses
	Cancellation of debts to the government
Over-compensation of VAT resulting from the flat rate system (mostly in agriculture, but also applies to VAT repayments to certain public corporations and other businesses that are specifically exempt from paying VAT)	Payments for damage to or loss of capital goods as a result of war, other political disturbance or national disaster
	Shares in corporate enterprises bought by general government
	Payments by general government agencies to market producers for goods and services provided directly to households to provide against 'social risks and needs' and to which the households have a legal right. The risks and needs in question include sickness, disability, old age, maternity, unemployment and 'general need'

Operating surplus or mixed income (B2 / B3)

5.29   These are alternative names for the balancing item in the generation of income account, measuring the surplus accruing from production before deducting payments or adding receipts of interest, rents and other property income. The balance is therefore unaffected by whether land is owned or rented, and by whether assets in general are financed by equity or loan capital. However the net measure of these balances does depend on the extent to which *fixed assets* are owned or rented. Renting reduces consumption of fixed capital by the lessee, but increases intermediate consumption by a larger amount because the lessor's operating and interest costs need to be covered, so the overall effect is to reduce net value added. There is a converse effect on the lessor's accounts, and the effect on aggregate accounts depends on whether the lessor and lessee are classified to the same sector.

5.30   The balance excludes property income, which is regarded by the SNA as a return to financial assets or land, rather than a reward for production. It is therefore recorded in the allocation of primary income account, rather than in the generation of income account (which relates only to productive activities). There is a special treatment of financial intermediation services indirectly measured (FISIM), which is described in paragraph 5 above and in Annex 1 to Chapter 20. The operating surplus generated by these services is included in the surplus of both the financial intermediation sector and (because these services are not charged to intermediate consumption) of the purchasing sectors. A notional sector is shown with offsetting negative operating surplus (the financial services adjustment) to give the correct operating surplus for the the UK economy as a whole.

5.31   The balancing item in the generation of income accounts of most sectors and institutional units is called the 'operating surplus' (B.2). 'Mixed income' (B.3) is the term used for unincorporated enterprises owned by households in which household members may work without receiving a wage or salary. (In practice it relates to all unincorporated enterprises except quasi-corporations and owner-occupiers.) It is 'mixed' because the total implicitly includes an element of remuneration for work done, alongside the surplus accruing from production. However, there is seldom enough information available on hours worked and appropriate rates of pay for the remuneration element to be accurately imputed. Furthermore, with unincorporated enterprises it is often impossible to distinguish between the assets of the enterprise and those of the owner in a personal capacity; that is, between goods purchased for intermediate consumption (in principle recorded in the production account) and those purchased for final consumption by household members (in principle recorded in the use of disposable income account).

Property income (D4)

5.32   This section deals with the income which is received by owners of financial assets, land and other tangible non-produced assets. These are known collectively as property incomes, their main components being interest, distributed income of corporations, and rents. Other items included are reinvested earnings on direct foreign investment and property income attributed to holders of insurance policies. All these are described in turn below.

5.33   An important preliminary distinction is that between rent – the component of property income, and rentals – treated as transactions in services.

5.34   The latter include rentals for:

operating leases
tenancies of buildings (including the land on which they stand)
royalty agreements for the use of intangible assets

and are considered to be the purchase of intermediate or final output.

5.35   On the other hand, recipients of property income such as rent merely place their assets at the disposal of producers and are not considered to be themselves engaged in productive activity. Another way to look at it is that operating leases, etc, apply to *produced* assets subject to capital consumption whereas property income relates to *non-produced* assets and no capital consumption is incurred.

Interest (D.41)

5.36   The first type of property income to be discussed is *interest*, which is defined as the amount that a debtor becomes liable to pay to a creditor over a given period of time without reducing the amount of principal outstanding, though it may not be due for payment until a later date. There are many kinds of financial instrument apart from traditional deposits, securities and loans and interest (not always described as such) may be paid in various different ways.

5.37   For economic accounts purposes interest is always recorded continuously on an accrual basis, not according to the amount paid or due to be paid. This applies even for bills and zero-coupon bonds which are issued at a discount to the redemption value where interest - the discount - has to be imputed over the life of the asset. For other bonds and debentures the interest is the guaranteed amount payable in each accounting period (known as the coupon payment) *plus* an appropriate share of any initial discount (i.e. the difference between the nominal or redemption value and the issue price). The methods by which this discount can be distributed over time vary. In the special case of index-linked securities, where the coupon payment is based on a specified price index, the full amount of this is treated as interest. Where it is the value of the principal that is index-linked the change in that value over the accounting period is treated as interest. In either event the interest accrued is then recorded as being effectively reinvested. Where reduced rate loans are given to employees, and the benefit included in income in kind, interest should be recorded at commercial rates.

5.38   Interest rate swaps are contractual arrangements between two institutional units who agree to exchange between themselves interest payments of different types (e.g. fixed rate payments swapped for variable rate ones). In this case only the *balance* of interest payments between the two parties is recorded here for economic accounts purposes. (This treatment is under review internationally.) The same principle applies to forward rate agreements (FRAs) whereby two parties seek to protect themselves against interest rate changes by agreeing in advance on an interest rate to be charged at a specified date on a notional principal. The only payment that takes place relates to the *difference* between the FRA rate and the market rate on the settlement date, and it is this which is recorded as interest.

5.39   Financial leases differ from other leases in that all the risks and rewards of ownership are transferred from the legal owner of the assets to the lessee. In the economic accounts this transaction is treated as if it were a purchase by the lessee funded by a loan from the lessor which is gradually paid off over the life of the lease. The 'rental' paid therefore incorporates both interest and repayment of principal, the first of which progressively reduces over time (assuming the total rental remains constant) and is recorded in the primary distribution of income account.

5.40   The amounts of interest payable to or receivable from financial intermediaries also include two elements: the true interest and the cost of providing services for which no explicit charges are made. The latter are regarded as sales of services, and recorded as such in the production accounts. The treatment of Financial Intermediation Services Indirectly Measured (FISIM) is outlined in Annex 1 to Chapter 20. Only the interest proper contributes to property income of the intermediaries, the imputed service charge having already been recorded in the production account.

5.41 The reserve funds of insurance companies and pension funds are treated as belonging to the policy holders or (prospective) pensioners. A payment is imputed here from financial corporations to the beneficiary sectors equivalent to the property income on these reserves. The same amounts are imputed as supplementary premium and contribution payments from these sectors to financial corporations in the secondary allocation of income account.

5.42 Finally as regards interest, it should be noted that the amounts recorded in the primary allocation of income account are all on a 'nominal' basis, i.e. in actual money values as distinct from the 'real' return which takes account of changes in purchasing power as a result of general inflation. It is possible to view actual payments as consisting of two elements, one to compensate for loss of purchasing power (calculated using a general price index) and the other representing real interest (which may be negative), but this approach is not followed in conventional economic accounting.

Distributed income of corporations (D.42)

5.43 A second type of property income is the distributed income of corporations (D.42), primarily dividends. These are the income to which shareholders are entitled as a result of placing funds at the disposal of corporations, though its amount is not fixed or predetermined. The term covers all distributions of profits to shareholders, by whatever name they are called.

5.44 Quasi-corporations (*see* Chapter 10) cannot pay dividends as such but from a conceptual point of view the withdrawal of entrepreneurial income from the business by their owners amounts to the same thing. Such withdrawals are separately identifiable because, in order to be treated as a quasi-corporation, an enterprise must have accounts of its own. They need to be distinguished from disposal of assets or liquidation of reserves, which are treated as financial transactions - withdrawals from the corporation's equity.

Re-invested earnings of direct foreign investment (D.43)

5.45 A direct foreign investment enterprise is one in which a foreign investor has made a direct investment as distinct from a portfolio investment (*see* Chapter 17). Its entrepreneurial income may be distributed in the form of dividends or (in the case of quasi-corporations) withdrawal of income, and these payments to foreigners are treated as international flows of property income, recorded in D.42. In addition, if earnings are retained they are treated as having been distributed and then reinvested, the distribution being recorded here in D.43. The rationale for this is that such enterprises are controlled by foreign investors so the decision to retain earnings must represent a deliberate decision on their part. If the enterprise is wholly owned by a single investor then all the retained earnings are deemed to have been remitted and reinvested: if only a part is foreign owned then the amount recorded is proportionate to the share of equity involved.

Property income attributed to insurance policyholders (D.44)

5.46 Technical reserves are held by insurance companies against outstanding risks in respect of life insurance policies. They are invested by the companies and yield investment income in the form of property income (from fixed assets and land) or operating surplus (from renting or leasing of buildings). However, the reserves are held in trust for policyholders and are therefore regarded as assets of theirs and liabilities of the companies. The income generated by the reserves (and retained by the companies) is shown in the primary allocation of income accounts as having been paid over by the insurance sector to the policy holders. It is then returned to the insurers in the form of premium supplements, additional to the actual premiums paid, recorded in the secondary distribution of income account (part of D.71 and D.611). The insurance companies' receipts from investment of technical reserves are recorded in the primary distribution of income account, so the balance of primary income and the disposable income of the companies are not affected. The income is allocated amongst policyholders in proportion to the actual premiums paid and shown under 'resources' in the allocation of primary income accounts. Similar treatment is accorded to the reserves of pension funds as supplements to contributions.

Rent (D.45)

5.47   Rent relates to income from non-produced assets, such as land and sub-soil assets. It is recorded on an accrual basis, so the amount shown for a particular accounting period is the accumulating amount payable as distinct from the amount due to be paid or actually paid. It may be denominated in cash or in kind, and in the latter case may not be fixed in advance (e.g. with share-cropping schemes). Rent on land also includes amounts payable to the owners of rivers for the right to exploit them (e.g. for fishing).

5.48   Taxes and maintenance costs incurred solely as a consequence of owning land are by convention treated as payable by the tenant and deducted from the rent that would otherwise be paid to the landlord. This net treatment avoids the need to create notional enterprises for landowners not already engaged in productive activity.

5.49   Another special situation arises where rentals for buildings (treated as purchasing services) and rents for land are subsumed into a single payment and cannot be separated. In this case the whole amount is to be allocated to whichever type of payment is believed to be the larger. In practice, virtually all such payments involving land with buildings are likely to be treated as rentals for buildings.

5.50   Rents are also paid in respect of subsoil assets (mineral deposits and fossil fuels). The owners of these – who may be either the owners of the land above or a government unit – grant leases permitting others to extract such deposits for a given period of time. The reciprocal payments are often described as royalties but are essentially rents, either of fixed amounts or related to the quantities extracted. Payments to landowners for the right to prospect for subsoil assets (e.g. by test drillings) are also treated as rents, even if no extraction takes place.

Secondary distribution of income account

5.51   This account shows how the balance of primary incomes for an institutional unit or sector is transformed into its disposable income by the receipt and payment of current transfers (excluding social transfers in kind). Transfers are defined as transactions in which one institutional unit provides a good, service or asset to another unit without receiving anything in return. Current transfers comprise those that do not relate to fixed assets or financial assets. Transfers payable are shown in the account as 'uses' while those receivable are under 'resources'. In accordance with the general rules of economic accounting the entries refer to amounts payable and receivable in the accounting period, not those actually paid or received. Social transfers in kind are defined later in this chapter.

5.52   The types of current transfer are shown in Table 5.6.

Table 5.6: Entries in the secondary distribution of income account

Taxes on income (D.51)

Other current taxes (D.59)

Social contributions (D.61):

Employers' actual social contributions (D.6111)

Employees' social contributions (D.6112)

Social contributions by self- and non-employed persons (D.6113)

Imputed social contributions (D.612)

Social benefits (D.62):

Social security benefits in cash (D.621)

Private funded social benefits (D.622)

Unfunded employee social benefits (D.623)

Social assistance benefits in cash (D.624)

Net non-life insurance premiums (D.71)

Non-life insurance claims (D.72)

Current transfers within general government (D.73)

Current international co-operation (D.74)

Miscellaneous current transfers (D.75)

5.53   The transfers fall into three broad groups:

- Current taxes on income, wealth, etc (D.5).

- Social contributions and benefits (D.6)

- Other current transfers (D.7)

5.54   Transfers are either 'capital' or 'current', and distinguishing between these is crucial both for prudent financial management and for meaningful measurement of income. A transfer in kind is 'capital' if it transfers ownership of an asset, other than inventories, and a transfer in cash is 'capital' if it is linked to the acquisition or disposal of an asset, other than inventories. In either case the asset may be a fixed asset or a financial asset. All other transfers are current, and directly affect disposable income. In practice capital transfers tend to be large. infrequent and irregular, and current transfers to be otherwise, but this is not a sufficient criterion to distinguish them. Some cash transfers may be regarded as 'capital' by one party to the transaction and 'current' by the other (e.g. households and government respectively in the case of inheritance tax). For economic accounting purposes these are treated as 'capital' in both sectors provided they relate to the transfer of an asset for at least one of the parties.

Current taxes on income, wealth, etc (D.5)

5.55   Taxes are a form of compulsory transfer, as the government provides nothing in return to the *individual* unit making the payment. Current taxes on income, wealth, etc, constitute a charge against income and are recorded in the secondary distribution of income account. Such taxes have traditionally been described as 'direct taxes' but this term is now no longer used in the economic accounts.

Changes from the former system

Taxes on income under the new definition includes capital gains tax (a tax on capital in the former system), local authority rates (a tax on expenditure) and the council tax (a tax *sui generis* in the former system). The Community Charge is classified to D.59. Dog and gun licences and motor vehicle licence duty paid by households, classified to D.59 were taxes on expenditure in the former system.

Taxes on income (D.51)

5.56   Taxes on income are listed in detail in Chapter 21. They include:

- Taxes on *individual or household income* (including pay-as-you-earn taxes and those on owners of unincorporated enterprises). These are levied on declared or presumed income from all sources, after deducting certain agreed allowances.
- Taxes on the *income of corporations* (including profits taxes and surtaxes). These are usually assessed on income from all sources, not only that generated by production.
- Taxes on *capital gains* (i.e. holding gains). These are usually payable on nominal rather than 'real' gains, and only when they have been realised.
- Taxes on *winnings* from lotteries and gambling.
- Taxes assessed on *holdings of property* or land where these are used as a basis for estimating the owners' income.

It is not always practicable to record tax on an accrual basis so some flexibility is allowed. For example, pay-as-you-earn taxes and regular prepayments may be recorded when paid.

Other current taxes (D.59)

5.57   Current taxes on capital are paid periodically, usually annually, on the net wealth of institutional units, and do not include taxes on assets used in production and taxes levied infrequently, at irregular intervals or in exceptional circumstances (e.g. death duties). They *do* include taxes on the use or ownership of land or buildings (including owner-occupied dwellings), on the value of fixed assets (*less* debts incurred on these) and on external signs of wealth (e.g. jewellery).

5.58   Current taxes not already mentioned include poll taxes (levied as specific amounts of money per person or per household), expenditure taxes (on the total spending of persons or households), payments by households for vehicle licences and licences to hunt, shoot or fish (but not other payments for licences which are regarded as purchases of services) and taxes on international transactions (e.g. travel abroad, foreign remittances and foreign investments).

Social insurance schemes (D.6)

5.59   Social insurance schemes may be organised privately or by government and provide benefits in cash or in kind in circumstances such as the following:

• Need for *medical treatment or long-term care*. Benefits are usually paid in kind, free or at prices which are not economically significant, or by reimbursing households' expenditure.
• Need to support *dependants*. Benefits are usually paid as regular cash allowances.
• *Not being able to work* (or to work full-time). Benefits are paid as regular allowances or by lump sum.
• *Death* of the main breadwinner. Benefits are paid as regular allowances or by lump sum.
• Free or subsidised *housing* or reimbursement of households' housing expenditure.
• Assistance with *education*. Expenses are normally met by allowances but services may occasionally be provided in kind.

However, coverage varies considerably from scheme to scheme and from country to country.

5.60   Such schemes are intended to cover beneficiaries and their dependents throughout their working lives and usually into retirement, provided the necessary contributions have been paid. Participation may be voluntary but is often obligatory (e.g. for employees of a particular firm). If a scheme is obligatory and has been set up to cater collectively for a particular group of workers with contributions from the employer then – provided it covers the sorts of eventualities listed above – it is regarded as a *social* insurance scheme, even if participants have individual policies. However, when individuals take out policies perhaps providing similar cover but on their own initiative and independently of their employers or government then the premiums/claims are treated not as social contributions/benefits but as current transfers in the secondary distribution of income account. Thus the nature of the benefit is not sufficient to establish the 'social' nature of the transaction.

5.61   Social insurance schemes can be of three types, the first covering the whole community (or large sections of it) and being imposed, controlled and financed by government (social security). The second type consists of private funded schemes, which may be managed either by separate institutional units or by the employer. In the latter case special reserves (called non-autonomous pension funds) are kept, segregated from the employer's other reserves and treated in the economic accounts as assets belonging to the beneficiaries. Thirdly there are unfunded schemes in which the employers pay social benefits out of their own resources without creating special reserves for the purpose. Schemes organised by the government for its own employees are classified as private schemes, not as social security.

5.62   The funds of government social security schemes are separate from other government funds, though they may receive transfers from them in addition to contributions by individuals or employers. The benefits received are not necessarily related to the contributions paid. Premiums for private funded social insurance schemes include a service charge which is recorded in the economic accounts as final consumption expenditure by households, except in the case of non-autonomous pension funds where it is assimilated into the employer's general production costs.

5.63   Social contributions and social benefits have various characteristics which affect their classification in the secondary distribution of income account and in the redistribution of income in kind account. These differences are explained in the following two sections.

Social contributions (D.61)

5.64   Employers' actual social contributions on behalf of their employees are usually paid directly to social security funds, insurance companies or pension funds. The underlying economic reality is that the contribution is part of the remuneration package. This is represented in the accounts by recording the amount firstly as part of compensation of employees (in the generation of income account) and secondly as a current transfer from the household (along with employee contributions) to the social security fund or whatever (in the secondary distribution of income account). The two transactions are recorded simultaneously, when the work is carried out which gives rise to the contributions. A distinction can be made between contributions that are compulsory by law and those that are not.

5.65   As regards employees' contributions, in the case of private funded social insurance schemes it is necessary to add to actual payments the contribution supplements reflecting income from the investment of reserves, and to deduct service charges (*see* the description of property income above, and Annex 2 to Chapter 20). Employees' contributions to social security funds, where they exist, are usually compulsory by law but voluntary contributions are sometimes made and may be separately recorded. All employees' contributions, like employers', are recorded simultaneously with the work giving rise to them.

5.66   A further row in the accounts shows social security contributions, compulsory and voluntary, by self-employed and non-employed people. Here, too, contribution supplements need to be added and service charges deducted.

5.67   Finally an entry is required in the secondary distribution of income account for imputed social contributions by employees where the employer is operating an unfunded social insurance scheme, to balance the imputed element of the employers' contributions in compensation of employees in the primary distribution of income account.

5.68   Payment of social insurance contributions and non-life insurance premiums may confer an entitlement to future benefits in certain circumstances but these are generally uncertain, unquantifiable or both, and bear no relation to the amount of contributions. The entitlement therefore cannot be regarded as an asset and any payments towards it have to be treated as transfers. However, life insurance policies taken out by households on their own initiative are a different matter. Here the policyholders themselves own the reserves, so premiums and benefits constitute acquisition and disposal of financial assets. In strict logic the same should apply to funded pension schemes but this would mean treating them differently from state pensions and go against the general perception of them as transfers. They are therefore recorded in the secondary distribution of income account as if they were current transfers.

Social benefits (D.62)

5.69   These benefits may be paid in kind as well as in cash, e.g. individual schemes may maintain their own clinics or retirement homes. When paid in cash unfunded benefits provided by employers (e.g. sickness or maternity pay) may be difficult to distinguish from ordinary wages and salaries, but this distinction does not affect balances on the accounts provided the identified benefits are balanced by imputed social contributions.

5.70   Social benefits provided by government units and by NPISH comprise:

- Social security benefits in cash, e.g. for sickness or invalidity, maternity, children or other dependents, unemployment, retirement and death, where eligibility depends on membership of the scheme;
- Social assistance benefits in cash, providing for circumstances or households not covered, or not covered sufficiently, by social security, but not including transfers in response to natural disasters;

However, social benefits in kind are recorded in the redistribution of income in kind account rather than the secondary distribution of income account.

5.71   Private social insurance benefits are divided into:

- Funded benefits, similar in type to social security but all in cash and included in the secondary distribution of income account;
- Unfunded benefits, which may be in cash or in kind and are recorded in the secondary distribution of income account in either event, e.g. education allowances and medical services not related to the employee's work.

Other current transfers (D.7)

5.72   This category includes a number of disparate transfers not so far described, of the following types:

- *Net non-life insurance premiums* (D.71). These relate to policies taken out by individual households on their own initiative, and include premium supplements (see para 5.43) as well as actual payments. 'Net' here means excluding service charges, this being the amount available to provide cover, which constitutes a current transfer to be recorded in the secondary distribution of income account. (*See* note on insurance in Chapter 20).
- *Non-life insurance claims* (D.72). Settlement of claims is regarded as a transfer to the claimant, always treated as 'current' even when very large (because lost assets are not necessarily replaced). The total of claims equals the total of net premiums because service charges are obtained by subtraction, the essential function of non-life insurance being to redistribute resources. 'Third party' claims are recorded as payable to the injured party's insurance company rather than the policyholder. (*See* note on insurance)
- *Current transfers within general government* (D.73). These include transfers between different levels of government and between government and social security funds. They exclude grants financing gross fixed capital formation and the paying-over of taxes collected by one government unit on behalf of another.
- *Current international co-operation* (D.74). This item covers transfers in cash or in kind between the governments of different countries (e.g. emergency aid and technical assistance) and between governments and international organisations (e.g. regular subscriptions for membership). Excluded are taxes payable to supra-national organisations (D.5) and capital grants (D.9).
- *Miscellaneous current transfers* (D.75). These include subscriptions and donations in cash and in kind to non-profit institutions serving households (NPISH), remittances home from family members working abroad for a year or more, fines and penalties imposed by courts of law, transfers between households in the form of lottery and gambling winnings (excluding the 'service charge' element in the original payments) and, finally, compensation from one institutional unit to another for injury or damage, excluding the payment of non-life insurance claims but including ex gratia payments from government and NPISH sectors following natural disasters.

Disposable income (B.6)

5.73 The balancing item of the secondary distribution of income account is disposable income, which may be recorded gross or net of consumption of fixed capital. Not all of it is available in cash as the account includes some transfers in kind (though not those from government and non-profit institutions to households). The recipients of these are, by convention, recorded as making imputed consumption expenditures on the goods and services in question, as if the transfers were in cash. Property income may also be in non-cash form, including that from the investment of insurance and pension reserves and imputed interest which is accrued but not yet paid, and is treated as automatically invested.

5.74 Disposable income defined in this way can be interpreted as the maximum amount that a household or other unit can afford to spend on consumption without reducing its net wealth.

Redistribution of income in kind account

5.75 From this point alternative presentations are possible in the accounts of general government, households and NPISH. The first takes disposable income into the *Use of disposable income account* and sets against it final consumption <u>expenditure</u> (and an adjustment for the equity of households in the reserves of pension funds) to arrive at the concept of saving for the sector. The second introduces a *Redistribution of income in kind account* which adjusts disposable income. The adjusted disposable income is carried forward into the *Adjusted use of disposable income account* and set against <u>actual</u> final consumption and the adjustment for the equity of households in the reserves of pension funds to arrive at the same concept of saving.

5.76 It is important to realise that this account relates to the transfer of goods and services from government and non-profit institutions serving households to the household sector. It does not have anything to do with the income in kind which is part of wages and salaries.

5.77 To understand the relationship between the two it is necessary to consider an aspect of consumption expenditure by government and NPISH. These can be classified according to whether they are on *individual* goods and services or *collective* services. The most important characteristic of an 'individual' service is that it is provided individually to one person or small group at a time, and its provision to some diminishes the amount available to others. The provider must therefore decide not only how much to spend in total but also how to allocate it. Examples are medical treatment or education. A collective service, in contrast, is provided simultaneously to the community as a whole (or distinct sections of it, such as those living in a particular area). Use of such a service is generally passive, not requiring the explicit agreement or active participation of those concerned, there is no rivalry in acquisition and individuals cannot be charged according to their usage or the benefit they derive. Examples are security and defence, law and order, legislation and regulation, maintenance of public health, protection of the environment, research and development.

5.78 Non profit making institutions serving households (NPISH), as their name suggests, exist to provide services for households; generally of the sort which their clients cannot easily obtain for themselves acting alone. In most cases the service is still essentially individual and by convention, as already noted, *all* NPISH services are treated as individual.

5.79   Finer distinctions need to be made within government. Public services are treated as collective to the extent that they are concerned with the formulation and administration of government policy, the setting and enforcement of public standards and the regulation, licensing and supervision of production. Individual public services, on the other hand, are those associated with the provision of services to particular individuals, such as hospitals, schools, etc. They also include central units providing common services to groups of those institutions but not those concerned with formulating government policy for these services.

5.80   Many government expenditures benefit both enterprises and households (e.g. provision of roads) and by convention these are treated as collective final consumption though, in principle at least, many are individual services which could be provided on a market basis (e.g. by charging tolls). Enterprises also benefit from genuinely collective services (e.g. police and fire services and research and development by non-market producers). Usage of these cannot easily be monitored so they too are treated as government final consumption.

5.81   In the alternative presentation, expenditure on individual goods and services is recorded as social transfers in kind in the *Redistribution of income in kind account*, paid by general government and NPISH to households. The balance on this account is the *adjusted disposable income*, which is carried forward into the *adjusted use of disposable income account*.

5.82   The latter account shows how households, government units and NPISH allocate their disposable income between actual final consumption and saving*. For households actual final consumption equals their final consumption expenditure plus social transfers from general government and NPISH (i.e. their individual consumption). It therefore measures their actual consumption of goods and services whether purchased by them or not. Actual final consumption of households, allowing for social transfers in kind, is a better indicator of living standards than final expenditure alone. In some countries it is described as 'enlarged' or 'total' consumption.

5.83   The following equations show the relationship between the two parallel *use of disposable income accounts*, one focusing on *expenditure* on consumption goods and services and the other on *acquisition*, whether by expenditure or through social transfers in kind. However, the essential purpose of both accounts is the same, namely to derive saving which is defined as:

Use of disposable income account

| Saving | = | Disposable income | *minus* | Final consumption expenditure | *plus* | expenditure adjustment |

Adjusted use of disposable income account

| Saving | = | Adjusted disposable income | *minus* | Actual final consumption | *plus* | consumption adjustment |

These two expressions are identical because the disposable income and adjusted disposable income differ from one another by the amount of social transfers in kind, as do final consumption expenditure and actual final consumption (but in the opposite direction). (These relationships are not meaningful for corporations which neither make nor receive social transfers in kind.) Like disposable income, saving can be measured gross or net of capital consumption.

5.84　The redistribution of income in kind account, then, shows how the disposable incomes of households, NPISH and government units are transformed into 'adjusted disposable incomes' by the receipt and payment of social transfers in kind. These include not only social *benefits* in kind but also transfers to households of individual non-market goods and services, most commonly education and health services. Some non-market services provided by NPISH have collective characteristics but most are individual in nature and all are treated as such, and therefore recorded as social transfers in kind.

5.85　Because social transfers in kind are payable only by government and NPISH and only to households it follows that adjusted disposable income is always lower than disposable income for the former and higher for the latter, whereas for the economy as a whole the two are identical. The adjusted disposable income of a household can be interpreted as the maximum *consumption* (as distinct from spending) that it can afford to undertake without reducing its net wealth. Conversely the adjusted disposable income of government is the maximum amount of collective services it can afford to provide on the same conditions. (*See* the discussions of individual and collective consumption, actual consumption and consumption expenditure in Chapters 3 and 4 and later in this chapter.)

Social transfers in kind (D.63)

5.86　These are, apart from balancing items, the only entries in the redistribution of income in kind account. They consist of individual goods and services supplied to households by government units (including social security funds) and NPISH, and have the advantage to policy-makers that the resources transferred can be targeted towards meeting specific needs and can only be consumed in the ways intended.

5.87　Social benefits in kind include reimbursement from social security funds of approved expenditure by households (e.g. for medicines or medical treatment). The household is regarded as an agent of the social security fund and the amount reimbursed is treated as expenditure by the fund, not as a cash transfer to the household. Only the difference, if any, between the purchase price and the reimbursement is recorded as household expenditure. Other social benefits in kind consist mainly of goods and services associated with health care. Services may be provided by either market or non-market producers and must be valued accordingly, any nominal payment by the household being deducted. Social *assistance* benefits in kind tend to be provided in similar circumstances to assistance in cash, as specified above.

5.88　The final type of social transfer in kind concerns individual non-market goods and services provided to individual households by government and the NPISH sector either free or at prices that are not economically significant. Services provided in this way may include education, health care, housing services, cultural and recreational services. However, the individual element of these services would not include expenditure on the formulation of government policy, general administration, regulation or research in these areas. The services are all designated as 'individual' to distinguish them from collective services provided to the community as a whole or large sections of it. They may come from market or non-market producers and are valued accordingly. Social transfers in kind account for the difference between final consumption expenditure and actual final consumption in the use of income account.

Use of disposable income account (and adjusted use of disposable income account)

5.89   This account and the parallel adjusted account show how households, government units and NPISH allocate their disposable income between final consumption and saving[*].

5.90   Disposable income, the balancing item carried forward from the secondary distribution of income account, is recorded in the use of income account, under 'resources', while final consumption expenditure, or actual final consumption is recorded under 'uses'.

Consumption expenditure and actual consumption

5.91   A consumption good or service is one that is used by households, NPISH or government for the direct satisfaction of needs and wants. Consumption expenditure attributes the acquisition of these goods and services to the sector which pays for them. Actual consumption attributes expenditure by government and NPISH sectors on individual goods and services to the household sector (*see* paragraphs 5.73–5.80 above).

5.92   Household expenditure on consumption goods and services includes tenants' rentals and the imputed value of housing services from owner-occupied dwellings, but excludes expenditure on dwellings as fixed assets and on valuables which do not deteriorate over time and are acquired primarily as stores of value even though their owners may also derive satisfaction from possessing them (e.g. works of art, precious stones and jewellery).

5.93   Specific conventions apply to certain situations and types of expenditure, as follows:

- *Households owning unincorporated enterprises.* All expenditure incurred for business purposes should be excluded from the final consumption expenditure of such households but this is difficult in practice if the same items are used in different ways, e.g. on farms or where vehicles are shared between business and personal use. In the latter case the expenditure on vehicles should be split between gross fixed capital formation and household final consumption expenditure.
- *Barter.* Household final consumption expenditure includes the value of consumption goods and services obtained through barter and, as a negative imputation, existing goods offered for barter.
- *Income in kind.* The value of goods and services provided is recorded both as income and as final consumption expenditure, since it is regarded as being simultaneously spent by the employees. The items concerned are goods and services that employees are able to use in their own time for their own direct satisfaction, as distinct from those they need for their work (which are part of the intermediate consumption of the enterprise), but inevitably there is something of a grey area in between.
- *Own-account production.* Households' final consumption expenditure includes the value of goods and services produced by unincorporated enterprises owned by households when these are retained for the households' own consumption (e.g. agricultural production by farmers and services provided by paid domestic staff). Values are imputed using the basic prices of similar goods sold on the market, by the costs of production or, in the case of domestic staff, by their compensation in cash and kind.
- *Payments for financial intermediation, insurance and pension fund services.* The treatment of these is described in Chapter 4 above.
- *Repairs.* Expenditure on repairs and maintenance of durables, including vehicles, together with that on materials for simple repairs and decoration of dwellings, is all treated as final consumption expenditure of households.

---

[*] For financial corporations the difference between disposable income and saving is the adjustment for the change in net equity of households in pension funds reserves. For non-financial corporations there is no difference

- *Licences and fees.* Payments to government units for licences, permits, certificates, passports, etc, are included in household consumption expenditure apart from vehicle licences and licences to hunt, shoot or fish (which are treated as taxes).
- *Hire purchase, other credit agreements and financial leases.* Expenditure on these is recorded at the time the goods are delivered. There is no legal change of ownership at this time but the purchaser acquires all the rights and responsibilities of ownership so a *de facto* change is deemed to have taken place.

5.94   Household expenditure is valued at purchasers' prices, including taxes and transport costs. Because it is common, especially with services, for different households to be charged different prices for the same item this may mean that the expenditure is not valued uniformly. Purchasers' prices do *not* include any interest or service charges for credit sales or late payments.

5.95   Household final consumption expenditure as defined in economic accounts is that by UK residents whether incurred within the country's economic territory or abroad. In practice this can be calculated by taking expenditure here by all households, adding expenditure by residents abroad and subtracting expenditure by non-residents within the economic territory.

5.96   Finally as regards the use of income it is necessary to describe the treatment of consumption expenditure not made by households at all, i.e. that incurred by government and NPISH. Most of this is spent on output of the government and private non-profit institutions themselves, which is provided free (or at prices which are not economically significant) to individual households or the community as a whole. It needs to be distinguished from the intermediate expenditures and other costs incurred by the government and private institutions as non-market producers producing those goods and services. In addition, government and NPISH purchase consumption goods and services from *market* producers for distribution to households as social transfers in kind, this expenditure being treated as final, not intermediate (as no further processing is undertaken).

Adjustment for net equity of households in pension funds reserves

5.97   Having entered disposable income and consumption in the use of income account it is not immediately possible to strike the balance and derive savings, as an adjustment needs to be made to reallocate some amounts between sectors. As explained above, contributions to and pensions from private funded schemes are treated as current transfers and affect disposable income, though logically they should be seen as acquisition and disposal of financial assets as the pension reserves are considered to be owned by the households with claims on them, as is recognised in the financial accounts and balance sheets. To reconcile these two approaches it is necessary to add back pension contributions to, and subtract pension receipts from, household disposable income, so that measured saving reflects any change in households' net equity in pension funds. Opposite adjustments need to be made to the use of income accounts of pension providers.

5.98   Once this adjustment has been made the balancing item representing saving can be calculated. It stands for that part of disposable income that is not spent on final consumption of goods and services. If positive it shows up as an increase in cash, acquisition of assets or reduction of liabilities, and if negative then the reverse. This saving provides the link between the current accounts and the subsequent accumulation accounts.

Annex 1

Expenditure, acquisition and use

5A.1 The distinction made in this chapter between final consumption expenditure and actual final consumption depends upon a more general distinction between expenditure on and acquisition of goods and services which will now be explained more fully. *Expenditure* is what a buyer pays to a seller and in economic accounts methodology the buyer of consumption goods and services is the unit which bears the cost, not necessarily the one that takes possession of the goods or even the one that makes the payment. As regards timing, expenditure is recorded when a liability is incurred, not when payment is made. Normally it is when ownership of a good is transferred or delivery of a service is satisfactorily completed, but these may be perceived differently by the two parties to the transaction.

5A.2 Liabilities may be discharged by forms of exchange other than cash, so values have to be imputed using appropriate prices taken from other transactions which *are* in cash. Where barter takes place different values may be imputed to the two sides of the deal, in which case a simple average is taken. It should be noted that the imputation refers to the value, not to the expenditure itself, though the term 'imputed expenditure' is sometimes used in the interests of brevity.

5A.3 'Existing goods' are ones which have already been supplied to a user (e.g. second-hand equipment) and their sale gives rise to negative expenditure. Durable goods may then have to be reclassified, e.g. sale of a company cars to a private buyer represents (negative) capital formation for the company but (positive) final consumption expenditure for the household.

5A.4 *Acquisition* by an institutional unit takes place when ownership of a good changes hands or delivery of a service is completed, including receipts in kind (less transfers in kind to other units). However, some transfers in kind are treated in economic accounts as if they were in cash, and recorded as expenditures by the units that acquire them. 'Acquisition' can refer to any kind of good or service, including investment goods, but special recognition is given to acquisition for purposes of final consumption or 'actual consumption'.

5A.5 A third concept, distinct from either expenditure or acquisition, is that of *use*. Goods and services are said to be used when they are employed in a process of production or for the direct satisfaction of human needs and wants. In the case of goods the distinction between acquisition and use is analytically important because it underlies that between durables and non-durables. A 'durable' good does not have to be hard-wearing in a physical sense but is one which can be used repeatedly or continuously over a period of more than a year, assuming a normal rate of usage.

5A.6 The repeated use of consumer durables by households to satisfy needs and wants is not easily measured. Doing so would mean postulating that, like fixed assets in a production account, they are gradually used up in a hypothetical process whose output consists of a stream of services acquired by households over time. In practice consumer durables are not treated in this way: household consumption is measured only by expenditure and acquisition.

5A.7 The distinction between acquisition and use has little practical relevance to services.

5A.8 Total expenditure on collective services is taken to be the government's actual final consumption (though of course it benefits the community rather than the government *per se*). Together with social transfers in kind it makes up the whole of government final consumption expenditure, and can therefore be deduced by deducting social transfers in kind from this total.

5A.9 For the economy as a whole the coverage of goods and services in total final consumption is the same whether viewed from the expenditure side or in terms of actual final consumption, and on a strictly practical level all the aggregates are derived from data on expenditure. For the two sides to balance it is only necessary to ensure that the treatment of the goods and services entering into social transfers in kind conforms, in valuation and timing, to that in the expenditure aggregates.

Chapter 6

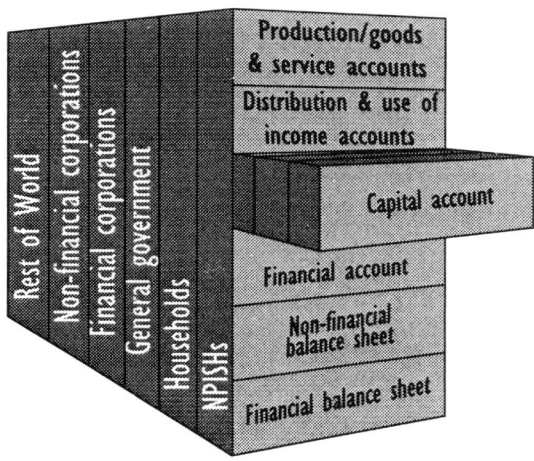

The Capital Account

Chapter 6 The capital account

Purpose and structure of the capital account

6.1 The capital account is the first of four 'accumulation accounts' dealing with changes in the value of assets. Specifically, it records the flows of *non-financial* assets acquired or disposed of. The other three elements are:

- The *financial account*, recording transactions in financial assets
- The *other changes in volume of assets account*, recording gains and losses not resulting from transactions, e.g. through discovery of new subsoil assets and destruction of existing assets by war or natural disaster
- The *revaluation account*, recording nominal holding gains and losses resulting from price changes.

Taken together, the four accounts wholly explain changes in the value of the *stock* of assets between the beginning and end of an accounting period, as recorded in the balance sheets.

6.2 In a similar way to the current accounts the top section of the capital account records the resources available for the accumulation of assets, *viz* saving and capital transfers (positive for those receivable and negative for those payable) while the lower part records changes in assets, consumption of capital and the balancing item, net lending or borrowing. The two parts are presented in the ESA as separate sub-accounts (numbered III.1.1 and III.1.2 respectively). Their contents are summarised in Table 6.1 below.

6.3 Capital transfers as recorded in the upper part are transactions in which:

- Ownership of an asset (other than cash and inventories) is transferred from one institutional unit to another, or
- Cash is transferred to finance acquisition of an asset, or
- Funds realised by the disposal of an asset are transferred.

Together with net saving and the external balance these constitute the resources available for the acquisition of assets as shown in the lower part of the account. Amongst the various types of acquisition gross fixed capital formation relates mainly to acquisition and disposal of produced assets which are used repeatedly in production processes, while consumption of fixed capital represents the value of fixed assets used up in production through physical deterioration, normal obsolescence or accidental damage.

6.4 The balancing item of the capital account – net lending or borrowing – is defined as:

Gross saving
+ current external balance
- Acquisitions of non-financial assets
+ Capital transfers receivable
+ Disposals of non-financial assets
– Capital transfers payable

Table 6.1: The capital account

Resources (saving and capital transfers)

Saving, gross (B.8g)
+ Current external balance (B.12)
+ Capital transfers, receivable (+) and payable (-) (D.9), including capital taxes (D.91) and investment grants (D.92)
= Changes in net worth due to saving and capital transfers (B.10.g)

Acquisition of non-financial assets

 Changes in net worth due to saving and capital transfers (B.10.g)
+ Consumption of fixed capital (K.1)
- Gross fixed capital formation (P.51), including acquisitions and disposals of fixed assets, both tangible (P.511) and intangible (P.512) and additions to the value of non-produced assets (P.513)
- Changes in inventories (P.52)
- Acquisitions *less* disposals of valuables (P.53) and non-produced assets (K.2)
= Net lending (+) or borrowing (-) (B.9)

Saving is that part of disposable income which is spent not on consumption but on the acquisition of assets (including cash) or repayment of liabilities. When negative it must be financed by disposing of assets or incurring new liabilities. Capital transfers are similarly concerned with the acquisition or disposal of assets. Thus net lending or borrowing, which encompasses both these elements, represents the resources finally remaining for purposes of lending by an institutional unit, or that need to be borrowed.

6.5 The capital account may be presented gross or net of consumption of fixed capital. In the gross presentation saving and fixed capital formation are shown gross, as in Table 6.1. In the net presentation, saving is shown net, and the consumption of fixed capital is deducted from gross fixed capital formation. The Blue Book presentation is essentially gross, but shows the capital consumption element of saving. The remainder of this chapter considers each of the constituents of the capital account in detail, together with their relationships to one another and to the rest of the integrated system.

Gross fixed capital formation (P.51)

6.6 Fixed assets are tangible or intangible assets which have been produced and are themselves used repeatedly or continuously in processes of production for more than a year. The value of such assets acquired during an accounting period, *less* disposals, constitutes the major part of gross fixed capital formation. The main activities involved, the various ways in which acquisitions and disposals take place are described in Chapter 3 (3.59–3.64).

6.7 Fixed assets which are acquired may be either new or existing assets (although the United Kingdom accounts are not able to make this distinction). New assets may not be complete units but renovations, reconstructions or enlargements that increase the capacity or extend the life of an existing asset. Disposals include items which are not used as fixed assets by their new owners, e.g. cars sold by businesses to households for their own use or equipment that is subsequently scrapped.

Table 6.2: Different types of gross fixed capital formation

Main categories	Subdivisions
Acquisitions *less* disposals of new or existing tangible fixed assets	Dwellings Other buildings and structures Machinery and equipment Cultivated assets (trees and livestock) used repeatedly or continuously to produce products such as fruit, rubber and milk
Acquisitions *less* disposals of new or existing intangible fixed assets	Mineral exploration Computer software Entertainment, literary or artistic originals Other intangible fixed assets
Major improvements to land and other tangible non-produced assets	
Costs associated with transfers of ownership of non-produced assets	

6.8   Capital formation is recorded at the time when ownership of the asset is transferred, and valued at purchasers' prices, including transport and installation charges and all costs and taxes incurred through the change of ownership. Conversely, sales of existing assets are valued after deducting any costs of transfer borne by the seller. Own-account production of fixed assets is valued at basic prices or, failing that, costs of production.

6.9   An existing asset is one whose ownership has passed from its producer to a user, and possibly between users, so that its value has already been included in gross fixed capital formation at least once. Sales and other disposals of such goods are recorded as negative expenditures or negative acquisitions. This negative value falls short of the positive one recorded by the buyer to the extent of the ownership transfer costs. Apart from these costs the positives and negatives cancel out in the case of transactions between resident producers. Even where the sale is actually to a non-resident it is sometimes deemed to be to a notional resident unit whose equity is then purchased by the non-resident as a financial asset. This convention applies to immovable fixed assets such as buildings but not to movable ones such as ships or aircraft.

6.10   Vehicles and some other durable goods may be classified as either fixed assets or consumer durables depending upon their owner and use. So if, for example, a car is sold to an enterprise by a household or non-resident unit then this is recorded as acquisition of a new fixed asset even though it is an existing good.

Improvements to existing assets

6.11   These are a special form of gross fixed capital formation which does not involve the creation of new assets that can be separately identified and valued. They need to be distinguished from ordinary repairs and maintenance, which constitute intermediate consumption. The criteria for doing so are given in Chapter 4 above but the choice is not always clear-cut as any repairs, like improvements, could be said to improve the performance or extend the life of an asset. Improvements, though, are characterised by the magnitude of the changes made and the fact that they go beyond the sort of change which takes place routinely in other assets of the same kind. **Improvements** so identified are classified with acquisitions of new fixed asset of the same type.

Additions to the value of non-produced assets (P513)

6.12   Another type of improvement affects tangible non-produced assets, which in practice means the treatment of land by way of reclamation from the sea, clearance of forests etc, drainage, irrigation and the prevention of flooding or erosion. Though such activities may involve the creation of substantial new structures, e.g. sea walls, these are not themselves used for production and are therefore not classified as fixed assets. Nor can they be shown as acquisitions of land itself, as this is a non-produced asset. They therefore require their own separate category of capital formation. Any decline in their value is treated as consumption of capital − a convention which reflects the need to write down the value of all gross fixed capital formation over time.

Costs of ownership transfer

6.13   As noted above, these form part of the cost of acquiring fixed assets. They include fees paid to lawyers, architects and surveyors, commission paid to estate agents and auctioneers, transport and installation costs and any taxes incurred (apart from capital gains tax liabilities of the seller). In common with the rest of gross fixed capital formation they are gradually written off unless and until the asset is sold on to another user, whereupon the first user may be unable to recover the transfer costs incurred on both the original purchase and subsequent sale. In this case a holding loss is recorded, equivalent to the difference between the depreciated value held in the balance sheet, and the proceeds of the sale.

6.14   Transfer costs apply equally to land as to fixed assets but, as land is not produced, there can be no gross fixed capital formation in it. The costs of ownership transfer therefore need to be recorded separately from the purchase and sale of land *per se*. The same applies to other non-produced assets such as mineral deposits that are used in production.

Special cases of acquisition or disposal of fixed assets

6.15   Fixed assets include both tangible items (such as structures, machinery, equipment and cultivated assets) and intangibles (such as computer software). Most acquisitions and disposals can be dealt with by a straightforward application of the definitions above but there are a number of cases where special treatment is called for. These are listed in Table 6.3.

Table 6.3: Examples of treatment of gross fixed capital formation

Item	Treatment
Small tools which are simple and relatively inexpensive	Treated as intermediate consumption if expenditure is fairly regular and small in relation to that on more complex machinery and equipment, but as gross fixed capital formation if they account for a significant part of the stock of durable producers' goods
Military equipment	
• Destructive weapons together with vehicles whose only function is to release them	Not treated as fixed assets as they are not concerned with production and (in the case of weapons) cannot be used repeatedly or continuously
• Structures used by the military, e.g. airfields, docks, roads and hospitals	Treated as fixed assets as they function in much the same way as civilian equivalents engaged in production and are often switched between military and civilian use
• Equipment used by the military which has alternative uses, e.g. transport, communications and hospital equipment	Treated as fixed assets unless expenditure cannot be separated from that on weapons and their delivery systems
• Light weapons and armoured vehicles used by non-military establishments engaged in internal security and policing	Treated as fixed assets
Buildings and other structures	
• Non-resident owners	All structures within UK economic territory are, by convention, deemed to be owned by UK residents. Where necessary a notional resident unit is created which is deemed to purchase the structure and sell its own equity (as a financial asset) to the non-resident.
• Dwellings, including mobile homes used as principal residences and any associated structures such as garages	All treated as fixed assets
• Major improvements to historic monuments and construction of new ones	Treated as gross fixed capital formation in dwellings, non-residential buildings or other structures as appropriate
• Consumption of fixed capital	Calculated on the assumption that all structures have finite service lives, though these may be very long. The life assumptions depend on eventual obsolescence as well as physical durability.

Table 6.3: *Cont....*

Buildings and other structures *Cont....*	
• Time of recording and valuation	Gross fixed capital formation takes place only when ownership of at least some part of the output is transferred to the eventual user. Until then it is recorded as work-in-progress or addition to inventories of finished goods. Acquisitions are valued at purchasers' prices (actual or estimated) *plus* costs of ownership transfer incurred by the purchaser, and disposals at prices receivable *minus* transfer costs incurred by the seller.
• Own-account construction	No change of ownership takes place so the whole output in each period is treated as gross fixed capital formation. In principle it should be valued as the appropriate share of the estimated basic price of the finished structure or, failing that, of production costs plus a mark-up. Any labour provided free should be valued using rates for comparable paid workers. Where own-account construction is transferred after completion to another sector a capital transfer and sale are imputed equal to the original gross fixed capital formation.
Machinery and equipment	
• Unsold items	New items which have not yet been sold, and imported assets, are not recorded as gross fixed capital formation until ownership is transferred to the eventual user.
• Acquisitions by households	Items acquired for final consumption are not included in gross fixed capital formation, though mobile homes, houseboats etc. used as principal residences are.
Cultivated assets (livestock or trees) used repeatedly or continuously for more than a year to produce goods or services	Breeding stock, dairy cattle, sheep reared for wool and draught animals are fixed assets but animals raised for slaughter are not. Similarly fruit trees and vines are fixed assets but trees grown for felling as timber are not.

Table 6.3: *Cont....*

Cultivated assets *Cont....*	
• Production which is incomplete at the end of an accounting period	For specialist producers (e.g. breeders and tree nurseries) this is treated as work-in-progress. However, immature animals, trees etc produced by farmers and others on their own account are treated as gross fixed capital formation, estimated by taking the basic price of the finished product and applying the share of total production costs incurred in the relevant period or, failing that, production costs themselves.
• Livestock cultivated for the products they yield year after year (e.g. dairy cattle and draught animals)	Gross fixed capital formation is the value of acquisitions <u>less</u> disposals, the latter including animals sold for slaughter or slaughtered by their owners but not exceptional losses due to disease or natural disasters.
• Plantations, orchards, etc	Acquisitions include those of immature trees produced on own account. Disposals consist of trees transferred to other units and those cut down before the end of their service lives but not exceptional losses due to drought or natural disasters.
Intangible fixed assets	Like tangible fixed assets these are produced themselves and go on to be used repeatedly or continuously in production. Major enhancements count as acquisitions.
• Mineral exploration	This is undertaken for commercial motives and the information obtained influences production activities in the future. It is therefore treated as acquisition of intangible fixed assets, the valuation of which is determined by the costs incurred, not the value of discoveries.
• Computer software	This includes the purchase or development of large databases and is treated as a fixed asset if used in production for more than a year. It is valued at purchasers' prices or, if developed in-house, at basic prices or costs of production
• Entertainment, literary or artistic originals	Acquisition of originals (manuscripts, tapes, films etc) is treated as gross fixed capital formation. If they are sold outright the value is the price paid. If not the valuation depends on the future benefits to be derived from them, but this may be difficult to estimate and in the absence of other information it may be necessary to use costs of production.

Changes in inventories (P.52)

6.16   As with other assets changes in inventories are measured by the difference between acquisitions and disposals, whether internal or external to the enterprise.

Acquisitions		*Disposals*
Purchase or barter		Sales or other uses
Internal transfer	INVENTORIES	Internal transfer
		Recurrent losses

To ensure consistent valuation at current prices all goods transferred into inventories are valued as if they were sold (or, in the case of work-in-progress, in proportion to the basic prices of the finished products) while all transfers out of inventories are valued at purchasers' prices. Inventories held abroad are excluded. As a consequence of the definition of residence they are regarded as being held by a quasi-corporate enterprise in the Rest of the World.

6.17   Inventories include *materials and supplies* which are to be used up in production (including losses through deterioration, accidental damage or pilfering). These need to be distinguished both from fixed assets and from valuables which are held as stores of value.

6.18   Also included in inventories is *work-in-progress*, especially where the production process is long or the accounting period short. The value of finished output (at basic prices) needs to be distributed between all the accounting periods in which it has been produced, taking account of the costs incurred in each period and, if necessary, price changes which have taken place during the process. When the whole process is complete the entire stock of work-in-progress is withdrawn from stock and transferred into finished product.

6.19   Work-in-progress includes cultivated assets in the form of single-use plants and livestock which have not been harvested or slaughtered by the end of the accounting period, though this should not be necessary if the accounts are annual and do not cut across the crop year. Trees and livestock that are for repeated or continuous use are treated as fixed assets when produced on own account, though work-in-progress may need to be calculated in the case of specialist breeders, e.g. of racehorses.

6.20   As already noted, structures and other fixed assets which take a long time to produce are only counted as work-in-progress as long as they are owned by the producer rather than the eventual user, which excludes own-account production and that under contracts of sale agreed in advance.

6.21   Inventories of *finished goods* can only be held by the enterprise that produced them. Something which is a finished product as far as its producer is concerned (e.g. coal or batteries) may subsequently become materials and supplies used as intermediate input to a further process of production. Entries to and withdrawals (or losses) from inventories of finished goods are valued at current basic prices.

6.22   A final type of inventory consists of goods acquired by enterprises – typically wholesalers or retailers – for the purpose of resale without further processing (apart from sorting, packing, etc). Such goods entering inventories are valued at purchasers' prices, including the cost of transportation other than that provided by the enterprise taking delivery. Reductions in inventories are valued at the purchasers' prices at which they could be replaced at that time, even if they are sold at a loss or not at all.

Acquisition and disposal of valuables (P.53)

6.23 Valuables, consisting of precious stones and metals (other than those used in production), works of art, jewellery and collections, are held primarily as stores of value, in the expectation that their prices relative to other goods and services will not decrease and may increase in the long run. Acquisitions are valued at the prices payable *plus* any associated costs of ownership transfer incurred by the buyer (e.g. for valuers, agents or auctioneers). Disposals are valued at the prices payable *minus* associated costs incurred by the seller. This means that, on aggregation, acquisition *less* disposals includes all dealers' margins and costs of ownership transfer.

Consumption of fixed capital (K1)

6.24 This is defined as the decline which takes place during an accounting period in the value of fixed assets owned by an enterprise as a result of physical deterioration, normal obsolescence and accidental damage. The value of an asset at any one time depends upon the benefits expected from using it in future, given by the present discounted value of the stream of receipts the owner could derive from renting it out, valued at the average prices of the period. The necessary estimates can be based on average service lives of assets and simple assumptions about the decline in their efficiency over time, taking account of information from markets in which new and existing assets are actually traded. (See Chapters 4, 15 and 16).

Non-produced non-financial assets (K2)

6.25 These comprise land; subsoil assets such as coal, oil, natural gas and minerals; non-cultivated biological resources; and below ground water resources. Land is defined in economic accounts as the ground itself, including soil and surface water but excluding structures built upon it, trees and crops growing on it, subsoil assets, non-cultivated biological resources and water resources below ground. The total stock of land is not fixed as it may be increased by reclamation, decreased by erosion, improved by clearance or irrigation and damaged by pollution or natural disasters. Activities that lead to major improvements in the quantity, quality or productivity or land are treated as a special category of gross fixed capital formation.

6.26 Acquisition and disposal of land are recorded at the same value (even if it is an imputation at current market value in a case of barter or capital transfer). Fees to estate agents, lawyers, etc. together with taxes, are treated as a separate category of gross fixed capital formation, whether paid by the purchaser or the seller. Where buildings or plantations are sold together with the land on which they stand, without separate valuations, the whole transaction should be classified to whichever component has the greater value, and if even this cannot be determined then it is classified, by convention, as purchase of a structure or plantation (i.e. as gross fixed capital formation) rather than of land.

6.27 Another important type of tangible non-produced asset consists of subsoil deposits such as coal, oil, gas, metallic ores and minerals, to the extent that ownership rights over these have been established (whether separate from or integral with the land below which they are located).

6.28 The capital account records only transactions in tangible non-produced assets when ownership passes from one institutional unit to another: not the depletion of reserves through extraction. Acquisitions and disposals are recorded in the same way as for land, all owners being deemed to be resident and ownership transfer costs being recorded separately.

6.28   The capital account also records, as sales of intangible non-produced assets, transactions in which the holder of a concession to extract subsoil assets sells this concession on to a third party. Similar treatment is accorded to the sale of leases on land or buildings. Other intangible non-produced assets include patents and purchased goodwill. In all cases the value of acquisitions includes ownership transfer costs incurred by the purchaser while disposals exclude the costs incurred by the seller.

Capital transfers (D.9)

6.30   The final component of the capital account is capital transfers, defined as transactions in which one institutional unit provides another with a good, service or asset without receiving anything in return. They may be in cash or in kind, the latter consisting of transfers of ownership, cancellations of liabilities or provision of services. An important distinction is made between:

- *Capital* transfers, which involve the disposal or acquisition of assets by one or both parties, and
- *Current* transfers, which reduce the income and consumption possibilities of one party and increase those of the other.

Some cash transfers may be regarded as capital by one party to the transaction but current by the other. For example, a payment of inheritance tax could be seen as a one-off capital transfer by the taxpayer but as a current receipt by the govern ment which collects many such payments. Similarly a large country making investment grants to a number of smaller countries might regard them as current payments even though they finance capital expenditure. In an integrated system of accounts any one transaction must be classified consistently so transfers are treated as capital for both parties even if only one of them acquires or disposes of assets. However, in other doubtful cases transfers should be treated as current rather than capital.

6.31   Non-financial assets are valued according to the price at which they could be sold on the market, including any costs of transportation, installation or ownership transfer incurred by the donor but not those of the recipient. The donation of a non-financial asset give rise to four entries in the accounts:

1. capital transfer paid by the donor
2. capital transfer received by the recipient
3. negative gross·fixed capital formation (i.e. an imputed sale) by the donor
4. positive gross fixed capital formation (i.e. an imputed purchase) by the recipient

Table 6.4: Examples of miscellaneous capital transfers (D.99)

Cancellation of debt by mutual agreement, including that owed by non-residents to residents and *vice versa*

Major payments of compensation for damage or serious injury not covered by insurance, e.g. by explosions, oil spillages or the side effects of prescribed drugs

Transfers from government units to publicly – or privately-owned enterprises to cover large operating deficits accumulated over two or more years

Transfers from central government to lower-level government units towards gross fixed capital formation or large expenditure deficits accumulated over two or more years

Legacies or large gifts, including those to non-profit instutions

Exceptionally large donations by households or enterprises to non-profit institutions to finance gross fixed capital formation, e.g. gifts to universities for new buildings

Taxes on capital (D.91)

6.32   Capital taxes are those levied at irregular and very infrequent intervals on the values of assets owned or transferred by institutional units. They include capital levies on assets owned by households or enterprises, which are treated as exceptional both by them and the government (e.g. betterment levies on agricultural land following the granting of planning permission). Other capital taxes are those on capital transfers, including inheritance tax and gifts taxes (on transfers made to avoid or minimise inheritance tax liability), but not taxes on sales of assets.

Investment grants (D.92)

6.33   Investment grants are capital transfers made by government to finance the acquisition of fixed assets, the recipients being obliged to use the proceeds for gross fixed capital formation, often on a specific project. They may include provision of equipment or buildings in kind, but exclude transfers of military weapons and their dedicated delivery systems.

Other capital transfers (D.93)

6.34   Finally there is a miscellaneous group of capital transfers which are neither taxes nor grants, some examples of which are listed in Table 6.5. There is a distinction between the cancellation of debt by mutual agreement between creditor and debtor, which is included, and the unilateral writing-off of debts by a debtor or creditor, which is not regarded as a transaction. If the creditor accepts such a write-off it will be recorded as a change in the volume of assets for each of the parties but neither provision for bad debts nor the unilateral repudiation of debts are recognised or recorded anywhere within the system of economic accounts. Debt cancellation gives rise to four entries in the accounts:

Chapter 7

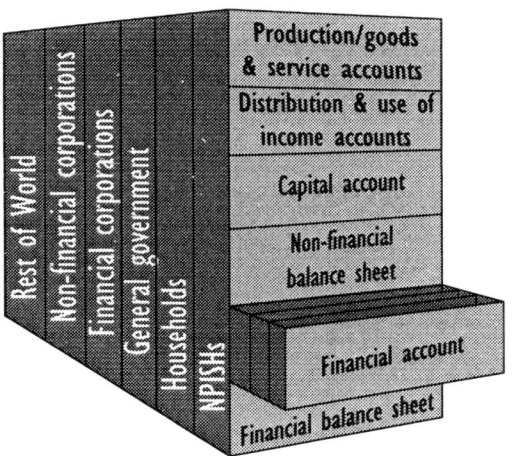

The Financial Account

Chapter 7 The financial account

Role of the financial account

7.1   The financial account is the final record of transactions between institutional units. It shows how the surplus or deficit on the capital account is financed by transactions in financial instruments. Thus the balance of the financial account (net incurrence of liabilities *less* net acquisition of financial assets) is equal in value but opposite in sign to net lending or borrowing, the balancing item of the capital account.

7.2   Some sectors are net lenders while others are net borrowers. When institutional units engage in financial transactions with one another the surplus resources of one sector can be made available for the use of others. The financial account indicates how net borrowing sectors obtain resources by incurring liabilities or reducing assets, and how net lending sectors allocate their surpluses by acquiring assets or reducing liabilities. The account also shows the contributions to these transactions of the various types of financial asset, and the role of financial intermediaries.

7.3   Most transactions involving the transfer of ownership of goods or assets or the provision of services have some counterpart entry in the financial account, even barter and transfers in kind, unless all elements of the transactions are completed simultaneously. The counterpart may take the form of a change in currency or transferable deposits, an account receivable or payable (e.g. a trade credit) or some other type of financial asset or liability.

7.4   Moreover there are many transactions which are recorded entirely within the financial account, where one financial asset is exchanged for another or a liability is repaid with an asset. New financial assets may be created through the incurrence of liabilities. Such transactions change the distribution of the portfolio of financial assets and liabilities and may change their total amounts but do not affect the *difference* between total financial assets and liabilities.

Nature of financial transactions

7.5   All financial transactions between institutional units are recorded in the financial account. Identifying them necessitates:

- Distinguishing financial from non-financial assets
- Distinguishing financial transactions from other changes that affect financial assets
- Distinguishing transactions in financial assets from operations involving contingent rather than actual assets.

Such issues have become more difficult with the proliferation of new and often complex financial instruments, some of which are tied to the prices of commodities, blurring the distinction between financial and non-financial transactions. International standardisation on these matters is further limited by variations between countries in the characteristics of their financial instruments and in national practices of classification and accounting.

7.6   Financial claims arise out of contractual obligations between pairs of institutional units. Many of these result in unconditional creditor/debtor relationships (e.g. with deposits, securities and loans) but in others the relationship is not unconditional and a liability is introduced only by convention (e.g. with shares and certain derivative instruments). A financial claim is created when a debtor accepts an obligation to make future payments to a creditor, normally of specified amounts in specified circumstances. Claims may also be created by force of law, e.g. to pay taxes and other compulsory levies. The construction of the accounts uses the identity that equates total transactions in assets for a given instrument to total transactions in liabilities across all sectors, including the rest of the world.

Table 7.1: Changes in financial assets involving other accounts

Item	Accounting treatment
Monetary gold and SDRs	Monetization and demonetization of gold and the allocation and cancellation of SDRs are recorded in the *other changes in volume of assets* account.
Valuation	Changes in the value of financial assets resulting from movements in prices or exchange rates are recorded in the *revaluation* account.
Debt forgiveness	Where debtor and creditor agree bilaterally that a financial claim no longer exists a capital transfer is recorded in the capital account and the extinction of the claim in the financial account.
Rescheduling of debt or change in assumptions, contractual terms or institutional sector of either party	These are regarded as constituting a new contract and are therefore reflected in the financial account.
Other write-offs and write-downs of debt	Recognition by a creditor that a financial claim cannot be collected is recorded in the *other changes in volume* account and in the balance sheet. Write-downs reflecting actual market values are recorded in the *revaluation* account. Write-offs and write-downs imposed to meet regulatory or supervisory requirements which do not reflect actual market values are not recorded in the economic accounts.
Debt repudiation	Unilateral cancellation by a debtor is not recognised in the economic accounts.
Defeasance, i.e. irrevocable pairing of assets and liabilities of equal value so that they can be removed from the debtor's own balance sheet	For economic accounts purposes the liabilities remain within the system though if they are transferred to a different institutional unit there will be a change in reporting arrangements.
Interest accrued but not paid	Recorded as property income as it accrues reinvested in the same financial instrument.

7.7 Certain changes in the volume and value of financial assets take place during an accounting period without involving any transactions between institutional units. These are excluded from the financial account but need to be recorded elsewhere. They are listed in Table 7.1.

7.8 Contractual financial arrangements which do not impose unconditional obligations but are dependent upon certain conditions being fulfilled are known as contingencies. They include:

- Guarantees of payment by third parties (in case the principal debtor defaults)
- Lines of credit (providing for funds to be made available if and when required)
- Letters of credit (promising payment when specified documents are presented)
- Underwritten note insurance facilities (guaranteeing that short-term securities issued by a potential debtor will be taken up).

7.9 In the economic accounts any fees for establishing such a contingent arrangement are treated as payments for services but the arrangement itself is not a current financial asset and is not recorded in the financial account until such an asset is actually created or changes ownership. However, where contingent positions are important for policy and analysis information about them may be collected and presented as supplementary to the accounts proper.

7.10 A borderline case of contingency is the banker's acceptance, which promises to honour certain drafts or bills of exchange by paying specified amounts on specified days. Though no funds are exchanged when the arrangement is made this is treated as an actual financial asset because the undertaking is unconditional. It appears in the balance sheet of the accepting financial institution.

Exceptions to general principles

7.11 The general principles described above are subject to certain exceptions as follows:

- *Financial claims arising from immovable non-financial assets*
  As described in Chapter 6, land and structures are all construed as being owned by residents. Owners who are actually non-residents are considered to have *financial* claims on notional resident units which own the land or structures.
- *Financial claims arising from unincorporated enterprises*
  An unincorporated enterprise operating in a different country from that in which its owner resides is considered to be a quasi-corporation resident in the country where it operates, the owner being deemed to own *financial* assets equal in value to the actual assets (both financial and non-financial) of the enterprise.
- *Financial leases*
  These are regarded as *de facto* changes of ownership, giving rise to financial claims which are treated as loans of the market value of the assets in question. Subsequent 'rental' payments are divided into interest (property income) and debt repayment (shown in the financial account).
- *Repurchase agreements (repos)*
  These involve sales of securities with a commitment to repurchase them at a fixed price at some future date. In most cases the securities do not change hands and the buyer is not entitled to sell them on, so it is doubtful whether a true change of ownership has taken place. The economic nature of the arrangement is similar to that of a collateralised loan from the purchaser to the seller and it is treated as such in the economic accounts, unless the instruments are short-term liabilities of monetary financial institutions, in which case they are classified as deposits.

- *Foreign exchange and gold swaps*

  These are a form of repurchase agreement undertaken by central banks with one another or with domestic banks. Swaps between central banks involve an exchange of deposits and, for each party, the acquisition of both a financial asset and a liability which should be recorded in the financial account. When a central bank acquires foreign exchange from a domestic bank in return for a deposit with a commitment to repurchase later the transaction is treated as a loan from the central bank and recorded as such in the financial account.

Financial derivatives

7.12   Many new financial instruments are linked to specific currencies, bonds, indices, interest rates or commodity prices and are therefore known as derivatives. As they are often devised for risk avoidance they are also referred to as hedging instruments. Various types can be distinguished:

- Giving rise to *contingent* assets and liabilities
- Giving rise to flows of property income but with no underlying transaction in a financial asset
- Giving rise to conditional rights like contingent assets but having a market value and being tradable.

7.13   The economic accounting convention here is that, to avoid asymmetrical treatment arising from conflicting perceptions of the various instruments, derivatives generally should be regarded as financial assets, and transactions in them as separate from any underlying deals to which they may be linked as hedges. Any explicit commissions to or from intermediaries for arranging derivative contracts are treated as payments for services, adjustment payments between parties to interest rates swaps (*see* following paragraph) are treated as interest payments while commodity-related contracts which (unusually) proceed to delivery are treated as transactions in commodities. All other payments made during the life of a derivative contract or when it is closed before maturity are treated as financial transactions and entered in the financial account.

7.14   Swaps are contractual arrangements to exchange streams of payments on the same amount of indebtedness over time, the most common types being:

- *Interest rate swaps* between interest payments of different types, e.g. fixed rate for floating rate
- *Currency swaps* between specified amounts of different currencies, including both interest and repayment flows.

Interest payments are treated in the economic accounts as property income while repayments are recorded in the financial account. Payments to third parties for arranging swaps are treated as being for services but the main parties to such transactions are not considered to be providing services to one another. The treatment of swaps is being reviewed internationally.

7.15   In forward rate options the parties agree in advance on an interest rate to be charged on the principal at a specified future date but the only payment which ever takes place relates to the difference between the agreed rate and the one actually prevailing at the time of settlement. Payments are treated as property income as the only asset involved is a notional amount of principal which is never exchanged.

7.16   Options are contracts that give their purchasers the right but not the obligation to buy (for a 'call' option) or sell (for a 'put' option) a particular financial instrument or commodity at a predetermined price (the 'strike' price) within a given time span (American option) or on a given date (European option). Many options, if exercised, are settled in cash rather than by delivery of the underlying assets or commodities. The buyer of the option pays a premium to the seller for the latter's commitment, which by convention is treated as a liability of the seller and recorded as such in its entirety (though in principle it could be considered to include a service element). Subsequent sales and purchases of options are also recorded in the financial account. Premiums are often called for in stages as 'margin payments', and are then recorded both as deposits and as transactions in securities.

7.17   Warrants are a form of option giving their holders the right to buy from the issuer (normally a corporation) a certain number of shares or bonds under specified conditions and for a specified period of time. There are also currency options based on the amount of one currency required to buy another. All warrants can be traded and have a market value.

Accounting rules for financial transactions

7.18   Transactions in financial assets are valued at the prices at which the assets are acquired or disposed of, excluding:

- Fees and commissions (recorded as payments for services)
- Tax on the transactions (recorded as tax on services).

When securities are issued at a discount the amount recorded in the financial account is the proceeds to the issuer at the time of sale, the difference between this and the face value being treated as interest accrued over the life of the instrument. When on the other hand securities are issued through underwriters or other intermediaries and then sold at higher prices to final investors the value recorded is that paid by the investors, the difference in this case being treated as service payments from the issuer to the underwriters.

7.19   When interest accrues on an instrument the amount accruing in the period is recorded as being re-invested in the same instrument.

7.20   As regards timing, financial transactions should be recorded by the two parties at the same time, aligned with any entries into non-financial accounts which give rise to the claims. If only the financial account is affected then the recording should take place when ownership of the asset is transferred or when the liability is incurred or redeemed as the case may be. In practice, however, there is often a lag between the making of payments and their receipt.

7.21   Reporting of financial assets and liabilities often entails a degree of netting-off of sales against purchases, particularly where transactions need to be inferred from balance sheet data. Various levels can be identified, from fully gross reporting through netting within specific assets (e.g. bonds) or categories of asset (e.g. all securities) right up to netting transactions in groups of liability categories against transactions in assets in the same groups. As the financial account is broken down by main categories of financial asset netting at this level is acceptable but for detailed flow of funds reporting information relating to specific assets is desirable.

7.22   Consolidation of the financial account is the process of offsetting transactions in assets against the counterpart transactions in liabilities for the same group of institutional units. It can be performed at any level of aggregation up to that of the whole economy and different levels are appropriate for different sorts of analysis. For example, consolidating by sector is sufficient to identify those which are net lenders and those which are net borrowers, but subsectoral detail provides much more information about financial intermediation. The ESA95 requires both consolidated and non-consolidated financial accounts and balance sheets for sectors and sub-sectors. Complete consolidation within each of these removes all transactions and balance sheet positions within the given sectors. Non-consolidated accounts are less easy to interpret, since the degree of de-consolidation may vary from sector to sector, possibly depending on the data sources and methods used.

Classification of financial instruments

7.23   Table 7.2 shows the economic accounts classification of transactions in and stocks of financial assets and liabilities. It is based on two criteria – liquidity and the form of the underlying creditor/ debtor relationship – and is designed to facilitate analysis and international comparisons. This is different from the old UK classification which defined transactions in terms of the counterpart sectors.

Table 7.2: Classification of financial assets and liabilities

Monetary gold and SDRs (F.1)

Currency and deposits (F.2)
   Currency (F.21)
   Transferable deposits (F.22)
   Other deposits (F.23)

Securities other than shares (F.3)
   Short-term (F.31)
   Long-term (F.32)

Loans (F.4)
   Short-term (F.331)   Derivatives (F.34)
   Long-term (F.332)

Shares and other equity (F.5)

Insurance technical reserves (F.6)
   Net equity of households in life insurance reserves (F.611)
   Net equity of households in pension funds (F.612)
   Payments of premiums and reserves against outstanding claims (F.62)

Other accounts receivable/payable (F.7)
   Trade credit and advances (F.71)
   Other (F.79)

7.24   It may be noted that the internationally-recognised system of economic accounts does not include any concept or measure of money itself. It is impossible (at least in present circumstances) to define either narrow or broad money in a way which is consistent and unambiguous for comparison between countries. However, this does not preclude the subdivision of economic accounts aggregates to square with currently-accepted conventions of monetary analysis.

7.25   Another concept which is downgraded in the general system is that of maturity, as innovations in financial markets have diminished the usefulness of simple distinctions here. Where differentiation between short-term and long-term instruments is still important the former are defined as being for one year or less (though individual countries can raise this to a maximum of two years if necessary to accommodate established national practices).

7.26   The classification includes some assets for which there are no corresponding liabilities, *viz* shares and other equity securities, certain financial derivatives, monetary gold and SDRs. Shares and other equity are treated as liabilities by convention - shares are the liabilities of the issuing corporations; other equity covers liabilities of quasi-corporations to their owners. Gold and SDRs form part of the country's reserve assets which are regarded as claims on non-residents although, strictly speaking, no sector or institution in the rest of the world bears the liability for them.

7.27   The contents of each of the main categories of financial asset distinguished in the classification can be summarised as follows:

Monetary gold and SDRs (F.1)

7.28   *Monetary gold* is gold owned by the monetary authorities or others subject to their effective control. (In the United Kingdom this is the central government itself.) It is held as a financial asset and as a component of the foreign reserves used for direct financing of international payment imbalances and intervention in the foreign exchange markets to affect the exchange rate. Monetisation and demonetisation of the authorities' gold stocks should be treated as reclassifications and shown in the *other changes in volume of assets* account rather than the financial account. Pledging of monetary gold as collateral does not affect the accounts though information about this is collected as it may have an impact on the gold's usability as a reserve asset.

7.29   *SDRs* (special drawing rights) are international reserve assets created by the International Monetary Fund and allocated to its members to supplement existing reserve assets, conferring an unconditional entitlement to obtain foreign exchange. Their value is determined daily on the basis of a basket of currencies and they are transferable between countries.

7.30   Monetary gold and SDRs do not appear as liabilities for any counterpart sector in the system. National foreign exchange reserves may also be held in other forms such as currency or securities; those are classified to the appropriate asset category below.

Currency and deposits (F.2)

7.31   *Currency* (national and foreign) comprises notes and coin in circulation that are commonly used to make payments.

7.32 *Deposits* as specified in the system are of two types: transferable and other. In the United Kingdom, deposits are generally the liabilities of monetary financial institutions. The distinction by type is of no real significance here and almost all deposits are therefore classified as transferable. The main exception is certain central government liabilities. A more relevant distinction is that between deposits and loans; deposits are generally initiated by the asset holder. Transferable deposits are defined in the system as those which are exchangeable on demand at par, freely transferable by cheque or giro and commonly used to make payments. Other deposits include savings accounts, term accounts and shares issued by building societies and others which are redeemable on demand or at short notice. Also included in non-transferable deposits are claims on the IMF that are part of international reserves (unless evidenced by loans), margin payments on options or futures contracts and very short-term repurchase arrangements (if considered part of broad money). For all deposits a distinction can be made according to whether they are denominated in national or foreign currency and whether they are liabilities of resident institutions or the rest of the world.

Securities other than shares (F.3)

7.33 This includes a wide range of instruments traded in the financial markets including bills, bonds and debentures giving holders the unconditional right to fixed or contractually-determined sums of money on specified dates (including preference shares in the UK as they generally do not give rights to participate in any distribution of company profits). Securities also include new assets created by 'securitisation' (i.e. repackaging of existing assets), derivatives, loans which have become negotiable *de facto*, preference stock (provided it does not confer a share in the value of the corporation) and bonds that are convertible into shares (provided the value of the conversion option is separated out). Other inclusions are:

- notional interest accrued on zero coupon or deep discount bonds which actually pay little or no interest during the life of the instrument, the amount being based on the difference between the discounted issue price and the price at maturity;

- indexation amounts due on index-linked securities for which either the periodic coupon payments or the principal are linked to a price or exchange rate index.

Loans (F.4)

7.34 Loans are created whenever creditors lend funds directly to debtors, whether the transaction is evidenced by non-negotiable documents or provides no security. They include hire-purchase, loans to finance trade credit, evidenced claims on the IMF, repurchase agreements (unless included in deposits) and financial leases. Cancellation of debt by mutual agreement between debtor and creditor is included. The counterpart is a capital transfer (D.99) or a transaction in equity (F.5).

Shares and other equity (F.5)

7.35 This category comprises all instruments conferring claims on the residual value of corporations and quasi-corporations. It is useful to compile data separately for shares that are and are not listed on the Stock Exchange (quoted and unquoted shares).

7.36 Besides shares in corporations the category includes investment in mutual funds and proprietors' capital additions to and withdrawals from their net equity in quasi-corporations. A quasi corporation is an entity which does not fulfil all the criteria to be a corporation but acts like one and produces accounts which enables them to be classified to the corporate sector in the economic accounts. An example would be a partnership. When the owner puts in or takes out capital this transaction would be classified to F.5. The counterpart of cancellation by government of debt of a public corporation about to be privatised is classified here as an increase in the government's equity in the corporation.

7.37   Also classified with *shares and other equity* are financial transactions related to immovable assets and unincorporated enterprises owned by non-residents. Unremitted profits of direct investment subsidiaries are treated as investment in unquoted shares.

Insurance technical reserves (F.6)

7.38   These comprise three elements, as follows.

- *Households' net equity in life insurance reserves (F.611)*: Life insurance companies hold and manage reserves that add to the maturity value of with-profits endowment policies. These are considered to be assets of the insured persons rather than the companies, and changes in them are recorded as such in the financial account.

- *Households' net equity in pension funds (F.612)*: These are similar to life insurance reserves and are treated in the same way.

- *Prepayments of insurance premiums and reserves for outstanding claims (F.62)*: Prepayments arise because insurance premiums are generally paid in advance and are partly intended to cover risks in the subsequent accounting period. They are considered to be assets of the policy holders. Reserves for outstanding claims are necessary because valid claims are held to be due for payment when the eventuality occurs which gives rise to them, however long they take to settle. Such reserves are treated as assets of the beneficiaries and liabilities of the insurance companies.

Other accounts receivable/payable (F.7)

7.39   A specific subdivision of this category (F.71) relates to trade credit for goods and services and advances for work that is in progress or to be undertaken, not including loans to finance trade credit. Other accounts receivable/payable are no longer intended to include interest which has accrued but not been paid or capitalised; that is included with the relevant financial instrument as far as possible. Accrued taxes not yet paid are included here. The category does not include statistical discrepancies.

Chapter 8

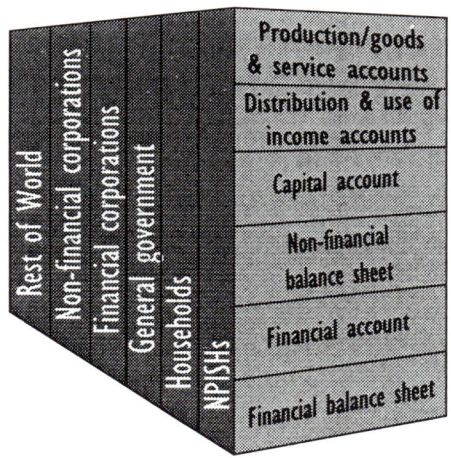

The *other changes in assets* account

Chapter 8 The *other changes in assets* account

Introduction

8.1   The preceding two chapters have described the recording of changes in the values of assets and liabilities resulting from transactions. This chapter deals with all other changes, which are of two quite distinct types:

- Actual changes in volume due to factors such as the discovery, depletion and destruction of assets; and
- Changes in value due to movements in the level or structure of prices, which are reflected in holding gains or losses.

These are to be recorded, respectively, in what are essentially two separate accounts: the *changes in volume of assets* account and the *revaluation* account, which are described in turn below. These accounts provide the link between transactions recorded in accounts described in earlier chapters and the change in balance sheets. Much of the analysis described is not yet possible for the United Kingdom and so the accounts will not be published.

Other changes in the volume of assets account

8.2   The *other changes in volume of assets* account has various functions:

- *To allow assets to enter and leave the accounting system in the normal course of events.* Whereas the capital and financial accounts relate mainly to changes in the ownership of existing economic assets as a result of interactions between institutional units, the *other changes* account deals with 'entrances and exits' (coming about through interactions between institutional units and nature).
- *To record the economic effects of exceptional unanticipated events.* Whereas the accounts already described allow for normal rates of loss (where appropriate) the other changes account includes highly destructive occurrences such as natural disasters and war.
- *To record the effect on the accounts of changes in classifications and institutional structure.*
- *To record changes in natural assets and environmental degradation.* These are not allowed for elsewhere in the national accounts (e.g. in consumption of fixed capital).

It is worth remembering the definition of economic assets within the system: they are entities over which ownership rights have been enforced and from which economic benefits may be derived. Thus for example environmental degradation refers to the degradation of assets within that definition. It would include, for example, the decrease in the value of farmed land due to pollution but would exclude similar effects on the value of unowned land.

8.3   For purposes of the *other changes* account nine new categories of change in assets and liabilities are introduced:

- Economic appearance of non-produced assets (K.3);
- Economic appearance of produced assets (K.4);
- Natural growth of non-cultivated biological resources (K.5);
- Economic disappearance of non-produced assets (K.6);
- Catastrophic losses (K.7);
- Uncompensated seizures (K.8);
- Other volume changes in non-financial assets (K.9);
- Other volume changes in financial assets and liabilities (K.10);
- Changes in classifications and structure (K.12).

Some of these may apply to any type of asset – produced, non-produced or financial – but most are specific to one or another, as indicated in Table 8.1.

Table 8.1: Components of the *other changes in volume of assets* account

Produced assets (AN.1)

Economic appearance of produced assets (K.4)
Catastrophic losses (K.7)
Uncompensated seizures (K.8)   Fixed assets (AN.11)
Other volume changes in non-financial assets (K.9)   Inventories (AN.12)
Changes in classifications and structure (K.12)   Valuables (AN.13)

Non-produced assets (AN.2)

Economic appearance of non-produced assets (K.3)
Growth of non-cultivated biological resources (K.5)
Economic disappearance of non-produced assets (K.6)
Catastrophic losses (K.7)   Tangible assets (AN.21)
Uncompensated seizures (K.8)   Intangible assets (AN.22)
Other volume changes in non-financial assets (K.9)
Changes in classifications and structure (K.12)

Financial assets and liabilities (AF)

Catastrophic losses (K.7)   Monetary gold and SDRs (AF.1)
Uncompensated seizures (K.8)   Currency and deposits (AF.2)
Other volume changes in financial assets and   Securities and other shares
liabilities (K.10)   (AF.3)
Changes in classifications and structure (K.12)   Loans (AF.4)
        Shares and other equity (AF.5)
        Insurance technical reserves
        (AF.6)
        Other accounts receivable/
        payable (AF.7)

8.4   Many of the entries in the *other changes* account are closely linked to entries in other accumulation accounts, which may provide estimates of value. For example, the 'economic appearance' of an asset may be associated with a market transaction, while a disappearance can be valued on the basis of the latest balance sheet.

8.5   The following sections describe each of the nine asset-change categories in turn.

Economic appearance of non-produced assets (K.3)

8.6   *Economic appearance* refers to a thing coming within the asset boundary to figure in the balance sheet for the first time, as distinct from assets resulting from processes of production. Some of the assets in question occur in nature while others are the result of human agency. Four sub-categories can be distinguished, as follows.

- *Gross additions to exploitable subsoil resources*
  The resources in question are coal, gas, oil, metals, minerals, etc, and 'exploitable' in this context means economically exploitable given current technology and relative prices. Reserves may therefore increase not only by virtue of new discoveries but also because formerly uneconomic deposits have been made viable by technological progress or price changes.
- *Transfers of other natural assets to economic activity*
  Economic assets are entities over which ownership rights are enforced and from which economic benefits may be derived. Wild or waste land does not necessarily qualify but may be brought within the asset boundary. The stock of land may also be increased by reclamation from the sea. For such natural assets economic appearance normally involves commercial exploitation, e.g. by forestry or diversion of groundwater.
- *Quality changes due to changes in economic use*
  In national accounting differences in quality are treated as differences in volume, and they may arise not from changes in the asset itself but from the way it is used, e.g. a change from agricultural to building land.
- *Appearance of intangible non-produced assets*
  Intangible non-produced assets are constructs of society, which make their appearance when, say, a patent grants legal protection to an invention or a transferable contract provides an economic benefit which can be passed on to a third party. A particular example is that of 'goodwill' when an enterprise is sold at a price that exceeds its net worth (i.e assets *minus* liabilities). Goodwill can only be considered an economic asset and reflected in the accounts if it is evidenced by a transaction so, prior to the sale, the appropriate value must be put into the enterprise's balance sheet via the *other changes in volume of assets* account so that it can be disposed of by the seller, acquired by the purchaser and remain thereafter as part of the enterprise's assets.

Economic appearance of produced assets (K.4)

8.7   Only two types of asset fall under this heading:

- *Historic monuments*
  Economic appearance occurs when the structure or site is first recognised as specially significant having formerly not been recorded in the balance sheet (perhaps because it had already been written-off).

- *Valuables (e.g. precious stones, antiques and works of art)*
  Until their high value or artistic significance is recognised these would be regarded as ordinary goods whose purchase should be included in households final consumption expenditure.

In each case an 'appearance' is recorded in the other changes in volume of assets account and immediately becomes the subject of a transaction in the capital account in which previously-recognised monuments and valuables are already recorded.

Natural growth of non-cultivated biological resources (K.5)

8.8   The resources in question here are natural forests, fishstocks, etc. These are economic assets but their growth is not under the direct control, responsibility and management of an institutional unit and is therefore not production but an economic appearance. In principle it should be recorded gross, depletion being regarded as an economic *dis*appearance but in practice the only measures available are likely to be net.

Economic disappearance of non-produced assets (K.6)

8.9   The various forms this can take include the following:

- *Depletion of natural assets*, e.g. subsoil assets, natural forests, fishstocks in the open sea and water resources;
- *Other reductions in exploitable subsoil resources* resulting from adverse changes in technology or relative prices, though these may be netted off against the additions mentioned above under 'economic appearance';
- *Quality changes due to changes in economic use*, which are also symmetrical with the economic appearances mentioned above, e.g. cultivated land given over to communal grazing;
- *Degradation due to economic activity*, whether anticipated or not, e.g. from erosion, improper farming practices, acid rain or agricultural run-off;
- *Write-off and cancellation of purchased goodwill, transferable contracts*, etc. and *exhaustion of patent protection*, which should each be recorded over an appropriate period, not all at once.

Catastrophic losses (K.7)

8.10   Accidental damage to fixed assets, recurrent losses from inventories and normal depletion or degradation of natural assets are all covered elsewhere in the accounts, as already described. This category relates to large-scale, discrete and recognisable events such as earthquakes, volcanic eruptions, tidal waves, hurricanes, drought, war, riots, toxic spillages and releases of radioactivity. All sorts of asset can be affected, e.g. damage to land and destruction of buildings, equipment, valuables, currency and bearer securities.

Uncompensated seizures (K.8)

8.11   These occur when government or other institutional units take possession of others' assets for reasons apart from payment of taxes, fines or similar levies and foreclosure or repossession of goods by creditors (the latter being treated as transactions because the agreement between debtor and creditor provides this recourse, either explicitly or by general understanding). The value of the seizure, *less* any compensation, should be recorded in the accounts of both parties.

Other volume changes in non-financial assets (K.9)

8.12   These arise from unexpected changes in the benefits derivable from assets, and especially from those not foreseen when allowances were determined for 'normal' consumption of fixed capital or inventory shrinkage. Examples are:

- *Unforeseen obsolescence* resulting from improved technology, e.g. a new model of the asset or an new process which renders the asset completely unnecessary;
- *Unforeseen damage to fixed capital*, which should be small for the economy as a whole but may be significant (and variable in sign since the damage might be greater or less than foreseen) for an individual unit;

- *Environmental degradation of fixed capital*, e.g. through the effects of acidity in rain or air on building surfaces or car bodies;
- *Abandonment of production facilities* where this occurs before they are complete and put into service, because they have already ceased to have an economic rationale;
- *Exceptional losses from inventories*, e.g. through fire damage, robbery or insect infestation;
- *Other changes*, e.g. assets lasting longer than expected (either economically or physically) and the effect of converting dwellings to commercial use or vice versa.

Other volume changes in financial assets and liabilities (K.10)

8.13   This category allows for the creation and extinction of assets for which there is no corresponding liability or no specific transaction, holding gain or loss or other balance sheet change already identified, as follows.

- *Allocation and cancellation of SDRs* (for which there is no IMF liability);
- *Writing-off of bad debts* when a creditor recognises that a financial claim can no longer be collected, e.g. due to bankruptcy (but not the cancellation of debt by mutual agreement, which is treated as a capital transfer);
- *Changes in actuarially-determined liabilities of 'defined benefit' pension plans* resulting from changes in benefit structure;
- *Miscellaneous changes* not covered elsewhere.

Changes in classifications and structure (K.12)

8.14   Two types of change are covered here:

- *Changes in sector classification and structure (K.12.1)*
  Reclassification of an institutional unit from one sector to another transfers its entire balance sheet and the effect of this is recorded in the *other changes in volume of assets* account. Corporate restructuring also needs to be taken into account since, when a corporation is absorbed and so ceases to be an independent institutional unit, all claims and liabilities between it and its new parent are eliminated. Conversely, when a corporation is split up new claims and liabilities may appear and need to be captured in the other assets account.

- *Changes in classification of assets and liabilities (K12.2)*
  The classification of an asset for purposes of the capital and financial accounts may be different from that in the opening balance sheet if the purpose for which it is used has changed, e.g. from non-monetary to monetary gold or from pasture to building plots. This discrepancy is accounted for in the *other changes* account.

Revaluation account: nominal, neutral and real holding gains (K.11)

8.15   The remainder of this chapter deals with the revaluation account which records holding gains on financial and non-financial assets and liabilities which accrue during an accounting period. Those on assets, whether positive or negative (i.e. losses) are shown on the opposite side of the account from those on liabilities. Table 8.2 lists in condensed form the components of the account.

8.16   The *nominal* holding gain is the benefit accruing as a result of a change in the monetary value of an asset, and is decomposed into:

- the *neutral* holding gain, which would accrue if the asset's price were to change in the same proportion as the general price level
- the *real* holding gain, which accrues as a result of any change in relative price.

Table 8.2: Components of the revaluation account

	Nominal holding gains/losses	of which:	
		Neutral holding gains/losses	Real holding gains/losses

Non-financial assets
  Produced assets
    Fixed assets
    Investments
    Valuables
  Non-produced assets
    Tangible non-produced assets
    Intangible non-produced assets			

Financial assets/liabilities
  Monetary gold and SDRs
  Currency and deposits
  Securities other than shares
  Loans
  Shares and other equity
  Insurance technical reserves
  Other accounts receivable/
  payable			

Changes in net worth due to holding gains/losses			

The algebraic sum of the nominal gains of an institutional unit (i.e. the balancing item of the account) is the change in net worth due to such gains, and it too may be decomposed into 'neutral' and 'real' elements. Holding gains are sometimes called 'capital gains' but include stock appreciation (gains on inventories) which is not a 'capital' item.

8.17 A nominal gain may accrue over any length of time – not just a complete accounting period – and is defined as the increase in the monetary value of the asset assuming it has not itself changed. This excludes:

- changes in the physical condition of assets, e.g. deterioration of stocks or normal obsolescence of capital equipment;
- changes in the value of seasonal products where storage is essentially an extension of the production process;
- changes in the market values of bills and bonds as they approach maturity, which may be attributable to the accumulation of accrued interest, representing growth in the asset itself;
- changes in amounts of assets and liabilities denominated in purely monetary terms (e.g. cash and deposits) since these are not measured in physical units, the 'quantity' is effectively the currency itself and *any* change is therefore a quantitative one.

8.18   Nominal holding gains are recorded in the revaluation account whether or not they have been 'realised', i.e. whether or not the asset in question has actually been disposed of. Whenever they take place acquisitions and disposals must be valued consistently with the capital and financial accounts. In the case of non-financial assets this means including the costs of ownership transfer, so if such an asset is purchased and subsequently resold at the same price then a nominal holding loss is recorded equal to the costs incurred on both purchase and resale.

8.19   The neutral holding gain needed to preserve intact the real value of quantity Q of an asset is given by:

$$\text{neutral gain} = P_0 Q (I_t / I_0 - 1)$$

where P denotes the unit price, $P_0 Q$ is the monetary value of the asset at time 0 and I is a general price index. The index used should cover as wide a range of goods and services as possible, e.g. all final consumption, but for short periods it may be necessary to use a more restricted indicator such as the retail prices index.

8.20   *Real* holding gains represent the arithmetic difference between the nominal and neutral holding gains and can be written as:

$$\text{real gain} = P_0 Q (P_t / P_0 - I_t / I_0).$$

These gains may lead to significant redistribution of net worth. For example,

- as already noted, there are no nominal gains on monetary assets and liabilities so in a period of inflation the neutral gains on them must be positive and the real gains negative, transferring real purchasing power from creditors to debtors (unless correspondingly higher nominal rates of interest are set);
- relative price changes may be very different for different types of asset so that some owners benefit from real holding gains while others experience real holding losses. These do not necessarily cancel out.

Real holding gains are therefore important economic variables in their own right and need to be taken into account along with income for purposes of analysing consumption and capital formation.

Measurement of holding gains

8.21   In order to calculate holding gains directly it would be necessary to record all the assets acquired and disposed of during an accounting period and their prices. Information in this detail is not normally available but, provided appropriate valuation methods have been used, nominal gains can be calculated indirectly as:

Difference between asset values values as shown in opening and closing balance sheets	} less {	Total value of all transactions and other volume changes

This derives from the following basic accounting identity, which itself derives from the way the various items have been defined and valued:

Value of stock in opening balance sheet	+	Value of quantities acquired or disposed of in transactions	+	Value of other volume changes	+	Nominal holding gain	=	Value of stock in closing balance sheet

Nominal gains may thus accrue on assets which do not appear in either the opening or closing balance sheets, and it is necessary to record these so that the accounts of transactions, inventories, etc, can be reconciled with the 'no change' situation in the balance sheet.

8.22   Assets acquired or disposed of during an accounting period also pose problems for the calculation of neutral holding gains since, to calculate them directly, it would be necessary to know precisely when the acquisitions and disposals all took place – information that is unlikely to be available. Again it may be possible to calculate neutral gains indirectly as a residual, but there is no simple method of doing so that can be used in all circumstances.

8.23   There may be situations (e.g. with inventories and certain financial assets) in which the *only* information available about assets is their value in the opening and closing balance sheets. In such cases the value of transactions and other volume changes and (hence) nominal holding gains can be deduced if assumptions are made about the path followed by both prices and quantities between the beginning and end of the accounting period. The simplest and most convenient assumption is a 'straight line' one, that both prices and quantities change at constant linear rates, in which case it can be shown that:

$$\text{Nominal gain} = (P_t - P_0) \cdot (Q_0 + Q_t) / 2$$

which can be decomposed into:

$$\text{Neutral gain} = (I_t / I_0 - 1) \cdot P_0 \cdot (Q_0 + Q_t) / 2$$

$$\text{Real gain} = (P_t / P_0 - I_t / I_0) \cdot P_0 \cdot (Q_0 + Q_t) / 2$$

8.24   The quality of these approximations depends on the realism of the underlying assumptions and the assumption that quantities change at a constant rate will be questionable in many cases, particularly where there are seasonal influences. However, approximations of this kind may serve as a useful check on estimates made by other methods.

Holding gains for fixed assets

8.25   Different types of asset present different problems and possibilities for the measurement of holding gains. In the case of fixed assets the accounting identity given in paragraph 8.21 above can be rewritten with 'Value of quantities acquired or disposed of in transactions' replaced by 'net fixed capital formation', i.e. gross fixed capital formation *minus* consumption of fixed capital. Nominal holding gains may then be obtained from:

$$\left.\begin{array}{l}\text{Difference between asset} \\ \text{values as shown in opening} \\ \text{and closing balance sheets}\end{array}\right\} - \left\{\begin{array}{c}\text{Net capital formation} \\ + \\ \text{Other volume changes}\end{array}\right.$$

This expression serves to emphasise that, when prices are increasing, the difference in value reflected in the balance sheets is not equivalent to capital formation. The former reflects the *change* in prices between the points at which the two balance sheets were drawn up (i.e. holding gains) while the latter is based on a single set of prices, namely those at which transactions took place during the period.

8.26 The change in balance sheet valuation should reflect inter alia changes in the purchasers' prices of new assets of the same kind provided these are still being produced and sold. However, if similar assets are no longer in production and their replacements have been much improved by technical progress then accurate valuation may be difficult.

Holding gains for inventories

8.27 For inventories nominal holding gains are given by:

$$\left.\begin{array}{l}\text{Difference between asset}\\ \text{values as shown as}\\ \text{opening and closing}\\ \text{balance sheets}\end{array}\right\} - \left\{\begin{array}{l}\text{Entries less withdrawals}\\ \text{(valued at prices prevail-}\\ \text{ing when they took place)}\\ +\quad\text{Other volume changes}\end{array}\right.$$

Other volume changes' are likely to consist of losses through exceptional events (e.g. natural disasters or major fires). Current losses (e.g. through regular wastage or pilfering) are included in withdrawals.

8.28 When records are not kept of the flow of entries and withdrawals and the prices attached to them it will be necessary to estimate changes from the opening and closing stocks by methods similar to those described above but, again, such methods are only as good as the assumptions underlying them.

8.29 For work-in-progress nominal holding gains are given by:

$$\boxed{\begin{array}{l}\text{Difference between assset}\\ \text{values as shown in opening}\\ \text{and closing balance sheets}\end{array}} - \boxed{\begin{array}{l}\text{Additions to work-in-progress}\\ \text{(as the production proceeds)}\\ \\ \textit{less } \text{withdrawals from w-i-p}\\ \text{(as the process is completed)}\end{array}}$$

Because some production processes take years to complete the whole of their output during this period may figure as additions to work-in-progress, which can generate substantial nominal holding gains even with only moderate inflation.

Holding gains for financial assets

8.30 Cash, deposits, loans, advances, credits, etc, which are denominated in currency cannot show *nominal* holding gains but, to calculate the neutral and real holding gains on them, it is necessary to record the times as well as the values of all transactions. Any estimates based on balance sheet data alone are likely to be unsatisfactory if these stocks are small in relation to the flows taking place within each accounting period.

8.31 In the case of bonds moving towards maturity there is a progressive growth in the principal outstanding as interest (measured by the difference between the issue price and redemption value) is accrued and simultaneously reinvested in the bond. This is essentially a quantity change rather than a price increase and does not generate any holding gains. The situation is similar to that of a good (e.g. wine) that matures while it is being stored.

8.32 However, bond prices change not only because of the passage of time but also (inversely) in response to changes in market rates of interest. These *are* price changes and give rise to holding gains or losses. The same is true of bills but, because these are short-term securities, the gains and losses produced are much smaller.

Foreign assets

8.33 The value of a foreign asset is measured by converting it to the currency of the country in which its owner is resident, using the current exchange rate. Nominal holding gains may therefore occur purely because of a change in the exchange rate. Their value can be calculated in the usual way by subtracting the value of transactions from the difference between the opening and closing balance sheet values of the asset, provided the various elements are converted from foreign to national currency using the appropriately-dated interest rate for each. Neutral and real holding gains can then be obtained, in national currency, in the same way as for any other asset.

Chapter 9

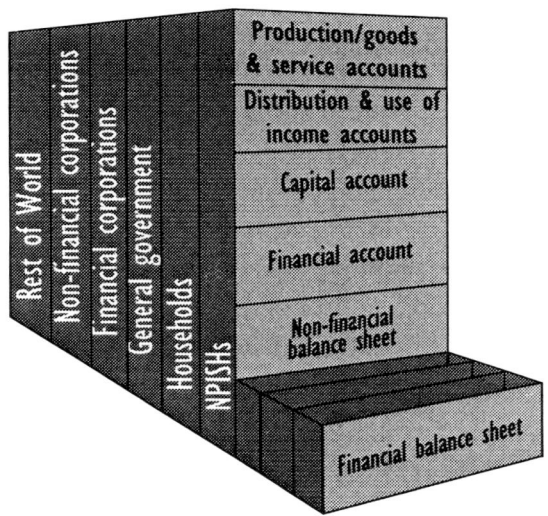

The balance sheet

Chapter 9 The balance sheet

Make-up and usage of the balance sheet

9.1   Balance sheets are statements of the value of assets and liabilities at a particular point of time, and can be drawn up for institutional units, sectors or the whole economy. The balancing item is 'net worth' or, in the case of the whole economy, 'national wealth' – the aggregate of non-financial assets and net claims on the rest of the world.

9.2   Moving forward through time each balance sheet is linked to the previous one by the accounting identity:

stock at the beginning of the accounting period

+   acquisitions *less* disposals of assets (financial and non-financial) by transactions taking place during the period

+   other changes in the volume of assets
e.g. through discovery or destruction during the period

+   holding gains *less* holding losses accruing during the period

=   stock at the end of the accounting period

9.3   Balance sheets can be used in various ways:

- for households, to analyse the effects of changes in net worth on consumption and savings patterns and the distribution of wealth;
- for corporations, to compare assets with the market value of shares and study liquidity and financing needs;
- for industries, to relate capital employed to production and broad measures of productivity;
- for the whole economy, to monitor the availability of resources and the overall financial position *vis à vis* other countries.

9.4   Economic assets are defined as entities over which ownership rights are enforced and from which economic benefits may be derived, and three types of asset are included in the balance sheet, namely

- *produced assets* (AN.1), which have come into existence as output from production processes. They are classified according to their role in production, as:

  fixed assets, which are used repeatedly or continuously: buildings and other structures, machinery and equipment, cultivated assets and certain intangible assets[1];
  inventories, which are either used up in production and intermediate consumption or disposed of (usually by being sold);
  valuables, which are not used for production but acquired and held primarily as stores of value.

- *Non-produced assets (AN.2)*, which are classified according to the way they came into existence, as:

  tangible assets which are natural resources coming within the asset boundary, including land, subsoil assets, non-cultivated biological resources and water resources but excluding the sea, air, etc.;
  intangibles, regarded as assets if evidenced by legal or accounting actions, e.g. patented entities, transferable contracts or purchased goodwill but not the *granting* of a patent.

- *Financial assets (AF)*, which are generally those for which there is a corresponding liability for some other institutional unit. They are classified according to their liquidity and the nature of the creditor/debtor relationship, into:

  Monetary gold and SDRs (for which, exceptionally, there is no counterpart liability)
  Currency and deposits
  Securities other than shares
  Loans
  Shares and other equity
  Insurance technical reserves
  Other accounts receivable/payable.

Only *actual* current liabilities are included, which rules out:

- contingency arrangements, unless they are unconditional on both sides and/or tradable; and
- any provisions for *future* liabilities or transactions.

Valuation of balance sheets

9.5   In order that they be consistent with the accumulation accounts all the items must be valued as if being acquired on the date to which the balance sheet relates, including costs of ownership transfer in the case of non-financial assets. The valuation may be undertaken in any of four ways, as shown in Table 9.1, the first of these being regarded as the ideal.

9.6   The practical application of these methods varies from one type of asset to another, as summarised in Table 9.2.

---

[1]   For reasons explained elsewhere most small tools, some military equipment, and all consumer durables acquired by households are not regarded as fixed assets and do not appear in the balance sheet. On the other hand, assets acquired under financial leases are included, as if they had been purchased outright with a loan from the lessor.

Table 9.1: Alternative methods of valuation for balance sheet assets

Method	Typical applications
Using prices actually observed in the market	Financial claims, e.g. securities quoted on the Stock Exchange Land and buildings Vehicles Crops and livestock Newly-produced fixed assets and inventories
Using estimates of what prices would have been (if no actual transactions have recently taken place)	Securities not quoted on the Stock Exchange, by analogy with similar securities which *are* quoted Tangible assets, comparing with observed prices for items which are not identical but are close substitutes (as is often done for insurance appraisals)
Using accrued acquisitions, *less* disposals, appropriately revalued, over the lifetime of the asset	Fixed assets, valued at current prices, adjusted for accumulated consumption of fixed capital (known as 'written-down replacement cost') Non-produced intangible assets (e.g. purchased goodwill and patented entities), typically valued by amortizing (progressively writing-off) the initial acquisition costs, appropriately revalued, over the expected life of the asset
Using the present (discounted) value of the future net benefits expected from the asset	Assets for which returns are either delayed (e.g. stands of timber) or spread over a long period of time (e.g. subsoil assets), the ultimate asset being valued at current prices and the discount factor reflecting transactions in the particular asset under consideration rather than a general interest rate

Net worth

9.7 Net worth is the difference between total assets and total liabilities at a particular moment of time. Corporations (*not* quasi-corporations) have a net worth – which may be positive or negative – even though they are wholly owned by their shareholders. Defined benefit pension funds increase or decrease the net worth of financial corporations (or, in the case of unsegregated funds, employers) according to whether they are overfunded or underfunded. Quasi-corporations, by convention, have a net worth of zero.

Table 9.2: Valuation methods for particular types of asset

Type of asset	Valuation
Tangible fixed assets	The stock of these is usually valued by the perpetual inventory method (PIM), using estimates of capital formation *less* capital consumption, classified by type of asset and year of acquisition, which have been accumulated and revalued over a long enough period to cover all fixed assets.
Dwellings and vehicles	Active second-hand markets exist for these items and prices observed for them may be used to supplement, though not generally to replace, written-down valuations.
Industrial plant and equipment	Observed prices may not be suitable as they are likely to be for items which have special characteristics, are obsolete or are being disposed of under duress.
Historic monuments	These are included only once their significance has been recognised by someone other than the owner, e.g. by a sale or formal appraisal.
Livestock used in produc tion year after year	These are valued on the basis of current prices for animals of a given age.
Trees cultivated for crops yielded year after year	Current prices are unlikely to be available so written-down values are used.
Intangible fixed assets	
Mineral exploration	Current exploration is valued on the basis of payments made to contractors or costs incurred on own account. The costs of past exploration which have not yet been written-off are revalued (which in this case may well *reduce* the value).
Entertainment, literary and artistic originals	These should be valued at their prices when last traded (or, for own-account production, costs). Alternatively the present value of expected future receipts is used.
Inventories	These are valued at the prices prevailing at the data to which the balance sheet relates, chosen as follows:
Material and supplies	Purchasers' prices
Finished products	Basic prices

Table 9.2: *Cont....*

Type of asset	Valuation
Goods for resale	Prices paid, excluding transport costs incurred.
Work-in-progress	Basic prices of finished products at balance sheet date multiplied by the proportion of total production costs incurred by then *or* Production costs plus a mark-up.
Single-use crops and livestock raised for slaughter	These can usually be valued using market prices.
Single-use timber	This is also work-in-progress, but it is valued by discounting future proceeds at current prices after deducting the expenses of bringing the timber to maturity.
Valuables	If well-organised markets exist these should be valued at actual or estimated purchase prices, including agents' fees and commissions. Otherwise insurance valuations may be used.
Non-produced assets	
Land	Land is valued at its current price, as would be paid by a new owner, taking account of its exact location and potential use which can affect the value enormously. The valuation explicitly includes written-down costs of ownership transfer and implicity incorporates the value of improvements brought about by gross fixed capital formation in the past.
Subsoil assets	The value of reserves is usually taken as the present value of expected net returns from commercial exploitation, though they are subject to uncertainty and revision.
Non-cultivated biological resources and water resources	Observed prices are unlikely to be available so the present value of expected future returns is usually used.
Intangible non-produced assets	When traded these should be valued at current prices. Otherwise the present value of expected future returns may be used. Goodwill is valued at cost of acquisition *less* accrued amortization, appropriately revalued.
Financial assets and liabilities	When regularly traded on organised markets these are valued at current prices excluding transaction costs. Otherwise the value is the amount the debtor must pay the creditor to extinguish the claim on the balance sheet data.

Table 9.2: *Cont....*

Type of asset	Valuation
Monetary gold	Prices are established in organised markets or bilateral arrangements between central banks.
SDRs	Values are determined daily by the IMF.
Currency	The balance sheet shows the nominal or face value.
Deposits	The balance sheet records the amount of principal that the debtor is contractually obliged to repay the creditor.
Securities other than shares	All securities, including derivatives, should be recorded at current market values but, in the absence of these, *short-term* bills and bonds can be valued at par (or issue price if discounted) *plus* accrued interest.
Loans	As for deposits
Shares and other equity	If regularly traded on organised financial markets these are valued at current prices. Otherwise an estimate is made using the prices of quoted shares that are comparable in earnings and dividend history and prospects, adjusted downwards if necessary to allow for the difference in marketability or liquidity. For quasi-corporations equity is equal to assets *less* liabilities.
Insurance technical reserves	The assets of which these are composed are valued at their actual or estimated current prices. Reserves against outstanding claims are the present value of the amount expected to be paid out in settlement.
Other accounts receivable	The balance sheet shows the principal which the debtor is contractually obliged to pay to the creditor.

Memorandum items with the balance sheet

9.8 Various types of memorandum item may accompany the balance sheet to show values which are not separately identified as assets in the main framework but are of analytical interest. These cover the following.

- *Consumer durables* The amount shown is the stock of durable goods used by households for final consumption (including cars), valued at current prices, net of consumption of fixed capital.
- *Direct foreign investment* In the balance sheet proper this is subsumed within the appropriate categories of asset (i.e. shares and other equity, loans or other accounts receivable/payable).
- *Net equity of households in unfunded pension schemes* There are no assets earmarked for the payment of benefits so there is nothing to show in the balance sheet but the present value of *promises* to pay pension benefits in future can be given separately.
- *Alternative valuations for long-term debt and equity* Examples could include face value for long-term bonds and revalued paid-in and equivalent value for corporate equity.

Changes in balance sheets account

9.9   For each group of assets and liabilities changes between the opening and closing balance sheets all result from entries in the accumulation accounts. For each item the change can be decomposed into:

- transactions in the item, e.g. acquisitions and disposals of non-financial assets, consumption of fixed capital, creation and extinction of claims;
- changes in the volume of assets; and
- nominal holding gains/losses, subdivided into 'neutral' and 'real' components.

The relationship between net saving and net worth can be shown in the form

Changes in net worth	=	Net saving	+	Net capital transfers	+	Other changes in volume of assets	+	Nominal holding gains or losses

This accounting identity can be traced for any type of asset, enabling the 'dynamics' of changes in net worth to be explored in detail.

Stocks of financial assets analysed by sector of debtor and creditor

9.10   The balance sheet, showing whether a sector is a creditor or debtor, presents a two-dimensional view of its financial position. In Chapter 7 above it was suggested that the similarly two-dimensional financial account could usefully be extended to show flows of funds between sectors and it is equally helpful to supplement the balance sheet with an analysis of who is financing whom, e.g.

- for the government, which sectors (or the rest of the world) have financed the liabilities it has built up
- for financial corporations, which sectors it has claims upon

Equally, debtor/creditor relationships within sectors can be analysed, e.g.

- between central and local government
- between the Bank of England and deposit institutions

Relationships across sectoral boundaries can also be examined. This more detailed approach serves particularly to spell out the role played by financial intermediation in mobilising resources.

9.11   It is also possible to compile financial balance sheets in consolidated or unconsolidated form. Consolidated balance sheets omit all assets with / liabilities to institutional units within the same sector or sub-sector. Non-consolidated balance sheets will include some or all of these asset and liability positions, but they should be matched by the counterpart positions in each instance, thus increasing the figures on both sides of the balance sheet by the same amount.

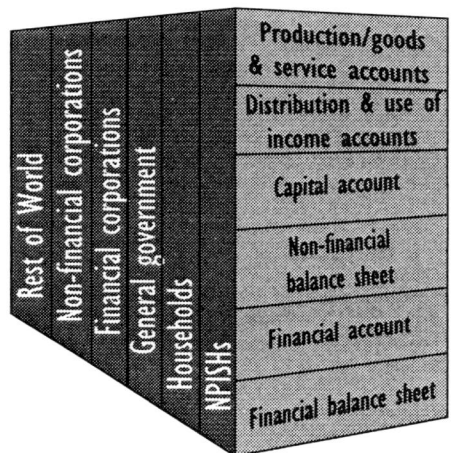

Chapter 10

Sectors

Chapter 10 Sectors

Introduction

10.1   In its analysis of the economy the system of national accounts identifies two kinds of institutional units: *consuming units* – mainly households – and *production units* – mainly corporations and non-profit institutions (NPIs) or government units. Each unit within the country is allocated to one of five sectors:

  • non-financial corporations
  • financial corporations (banks and building societies, other financial intermediaries, financial auxiliaries, insurance companies and pension funds)
  • general government (covering central government and local authorities)
  • households
  • non-profit institutions serving households (NPISH).

Units not in the country are allocated to a further sector which is called the rest of the world sector. These sectors may be subdivided further in certain parts of the accounts. The list of ESA sectors used in the UK is shown in Table 10.1. This chapter describes the principles of sectorisation and how they have been applied to the UK economy.

10.2   Dividing the economy up into sectors allows us to summarise the transactions of those sectors and thus consider the UK economy in greater detail. The sector approach to the economy also provides an alternative view to the industrial breakdown shown in the tables describing the goods and services account and the production account and the input-output tables.

10.3   The sector classification differs from the industrial analysis (considered further in Chapter 13) in that the latter groups together economic units according to their main productive activity, without reference to ownership or who operates them. Sector classification divides the economy into a relatively small number of institutional sectors according to their control, ownership and functions. This is particularly of use when the behavioural relationship between economic agents is of interest and thus the flows affecting income, capital and financial transactions and balance sheets are studied.

10.4   Sector accounts provide information on the contribution to the national aggregates by each sector and show the resources employed by each sector. These accounts also record the movement of resources between sectors by way of taxes, grants and interest, as well as transactions in financial assets; and balance sheets showing how each sector holds its wealth.

10.5   The set of accounts summarising the transactions of individual sectors of the UK economy is thus an important feature of the national accounting system. They show the relationships between different parts of the economy and between different types of economic activity in a way which is completely hidden within the statistics and tables of aggregate national income, expenditure, and wealth.

10.6   In the United Kingdom an important aspect of the classification is the separation of public and private sectors. The size and function of the public sector (which includes public corporations as well as general government) in the UK economy is of interest both for economic analysis and because of the crucial role it has played in the policies of many governments. Internationally consistent definitions of these measures are therefore important. For EU member states the Maastricht Treaty's economic convergence criteria include two which cover government sector deficit and debt. In the United Kingdom, government sector spending and the borrowing of the whole public sector have targets in domestic economic policy.

10.7   The framework of sectoral classification provided by the SNA and ESA assists clear and consistent classification of enterprises and other bodies. For example, the system specifies rules or guidelines to determine what is a significant level of government control and influence which should lead to a unit being classified to the government sector. In comparison, the Standard Industrial Classification classes units by activity. Thus National Health Service hospitals, National Health Service Trust hospitals and private hospitals are all classified to the same industry but will all be in different sectors of the UK economy.

Principles of sectorisation

10.8   The sectors are formed by grouping together institutional units which have common characteristics likely to affect economic behaviour. It is essential therefore to define what we mean by an institutional unit. The next step is to decide whether the unit is part of the UK economy, or in the Rest of the World. The final allocation to sectors depends on the economic function of the unit. Each of these three 'foundation stones' is considered in turn below.

The institutional units

10.9   An institutional unit is defined in the ESA as:

'an elementary economic decision-making centre characterised by uniformity of behaviour and decision-making autonomy in the exercise of its principal function. A resident unit is regarded as constituting an institutional unit if it has decision-making autonomy in respect of its principal function and either keeps a complete set of accounts or it would be possible and meaningful, both from an economic and legal viewpoint, to compile a complete set of accounts if they were required.'

[ESA 1995 paragraph 2. 12]

10.10   The first characteristic is that the unit must have autonomy of decision as far as its principal function is concerned. This means that it must:

- be able to own goods or assets in its own right (and thus be able to exchange them in transactions with others);
- be entitled to take economic decisions and carry out economic activities for which it is held be directly responsible and accountable at law;
- be able to incur liabilities on its own behalf, to take on other obligations or further commitments and to enter into contracts.

10.11   The second condition is that a unit keeps a complete set of accounts. In order to meet this, an entity must keep accounting type records which cover all the economic and financial transactions it has made during the accounting period, as well as a balance sheet of assets and liabilities. However, as described below, this condition is relaxed in the case of households.

10.12   There are therefore two types of units that may qualify as institutional units: persons, or groups of persons, in the form of households, and legal or social entities whose existence is recognised by law or society independently of the persons or other entities that own or control them.

10.13   The individual members of multi-person households are not treated as separate institutional units. This is for a number of reasons. Many assets are owned, or liabilities incurred, jointly by two or more members of the same household. Similarly, some or all of the income received by individual household members may be pooled for the benefit of all members. Also, many spending decisions, especially ones about the consumption of food or the housing of the household, may be made collectively for the household as a whole. It may therefore not be possible to draw up meaningful accounts (balance sheets etc.) for individual members of a household. A household is thus treated as the institutional unit. An unincorporated enterprise that is entirely owned by one (or more) members of the same household is treated as an integral part of that household, and not as a separate institutional unit, except when the enterprise is regarded as a quasi-corporation.

10.14   The second type of institutional unit is a legal or social entity that engages in economic activities and transactions in its own right. Examples include companies, charities, government departments, etc. These bodies are responsible and accountable for the economic decisions or actions they take, though these may be influenced to some extent by other units – for example, the shareholders in the case of a corporation. Unincorporated enterprises are in general classified with the household or other unit that owns them. However, some unincorporated enterprises belonging to households or government units behave in much the same way as corporations. When they have complete sets of accounts and autonomy of decision-making to an extent that makes them separable from the owners, they are known as quasi-corporations, and classified to the corporate sectors.

10.15   Certain principles may be used whenever one or other of the parts of the definition of an institutional unit are not clearly met. These are as follows:

a) households always have autonomy of decision and so must be institutional units, even though they do not keep complete sets of accounts;

b) entities which do not keep a complete set of accounts, and for which it would not be possible or meaningful to compile a complete set of accounts, are combined with the institutional units into whose accounts their partial set is integrated;

c) if an entity has a complete set of accounts but has no autonomy of decision in the exercise of its principal function, it is combined with the unit which controls it;

d) if accounts are not published for an entity, but otherwise it satisfies the definition of an institutional unit, it will be treated as such. One example of this is privately owned companies in the UK;

e) where a holding corporation is responsible for the general direction of a group of companies the holding corporation is regarded as an institutional unit distinct from the units it controls, unless b) above applies. Thus as long as the production entities which the holding corporation partially controls have a complete set of accounts they will be regarded as separate institutional units. This principle applies across all groups of corporations – and to their subsidiaries at all levels, as long as b) above is not applicable;

f) quasi-corporations are institutional units which complete a set of accounts but have no independent legal status. However, their behaviour is different from that of their owners and similar to that of corporations. Hence they are said to have autonomy of decision and thus meet the criteria for definition as an institutional unit. In the UK, all partnerships are considered to meet these criteria.

g) The UK sectors also contain *notional resident units*. These units are:

- those parts of non-resident units which have a centre of economic interest on the economic territory of the UK;
- non-resident units where they are owners of land or buildings on the economic territory of the UK – but only in respect of transactions affecting such land or buildings.

Notional resident units, although they may only keep partial accounts and may not have autonomy of decision, are regarded as institutional units.

10.16   Institutional units in the UK are therefore deemed to be:

a) units which have a complete set of accounts together with autonomy of decision:

i) private and public corporations;
ii) co-operatives or partnerships recognised as independent legal entities;
iii) public producers which by virtue of special legislation are recognised as independent legal entities;
iv) agencies of general government;
v) non-profit institutions recognised as independent legal entities;

b) units which have a complete set of accounts and which are deemed to have autonomy of decision:
quasi-corporations

c) units which may not keep a complete set of accounts, but which by convention are deemed to have autonomy of decision, these are:

i) households;
ii) notional resident units (see 10.15 above).

The limits of the UK economy

10.17   For national accounting (including balance of payments) purposes, the United Kingdom is defined as England, Scotland Wales and Northern Ireland. It does not include the Channel Islands and Isle of Man. The units which form the economy of the United Kingdom, and whose transactions are recorded in the UK sector are those which have a 'centre of economic interest' on the 'economic territory' of the UK. They are therefore termed 'resident units'. These terms and the concept of residence are described in detail in Chapter 24 (24.03–24.10). Resident units may or may not have UK nationality and may not actually be legal entities; indeed, such units may not even be present on the UK economic territory when they carry out a transaction. Thus, residence is not based on nationality or legal criteria, although it is similar to the concepts of residence used for exchange control and tax purposes in many countries.

10.18   Where a unit owns land or buildings in the UK it is, as far as this ownership is concerned, deemed by the system to be a resident of the UK. The foreign owner is recorded as having a financial investment in the UK. Similarly, the UK owners of homes etc. abroad are deemed, in respect of that ownership, to be resident in the foreign country concerned, and the UK owner is recorded as having a financial investment in the foreign resident unit.

Changes to the definition of the UK

In order to comply with the new ESA, the definition of the UK no longer includes the Channel Islands or the Isle of Man. These offshore islands are now treated as part of the Rest of the World sector (other than EU). Surveys run by both the ONS and the Bank of England have been amended to exclude these offshore islands. Estimate of trade in goods, received from HM Customs and Excise, have also been amended to reflect this change.

10.19   Corporations and non-profit institutions (NPI) are normally expected to have a centre of economic interest in the country in which they are legally constituted and registered. However, corporations may be resident in countries different from their shareholders and subsidiary corporations may also be resident in countries other than that of their parent corporation. When a corporation, or unincorporated enterprise, maintains a branch, office or production site in another country in order to engage in a significant amount of production over a long period of time (but without creating a subsidiary corporation for the purpose) the branch, office or site is deemed to be a 'quasi-corporation' resident in the country in which it is located. It is thus treated as a separate institutional unit for the purpose of the national accounts.

10.20   Households which have a centre of economic interest in the UK are deemed to be resident units, even if they go abroad for short periods. (A short period is one of less than a year.) In addition, the residence of individual persons is determined by that of the household of which they form part and not, for example, their place of work. UK households include those in the following circumstances:

a) border workers who cross the border each day to work in a neighbouring country -this is, of course, far more common in mainland Europe than it is in the UK;

b) seasonal workers who leave the country for several months to work in another in sectors where extra workers are needed periodically;

c) tourists, patients, students, visiting officials, businessmen, salesmen, artists and crew members who travel abroad. In fact, students are always treated as residents of their home country, however long they study abroad;

d) locally recruited staff working in the extra-territorial enclaves of foreign governments;

e) the staff of institutions of the EU and of civilian or military international organisations which have their headquarters in extra-territorial enclaves;

f) the official, civilian or military representatives of the government of the UK established in territorial enclaves abroad. This includes the families and households of such UK representatives.

10. 21   Unincorporated enterprises that are not quasi-corporations, i.e. sole traders, are not separate institutional units from their owners and so they have the same residence as their owners.

Economic functions and behaviour

10.22   Corporations, non-profit institutions (NPI), government units and households are intrinsically different from each other. Their economic objectives, roles and behaviour are also different in many ways.

Table 10.1: ESA sectors used in the UK national accounts

Sector and subsector	Reference Number
Non-financial corporations	S.11
of which:	
Public	S.11001
Private	S11002/3
Financial corporations	S.12
of which:	
Monetary financial institutions	S.121/2
Other financial intermediaries and financial auxiliaries	S.123/4
Insurance companies and pension funds	S.125
General government	S.13
of which:	
Central government	S.1311
Local government	S.1313
Households and non-profit institutions serving households	S.14 and S.15
Rest of the world	S.2
of which:	
European Union (EU)	S.21
Member countries of the EU	S.211
Institutions of the EU	S.212
Other countries and international institutions	S.22

10.23 *Corporations* are institutional units set up for the purpose of producing goods or services to sell. They will usually be a source of profit to the units that own them. *NPI* however, whilst similarly being created to produce or distribute goods and services, do not have the purpose of income or profit generation for the units that control or finance them. In contrast to corporations, NPI may consume goods or services in order to provide collective benefits to society as a whole, or to specific groups of individuals, or to other economic units. Their resources often come from voluntary donations or government grants rather than sales of goods or services. NPI are included with the sector that owns and controls them, apart from those serving households, which constitute a separate sector in the system.

10.24 The economic aims, functions and behaviour of *government units* are quite distinct. They organise and finance the provision of non-market goods and services (*see* Chapter 3), including both individual and collective services to households and the community. They are also involved with the distribution or re-distribution of income and wealth through taxation and other transfers. They may engage in market production themselves.

10.25 The main activity of *households* is as consumers. However, they can engage in any kind of economic activity. They supply labour to enterprises and also run their own economic units in the form of unincorporated enterprises.

10.26 The economy is divided into sectors by grouping together institutional units according to these characteristics. The institutional units are grouped together according to the type of producer they are and depending on their main activity and function; these are taken to indicate their economic behaviour. Thus banks, building societies, insurance corporations and other financial type units are grouped together in the financial corporations sector. Any further sub division of sectors is according to particular criteria relevant to that sector: which allows a more precise description of the economic behaviour of the units. So, within the financial corporations sector, the central bank, other banks and building societies are grouped together and called 'monetary financial institutions'.

10.27 When the principal function of the institutional unit is to produce goods and services the first criterion used in deciding which sector it should be allocated to is to that of the type of producer it is. There are three types of producers defined in the ESA:

- private and public 'market producers';
- private producers 'for own final use';
- private and public 'other non-market producers'.

10.28 The first distinction is between public and private producers. A *public producer* is one that is controlled by the general government. All other producers are private producers. Control of a producer is defined as the ability to determine the general (corporate) policy or programme of an institutional unit by appointing appropriate directors or managers if necessary. Owning more than half of a corporation's voting shares is a sufficient, but not a necessary, condition for control. For non-profit institutions (NPI), a public producer is an NPI that is controlled and mainly financed by the general government.

10.29 Private producers are found in all sectors except the general government sector. Public producers, however, are only found in the corporate sectors and in the general government sector. As there are currently no public corporations classified to the financial corporations sector in the UK, public corporations are only found in the UK non-financial corporations sector.

10.30 Having decided whether the unit is a private or public producer the next question to consider concerns its output. *Market output* is output that is sold at prices that are economically significant or otherwise disposed of on the market. Prices are said to be economically significant when they have a significant influence on the amounts the producers are willing to supply and purchasers are willing to buy. Market producers are ones for which the major part of output is market output. Institutional units that are market producers may be classified as non-financial corporations (S.11), financial corporations (S.12) or households (S.14).

10.31 *Producers for 'own final use'* are units the major part of whose output is retained for their own final use within the same institutional unit. As corporations have no final consumption, output for own final consumption is produced only by unincorporated enterprises: for example agricultural goods produced and consumed by members of the producing household. The household sector (S.14) includes any unincorporated enterprises owned by households, which are a specific category of private producers. They are always market producers or producers for own final use. The main example for production for own final use in UK households apart from agriculture is the production of owner occupied dwelling services. Where unincorporated enterprises are owned by households (e. g. sole-proprietor businesses or partnerships) these are market producers and are classified as quasi-corporations if possible. In practice, this can be done only for partnerships, which are classified to either the non-financial or financial corporations sectors as appropriate. Sole proprietor businesses are generally classified with households.

10.32 *Other non-market output* consists of goods and individual or collective services produced by non-profit institutions serving households (NPISH, S.15) or government (S.13). They are non-market in that they are supplied free (or at prices that are not economically significant) to other institutional units or the community as a whole. These services are produced without a market price either because it is impossible to charge for them or because the Government makes a policy decision not to charge. Military defence is an example of the former while the National Health Service is one of the latter.

The rest of the world (S.2) is a grouping of institutional units which is not characterised by similar objectives and types of behaviour; it groups together non-resident institutional units insofar as they carry out transactions with resident institutional units.

10. 33 Producers are allocated to market or non-market categories according to their main output. Many enterprises may produce both market and non-market output.

Legal or social entities

10.34 Before moving on to consider each sector of the economy individually it may be helpful to consider further those institutional units which are legal or social entities. These are of three main types: corporations, quasi-corporations and non-profit institutions. Units of government are described separately.

Corporations

10.35 This name is used to describe any legal entity set up for the purpose of producing goods or services for the market so that its owners may make a profit from it. It is collectively owned by its shareholders who are able to appoint directors to look after its general management. The term corporation thus covers not only those units called corporations but also unincorporated enterprises, public limited companies ('plcs'), public corporations, private companies, limited liability partnerships, etc.

Table 10.2: The type or producer, the principal activities and functions classified by sector

Sector	Type of producer	Principal activity and function
Non-financial corporations (S.11)	Market producer	Production of market goods and non-financial services
Financial corporations (S.12)	Market producer	Financial intermediation including insurance Auxiliary financial activities
General government (S.13)	Public other non-market producer	Production and supply of other non-market output[1] for collective and individual consumption and carrying out transactions intended to redistribute national income and wealth.
Households (S.14) - as consumers - as entrepreneurs	Market producer or private producer for own final use	Consumption Production of market output and output for own final use.
Non-profit institutions serving households (S.15)	Private other non-market producer	Production and supply of other non-market output[1] for individual consumption.

[1]   non-market output other than output for own final use

10.36   In the system of national accounts, a corporation cannot be a final consumer- it cannot incur final expenditure for the benefit of households. When a corporation provides goods or services to its employees these must be recorded as either 'compensation of employees paid in kind' or 'intermediate consumption' (*see* 4.31). A further characteristic of corporations is that their profit ultimately benefits other institutional units, their shareholders. These two characteristics differentiate a corporation from a non-profit institution since the latter can incur final expenditure for the benefit of households and cannot generate income that other institutional units can take.

Ownership and control of corporations

10.37   A corporation is owned by its shareholders. Although a single institutional unit (e.g. government or an individual) may own all the shares in a corporation, ownership is usually spread among a number – and this number may run to hundreds of thousands, for example in the case of privatised corporations which were formerly publicly owned. This has occurred in the United Kingdom with corporations such as British Telecom (BT) and British Gas (BG).

10.38   If any one institutional unit owns more than half the voting shares of a corporation it can control its policy and operations. Likewise a small group of shareholders will generally have such control if their combined shareholdings exceed 50 per cent. The extent to which the board of directors and managers of a corporation have freedom of actions depends on the degree to which ownership of its shares is concentrated in a small number of other units (whether these are other corporations, households or government units). The general rule is that institutional units have to be responsible, and accountable, for the decisions and actions they take even it they are not, strictly, autonomous.

10.39   A small number of shareholders may be able to obtain control of a corporation even though they control considerably less than half the total shares – for example, perhaps as few as 10 per cent. Thus it is not possible to specify a minimum shareholding below 50 per cent which will guarantee control. The minimum varies according to factors including the total number of shareholders, the distribution of shares between them and the extent to which small shareholders take an active interest.

10.40   The national accounts distinguish between the activities of corporations and those of their owners, which may be in other sectors.

10.41   The sub-sectors of the national accounts system allow for a separation of private corporations from public ones which are subject to control by government units. A further division of private corporations is to separate out those controlled by non-resident units. For practical reasons, the only subdivision in the UK accounts is between public and private corporations. For this purpose, control is defined as the ability to determine general corporate policy by appointing appropriate directors, if necessary. If a government is able to control a corporation because of special legislation the corporation should be treated as a public one, whatever the disposition of its shares.

Subsidiary corporations and associated corporations

10.42   Each subsidiary or associated corporation should be treated as a separate institutional unit whether or not it forms part of a large group (with the exception of ancillary corporations which are covered below). Even subsidiaries which are wholly owned by other corporations are separate legal entities which are required to produce a complete set of accounts and they remain responsible for the conduct of their own production activities. A further reason for not treating groups of organisations as single institutional units is that groups are not always well defined, stable or easily identified in practice. Where the activities of a group are not closely integrated it may be difficult to obtain data. To conclude on this point, many conglomerates are far too large and heterogeneous for them to be treated as single units; their size and composition may be continually changing due to mergers and take-overs.

Ancillary corporations

10.43   These are wholly owned subsidiary corporations whose productive activities are strictly confined to providing services to the parent corporation, or to other ancillary corporations owned by the same parent corporation.

10.44   Ancillary activities are those whose sole function is to produce one or more common types of services for intermediate consumption within the same enterprise. These services typically include transportation, purchasing, sales and marketing, various types of financial or business services, computing and communications, security, maintenance and cleaning. Neither the inputs to, nor the outputs from, ancillary activities are recorded separately from those of the principal or secondary productive activities.

10.45   Corporations may create such subsidiary ancillary corporations for tax or other reasons. One example is where ownership of land, buildings or equipment is transferred and the ancillary's sole function is to lease them back again to the parent corporation. In the UK, financing subsidiaries of non-financial corporations are considered to be ancillary corporations despite not strictly fulfilling these criteria. There would be considerable practical difficulty in separating them out. Some corporations set up 'dormant' subsidiaries, which are not actually engaged in any production but which may be activated at the convenience of the parent corporation.

10.46   Ancillary corporations are not treated as separate institutional units within the national accounts. This is because they can be regarded as artificial units created to secure taxation, legal or administrative advantages. Ancillary corporations are combined with the parent corporation and viewed as a single institutional unit.

Co-operatives, limited liability partnerships, etc.

10.47 This heading is a 'catch all' one for legal units which have been set up to produce for the markets for profit but which do not call themselves 'corporations' or 'companies'. Such institutions may be described differently because they have somewhat specialised functions, however, they are categorised as corporations for the purposes of national accounts.

10.48 As the heading suggests, this covers co-operatives formed to market the collective output of producers such as farmers. The profits of co-operatives are distributed according to rules agreed (which may not be proportionate to shares held) and they behave like corporations. Limited liability partnerships are similarly separate legal units which act like corporations, the partners being both shareholders and managers at the same time.

10.49 All units, whatever their description or name, which meet the three criteria of:

- having been set up in order to engage in market production;
- being capable of generating a profit or other financial gain for their owners;
- being recognised at law as separate legal entities from their owners who enjoy limited liability,

are classified as corporations in the national accounts system.

10.50 Some legal units which are actually non-profit institutions are sometimes called 'corporations'. The status of an institutional unit cannot necessarily be inferred from its name; it may be necessary on occasion to look at its objectives and functions.

Quasi-corporations

10.51 These are unincorporated enterprises that operate as though they are corporations. A quasi-corporation may be in one of two categories:

- an unincorporated enterprise owned by a resident institutional unit that is operated as if it were a corporation and whose *de facto* relationship to its owner is that of a corporation to its shareholders: such a unit must, of course, keep a complete set of accounts; or
- an unincorporated enterprise owned by a non-resident institutional unit that is deemed to be a resident unit because it engages in a significant amount of production in the UK economic territory over a long (one year or more) or indefinite period of time, or owns land and buildings in the UK.

10.52 Quasi-corporations are sectorised separately from their legal owners. Although owned by other sectors, they are allocated to the non-financial corporations or financial corporations sectors as appropriate. There are three main types of quasi-corporations:

- unincorporated enterprises owned by government units which are engaged in market production and which are operated in a similar way to publicly owned corporations;
- unincorporated enterprises (including partnerships), owned by households which are operated as if they were privately owned corporations;
- unincorporated units which belong to foreign institutional units. Examples of these are the permanent branches, or offices of foreign corporate or unincorporated enterprises, or of production units belonging to enterprises resident abroad which carry out significant amounts of production within the economic territory over long or indefinite periods of time (such as

those constructing bridges or other large structures).

10.53 The term quasi-corporation has been coined for the purpose of bringing into the corporate sectors unincorporated enterprises which are sufficiently self-contained and independent that they operate like corporations. To behave like corporations they must have complete sets of accounts. In fact, such accounts (covering balance sheets as well as value added, saving, etc.) have to exist for a quasi-corporation to be separable statistically. It should also be possible to identify any flows of income and capital between the quasi-corporation and its owner. Income withdrawn from a quasi-corporation is decided by the owner and is equivalent to a corporation's payment of a dividend to its shareholders. It is assumed that the owner's net equity in a quasi-corporation equals the difference between the value of its assets and the value of its other liabilities so that the net worth of the quasi-corporation is always zero.

10.54 It is often difficult to distinguish enterprises owned by households from their owners. Even a large enterprise, if it is not in effect operated like a corporation and does not have a complete set of accounts, would be classified with its owner as a household. In the UK is is expected that only partnerships will be sufficiently separable from their owning households to be classified as quasi-corporations; sole-proprietor businesses will not be.

Non-profit institutions (NPI)

10.55 Institutions set up to produce goods and services but whose status does not allow them to be a source of income or profit etc. for the units which establish, control or finance them are non-profit institutions (NPI). They will produce either a surplus or deficit as a result of their production activities but any such surpluses cannot be used by other institutional units. NPI are created using articles of association which state that their controlling or financing units are not entitled to a share in any profit or other income they receive. As a result, they are often exempt from various taxes which apply to corporations and others.

10.56 NPI are set-up for many different reasons. These include the provision of services to benefit the people or corporations who control or finance them, or from charitable or philanthropic motives, to give goods or services to other persons in need of them. They may also be created to act as pressure groups in politics etc. or to provide education or health services for a fee, but not for a profit.

The main features of NPI

10.57 The characteristics of NPI are as follows:

- most are legal entities whose existence is recognised independently of the units (persons, corporations or government) that establish, finance, control or manage them. The articles of association, or a similar document drawn up when it is established, generally give the purpose of the NPI;
- many NPI are controlled by associations whose members have equal voting rights on all important decisions which affect the operation of the NPI. Members have limited liability as far as the affairs of the NPI are concerned;
- there are no shareholders with claims on the profits or equity of the NPI. Members are not entitled to a share in any such profits etc. which the NPI may generate, all profits are retained within the NPI;
- a group of officers, such as an executive committee elected by a simple majority of the members, control the direction of the NPI. They are the equivalent to a corporation's board of directors;
- the term 'non-profit institution' comes from the fact that those controlling the NPI are not allowed to gain financially from its operation and cannot appropriate any surplus it makes.

The NPI may, however, make an operating surplus on its production.

NPI as market and non-market producers

10.58 NPI may engage in both non-market and market production. The distinction between these is used to allocate them to a sector of the economy.

NPI engaged in market production

10.59 Market producers are ones which sell most, if not all, of their output at prices that are economically significant. Schools, colleges, clinics, hospitals, etc. constituted as NPI are market producers when they charge fees which are based on their production costs and which are sufficiently high to have a significant influence on the demand for their services. Their production activities must generate an operating surplus or loss. Any surpluses they make must be retained within the institutions as their status prevents them from distributing them to others. At the same time, as they are 'non-profit institutions' they may also raise additional funds by appealing for donations. Hence they may be able to acquire assets which generate significant property income in addition to their revenues from fees. This lets them charge fees that are less than the average costs they incur. However, they must continue to be classified as market producers as long as their fees are determined mainly by their production costs and are high enough to have a significant effect on demand. Market NPI are classified to the non-financial or financial corporate sectors, as appropriate.

Market NPI serving businesses

10.60 Groups of businesses in an industry, town or region may set up associations to promote their interests. They consist of chambers of commerce, agricultural, manufacturing or trade associations and employers' organisations. Research or testing laboratories or other organisations or institutes which carry out activities which are of mutual interest or benefit to the group of businesses that control and finance them also fall into this category. These NPI often speak and act publicly for the group and also lobby politicians or give advice or assistance to individual members when they are in difficulty. The finance for such NPI generally comes from subscriptions or contributions from the group of businesses. The system treats the subscriptions as payments for services rendered (rather than transfers) and therefore these NPI are said to be market producers. There is an exception to this: when government units control and mainly finance NPI such as chambers of commerce they are classified as non-market NPI and allocated to the general government sector. This is enlarged upon below.

NPI engaged in non-market production

10.61 Many NPI are actually non-market producers; they provide most of their output to others free, or at prices which are not economically significant. The prices do not have a significant influence on either the amounts these NPI are prepared to supply or on the amounts the 'purchasers' wish to buy. Thus, non-market NPI may be distinguishable not only because they are barred from providing financial gain to the units which control or manage them but also because they must rely mainly on funds other than those earned from sales to cover their costs of production or other activities. Their main source of finance may be regular subscriptions paid by the members of the association that controls them or transfers or donations from third parties, including government.

10.62 NPI engaged in non-market production may be divided into two groups:

- NPI controlled and mainly financed by government;
- those NPI providing non-market goods and services to households financed mainly by transfers from non-government sources (households, corporations or non-residents).

It is this second group of non-market NPI which are described within the system as 'NPI serving households (NPISH)' and these form a separate sector in the system.

NPI controlled and mainly financed by government

10.63   To be classified here a NPI must be a properly constituted legal entity which exists separately from government. Control is the ability of government to determine the general policy or programme of the NPI by having the right to appoint the officers managing the NPI. Such NPI may be engaged in research or development, for example, for the benefit of certain groups of producers. They may also have been set up to set or maintain standards in areas such as health, safety, the environment, accounting, finance, education, etc. for the benefit of both enterprises and households. This enables such NPI to be seen as detached and objective and not subject to political pressures, which would not be the case if they had been created as government agencies. NPI controlled and financed by government are allocated to the general government sector, irrespective of the sorts of institutional units that mainly benefit from their activities.

10.64   In some countries, certain legal entities created by government units may have the characteristics of, and behave like, NPI controlled and mainly financed by governments but be formally described as 'corporations'. These entities are viewed by the system as NPI whatever their names. Thus it is necessary to take account of the functions and purpose of a legal entity and not to classify it merely by its name.

Non-profit institutions serving households (NPISH)

10.65   These are a category of NPI which provide goods or services to households either free or at prices that are not economically significant. There are two main types:

   •   those created by associations of persons to provide goods or, more often, services mainly for the benefit of the members themselves;
   •   charities, relief or aid agencies, set up for philanthropic purposes.

10.66   The former group, set up mainly to benefit their members usually provide services free and are financed by regular membership subscriptions. They include professional societies, trade unions, churches or religious societies, and social or sports clubs.

10.67   Philanthropic institutions serving households comprise charities, relief and aid agencies etc. They provide goods and/or services on a non-market basis to those households most in need, including those affected by natural disasters or war. They are financed mainly by donations from the general public, government or corporations.

10.68   Non-profit institutions serving households are covered in greater detail in the section on the NPISH sector towards the end of this chapter.

The institutional sectors: an overview

10.69   Having considered the foundations upon which a classification of the economy is based it is now possible to look at the actual sectors themselves.

10.70   Individual institutional units are combined into groups called institutional sectors – or, more simply, just sectors. These can be further divided into sub-sectors. Each sector brings together institutional units which have a similar type of economic behaviour. Definition of each sector is according to the type of producer an institutional unit is and depends on its main activity and function: these are taken as indicators of their economic behaviour. The further division of certain sectors into sub-sectors is according to criteria relevant to the particular sector concerned, thus producing a more exact description of the economic behaviour of the units. All of the activities of a sector (and sub-sector) are recorded in the accounts for that sector, whether they relate to the main activity or to other, secondary, ones. Each institutional unit is a member of only one sector and sub-sector.

10.71 The six sectors specified by the ESA are as follows: non-financial corporations; financial corporations; general government; households; non-profit institutions serving households; rest of the world. In summary, these correspond broadly to the sectors under the previous ESA as follows:

Table 10.3

ESA 1995	Former UK system
S.11 non-financial corporations	companies and financial institutions (which was split into: industrial and commercial companies; financial companies and institutions); and public corporations
S.12 financial corporations	
S.13 general government	general government
S.14 households	personal sector
S.15 non-profit institutions serving households	
S.2 rest of the world	overseas

10.72 Chapters 19 to 24 of this publication deal with the data sources and methods for each of these six sectors. The coverage of these sectors will vary over time, for example, as a result of privatisation or nationalisation. Time series published in the UK national accounts reflect the changing composition of sectors. The new coverage of a sector is only adopted for the periods after the date on which a change occurred; data for earlier years continue to reflect its previous composition.

10.73 Having discussed the principles of sectorisation and briefly outlined the six sectors above, we now move on to consider in more detail the definitions of each sector in turn.

Non-financial corporations (S.11)

10.74 This sector consists of:

'institutional units whose distributive and financial transactions are distinct from those of their owners and which are market producers, whose principal activity is the production of goods and non-financial services.'

[ESA 1995, para 2. 21]

The sector also includes non-financial quasi-corporations.

10.75 The term 'non-financial corporations' describes all bodies which are independent legal entities and are market producers and whose principal activity is the production of goods and non-financial services. The types of institutions classified to this sector are as follows:

a) private and public corporations which are market producers mainly engaged in the production of goods and non-financial services;

b) co-operatives and partnerships recognised as independent legal entities which are market producers principally engaged in the production of goods and non-financial services;

c) public producers which, by virtue of special legislation, are recognised as independent legal entities and which are market producers mainly engaged in the production of goods and non-financial services. In practice in the UK these are bodies subject to an external finance control regime by the Treasury;

d) non-profit institutions or associations serving non-financial corporations, which are recognised as independent legal entities and which are market producers principally engaged in the production of goods and non-financial services;

e) holding corporations controlling a group of corporations where the dominant activity of the group as a whole is production of goods and non-financial services;

f) private and public quasi-corporations which are market producers of goods and non-financial services.

10.76 Some non-financial corporations may have secondary financial activities such as providing consumer credit to their customers. However, such corporations are classified on the basis of their main activity in the non-financial corporations sector. Since sectors are groups of mutually exclusive units, it is not possible for a corporation (or quasi-corporation) to be classified to more than one sector.

10.77 This sector is divided by the ESA into three sub-sectors. The division is on the basis of the sector which controls the corporations, quasi-corporations or market NPI. Control is said to be achieved when a unit (or group of units) own, whether directly or through a subsidiary, more than 50 per cent of the voting shares of a corporation, or when there is other evidence that control is exercised.

10.78 The sub-sectors specified for non-financial corporations are:

- public non-financial corporations;
- national private non-financial corporations;
- foreign controlled non-financial corporations.

The last two are combined in the UK national accounts.

Public non-financial corporations (S.11001)

10.79 These are all resident non-financial corporations and quasi-corporations that are subject to control by general government units. Control may be achieved by owning more than half the voting shares or, as a result of special legislation, being able to define corporate policy or appoint the directors. Public corporations may be subsidiaries of other public corporations; public quasi-corporations are quasi-corporations owned directly by government units. However, public corporations do not include any non-market NPIs controlled and financed by government units – these are included in the general government sector. In practice UK public non-financial corporations are bodies subject to an external finance control regime by HM Treasury.

10.80 The composition of this sub-sector may change over time as governments act to nationalise and also to privatise (or de-nationalise) corporations and quasi-corporations.

10.81 In the United Kingdom this sub-sector currently includes public corporations such as the BBC, the Civil Aviation Authority, the Land Authority for Wales, and the United Kingdom Atomic Energy Authority. A full list of them is given in the *Sector Classification for the National Accounts*.

10.82 As part of the move to the ESA 1995 the classification of trading bodies within the central government sector was examined. As a result, most of these bodies are now being shown as quasi-corporations within this sub-sector.

10.83 The Export Credits Guarantee Department is a central government trading body which remains classified to the central government sector. *See* paragraphs 10. 176 to 10. 181.

National private and foreign controlled non-financial corporations (S. 11002 and S. 11003)

10.84 These comprise all resident corporations and quasi-corporations that are not controlled by the UK government. These corporations may, or may not, be controlled by other institutional units resident or non-resident in the United Kingdom. This sub-sector also includes market NPIs producing goods or non-financial services – for example, those engaged in providing education or health services on a fee-paying basis, or trade associations serving enterprises.

10.85 The foreign controlled element of this subsection includes:

- all non-financial UK subsidiaries and associates of non-resident corporations;
- all corporations controlled by non-resident units that are not themselves corporations – such as foreign governments. It also includes corporations controlled by a group of non-resident units acting together;
- all branches or other unincorporated agencies of non-resident corporate or unincorporated enterprises that have significant amounts of production on the United Kingdom's economic territory on a long-term basis. These are thus classed as resident quasi-corporations.

Financial corporations (S.12)

10.86 This sector consists of:

'all corporations and quasi-corporations which are principally engaged in financial intermediation (financial intermediaries) and/or in auxiliary financial activities (financial auxiliaries). '
[ESA 1995 paragraph 2. 32]

10.87 The distinction between financial and non-financial corporations is made because financial intermediation is inherently different from most other types of productive activity.

Financial intermediation

10.88 This is defined as:

'a productive activity in which an institutional unit incurs liabilities on its own account for the purpose of acquiring financial assets by engaging in financial transactions on the market.'
[SNA paragraph 4. 77].

Financial intermediaries channel funds from lenders who have excess funds to borrowers by intermediating between them. They collect funds from lenders and transform, or repackage, them in order the suit the requirements of borrowers. Such repackaging may take the form of altering the maturity or scale or risk of the funds.

10.89 A financial intermediary does not, however, merely act as an agent between different parties. It also places itself at risk by acquiring or incurring liabilities on its own account. It may obtain funds by taking deposits, or by issuing bills, bonds or other securities. It uses these funds to acquire financial assets, mainly by making loans to others but also by buying bills, bonds and other securities.

10.90 Financial intermediation is usually restricted to financial transactions on the market, that is, acquiring assets and incurring liabilities with the general public or specified and relatively large sub-groups of them. Financial intermediation does not normally include the transactions of institutional units providing treasury services to a company group. Such units are allocated to a sector depending on the main function of the company group. However, where the unit providing the treasury services is subject to supervision by financial authorities in the UK, the convention is that it be classified in the financial corporations sector.

10.91 There are some exceptions to the rule that financial intermediation is limited to financial transactions on the market. Examples of them are municipal credit and savings banks, which rely heavily on the municipality involved, or financial leasing corporations which depend on a parent group of companies in acquiring funds. However, for them to be classified as financial intermediaries, their lending or their acceptance or savings should be independent of the municipality or parent group involved.

Financial enterprises

10.92 Financial enterprises are those which are mainly involved in financial intermediation or in closely related auxiliary financial activities. This sector thus includes enterprises that do not actually engage in financial intermediation themselves but whose principal function is to facilitate financial intermediation without incurring liabilities themselves. Financial enterprises in total comprise all those classified under Divisions 65, 66 and 67, i.e. financial intermediation, insurance and pension funding, and activities auxiliary to financial intermediation, of the International Standard Industrial Classification of All Economic Activities (ISIC) Rev. 3 and the corresponding EU classification.

Auxiliary financial services

10.93 Auxiliary financial services may be carried out as secondary financial activities of financial intermediaries or they may be provided by specialist agencies or brokers. Examples of the such agencies are securities brokers, flotation companies, loan brokers and managers of mutual or pension funds. There are also other agencies whose main role is to guarantee, by endorsement, bills or other similar instruments intended for discounting or refinancing by financial enterprises and also institutions that arrange hedging instruments (swaps, options and futures etc.) which have evolved. These enterprises also provide services which are very close to financial intermediation without necessarily being it (as the enterprises may not put themselves at risk by incurring liabilities on their own account). The boundary between financial intermediation and many services which are auxiliary to it has become somewhat blurred due to continuous innovation in financial markets. It has become increasingly difficult to make a clear distinction between true intermediation and certain other financial activities.

Mutual funds

10.94 Mutual funds mainly incur liabilities through the issue of shares, or units. In the UK these include unit trusts and investment trusts. They transform these funds by acquiring financial assets and/or real estate. They are thus classified as financial intermediaries. As with other corporations, any change in the value of their assets and liabilities other than their own shares is reflected in their own funds. Because the amount of the funds usually equals the value of the mutual fund's shares, changes in the value of the fund's assets and liabilities will be mirrored in the market price of these shares. Mutual funds investing solely in real estate (such as property unit trusts) are also treated as financial intermediaries.

Unincorporated financial enterprises

10.95 Financial intermediation generally is limited to transactions on the market. However, individuals or households may lend money to others or buy and sell foreign currency. Such unincorporated financial enterprises of this kind are included in the financial corporations sector only if they qualify as proper financial intermediaries or auxiliaries and as quasi-corporations. To do this, they must have complete sets of accounts that are separable from those of their owners – so money lenders, currency traders and others operating in financial activities on a small scale are unlikely to qualify and so will not be included in the financial corporations sector. However, large unincorporated financial enterprises may be subject to government regulation and control and thus obliged to keep accounts: they would therefore fall into this sector.

The institutional units included in the financial corporations sector

10.96 These are the following:

a) private or public corporations which are principally engaged in financial intermediation and/ or auxiliary financial activities;
b) co-operatives and partnerships recognised as independent legal entities which predominantly carry out financial intermediation and/or auxiliary financial activities;
c) public producers, which by virtue of special legislation are recognised as independent legal entities, which are predominantly engaged in financial market intermediation and/or in auxiliary market financial activities;
d) NPIs recognised as independent legal entities which are mainly engaged in financial intermediation and/or auxiliary market financial activities, or which are serving financial corporations;
e) holding corporations if the group of subsidiaries within the economic territory as a whole is principally engaged in financial intermediation and/or in auxiliary financial activities;
f) unincorporated mutual funds composed of investment portfolios owned by groups of participants, and whose management is usually undertaken by other financial corporations. By convention these funds are regarded as institutional units separate from the managing financial corporation;
g) financial quasi-corporations.

Amongst the types of institutional unit included in the list above of financial corporations are some which do not currently exist in the United Kingdom. For example, there are no public producers ((c) above) classified to the financial corporations sector.

Sub-sectors of the financial corporations sector and changes in the new ESA

10.97 Under the new system the financial corporations' sector (S. 12) is split between:

- the central bank (S.121)
- other monetary financial institutions (other banks and building societies) (S.122)
- other financial intermediaries, except insurance corporations and pension funds (S.123)
- financial auxiliaries (S.124)
- insurance corporations and pension funds (S.125).

10.98 The other monetary financial institutions sub-sector is the ESA form of the 'other depository corporations' sub-sector as defined in the SNA 93, paragraphs 4.88–4.94. The sub-sectors S.121 and S.122 used in the ESA 95 correspond to monetary financial institutions for statistical purposes as defined by the European Central Bank (ECB).

10.99 Holding corporations which control and direct a group of subsidiaries principally engaged in financial intermediation or in auxilliary financial activities are classified to the financial corporations sector. If the holding corporation itself is engaged in such activities it is classified to the appropriate subsector. If not, it is classified to the sub-sector 'other financial intermediaries except insurance corporations and pension funds' (S.123). This latter is a deviation from the SNA 1993 (paragraph 4.100) which advocated classifying the holding corporation of the sub-sector of the majority of the group. It was agreed for the ESA 95 in order to maintain consistency with the monetary financial institutions for statistical purposes as defined by the ECB and with official statistics on insurance corporations. Where NPI exist which are recognised as independent legal entities serving financial corporations, but not themselves carrying out financial intermediation or auxiliary financial activities, they are allocated to the financial auxiliaries sub-sector (S.124).

10.100 Financial corporations may also be subdivided according to whether they are subject to public, private or foreign control (using the same criteria as described for non-financial corporations above).

10.101 The first level of sub-sectors for financial corporations is the five categories listed above, although EU legislation implementing the ESA 1995 does not actually specify separate figures for the central bank (S.121). Instead the EU legislation requires and the United Kingdom publishes figures for 'monetary financial institutions' which are the combined data for the central bank and other monetary financial institutions (S.121 and S.122) sub-sectors.

Change from the former system of accounts

Under ESA 79 the financial account and balance sheet showed the financial corporations and institutions sector split between banks, building societies, life assurance and pension funds, and other financial institutions. This split was not however made for the current and capital accounts. It is now part of the 'ESA 95' system, though still not required by Regulation.

10.102 Subdivision between public, private and foreign controlled corporations is not done in the UK accounts.

The central bank (S.121)

10.103 The ESA 1995 (paragraph 2.45) defines this as:

'all financial corporations and quasi-corporations whose principal function is to issue currency, to maintain the internal and external value of the currency and to hold all or part of the international reserves of the country.'

10.104 This sub-sector includes the national central bank plus any other central monetary agencies of essentially public origin which keep a complete set of accounts and can make decisions independent of central government. In the UK this means the Issue and Banking Departments of the Bank of England. Decisions on the UK's official reserves are handled within central government so this does not form part of the central bank's role. This sub-sector does not include agencies and bodies (other than the central bank) which regulate or supervise financial corporations or financial markets. These are classified to the financial auxiliaries sub-sector (S.124). This is an instance where the ESA 1995 deviates from the 1993 SNA so that consistency with the European Central Bank's definition of monetary financial institutions is maintained.

Other monetary financial institutions (S.122)

10.105 This sub-sector comprises:

'all financial corporations and quasi-corporations, except those classified in the central bank sub-sector, which are principally engaged in financial intermediation and whose business is to

receive deposits and/or close substitutes for deposits from institutional units other than monetary financial institutions, and, for their own account, to grant loans and/or make investments in securities. '

[ESA 1995 paragraph 2. 48]

10.106 Monetary financial institutions (MFIs) consist of the central bank (S.121) and those other institutions classified to S.122. This is the same as the 'Monetary Financial Institutions for statistical purposes' defined by the ECB. The discussion of the Central Bank sub-sector (S.121) above explains how the definition differs from that set out in the SNA.

10.107 In the United Kingdom, banks and building socities are MFIs. More generally, other monetary financial institutions may include some corporations (such as building societies in the UK) which do not call, or are not allowed to call, themselves banks. In some countries, other institutions described as 'banks' may not actually be MFIs. In the UK all institutions allowed to call themselves 'banks' are monetary financial institutions.

10.108 The following intermediaries are usually allocated to this sub-sector (S.122) in countries covered by the ESA 95:

- commercial banks, 'universal' banks, 'all-purpose' banks;
- savings banks (including trustee savings banks and savings banks and loan associations);
- post office giro institutions, post banks, giro banks;
- rural and agricultural credit banks;
- co-operative credit banks and credit unions;
- specialist banks such as merchant banks, issuing houses and private banks.

10.109 In the UK credit unions are very small and are treated as other financial intermediaries (S.123) rather than monetary financial institutions. Also, some of the above types of intermediaries, whilst common in certain other EU member states, do not exist in the UK.

10.110 Where other financial intermediaries exist to receive repayable funds from the public, whether as deposits or the continuing issue of bonds and other similar securities, they are also be classified in this sub-sector. Such institutions are corporations which grant mortgages (including building societies and mortgage credit institutions), mutual funds and municipal credit institutions. In the UK, only a few small unit trusts, known as money market funds, currently fulfil the requirements of this category. Where such intermediaries do not receive repayable funds from the public they are allocated to the 'Other financial intermediaries, except insurance corporations and pension funds' sub-sector (S.123). UK investment trusts, units trusts, etc. are classified to sub-sector S.123.

10.111 Sub-sector S.122 does not include:

- holding corporations which only control and direct a group consisting mainly of other monetary financial institutions, but which are not other monetary financial institutions themselves. Such institutions are allocated to sub-sector S.123: 'other financial intermediaries, except insurance corporations and pension funds';
- non-profit institutions recognised as independent legal entities serving other monetary financial institutions which are not themselves engaged in financial intermediation. They are classified in sub-sector S.124 as financial auxiliaries.

10.112 There are currently no public monetary financial institutions other than the Bank of England in the UK, although Girobank was classified here prior to privatisation in 1990.

Other financial intermediaries, except insurance corporations and pension funds (S.123)

10.113 This sub-sector can be described as follows:

'all financial corporations and quasi-corporations which are principally engaged in financial intermediation by incurring liabilities in forms other than currency, deposits and/or close substitutes for deposits from institutional units other than monetary financial institutions, or insurance technical reserves.'

[ESA 1995 paragraph 2.53]

10.114 It includes various types of financial intermediaries, especially those which are mainly involved in long-term financing such as investment corporations and hire purchase corporations. This predominant maturity generally provides the basis for distinguishing them from the other monetary financial institutions described above. The distinction between S.123 and S.125 'Insurance corporations and pension funds' is based on the fact that the former institutions do not have liabilities in the form of insurance technical reserves.

10.115 The corporations classified in this sub-sector, S.123, are thus the following unless they are MFIs:

- those engaged in financial leasing;
- hire purchase and the provision of personal or commercial finance;
- factoring;
- securities and derivatives dealers (on own account);
- specialised financial corporations such as venture and development capital companies, export/import financing companies;
- financial vehicle corporations, created to be holders of securitised assets;
- financial intermediaries which receive deposits and/or close substitutes for deposits from MFIs only;
- holding corporations which only control and direct a group of subsidiaries mainly engaged in financial intermediation and/or auxiliary financial activities, but which are not financial corporations themselves;
- mortgage credit unions;
- mutual funds other than monetary market funds, including most unit trusts and other collective investment schemes, e.g. undertakings for collective investment in transferable securities (UCITS);
- mortgage lenders;
- credit card issuers.

10.116 Non-profit institutions (NPI), recognised as independent legal entities, serving the bodies in this sub-sector but not themselves engaged in financial intermediation are excluded from S.123 – they are classified as financial auxiliaries, S.124.

Financial auxiliaries (S.124)

10.117 These are defined as:

'all financial corporations and quasi-corporations which are principally engaged in auxiliary financial activities, that is to say activities closely related to financial intermediation but which are not financial intermediation themselves.'

[ESA 1995, paragraph 2.57]

10.118 This sub-sector consists of corporations and quasi-corporations such as:

- insurance brokers, salvage and average administrators, insurance and pension consultants, etc.

- securities brokers, loan brokers, investment advisers, etc.
- flotation corporations that manage the issue of securities;
- corporations whose main role is to guarantee (by endorsement) bills and similar instruments;
- corporations which arrange (but do not issue) derivative and hedging instruments such as swaps, futures and options;
- corporations providing infrastructure for financial markets;
- central supervisory authorities of financial intermediaries and financial markets (when they are separate institutional units);
- managers of pension funds, mutual funds, etc. ;
- corporations providing stock exchange and insurance exchange;
- NPIs recognised as independent legal entities serving financial corporations but which are not directly involved in financial intermediation or auxiliary activities themselves.

10.119 This sub-sector excludes holding corporations which merely control and direct a group predominantly comprising financial auxiliaries, but which are not financial auxiliaries themselves. These are categorised to sub-sector S.123 instead.

10.120 Institutions in this sub-sector will initially be covered elsewhere in the accounts. The United Kingdom has derogations under the European Union Council regulation regarding this sub-sector. One derogation allows the United Kingdom to include the financial transactions of financial auxiliaries with those of non-financial corporations in the financial accounts until the year 2002. There is a similar derogation for the balance sheets of financial assets and liabilities of these particular institutions. These are necessary because work on identifying them is still in progress. Many such units are small and unincorporated and the lists of them are consequently incomplete. Most unincorporated businesses that should be classified here were previously allocated to unincorporated non-financial businesses. However, some, such as the Corporation of Lloyd's and supervisory bodies such as IMRO, were previously included in the 'other financial institutions' sector under the UK's application of the 1979 ESA. The data for them will, for the time being, be recorded in the 'other financial institutions except insurance corporations and pension funds' (S.123) sub-sector (rather than with non-financial corporations, S.11) whilst work continues to develop a data set for all the financial auxiliaries sector in the UK.

10.121 Examples of the UK institutions which will in future fall in the financial auxiliaries sub-sector are as follows:

- Bureaux de Change (previously classified to other financial institutions); clearing houses (e.g. the London Clearing House, Crest Co.); insurance exchange corporations -the Corporation of Lloyd's; Lloyd's brokers; the International Petroleum Exchange; and the London Stock Exchange.

- Various regulatory organisations (previously 'other financial institutions'), will be classified as financial auxiliaries. The new Financial Services Authority (FSA) is expected to be part of central government.

Insurance corporations and pension funds (S.125)

10.122 These are defined as follows:

'all financial corporations and quasi-corporations which are principally engaged in financial intermediation as the consequence of the pooling of risks.'

[ESA 1995. paragraph 2.60]

10.123 The insurance contracts administered by these institutions may relate to individuals or groups, whether or not there is a government imposed obligation for them to participate in such contracts.

10.124 This sub-sector includes both captive insurance corporations and reinsurance corporations.

10.125 This sub-sector does not include 'social security funds'. Those funds are ones in which certain groups of the population are legally obliged to participate and which general government is responsible for managing independently of its role as supervisory body or employer. So this sub-sector does not include the UK national insurance scheme, which is operated by central government, S.1311.

10.126 The insurance corporations and pension funds sub-sector does not include holding corporations which only control and direct a group of these (S.125) units without being insurance corporations and pension funds themselves. Such holding corporations are to be found in sub-sector S.123 (see above). NPI which serve pension funds or insurance corporations, without being engaged in financial intermediation, are classified as financial auxiliaries (S.124).

10.127 This sub-sector may be further divided, into:

- insurance corporations;
- (autonomous) pension funds.

Insurance corporations

10.128 These are incorporated, mutual and other entities whose principal function is to provide life, accident, sickness, fire or other types of insurance to individual institutional units or groups of units. The insurance contracts administered might relate to individuals and/or groups, and their participation may, or may not, be due to an obligation imposed by government. For example, there is usually a legal obligation for third party motor vehicle insurance. Social insurance contracts (described in Chapter 5) are sometimes a large part of the contracts administered.

10.129 Risks concerning individuals or groups could both be included in the activities of life and non-life insurance corporations. Some insurance corporations may restrict their activities to group contracts only.

Pension funds

10.130 Pension funds are institutions which insure group risks relating to social risks and needs of the insured persons. So participants may be employees of a single enterprise, employees of a certain industry or profession. The benefits included in the insurance contract may include death benefits, retirement benefits and benefits paid on early retirement on medical grounds. In some countries, including the United Kingdom, all these types of risk could be equally well insured by life insurance corporations. In other countries some of these types of risk have to be insured through life insurance corporations. In contrast to life insurance corporations, pension funds are restricted, by law, to specific groups of employees and the self-employed.

10.131 The pension funds included here, in sub-sector S.125, are those which are *autonomous*. That is, they are separate institutional units from the units (such as corporations) which create them. They are set up to provide benefits on retirement for specific groups of employees and have their own assets and liabilities. They also make their own transactions in financial assets in the market. These funds are organised, and directed, by individual private or government employers, or jointly by individual employers and their employees. The employers and/or employees make regular

© National Accounts Concepts, Sources and Methods

contributions to these funds. Examples of such pension funds in the United Kingdom are the many occupational pension schemes established by employers. In many EU member states autonomous pension schemes are rare, non-autonomous schemes being much more common.

10.132 *Non-autonomous pension funds* are those where the pension arrangements for the employees of government or private sector entities do not include a separately organised fund. They also include arrangements organised by a non-government employer in which the reserves of the fund are simply added to that employer's own reserves or invested in securities issued by that employer. These schemes are not classified to this sub-sector but instead to that of the employer. The main non-autonomous pension schemes in the UK are for government sector employees (such as the civil service pension scheme). However, in other EU countries like Germany non-autonomous schemes are common.

Changes from the former system of accounts

Under the previous ESA 'casualty insurance', which is now more usually called non-life insurance in the UK, was classified to 'other financial institutions', as were insurance holding companies. Non-life insurance is also sometimes referred to as general, or 'other than long-term' insurance in the UK. The old ESA also treated life insurance and pension funds differently - as part of the personal (now households) sector above the net borrowing or lending line as far as their income and expenditure recorded in the current account were concerned; but in the capital and financial accounts they formed the life assurance and pension fund sub-sector of financial corporations. However, in the 1995 ESA all insurance and pension funds data are brought together into this sub-sector, 'insurance corporations and pension funds', S.125. Actual and imputed transactions with other sectors then show the redistribution of income to policy holders.

General government (S.13)

10.133 This sector consists of:

> '*all institutional units which are other non-market producers whose output is intended for individual and collective consumption, and mainly financed by compulsory payments made by units belonging to other sectors, and/or all institutional units principally engaged in the redistribution of national income and wealth.*'
>
> [ESA 1995 paragraph 2.68]

Governments as institutional units

10.134 Government units can be considered to be unique kinds of legal entities established by political processes which have legislative, judicial or executive authority over other institutional units within a certain area. So the principal roles of government involve responsibility for providing goods and services to the community or to individual households and paying for this out of taxation or other incomes; redistributing income and wealth by means of transfers; and engaging in non-market production.

10.135 A government unit usually has the authority to raise funds by collecting taxes or compulsory transfers from other institutional units. To be an institutional unit it must also have funds of its own, (either raised by taxing or received as transfers from other government units) and the authority to distribute some, or all, of such funds towards its policy objectives. It may also be able to borrow funds on its own account.

10.136 Government units will usually make three different types of outlays:

- actual or imputed expenditures on providing free to the community collective services like public administration, defence, law enforcement, public health, etc. These types of services have to be organised collectively by government and financed by taxation or other income because of the inability of the market to provide them;
- the expenditures on providing goods and services free (or at prices that are not economically significant) to individual households. These are deliberately made and financed out of taxation etc. by government in the pursuit of its policy objectives, even though individuals could be charged according to their use of them;
- the third type of outlay that government units make is transfers paid to other institutional units, mostly households, to achieve some redistribution of income or wealth.

10.137 In a single country there may be many separate government units when there are different levels of government – central, state or local government. In the UK the state level does not exist at present; there is only central government and local government -the latter carried out by local authorities. As well as these, social security funds are also viewed as government units.

Government units as producers and when they are quasi-corporations

10.138 Government units may own and operate unincorporated enterprises that produce goods or services. Depending largely upon political choice, government units may produce goods and services themselves or may purchase them from market producers. It is not necessary for a government itself to produce the goods and services which it may provide free (or at insignificant prices) to households and others – it is only required to assume responsibility for organising and financing their production. It is common, however, for government units to be involved in producing a wide range of goods and services. The extent of this varies greatly from one country to another. Thus, apart from collective services like defence and public administration it is difficult to categorise other types of production (such as education or health services) as intrinsically government production.

10.139 The government has three different ways of intervening in the production arena:

- it may create a public corporation whose corporate policy, including pricing and investment, it is able to control;
- it may set up a NPI that it controls and at least mainly finances;
- it may produce the goods or services itself, in an establishment it owns and which does not exist as a separate legal entity.

10.140 However, if a government establishment can be described as follows it should be treated as a *quasi-corporation*:

- it charges economically significant prices for its outputs;
- it is operated and managed in a similar way to a corporation;
- it has a complete set of accounts from which its operating surpluses, savings, assets and liabilities can be calculated.

Such quasi-corporations are market producers that are treated as separate institutional units from the government units that own them. They are classified to the corporate sectors in the same way as public corporations.

10.141 The management of an enterprise must have a large amount of discretion in relation to both the management of the production process and also the use of funds if it is to be treated as a quasi-corporation. For example, it must be able to finance at least some of its capital formation out of its own savings, depreciation reserves or borrowing. The net operating surplus of a government quasi-corporation is not a component of government revenue, and the accounts for government record only the actual or imputed flows of income and capital between the government and the quasi-corporation.

10.142 The producer units which remain integrated with the government units which own them, and thus stay in the general government sector, are ones which cannot be treated as quasi-corporations. They will be mainly non-market producers: producers whose output is supplied to others free or at prices that are insignificant economically. Such units may include government producers supplying non-market goods or services to other government units, such as government printing offices, transport agencies, computer or communication agencies. In principle, government units can also be market producers.

10.143 There are usually no suitable markets whose prices can be used to value government non-market output. The international convention is that such output is valued at its production costs.

Social security funds

10.144 These are special kinds of institutional units which may be found at any level of government. In practice, they are not recognised as separate units in the United Kingdom. They stem from social security schemes, which are one form of *social insurance schemes*. (The others are private funded insurance or pension schemes; and unfunded schemes run by employers for their employees.)

10.145 *Social security schemes* are set up to provide social benefits (including pensions) to members of the community, or to groups of individuals (such as the employees of an enterprise and their dependants) out of funds derived mainly from social contributions. They cover the community as a whole, or at least large sections of it, and are imposed and controlled by government units. Usually employees or employers, or both, have to make compulsory contributions to these schemes; the terms on which the benefits are paid to recipients are determined by government units. The schemes cover a wide range of programmes, giving benefits in cash or in kind for old age, invalidity or death, survivors, sickness and maternity, work injury, unemployment, family allowance, health care, etc. There is usually no direct link between the amount of the contribution paid by an individual and the risk to which he or she is exposed. Social security schemes, operated by government, are different from pension schemes, or other social insurance schemes set up by individual employers by mutual agreement with their employees, where the benefits are linked to contributions. Those employee pension schemes are all regarded as private and, if they are funded, are classified to the insurance corporations and pension funds sub-sector, S.125, within financial corporations. The government's own pension schemes for its employees are similarly classified as social insurance schemes, not social security schemes. Those that are unfunded (such as the UK Principal Civil Service Pension Scheme) are not recognised as institutional units separate from government.

10.146 *Social security funds* may be distinguished by the fact that they are separately organised from the other activities of government units and hold their assets and liabilities separately from the latter. They are then separate institutional units because they are autonomous funds, they have their own assets and liabilities and engage in financial transactions on their own account. However, arrangements for social security vary from country to country and in some countries, such as the United Kingdom, they are so closely integrated with the other finances of government that they cannot be treated as a separate sub-sector. The amounts raised, and paid out, in social security contributions and benefits may be deliberately varied in order to achieve government policy objectives that have no direct connection with the concept of social security as a scheme to provide social benefits to members of the community.

The general government sector

10.147 This consists of groups of resident institutional units:

- all units of central, state or local government;
- all social security funds at each level of government;
- all non-market NPI that are controlled and mainly financed by government units.

10.148 The institutional units classified here are as follows:

- general government departments and agencies which administer and finance a group of activities, mainly providing non-market goods and services, intended for the benefit of the community;
- NPIs recognised as independent legal entities which are other non-market producers and which are controlled and principally financed by general government;
- autonomous pension funds where these can be classified as social security funds;
- non-autonomous (unfunded or notionally funded) pension schemes for government employees;
- market regulatory organisations which are either exclusively or principally simple distributors of subsidies even if they are engaged in buying, holding and selling agricultural or food products.

10.149 This sector does not include public corporations, even when all their equity is owned by government units. It also excludes quasi-corporations that are owned and controlled by government units, when these are classified to the non-financial or financial corporations' sectors. However, unincorporated enterprises owned by government units that are not quasi-corporations remain an integral part of those units and so must be included in the general government sector.

10.150 There is some discussion below of the UK public enterprises that remain in this sector. A number of government owned trading bodies have moved from the central government sector to the public non-financial corporations sector because of the move to the new ESA.

Sub-sectors of the general government sector

10.151 The general government sector is divided into four sub-sectors in the ESA:

- central government (S.1311);
- state government (S.1312);
- local government (S.1313);
- social security funds (S.1314).

10.152 The sub-sector state government (S.1312) is not used in the UK national accounts as it is not a level of government that exists currently in the UK. There is no social security funds sub-sector for the UK since the arrangements for social security here are so closely integrated with other finances of government that they cannot be treated as separate. Although these two subsectors are not currently applicable in the UK context they are both described below.

Central government (S. 1311)

10.153 This is defined as including:

'all administrative department of the State and other central agencies whose competence extends normally over the whole economic territory, except for the administration of social security funds. '
[ESA 1995 paragraph 2.71]

10.154 As well as government departments, (including, in the United Kingdom, the Scottish and Welsh Offices and Northern Ireland department) this sub-sector also includes those NPI which are controlled and predominantly financed by central government and whose competence also extends over the whole economic territory.

10.155 In the UK the administration of social security funds is an integral part of central government for both its funding and decision-making, and so cannot be separately classified as social security funds (S.1314).

10.156 The political authority of central government extends over the entire territory of the country. It thus has the authority to impose taxes on all resident and non-resident units engaged in economic activities in the country. Its political responsibilities include national defence and relations with foreign governments. It also uses legislation and regulation to try to ensure the efficient working of the social and economic system and the maintenance of law and order. Its responsibilities include providing collective services for the benefit of the community as a whole – hence expenditure on defence and public administration. Central government may also pay for services like education and health to be provided, primarily to benefit individual households. A final function of central government is that it may make transfers to other institutional units, including other levels of government.

10.157 As in most countries this sub-sector is large and complex. It comprises a central group of departments or ministries, which form a single institutional unit. The departments are responsible for large amounts of expenditure but they are not separate institutional units since they do not own assets, engage in transactions, etc. independently of central government as a whole. Central government includes other institutional units such as agencies created to carry out specific functions such as road construction or the non-market production of services such as health or education.

10.158 Central government departments may be located in different parts of the country. However, they are still regarded as part of central government, although in order to produce production accounts by type of productive activity, the establishment is often used as the statistical unit, and the producer units in different regions are treated as different establishments, despite being part of a single institutional unit.

10.159 This sub-sector may include units which make financial transactions that in another country would be carried out by the central bank; in particular, managing international exchange reserves, operating exchange stabilisation funds and transacting with the International Monetary Fund (IMF). Those monetary authority functions carried out by central government are, of course, recorded in the government sector. In view of possible variations between countries, international comparisons should be made with caution.

10.160 In the UK, central government includes the Exchange Equalisation Account handling the official reserves. However, the Banking Department and the Issue Department of the Bank of England form the UK's central bank sub-sector (S.121).

Central government trading bodies

10.161 As discussed above, some UK central government trading bodies are, under the ESA 1995 system, now classified as public corporations (and are within the non-financial corporations sector). They have been reclassified since they have sufficient independence to control their own affairs day to day and their finance is not controlled by parliamentary 'votes' of money. These include the following: the Central Office of Information, Companies House, Horserace Totalisator Board, Her Majesty's Land Registry and Remploy Ltd.

10.162 Those that are not sufficiently independent are deemed to be central government market bodies. These include the Export Credits Guarantee Department. Other bodies classified to central government include: Regional Health Authorities; NHS hospitals other than Trusts; the Driving Standards Agency; the Housing Corporation; and Housing for Wales.

Intervention Board Executive Agency

10.163 The classification of this body was reviewed as part of the move to the new ESA because of a footnote to paragraph 269 in ESA 1995 that effectively puts intervention activities in agriculture in the corporate sector and those redistributing subsidies in the government sector. The Intervention Board is predominantly involved in redistributing subsidies although about 10 per cent of its activities involve intervention buying. The UK consulted some other countries to establish what they were doing with similar bodies: no clear consensus was found: It was decided that this agency should be classified to central government in the UK.

Export Credits Guarantee Department (ECGD)

10.164 The classification of this body in the central government sector was examined during the move to the 1995 ESA since much of its activity is analogous to that of financial corporations. There had been changes in the functions and accounting for ECGD after the passing of the Export and Investment Guarantees Act 1991. The comments below describe the position of this body from April 1991.

10.165 The ECGD's main activity is credit insurance support for UK capital goods and project exports, typically sold on medium or long terms of payment. Support for short-term exports is also important and ECGD works to ensure that adequate short-term credit insurance is available for exporters by entering into reinsurance arrangements with the private sector. ECGD also has some non-trading activity, which is treated as general government expenditure in the national accounts and is not covered by the comments here.

10.166 The ECGD is a department of the Secretary of State for Trade and Industry. ECGD's statutory authority comes from the Export and Investment Guarantees Act 1991. Its main role under this Act is to facilitate exports of goods and services through the provision of guarantees and insurance (Section 1 of the Act). Section 2 enables ECGD to provide overseas investment insurance. ECGD is able to make any arrangements considered to be in the interests of the proper management of the ECGD portfolio on the terms and conditions it considers appropriate, providing the consent of the Treasury in obtained. The ECGD has to consult the Export Guarantees Advisory Council when determining if there is a national interest case for providing reinsurance support.

10.167 ECGD is required to operate its trading activities at no overall cost to public funds. However, since 1980–81 the need for many of the UK's overseas trading partners to reschedule debt has caused claims to exceed income, resulting in net expenditure on the ECGD vote. This net expenditure is regarded in the ECGD trading accounts as notional borrowing from the Consolidated Fund and interest on the outstanding notional debt is charged to the Trading Account. A trading account and balance sheet are available in the ECGD's annual report.

10.168 For the ECGD, as an agency of general government, to be considered an institutional unit under the 1995 ESA it must not only have a complete set of accounts but also have autonomy of decisions. The following points were relevant:

a) ministerial authority is required to write off debt, except where it is not cost-effective to pursue;
b) the legislation for the ECGD is couched in terms of the Secretary of State carrying out the function through the agency of ECGD;
c) ECGD carries out certain non-trading functions on behalf of the Secretary of State;
d) ECGD when providing credit guarantees is providing a market service;
e) All ECGD expenditure has to be voted, and all receipts paid back to the consolidated fund; thus ECGD keeps no reserves.

10.169 On balance it was considered that ECGD's ties to government were too close to reclassify it as a financial corporation, despite much of its activity being analogous.

State government (S. 1312)

10.170 The state government sub-sector consists of:

'state governments which are separate institutional units exercising some of the functions of government at a level below that of central government and above that of the governmental institutional units existing at local government level, except for the administration of social security funds.'

[ESA 1995, paragraph 2.72]

10.171 A state government usually has the power to levy taxes on institutional units which are resident, or engage in economic activities or transactions within, its area. It must also be entitled to spend some or perhaps all of the income it received as it chooses (within the general laws of the country). To be recognised as an institutional unit it must also be able to appoint its own officers, independently of external administrative control.

10.172 There are no institutions in the UK at present which satisfy the criteria for classification as state government.

Local government (S. 1313)

10.173 The sub-sector 'local government' includes:

'those types of public administration whose competence extends only to a local part of the economic territory, apart from local agencies of social security funds.'

[ESA 1995, paragraph 2.73]

10.174 This sub-sector includes any NPIs which are controlled and mainly financed by local governments.

10.175 These units are in principle those whose fiscal, legislative and executive authority extends over small geographical areas distinguished for administrative and political purposes. However, the authority of local governments is generally much less than that of central or regional/state governments and they may not necessarily have the fiscal authority to levy taxes on institutional units which are resident in their areas. Grants or transfers from higher levels of government are often important components of their income and part of their function may be to act as agents for these higher levels of government to some extent. However, the fact that they may also act as such agents does not prevent them from being treated as separate institutional units as long as they are also able to raise and spend some of their funds on their own initiative and own responsibility.

10.176 Local government units often provide a wide range of services to local residents. The rules about the treatment of the production of goods and services by the other levels of government also apply to them. Hence, municipal theatres, museums, swimming pools etc. which supply goods or services on a market basis should be classified as quasi-corporations whenever appropriate. However, UK local authority trading bodies do not have sufficient independence to warrant classification as public corporations. As trading bodies they stay within the local government sector as producers of market output. This sub-sector thus covers local authorities throughout the UK and a selection of trading bodies. The latter include: the Barbican Centre; Bath Library Co. Ltd. ; Enniskillen Aerodrome; Motherwell District Slaughterhouse; municipally owned industrial and trading estates, fishing harbours, ports and piers, restaurants, theatres, toll bridges, etc. Units supplying services on a non-market basis (such as education and health) stay part of the local government unit to which they belong.

10.177 'Local government' does not include the local agencies of central government such as the National Health Service.

10.178 A full list of bodies within this subsector is published in the *Sector Classification Guide for National Accounts*.

Local authority schools

10.179 These schools are recorded within this sub-sector, as are grant-maintained schools. Further and higher education establishments and sixth form colleges are also recorded as NPISH, with the exception of Buckingham University which is a private corporation.

Social security funds (S.1314)

10.180 These are defined as:

> 'all central, state and local institutional units whose principal activity is to provide social benefits and which fulfil each of the following two criteria:
>
> • by law or regulation certain groups of the population are obliged to participate in the scheme or to pay contributions;
> • general government is responsible for the management of the institution in respect of the settlement or approval of the contributions and benefits independently from its role as supervisory body or employer.'

[ESA 1995, paragraph 2.74]

10.181 There is usually no direct link between the amount of the contribution paid by an individual, to social security funds, and the risk to which that individual is exposed.

10.182 If they were separately funded, this sub-sector would cover the Basic State Pension Scheme, the National Insurance Scheme, the Maternity Fund in the UK. However, the administration of social security funds in the UK is an integral part of central government as far as both funding and decision making is concerned, and so they cannot be classified separately as social security funds. Hence the Maternity Pay Fund, the National Insurance Fund and the Social Fund have been classified to central government under the ESA 1995 and are no longer shown as a separate Social Security Funds sub-sector.

Households (S.14)

10.183 The households sector covers:

> 'individuals or groups of individuals as consumers and possibly also as entrepreneurs producing market goods and non-financial and financial services (market producers) provided that, in the latter case, the corresponding activities are not those of separate entities treated as quasi-corporations. It also includes individuals or groups of individuals as producers of goods and non-financial services for exclusively own final use.'

[ESA 1995 paragraph 2.75]

In the previous system households were merged with the NPISH sector in the personal sector. Until data problems have been resolved the 'Households' sector will continue to include non-profit institutions serving households in the UK national accounts.

Households as institutional units

10.184 To define households the ESA 1995 uses the SNA 1993 definition:

'Households as consumers may be defined as small groups of people who share the same living accommodation, who pool some, or all, of their income and wealth and who consume certain types of goods and services collectively, mainly housing and food.'

[SNA 1993, paragraph 4.132]

10.185 Households are often families; however, members of the same household do not have to belong to the same family as long as there is some sharing of resources and consumption between them. Each member of a household should normally have some claim on its combined resources and, similarly, at least some of the spending decisions should be taken by the household as a whole.

10.186 The definition adopted for the 1991 census in the UK was that:

'a household comprises either one person living alone or a group of people (not necessarily related) living at the same address with common housekeeping – that is, sharing at least one meal per day or sharing a living room or sitting room.'

This definition closely corresponds to the concept of a household as defined in the national accounts system

10.187 Servants or paid domestic employees who live on the same premises as their employer do not form part of their employer's household, even though they may receive accommodation and meals as remuneration in kind. This is because paid employees have no claim upon the collective resources of their employer's households and the accommodation and food they consume are not included with their employer's consumption. Thus they are seen as belonging to separate households from their employers.

10.188 Households may be institutionally based: people who live permanently in an institution are treated as forming a single institutional household. Examples of such institutions are prisons, retirement homes and monasteries: in all of which those residing have little or no autonomy of action or decision in economic respects. Persons expected to live for a long time in such institutions (for example, long-term patients in hospitals), although they themselves may not regard it as their permanent residence, are included.

10.189 People who enter institutions such as hospitals, clinics, convalescent homes, religious retreats, etc. for short periods, or who attend boarding schools are treated as members of the individual households to which they usually belong. This also applies to those who live away from home while at university, or who serve short prison sentences.

10.190 Households' economic behaviour may be more varied than that of members of other sectors since they may carry out any kind of economic activity and not just consumption. This is particularly seen when members of households act in the production process either by operating their own unincorporated businesses or by supplying labour to other unincorporated or corporate enterprises by working as employees. They also lend and borrow funds, and so on.

Households as producers

10.191 Most working people work as employees for corporations, quasi-corporations or government – so the production to which they contribute takes place outside the household sector, in one of the two corporations sectors or the government sector. Production within the household sector does occur, within enterprises that are directly owned and controlled by households, either individually or in partnership with others. However, independent partnerships producing market goods and/or services are classified as quasi-corporations in the UK.

10.192 The producer units in the household sector are called 'unincorporated enterprises', they are not incorporated as legal units separate from the households themselves. The fixed and other assets used in unincorporated businesses belong to the owners of the businesses. The enterprises themselves cannot enter into contractual relationships with other units nor incur liabilities on their own behalf. The owners of these businesses are themselves personally liable, without limit, for any debts or obligations incurred during the production process.

10.193 The owner of a household unincorporated enterprise generally has two different functions:

- as the entrepreneur who is responsible for the creation and management of the enterprise;
- as a worker who contributes his or her labour.

Setting up an unincorporated business requires initiative, enterprise and capital equipment. The owners of such businesses have to raise the finance required at their own risk and on their own personal security. They must also find suitable premises, and buy or hire the capital equipment or materials they need. Owners may also have to hire and supervise paid employees. It can therefore be seen that in some cases the main function of the owner is to act as an entrepreneur, innovator and risk-taker: hence the surplus from production which eventually accrues to the owner represents predominantly a return to entrepreneurship. On the other hand, the principal role of the owner may be to provide labour, often highly skilled professional labour, in which case most of the surplus could be described as remuneration for work done – an example of this latter role is an owner acting as a self-employed accountant.

10.194 Since the surplus produced by the productive activities of a household's unincorporated enterprise can commonly be seen as mixture of two different types of income it is described as 'mixed income' instead of 'operating surplus' within the national accounts system (the only exception is the surplus from the production of own-account housing services, which is considered further in Chapters 13 and 22). The discussion in the preceding paragraph showed that the balance between return to entrepreneurship and remuneration for work done will vary significantly between the many different types of household unincorporated enterprises and that it is even difficult to make a distinct separation conceptually.

Households' unincorporated market enterprises

10.195 These enterprises are set up to produce goods or services for sale (or barter) on the market. They can cover more or less any type of activity: agriculture, mining, manufacturing, construction, retail distribution or other kinds of services. They can be individuals working as street traders with virtually no capital or premises of their own or large manufacturing, construction or service enterprises employing many other people.

10.196 The SNA acknowledges that it is extremely difficult to separate unincorporated businesses from their owners, since the owners are entitled to use the assets of such businesses in any way they choose. Thus some of the outputs of unincorporated market producers may be retained for consumption by the owner or other members of the household to which he or she belongs. Such goods or services should be included in the outputs of the enterprises and in the final consumption of the households. However, it may be difficult to get the data needed for this, especially if there is an element of tax avoidance or evasion involved! Similarly, buildings or capital equipment may be used partly for production and partly for consumption (that is the owner's household may live on the same premises). Hence the difficulties in applying the national accounts standards to households in these respects.

Household enterprises producing for own final use

10.197 These are household enterprises operated mainly to produce goods or certain services for their own final consumption or own gross fixed capital formation. The value of their output is imputed using the prices of similar goods or services sold on the market.

10.198 Household unincorporated enterprises producing goods for their own final use are:

- subsistence farmers, crofters or others engaged in the production of agricultural goods for their own final consumption;
- households engaged in the production of their own houses or other structures for their own final use, or on structural improvements or extensions to their existing homes;
- households producing other goods for their own consumption, for example, cloth, clothing, furniture, foodstuffs (but excluding the preparation of meals for immediate consumption), etc.

These enterprises may sell any of their surplus output – but if they regularly sell most of it they should instead be treated as market enterprises. Where groups of households carry out communal construction of buildings etc. for their own individual or community use, they should be treated as informal partnerships engaged in non-market production.

10.199 The national accounts system recognises only two types of services produced by households for their own final consumption:

- services of owner occupied dwellings: owner occupiers are said to own household unincorporated businesses that produce housing services for their own consumption;
- domestic services produced by employing paid staff: households are deemed to own household unincorporated businesses in which they employ paid staff -servants, cooks, gardeners, etc. – to produce services for their own consumption.

10.200 The production of these services does not generate mixed income, however. This is because there is no labour input into the production of the services of owner-occupied dwellings so any surplus arising is operating surplus. There is no surplus generated by employing paid staff since national accounting conventions state that the output produced by employing paid staff is valued as the costs of the compensation of employees (wages and salaries etc.) paid to the staff, no other inputs being recognised.

Sub division of this sector

10.201 The households sector is not subdivided in the UK national accounts, but the ESA provides for division into into six sub-sectors reflecting the main sources of household income:

a) employers (including own-account workers) (S.141 + S.142);
b) employees (S.143);
c) recipients of property incomes (S.1441);
d) recipients of pensions (S.1442);
e) recipients of other transfer incomes (S.1443);
f) others (S.145).

Changes from the previous ESA

The household sector is similar to, but not exactly the same as, the old 'personal sector'. It includes individuals and private trusts, although the latter are included independent legal status. Sole traders as unincorporated businesses continue to be recorded here since their accounts are not separable from those of households. Partnerships, formerly in the personal sector, are now classified to non-financial corporations (S.11) or financial corporations (S.12). Individual Lloyd's Underwriters continue in the household sector for the time being. UK students abroad are still included. However, private non-profit making bodies serving persons - previously part of the personal sector - are now allocated their own sector as 'Non-profit institutions serving households' (NPISH) (S.15). Life assurance and pension funds' income and expenditure were previously recorded in the personal sector. They are now recorded with the rest of their data in the 'Insurance corporations and pension funds' sub-sector (S.125) of the financial corporations sector.

Non-profit institutions serving households (NPISH) (S.15)

10.202 This sector consists of:

'non-profit institutions which are separate legal entities, which serve households and which are private other non-market producers. Their principal resources, apart from those derived from occasional sales, are derived from voluntary contributions in cash or in kind from households in the capacity as consumers, from payments made by general governments and from property income.'

[ESA 1995 paragraph 2.87]

10.203 These institutions were previously part of the personal sector under the 1979 ESA. Now they have been allocated a separate sector under the 1995 ESA rather than continuing in the renamed 'Household' sector. (However, in the UK accounts, until data problems have been resolved the sector remains combined with Households.)

10.204 The NPISH sector can thus be defined as the set of all resident non-profit institutions (NPIs) except:

- NPIs that are market producers; and
- non-market NPIs that are controlled and mainly financed by government units.

10.205 Where NPISH do not have legal status they are indistinguishable from households and so they are included in the household sector (S.14).

10.206 This sector includes charities, relief or aid agencies serving non-resident units but excludes entities where membership gives right to a predetermined set of goods and services. NPISH are described as NPI which provide goods or services to households either free or at prices that are not economically significant. They are of two main types:

- those created by associations of persons to provide goods or, more often, services mainly for the benefit of the members themselves;
- charities, relief or aid agencies, set up for philanthropic purposes.

10.207 Those set up mainly to benefit their members usually provide services free and are financed by regular membership subscriptions. They include trade unions, professional or learned societies, consumers' associations, political parties, churches or religious societies, social, cultural, recreational and sports clubs. They do not include bodies serving similar functions that are controlled and mainly financed by government units (except that churches are always treated as serving households even when, as in some countries, mainly financed by government units).

10.208 In some countries NPISH exist which do not possess any legal status or any formal articles of association. In the system they should be treated as NPISH when they carry out similar roles to the societies, parties, unions etc. described above, although they are not legally constituted. However, when groups of households collaborate on communal construction projects (such as construction of buildings, bridges, ditches, etc.) they are regarded as informal partnerships engaged on own-account construction, rather than NPISH. NPISH should normally have a continuing function and not be created solely for a single project of limited duration.

10.209 Philanthropic non-profit institutions serving households comprise charities, relief and aid agencies, etc. They provide goods and/or services on a non-market basis to those households most in need, including those affected by natural disasters or war. They are financed mainly by donations from the general public, government or corporations. Such donations may be in cash or in kind. They may also be provided by transfers from non-residents – including similar kinds of NPISH in other countries.

10.210 A list of NPISH is included in the *Sector Classification Guide for National Accounts* published by ONS. The wide range of activities of this sector is illustrated by the following brief selection, the Association of Corporate and Certified Accountants; Barnardos; Boy Scouts; British Society for the Advancement of Science; Civil Service Sports Council; the National Trust; Oxfam and the Salvation Army.

Educational institutions

10.211 *Universities* are diverse in their operations, but were reviewed en bloc because the main elements of their activities are the same and it is helpful to have consistency of statistical treatment between them. The only clear exception to this is Buckingham University which is explicitly a profit-making institution and has complete autonomy from government. Since Buckingham does not rely on, and is completely separate from, the traditional funding arrangements for universities it was, and continues to be, classified to the corporate sector.

10.212 Other universities are non-profit making, have responsibility for all their financial and day-to-day activities and compile full accounts. It can be argued that these bodies are active in a market because they compete freely with other institutions for business. However, governments control the whole higher education market either by funding or by selecting the participants, although this does not necessarily stop the market being viewed as a true one. The government is a key operator in many markets (eg defence) where, were it to withdraw, many private firms would subsequently cease trading; the government's ability to select the participants could be viewed as market control.

10.213 The most important question for universities is whether the price they charge is significant and is set by the market. The income of universities largely depends on the number of students they get. However it is considered that the price is not significant because the block grant from government is effectively a transfer to universities and the tuition fees are paid on a blanket basis. Thus these establishments are not able to alter the demand of government or students; they are not market producers.

10.214 Universities are not controlled by government since it cannot appoint the majority of officers of the governing bodies nor does it control the programme of the principal function, education. Therefore they should be considered as private institutions.

10.215 Universities other than Buckingham have therefore been classified as NPISH.

10.216 *Further education institutions* were also considered en bloc. Prior to April 1993 they were owned and effectively controlled by Local Education Authorities and classified to the Local Authority Sector (now the Local Government sub-sector). From April 1993 they acquired the same financial and educational freedoms as higher education establishments and are therefore viewed as institutional units. Concerning the classification of these units as market producers or not, the same arguments apply as have been set out above for universities. These bodies are not controlled by government for the same reasons as for universities and they are therefore classified as NPISH from April 1993.

10.217 Since 1989 many *Local Authority schools* have become *grant maintained*. However, following discussions with Eurostat it has been decided that they are still essentially financed and controlled by local authorities and they therefore remain within the local government sub-sector of general government (S.1313).

Rest of the World (S.2)

10.218 This sector is:

> '*a grouping of units without any characteristic functions and resources; it consists of non-resident units insofar as they are engaged in transactions with resident institutional units, or have other economic links with resident units. Its accounts provide an overall view of the economic relationships linking the national economy with the rest of the world.*'

> ESA 1995 paragraph 2.89]

10.219 The rest of the world is not a sector for which complete accounts have to be kept or compiled, although it is often helpful to describe the rest of the world as though it were a sector. The sectors of the national economy (S.11 through S.15) are derived by disaggregating the total economy to obtain more homogeneous groups of resident institutional units which are similar in their economic behaviour, their objectives and roles. However, this is patently not the case for the rest of the world sector (S.2) where the transactions and other flows of all sectors within all non-UK economies are shown only to the extent that they involve units in the UK economy. The subject of the balance of payments between the United Kingdom and the rest of the world forms Chapter 24 in this publication and provides another and more detailed view of this area.

10.220 The rule that states that the accounts for the rest of the world include only transactions carried out between resident institutional units and non-resident units is subject to the following exceptions:

- certain transport services in respect of imported and exported goods, for consistency with the f.o.b. valuation (*see* Table 3.6);
- transactions in foreign assets between residents belonging to different UK sectors are shown in the detailed financial accounts for the rest of the world; although they do not affect the UK's financial position in relation to the rest of the world, they do affect the financial relationships of individual sectors with the rest of the world;
- transactions in the UK's liabilities between non-residents belonging to different geographical zones are shown in the geographical breakdown of the rest of the world accounts. Although these transactions do not affect the UK's overall liability to the rest of the world, they affect its liabilities to different parts of the world.

10.221 For the UK, and other countries to which the ESA applies, this sector is subdivided into:

- the European Union (S.21);
  - the member countries of the EU (S.211);
  - the Institutions of the EU (S.212);
- third countries and international organisations (*see* next paragraph) (S.22).

International organisations

10.222 Some international organisations have all the attributes needed by institutional units. The special characteristics of an 'international organisation' as used here are summarised below.

a) The members of an international organisation are either national states or other international bodies whose members are national states; they therefore obtain their authority either directly from the national states which are their members or indirectly from them via other international organisations.

b) They are entities established by formal political agreements between their members that have the status of international treaties; their existence is recognised by law in their member countries.

c) Since they are established by international agreement, they are accorded sovereign status: that is, they are not subject to the laws or regulations of the country, or countries, in which they are located.

d) International organisations are set up for many purposes, including the following types of activities:

i) the provision of non-market services, of a collective nature, for the benefit of their members;

ii) financial intermediation at an international level – that is, channelling funds between lenders and borrowers in different countries; such an organisation may also act as a central bank to a group of countries.

10.223 Formal agreements concluded by all the member states of an international organisation sometimes have the force of law within those countries. Most international organisations are financed wholly or partly by contributions (transfers) from their member countries, but some may raise funds in other ways – for example by borrowing on financial markets.

Chapter 11

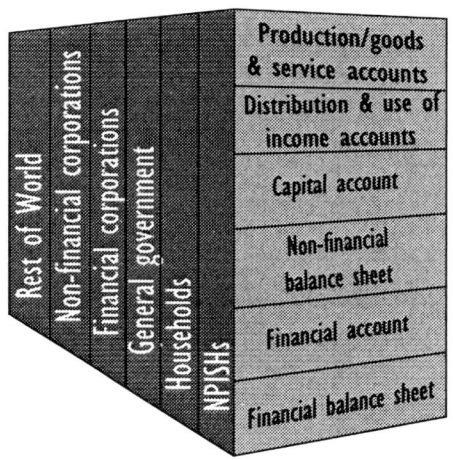

The United Kingdom economic accounts

Chapter 11 The United Kingdom economic accounts

11.1   In the United Kingdom the economic accounts are the accounts of the nation. These accounts, which comprise a great wealth of economic series, are compiled by the Office for National Statistics (ONS) and calculated according to international standards. They record and describe economic activity in the United Kingdom and as such are used to support the formulation and monitoring of economic and social policies.

11.2   A general discussion of the economy and its measurement and a detailed description of the conceptual structure of the accounts, as defined according to the 1995 ESA, has been provided in the first part of this book. The aim of this chapter however, is to provide an overview of the United Kingdom economic accounts and describe the estimation of the main aggregates, in particular gross domestic product (GDP). The subsequent Chapters (12–18) then give a detailed description of the series which comprise the various accounts or groupings of economic units that make up GDP; that is the current accounts, the accumulation accounts and the balance sheets. These are followed by Chapters 19–24 which deal separately with the six sectors of the accounts and the sources and methods used in the estimation of these sector accounts.

Overview of the accounts

11.3   The full system of the UK accounts reveals a highly articulated system which embraces a vast amount of economic information: the production processes revealed in the input-output framework, the sector accounts showing, for example, the income, expenditure, saving and financial transactions and balance sheets of each sector, and estimates of gross domestic product.

11.4   In the United Kingdom the general approach essentially begins with the quarterly economic accounts and the production of a single estimate of GDP with its income, output and expenditure components. The income analysis is available at current prices, expenditure is available at both current and constant prices and output is in constant prices only. Income, capital and financial accounts are also produced for each of the institutional sectors: non-financial corporations, financial corporations, general government, households and (in due course) non profit institutions serving households. The accounts are fully integrated, but with a statistical discrepancy shown for each sector account (which reflects the mismatch between the sector financial surplus or deficit and the identified borrowing or lending in the financial accounts). Accounts are also produced for the Rest of the World sector in respect of its dealings with the United Kingdom. Financial balance sheets are also produced.

11.5   In the weeks after the end of each quarter quarterly economic accounts statistics are produced as follows:

- 3 weeks after the end of the quarter : a preliminary estimate of the growth in the volume of GDP (which reflects the growth in the quantity of output produced rather than an increase in the price level) from the previous quarter, based on output information.

- 8 weeks after the end of the quarter: expenditure, output and income breakdowns of GDP showing the growth in the latest quarter.

- 12 weeks after the end of the quarter: full set of economic accounts for recent quarters, including gross national product and accounts for the institutional sectors.

11.6   ONS aims to publish estimates which are timely, reliable and internally consistent. However, because these objectives can conflict with one another, the earlier, more timely estimates, for example the estimates made after 3 and 8 weeks, are necessarily based on less information than later estimates. Further details on these quarterly estimates are given in paragraphs 108–132 below.

11.7   The quarterly accounts are fully integrated with the annual accounts. Developments in recent years have improved the coverage and quality of quarterly estimates and strengthened this method of production, although some quarterly surveys remain benchmarked on annual surveys (e.g. benchmarking the quarterly estimates of capital expenditure surveys on the annual census of production). However, some series are only available some time after the end of the year to which they refer and quarterly indicators are used to project forward the last annual estimate and so provide quarterly estimates. For example, the Inland Revenue tax based estimates of compensation of employees are projected forward using employment and average earnings estimates.

11.8   Following the end of the calendar year annual economic accounts estimates are published as follows:

- 12 weeks after the end of the year: production of the first full set of economic accounts for the fourth quarter, allowing calculation of the first full set of annual accounts (though earlier estimates of the annual growth rate and estimates of income, output and expenditure will have been available with the earlier estimates for the fourth quarter).

- 1st annual estimate: 6 months after the end of the year, including all of the main annual series with detailed annual breakdowns.

- 2nd annual estimate: 18 months after the end of the year, including summary input-output balances.

11.9   The full annual accounts, first produced six months after the latest year to which they refer, are published each August in the ONS publication *'The United Kingdom national accounts' (The Blue Book)*. Further details on the production and publication of these annual estimates are provided in paragraphs 11.132–11.137.

Contents of Chapter 11

11.10   In order to help readers find their way around this chapter, after the exposition of some basic definitions information has been presented around the production and publication of the successive quarterly and annual estimates. Descriptions of the particular techniques used in the production of the accounts have been given at that point in the process when these techniques are used. For example the balancing of the quarterly output, income and expenditure estimates is given in the section on the income, output and expenditure estimates – published eight weeks after the end of the quarter – when the technique is used. Other topics which are not specific to any particular point in the production cycle but have general relevance (for example seasonal adjustment) are then explained in the sections that follow.

The main sections of the rest of this chapter are:

- *Definitions (in the UK context)*

- *The economic accounts framework in the UK*

- *Sources and methods: GDP at current and constant prices*

- *The economic accounts publication and processes*

- *Current prices, constant prices, index numbers and rebasing*

- *Accuracy and bias*

- *The seasonal adjustment of the accounts*

- *The satellite accounts*

Some Definitions

11.11   Before providing an in-depth description of the framework used in the United Kingdom, and the sources and methods used in the estimation of the UK economic accounts, it is sensible to begin with an explanation of some of the basic concepts and their 'UK specific' definitions, namely:

- the limits of the UK national economy: economic territory, residency and centre of economic interest (see paragraphs 11.12–11.18)

- economic activity: what production is included - the production boundary (see paragraphs 11.19–11.25)

- what price is used to value the products of economic activity (see paragraphs 11.26–11.28)

- estimation or imputation of values for non-monetary transactions (see paragraphs 11.29–11.31)

- the rest of the world: national and domestic (see paragraphs 11.32–11.36)

A full description of the rules of accounting is provided in Chapter 2 (2.21 *et seq*).

The limits of the national economy: economic territory, residence and centre of economic interest

11.12   The economy of the United Kingdom is made up of institutional units (*see* Chapter 10) which have a centre of economic interest on the UK economic territory. These units are known as resident units and it is their transactions which are recorded in the UK economic accounts. The definitions of these terms are given below.

11.13   The *UK economic territory* is made up of:

Great Britain and Northern Ireland (the geographic territory administered by the UK government within which persons, goods, services and capital move freely)

any free zones, including bonded warehouses and factories under UK customs control

the national airspace, UK territorial waters and the UK sector of the continental shelf

but *excludes* the offshore islands - Channel Islands and the Isle of Man - which are not members of the European Union and are therefore not subject to the same fiscal and monetary authorities as the rest of the United Kingdom.

11.14   Within the ESA (paragraphs 2.04–2.07) the definition of economic territory also includes:

territorial enclaves in the rest of the world (like embassies, military bases, scientific stations, information or immigration offices, aid agencies, etc, used by the British government with the formal political agreement of the governments in which these units are located).

but excludes:

any extraterritorial enclaves (i.e. parts of the UK geographic territory - like embassies and US military bases - used by general government agencies of other countries, by the institutions of the European Union or by international organisations under treaties or by agreement).

11.15   The UK complies with this definition by obtaining the relevant information (for inclusion and exclusion from the relevant economic accounting estimates) in an *ad hoc* way from the agencies involved. For example, the Foreign and Commonwealth Office and the Ministry of Defence provide information on the expenditure of UK embassies and bases overseas and, on an annual basis, inquiries are sent to the embassies of foreign governments and international organisations situated in the United Kingdom. The transactions of these units will be reflected within the import and export of services and compensation of employees.

Centre of economic interest and residency

11.16   An institutional unit has a centre of economic interest and is a resident of the UK when, from a location (for example a dwelling, place of production or premises), within the UK economic territory, it engages and intends to continue engaging (indefinitely or for a finite period; one year or more is used as a guideline) in economic activities on a significant scale. It follows that, if a unit carries out transactions on the economic territory of several countries, it has a centre of economic interest in each of them (for example BP will have an interest in many countries where it is involved in the exploration and production of oil and gas). Ownership of land and structures in the UK is enough to qualify the owner to have a centre of interest.

11.17   Within the definition given above, resident units are basically households, legal and social entities, such as corporations and quasi corporations (for example branches of foreign investors), non profit institutions and government. Also included here, however, are so called 'notional residents' (see paragraph 10.15g).

11.18   Travellers, cross border and seasonal workers, crews of ships and aircraft and students studying overseas are all residents of their home country and remain members of their households. However an individual who leaves the UK for a year or more (except students and patients receiving medical treatment) ceases to be a member of a resident household and remains a non-resident on home visits.

Economic activity: what production is included?

11.19 As GDP is defined as the sum of all economic activity taking place in UK territory, having defined the economic territory it is important to be clear about what is defined as economic activity. In its widest sense it could cover all activities resulting in the production of goods or services and so encompass some activities which are very difficult to measure.

11.20 In practice a production boundary is defined, inside which are all the economic activities taken to contribute to economic performance. This economic production may be defined as activity carried out under the control of an institutional unit that uses inputs of labour or capital and goods and services to produce outputs of other goods and services. These activities range from agriculture and production through service-producing activities (for example financial services and hotels and catering) to the provision of health, education, public administration and defence: they are all activities where an output is owned and produced by an institutional unit for which payment or other compensation has to be made to enable a change of ownership. (This omits purely natural processes.)

11.21 Basically the decision whether to include a particular activity in the production boundary takes into account the following:

- does the activity produce a useful output?

- is the product of the activity marketable and so have a market value?

- if the product does not have a meaningful market value can a meaningful market value be assigned? (i.e. can we impute a value)

- would exclusion (or inclusion) of the product of the activity make comparisons between countries or over time more meaningful?

11.22 In practice under the ESA 95 the production boundary can be summarised as follows:

the production of all **goods** whether supplied to other units or retained by the producer for own final consumption or gross capital formation and,

services only as far as they are exchanged in the market and/or generate income for other economic units

11.23 For household this has the result of including the production of goods on own-account (for example the produce of farms consumed by the farmer's own household). However, in practice produce from gardens or allotments has proved impossible to estimate in the United Kingdom so far). Production excludes services for own final consumption (household domestic and personal services like cleaning, cooking, ironing and the care of children and the sick or infirm). Although the production of these services does take a considerable time and effort, the activities are self contained with limited repercussions for the rest of the economy and, as the vast majority of household domestic and personal services are not produced for the market, it is very difficult to value the services in a meaningful way.

11.24 Member states of the European Union will be required to produce current market price gross national product on the previous ESA definition, for use by the European Commission in the calculation of 4th resource contributions to the European budget. This will be a requirement for all member states until the basis of the 4th resource contributions has been renegotiated in ESA95 terms.

11.25   The list below summarises the changes from the previous European system of accounts which will affect the calculation of GNP. It should be remembered that the previous UK system of accounts was not wholly in line with the previous ESA. Differences from the previous UK system are indicated as appropriate in subsequent chapters.

1.   Residence criteria
Different treatment of students, installation of equipment and construction activities

2.   Financial intermediation services indirectly measured (FISIM)
Allocation of FISIM to users - see Annex 1 to Chapter 20.

3.   Insurance
Output of insurance redefined - see Annex 2 to Chapter 20.

4.   Direct investment earnings
Re-invested direct investment earnings to be included in property income.

5.   Interest income
Interest to be recorded as it accrues rather than when it becomes due.

6.   Cultivated natural growth of plants
Output of these to be recorded as they grow (work in progress) rather than when they are harvested.

7.   Computer software and large databases
These to be recorded as intangible fixed assets; ESA79 gave no clear guidance.

8.   Military equipment and vehicles, other than weapons
To be treated as fixed assets rather than consumption.

9.   Work in progress on services
Output of these to be recorded as they are produced (work in progress) rather than when the finished service is delivered to the user.

10.   Mineral exploration expenditures
To be treated as fixed assets rather than intermediate consumption.

11.   Consumption of fixed capital on roads, bridges, etc.
Previously deemed to have an infinite life (and thus no capital consumption), these assets are now to have capital consumption estimated for them.

12.   Government licences and fees
A distinction is now drawn between business licences used as intermediate consumption of producers, and licences acquired by households as consumers.

13.   Valuation of output for own final use and output from voluntary activity
An element of operating surplus to be included in the valuation of own account production; and in the valuation of construction of fixed assets by voluntary activity an estimate of the value of the labour used to be included. Neither of these required by ESA79.

14.   Value threshold for capital goods
Threshold for inclusion of durable goods of small value in gross fixed capital formation raised in real terms from 100 ECU at 1970 prices to 500 ECU at 1995 prices.

15.   Market / non-market criteria
Rules for classifying market or non-market have been changed, with consequences for the valuation of output.

16.   Subsidies
Rules for distinguishing between social benefits and subsidies changed. Also, non-market producers allowed to receive subsidies in certain circumstances.

17.   Entertainment, literary and artistic originals
The production of these to be within the production boundary (part of gross fixed capital formation)

18.   Services associated with 17.
Payments for rights to use these originals to be recorded as purchases and sales of services; previously recorded as property income.

19. Garages

    Garages used by the owner of a dwelling for consumption purposes, even if separate from the dwelling, to be included in the imputed output of dwelling services. ESA79 gave no guidance.

20. Car registration taxes paid by households

    Classified as taxes on products, rather than miscellaneous current transfers.

21. Wages and salaries in kind

    ESA95 includes the provision of sports and recreation facilities here, rather than as intermediate consumption. Also, pay in kind to be recorded at basic prices rather than the sum of the costs of production.

22. Licences for the use of intangible non-produced assets

    Payments and receipts for the use of such assets (patents, copyrights, trade marks etc) to be recorded as consumption and output rather than property income.

23. Stamp taxes

    To be included in taxes on products rather than other taxes on production (producers) or miscellaneous current transfers (consumers).

24. Finance leasing

    To be treated as a form of borrowing. The acquisition of the asset is gross fixed capital formation of the lessee, financed by a loan from the lessor to the lessee.

25. Illegal activity

    To be included where it takes the form of transactions between consenting parties (see Chapter 2)

What price is used to value the products of economic activity?

11.26 In the UK there are a number of different prices used to value inputs and outputs depending on the treatment of taxes and subsidies on products and trade and transport margins. These prices - purchasers' prices (or market prices), basic prices and producers' prices - are looked at in turn below. Although the factor cost valuation (see paragraphs 11.85) is not required under SNA or ESA, ONS will continue to provide figures for GDP at factor for as long as customers continue to find this analysis useful.

11.27 The 'market price', the price agreed and paid by transactors, is the basic reference for the valuation of transactions in the accounts. However the market prices of products will include indirect taxes (for instance VAT) paid to the government and subsidies paid to producers by the government. As a result the producer and user of a product will usually perceive the value of the product differently. This has resulted in two basic distinctions in the valuation of products: output prices received by producers, and prices paid as products are acquired.

Basic prices

11.28 These prices are the preferred method of valuing output as they reflect the amount received by the producer for a unit of goods or services, minus any taxes payable, and plus any subsidy receivable on that unit as a consequence of production or sale (i.e. the cost of production including subsidies). As a result the only taxes included in the price will be taxes on the output process - business rates and vehicle excise duty - which are not specifically levied on the production of a unit of output. (Currently there are no relevant subsidies included in the basic price of any goods or services produced in the UK.) Basic prices exclude transport charges invoiced separately by the producer.

When a valuation at basic prices is not feasible producers' prices may be used where:

Producers' prices

= basic prices *plus* those taxes paid (other than VAT or similar deductible taxes invoiced on the output sold) per unit of output *less* any subsidies received per unit of output.

Purchaser's prices

Essentially these are the prices paid by the purchaser and include transport costs, trade margins and taxes (unless the taxes are deductible by the purchaser), i.e.

Purchasers' price = producers' price **plus** any non deductible VAT or similar tax payable by the purchaser **plus** transport prices paid separately by the purchaser and not included in the producers' price.

Market prices

The term 'purchasers prices' is used in preference to 'market prices' although the terminology 'GDP at market prices' has been retained and equates to 'GDP at purchasers' prices'.

Estimation or imputation of values for non monetary transactions.

11.29   When goods or services are exchanged for money (a monetary transaction) the value of the transaction - the purchaser's price - is directly available. However, for those transactions that do not involve the exchange of money (non-monetary transactions), values have to be assigned. This can be done in two ways:

- estimating or imputing a value by reference to market prices for similar goods and services, or,

- estimating or imputing a value by reference to the costs incurred.

11.30   For example market price can be estimated for the provision of remuneration in kind to employees (e.g. subsidised canteens, provision of a company car) by reference to the cost to the employer; and the payment of inheritance tax through the donation of paintings or other valuables by reference to their market value.

11.31   For other non-monetary transactions when goods are retained for own use, the transaction itself has to be constructed, or imputed, before values can be estimated. For example the provision of owner-occupied housing (by the owner for their own use) and the non trading use of fixed assets owned by the government and by private non-profit-making bodies (for example, school buildings).

The rest of the world: National and Domestic

11.32   *Domestic product* (or income) includes production (or primary incomes generated and distributed) resulting from all activities taking place 'at home' or inside the UK domestic territory. This will *include* production by any foreign owned company in the United Kingdom but *exclude* any income earned by UK residents from production taking place outside the domestic territory. Thus Gross domestic product is also equal to the sum of primary incomes distributed by resident producer units.

11.33   The definition of gross *national* income can be introduced by considering the primary incomes distributed by the resident producer units above. These primary incomes, generated in the production activity of resident producer units, are distributed mostly to other resident institutional units; however part of them may go to non-resident institutional units. For example, when a resident producer unit is owned by an overseas company, some of the primary incomes generated by the producer unit are likely to be paid overseas. Symmetrically, some primary incomes generated in the rest of the world may go to resident units. Thus when looking at the income of the nation it is necessary to exclude that part of resident producers' primary income paid overseas, but include the primary incomes generated overseas but paid to resident units; i.e.

Gross *domestic* product (or income)

less

primary incomes payable to non-resident units

plus

primary incomes receivable from the rest of the world

equals

Gross *national* income

11.34 Thus gross national income (GNI) at market prices is the sum of gross primary incomes receivable by resident institutional units/sectors.

11.35 It is worth noting that although GNI at market prices was previously called gross national product (GNP), in contrast to gross national product, gross national income is not a concept of value added. GNI is an income (primary income) concept.

11.36 National income includes income earned by residents of the national territory, remitted (or deemed to be remitted in the case of direct investment) to the national territory, no matter where the income is earned.

Gross domestic product: the concept of net and gross

11.37 The term *gross* refers to the fact that when measuring domestic production we have not allowed for an important phenomenon: capital consumption or depreciation. Capital goods are different from the materials and fuels used up in the production process because they are *not* used up in the period of account but are instrumental in allowing that process to take place. However, over time capital goods do wear out or become obsolete and in this sense gross domestic product does not give a true picture of value added in the economy. In other words, in calculating value added as the difference between output and costs we should include as a current cost that part of the capital goods used up in the production process, that is, the depreciation of the capital assets. *Net* concepts are net of this capital depreciation, for example:

Gross domestic product

minus

consumption of fixed capital

equals

Net domestic product

11.38 However, because of the difficulties in obtaining reliable and timely estimates of the consumption of fixed capital (depreciation), gross domestic production remains the most widely used measure of economic activity. Capital consumption and its estimation are discussed in more detail in Chapters 15 and 16.

The effects of inflation: How are changes in real output estimated?

11.39 When looking at the change in the economy over time the main concern is usually whether more goods and services are actually being produced now than at some time in the past. How far are the changes "real" and how far are they the result of inflation? This is not as easy as making a simple comparison of the current price figures. Most economic accounting data are shown in current prices which combine the effects of the changes in both prices and quantities.

11.40 However, comparisons over time can be more useful analytically when the separate effects of movements in price and volume can be separated. The expenditure approach and the output approach to GDP (discussed below) provide estimates at constant prices, that is prices which have been revalued to a constant price level. Current prices, constant prices and index numbers are all discussed further in paragraphs 11.160–178.

Figure 11.1 Synoptic presentation of the accounts, balancing items and main aggregates

Accounts			Balancing items	Main aggregates 1/
Full sequence of accounts for institutional sectors				
Current accounts	I. Production account 2/		B.1 Value added	Domestic product (GDP/NDP)
	II. Distribution and use of income accounts	II.1. Primary distribution of income accounts — II.1.1. Generation of income account 2/	B.2 Operating surplus; B.3 Mixed income	
		II.1.2. Allocation of primary income account	B.5 Balance of primary incomes	National income (GNI,NNI)
		II.2. Secondary distribution of income account	B.6 Disposable income	National disposable income
		II.3. Redistribution of income in kind account	B.7 Adjusted disposable income	
		II.4. Use of income account		
		II.4.1. Use of disposable income account	B.8 Saving	National saving
		II.4.2. Use of adjusted disposable income account		
Accumulation accounts	III. Accumulation accounts	III.1. Capital account	B.10.1 (Changes in net worth, due to saving and capital transfers) 3/; B.9 Net lending/Net borrowing	
		III.2. Financial account	B.9 Net lending/Net borrowing	
Balance sheets	IV. Financial balance sheets	IV.3. Closing balance sheet	B.90 Financial net worth	
Transaction accounts				
	0. Goods and services account			National expenditure
Rest of the world account (external transactions account)				
Current accounts	V. Rest of the world account	V.I. External account of goods and services	B.11 External balance of goods and services	External balance of goods and services
		V.II. External account of primary income and current transfers	B.12 Current external balance	Current external balance
Accumulation accounts		V.III. External accumulation accounts — V.III.1. Capital account	B.10.1 (Changes in net worth due to current external balance and capital tranfers) 3/; B.9 Net lending/Net borrowing	
		V.III.2. Financial account	B.9 Net lending/Net borrowing	Net lending/Net borrowing of the nation
Balance sheets	V.IV. External assets and liabilities account	V.IV.3. Closing balance sheet	B.90 Net worth	
Balance sheets			B.10 Changes in net worth; B.90 Net worth	

1/ Most balancing items and aggregates may be calculated gross or net.
2/ Applies also to industries.
3/ Not a balancing item, but plays a similar role.

The basic framework in the UK

What is an account? What is its purpose?

11.41   As discussed in Chapter 2 an account records and displays all of the flows and stocks for a given aspect of economic life. The sum of resources is equal to the sum of uses with a *balancing item* to ensure this equality. Normally the balancing item will be an economic measure which is itself of interest.

11.42   By employing a system of economic accounts we can build up accounts for different areas of the economy which highlight, for example, production, income and financial transactions. In many cases these accounts can be elaborated and set out for different institutional units and groups of units (or sectors). Usually a balancing item has to be introduced between the total of assets and total liabilities of these units or sectors and, when summed across the whole economy, these balancing items constitute significant aggregates (for example GDP).

The integrated economic accounts

11.43   The integrated economic accounts of the UK provide an overall view of the economy and sit at the centre of the accounting framework. The sequence of the integrated economic accounts is shown in Figure 11.1 below. Figure 11.1 presents a summary view of the accounts, balancing items and main aggregates and shows how they are expressed.

11.44   The accounts are grouped into four main categories: goods and services account, current accounts, accumulation accounts and balance sheets.

11.45   *Current accounts* deal with production, distribution of income and use of income. The *accumulation accounts* cover changes in assets and liabilities and changes in net worth (the difference between assets and liabilities). *Balance sheets* present stocks of assets and liabilities and net worth.

The goods and services account

11.46   The accounting structure is uniform throughout the system and applies to all units in the economy, whether they are institutional units, sub sectors, sectors or the whole economy. However, some accounts (or transactions) may not be relevant for some sectors and these details will be covered in subsequent chapters.

11.47   The goods and services account is a transactions account which balances total resources, from output and imports, against the uses of these resources in consumption, investment, inventories and exports. This account records the flow of goods and services, in contrast to the production account which records the counterpart monetary flows. Because the resources are simply balanced with the uses, there is no balancing item. The goods and services account is discussed in detail in Chapters 3 and 12.

Current accounts

Production account (Account I)

11.48   The production account displays the transactions involved in the production of goods and services. In this case the balancing item is *value added*. For the nation's accounts the balancing item, the sum of value added, is gross domestic product (GDP) or net domestic product when measured net of capital consumption. The production accounts are discussed in detail in Chapters 4 and 13.

Distribution and use of income accounts (Account II)

11.49   The distribution and use of income account is a very important account for any country and shows the distribution of current income (in this case value added or GDP) carried forward from the production account, and has as its balancing item, saving, which is the difference between income (disposable income) and expenditure (or final consumption). There are three sub-accounts which break down the distribution of income into the primary distribution of income, the secondary distribution of income and the redistribution of income in kind. The primary distribution account is divided into two sub accounts (the generation and allocation of primary incomes), but the ESA's further breakdown of the allocation of primary income account into an entrepreneurial income account and an allocation of other primary income account has not been adopted in the United Kingdom. A further two sub accounts - the use of disposable income and the use of adjusted disposable income - look at the use of income for either consumption or saving.

11.50   These accounts are examined in detail in Chapters 5 and 14.

11.51   Aggregated across the whole economy: the balance of the primary distribution of income provides *national income* (which can be measured net or gross), the balance of the secondary distribution of income provides *national disposable income*, and the balance of the use of income accounts provides *national saving*. These are shown in Figure 11.1 above

The accumulation accounts

11.52   The accumulation accounts cover all changes in assets, liabilities and net worth (the difference for any sector between its assets and liabilities). The accounts are structured to allow various types of changes in these elements to be distinguished.

11.53   The first group of accounts covers transactions which would correspond to all changes in assets/ liabilities and net worth which result from transactions (eg savings and voluntary transfers of wealth (capital transfers). These accounts are the *capital account* and *financial account* which are distinguished in order to show the balancing item net lending/borrowing. The second group of accounts relate to changes in assets, liabilities and net worth due to other factors (for example the discovery or re-evaluation of mineral reserves, or the reclassification of a body from one sector to another). This second group, the 'other changes in assets accounts' has not been implemented in the United Kingdom.

Capital account (Account III.1)

11.54   The capital account concerns the acquisition of non-financial assets, (some of which will be income creating and others which are wealth only) such as fixed assets or inventories, financed out of saving and capital transfers involving the redistribution of wealth. Capital transfers include, for example, capital grants from private corporations to public corporations (eg private sector contributions to the extension to the Jubilee line). This account shows how saving finances investment in the economy. In addition to new gross fixed capital formation and changes in inventories, it shows the redistribution of capital assets between sectors of the economy and the rest of the world. The balance on the capital account if negative, is designated net borrowing, and measures the net amount a unit or sector is obliged to borrow from others; if positive the balance is described as net lending, the amount the United Kingdom or a sector has available to finance others. This balance is also referred to as the financial surplus or deficit and the net aggregate for the five sectors of the economy equals net lending/borrowing from the rest of the world.

Financial account (Account III.11)

11.55   The financial account shows how net lending and borrowing are achieved by transactions in financial instruments. The net acquisitions of financial assets are shown separately from the net incurring of liabilities. The balancing item is again net lending or borrowing.

11.56 In principle net lending or borrowing in the capital account should be identical to net lending or borrowing on the financial account. However in practice, because of errors and omissions, this identity can be very difficult to achieve for the sectors and the economy as a whole. The difference is known as the *statistical* discrepancy (previously known as the balancing item).

The balance sheets (Account IV)

11.57 The second group of accounts within the accumulation accounts completes the full set of accounts in the system. These include the balance sheets and a reconciliation of the changes that have brought about the change in net worth between the beginning and the end of the accounting period.

11.58 The opening and closing balance sheets show how total holdings of assets by the UK or its sectors match total liabilities and net worth (the balancing item). In detailed presentations of the balance sheets the various types of asset and liability can be shown. Changes between the opening and closing balance sheets for each group of assets and liabilities result from transactions and other flows recorded in the accumulation accounts, reclassifications and revaluations, and changes in net worth equal changes in assets less changes in liabilities.

11.59 However, in the United Kingdom at present only the end-period financial balance sheets are published. Neither does the United Kingdom at present publish the *Other changes in assets* accounts, which reconcile the transactions in assets and liabilities with the changes between balance sheets.

Rest of the world account (Account V)

11.60 This account covers the transactions between resident and non-resident institutional units and the related stocks of assets and liabilities. The 'rest of the world' plays a similar role to an institutional sector and the account is written from the point of view of the rest of the world. This account is discussed in detail in Chapter 24.

Satellite accounts

11.61 Satellite accounts are accounts which involve areas or activities not dealt with within the central framework above, either because they add unnecessary detail to an already complex system, or because they actually conflict with the conceptual framework. The UK has begun work on a number of satellite accounts and one such account - the UK environmental account - aims to give an industrially disaggregated assessment of pressures on the environment from economic activity. This development work includes the publication of *UK Environmental Accounts 1998*.

11.62 Corresponding to the components of the model described above, the integrated economic accounts of the United Kingdom fall into three major parts:

- GDP and its components;
- the sector accounts;
- the balance of payments.

GDP and its components and the sector accounts are dealt with as follows below:

Gross domestic product at current and constant prices

The three approaches and the need for balancing

11.63 An understanding of this system in the United Kingdom can begin with a brief discussion of gross domestic product (GDP). GDP is arguably the most important aggregate or summary indicator for purposes of economic analysis and comparisons over time. It measures total domestic activity and can be defined in three different ways:

- GDP is the sum of gross value added of the institutional sectors or the industries plus taxes and less subsidies on products (which are not allocated to sectors and industries). It is also the balancing item in the total economy production account.

- GDP is the sum of final uses of goods and services by resident institutional units (actual final consumption and gross capital formation), plus exports and minus imports of goods and services.

- GDP is the sum of uses in the total economy generation of income account (compensation of employees, taxes on production and imports less subsidies, gross operating surplus and mixed income of the total economy).

11.64 This is also how we approach the task of estimating GDP. By using three different methods which, as far as possible, use independent sources of information we are not relying solely on one source and so can be more confident of the overall estimation process.

11.65 The resulting estimates however, like all statistical estimates, contain errors and omissions; we obtain the best estimate of GDP (ie the published figure) by reconciling the estimates obtained from all three approaches. On an annual basis this reconciliation is carried out through the construction of the input-output balances for the years when an input output balance is available, and for subsequent periods by carrying forward the level of GDP set by the annual balancing process by using the quarterly movements in output, income and expenditure totals. The annual balancing process is described in more detail in paras 11.138 to 11.159 and the quarterly balancing process is described in paragraphs 11.119 to 11.132.

11.66 For years in which no input-output balance has been struck a statistical discrepancy exists between estimates of the total of expenditure components of GDP at factor cost and the total income components of GDP after the balancing process has been carried out. This statistical discrepancy is made up of two components which are shown in the accounts, namely:

the *statistical discrepancy (expenditure adjustment)*, which is the difference between the sum of the expenditure components and the definitive estimate of GDP, *plus*

the *statistical discrepancy (income adjustment)*, which is the difference between the sum of the income components and the definitive estimate of GDP *(with sign reversed)*.

11.67 As outlined in the framework above, the different approaches to the measurement of GDP provide a breakdown into different component parts and give different perspectives on the data. These approaches, with each of their separate sources, are described in more detail in turn below.

The income approach

11.68   The income approach provides estimates of GDP and its 'income' component parts at current market prices. The sources and methods of this approach are described in detail in Chapter 14.

11.69   As it suggests, the income approach adds up all income earned by resident individuals or corporations in the production of goods and services and is therefore the sum of uses in the generation of income account for the total economy (or alternatively the sum of primary incomes distributed by resident producer units).

11.70   However some types of income are not included - these are transfer payments like unemployment benefit, child benefit or state pensions. Although they do provide individuals with money to spend, the payments are made out of, for example, taxes and national insurance contributions. Transfer payments are a *redistribution* of existing incomes and do not themselves represent any addition to current economic activity. To avoid double counting, these transfer payments and other current transfers (for example taxes on income and wealth) are excluded from the calculation of GDP although they are recorded in the secondary distribution of income account.

11.71   In the UK the income measure of GDP is obtained by summing together:

   gross operating surplus,
   mixed income,
   compensation of employees (wages and salaries and employers' social contributions),
   taxes on production and imports,
   less any subsidies on production.

11.72   Mixed income is effectively the operating surplus of unincorporated enterprises owned by households which implicitly includes remuneration for work done by the owner or other members of the household. This remuneration cannot be identified separately from the return to the owner as entrepreneur.

11.73   As most of these incomes are subject to tax the figures are usually obtained from data collected for tax purposes by the Inland Revenue. However, because there is some delay in providing good quality estimates by this method, other methods are used to provide initial estimates.

11.74   Although in the old system a 'stock appreciation adjustment' (as it was known) was required to remove the effects of holding gains on inventories and fixed assets resulting from revaluation, this adjustment is no longer required. Under ESA95 the operating surplus and mixed income are measures of profit that exclude any holding gains. (Holding gains result to the extent that the value of assets has increased as a result of an increase in their prices rather than from production.)

11.75   Where no annual input-output balance is struck (see paragraph 11.67) a *statistical discrepancy* or *income adjustment* is used to bring the income measure and expenditure measure into line. (As with all estimates there will be some errors and omissions in both measures. We adjust to give one 'best' estimate).

11.76   It should also be noted that within each of the figures for compensation of employees, operating surplus and mixed income, we have included an evasion adjustment. This adjustment is added to the initial estimate of income to allow for income earned but not declared to the Inland Revenue. (Further information can be found in Chapter 14.)

11.77 Although the income approach cannot be used to calculate constant price estimates directly (because it is not possible to separate income components into prices and quantities in the same way as for goods and services) some estimates are obtained indirectly. The expenditure-based *GDP deflator at factor cost* (also known as the *index of total home costs*) is used to deflate the current factor cost estimates to provide a constant price version of the total income component of GDP.

11.78 Data on the income components can be found in Blue Book table 1.2.

The expenditure approach

11.79 The expenditure approach measures total expenditure on finished or final goods and services produced in the domestic economy or, alternatively, the sum of final uses of goods and services by resident institutional units less the value of imports of goods and services.

11.80 The total is obtained from the sum of final consumption expenditure by households, government and non profit institutions serving households on goods and services, gross capital formation (capital expenditure on fixed and intangible assets, changes in inventories and acquisitions *less* disposals of valuables), and net exports of goods and services. This approach can be represented by the following equation:

$$GDP_{mp} = C + G + I + X - M$$

Where: C = final consumption expenditure by households and NPISH sectors, G = government consumption expenditure, I = investment or gross capital formation, X = exports and M = imports.

11.81 These categories are estimated from a wide variety of sources including expenditure surveys, the government's internal accounting system, surveys of traders and the administrative documents used in the importing and exporting of some goods.

11.82 To avoid double counting in this approach it is important to classify consumption expenditures as either final or intermediate. *Final consumption* involves the consumption of goods purchased by or for the ultimate consumer or user. These expenditures are final because the goods are no longer part of the economic flow or being traded in the market place. *Intermediate consumption* on the other hand is consumption expenditure on goods which are used or consumed in the production process.

11.83 Exports include all sales to non-residents, and both exports of goods and services have to be regarded as final consumption expenditure, since they are final as far as the UK economy is concerned.

11.84 Imports of goods and services are deducted because although they are included directly or indirectly in final consumption expenditure they are not part of domestic production. What remains is what has been produced in the United Kingdom:- gross domestic product from the expenditure side.

11.85 This approach provides estimates of GDP and its component expenditure parts at *current market (or purchasers') prices, current basic prices* and *current factor cost.*

Where:

GDP at market prices is shown as

$$GDP_{mp} = C + G + I + (X\text{-}M)$$

GDP at basic prices (i.e gross value added) is

$$GDP_{bp} = GDP_{mp} \text{ - taxes on products + subsidies on products}$$

GDP at factor cost

$$GDP_{fc} = GDP_{mp} \text{ - (taxes } less \text{ subsidies on production)}$$

Hence GDP at basic prices includes those taxes on production, such as business rates (NNDR), which are not taxes on products.

The expenditure approach at constant prices

11.86   As well as GDP at current prices the expenditure approach is used to provide information on expenditure at constant prices. In constant price series the transactions are revalued for all years to a fixed price level; that is, at the average prices of a selected year (known as the base year).

11.87   There are therefore two series of GDP at market prices. One series is at current prices (ie the price prevailing in that year) and the other is 'at constant prices'. The constant price series shows the change in GDP *after* the effects of inflation have been removed.

11.88   Because the constant price figures are shown at 1995 prices 1995 is called the base year. (Constant price series are discussed in more detail in paragraphs 11.160 *et seq* below.)

11.89   How do we remove the effects of inflation to obtain this constant price series? Information obtained on price changes for particular goods and services - such as that collected for the retail prices index (RPI) or the producer prices index (PPI) - is used to 'deflate' the current price series.

11.90   Constant price versions of GDP and its main expenditure components are given in Blue Book table 1.4.

The Output approach

11.91   The output approach to the estimation of GDP looks at the contribution to production of each economic unit; that is the value at basic prices of their total output *less* the value of the inputs used up in the production process. The sum of this gross value added, plus taxes and less subsidies on products for all producers, is GDP at market prices, the production account balancing item. The methodology is discussed in detail in Chapter 13, and the following paragraphs give a brief overview. However, the output approach concentrates on the basic price concept, gross value added (GVA).

11.92   GVA measured by this approach is presented in seasonally adjusted index number form at constant prices. (The input-output balance compilation allows the calculation of this measure at current basic prices although this information is only available for those years for which the balances are produced.)

11.93   In theory value added at constant prices should be estimated by double deflation; that is, deflating separately the gross inputs and the gross outputs of each economic unit (valued at constant prices) and then subtracting one from the other. But, because it is hard to get reliable information from companies which would make this calculation possible on a timely basis, double deflation is only used in the estimation of output for the agriculture and electricity industries.

11.94   In practice, for the estimation of the value added of most other economic units in the economy, a simplifying assumption is made: at constant prices, changes from the base year in value added are assumed to be proportional to corresponding changes in the output produced. Movements in the value added by these industries at constant prices are then estimated by the use of output series. For industries whose outputs are goods output can be estimated from the physical quantities of goods produced or from the value of output deflated by an index of price.

11.95   Apart from output which accounts for around 80 per cent of the total of the output measure a number of other kinds of indicator might be used as a proxy for the change in value added. For example, they may be estimated by changes in inputs, where the inputs chosen may be materials used, employment or some combination of these.

11.96   In the short-term it is reasonable to assume that movements in value added can be measured this way. However, changes in the ratio of output and inputs to value added can be caused by many factors: new production processes, new products made and inputs used and changes in inputs from other industries will all occur over time. Aggregated over all industries the impact of these changes will be lessened. In the longer term all indicators are under constant review, with more suitable ones being used as they become available. In addition, the ratio of the proxy series to value added is re-established every five years when the output measure is rebased.

11.97   The estimate of value added for all industries (GDP) is finally obtained by combining or 'weighting together' the estimates for each industrial sector according to its relative importance (as established in the input-output balance) in the base year.

Other aggregates

Gross national disposable income and real national disposable income

11.98   In the discussions so far we have yet to consider the measure which represents the total *disposable income* of the country's residents.

11.99   Gross national income (GNI) represents the *total income* of the country's residents and is derived from GDP by adding net employment income and net property income from the rest of the world. However there are two other areas which affect the country's residents' command over resources:

• There are flows into and out of the country which are not concerned with economic production. These are current transfers from abroad and current transfers paid abroad. They include transactions with the European Union, overseas aid and private gifts. An estimate of gross national disposable income (GNDI) is reached by adjusting GNI by the net amount of net income received. GNI and GNDI are shown in Blue Book table 1.1.

• Second, disposable income is affected by the terms of trade effect. Some of the expenditure by UK residents is on imported goods and services; some of the income earned by residents is from exports of goods and services. If UK export prices fell relative to the price of imports then the terms of trade effect would move against the UK; that is, residents would have to sell more exports to be able to continue to buy the same amount of imports. The purchasing power of UK residents would be diminished to this extent. Similarly, if the UK export prices rose relative to prices of imports then the effect would be opposite: the purchasing power of residents would rise. An adjustment is made specifically for the terms of trade effect in calculating real national disposable income (RNDI), which is the constant price version of GNDI, also shown in Blue Book Table 1.1.

The sector accounts

11.100 The division of the economy into institutional sectors and the presentation of the relationships and transactions between the different sectors is a primary feature of the SNA and the UK economic accounting framework. This framework - the current accounts, accumulation accounts and balance sheets -is used to record the activities of particular groups of institutions or people in the economy, showing how income is distributed (and redistributed), and how savings are used to add to wealth, in a way in which the aggregates statistics of national income cannot. Indeed this is the framework within which the UK economic accounts are presented in the Blue Book and elsewhere.

11.101 A full description of the conceptual definition of the sector accounts is given in Chapter 10. The UK approach and the data sources of the individual sectors and subsectors are described in detail chapters 19–24. A brief description of the approach is provided below.

The framework

11.102 As can be seen in table 1.8 of the Blue Book the UK sector accounts can be used to show the economic accounting framework in considerable detail by elaborating the accounts on three different dimensions:

the institutional sectors,
the types of transactions and
the national and sector balance sheets.

The institutional sectors

11.103 The first dimension involves the breakdown of the current account into institutional sectors grouped broadly according to their roles in the economy. Examples of these roles are : income distribution, income redistribution, private consumption, collective consumption, investment, financial intermediation etc. Most units have more than one role but a natural classification is to distinguish between corporations, government and households. The rest of the world sector is also identified as having a role although it is obviously not part of the domestic economy.

11.104 A summary of the UK institutional sectors is provided below. Full definitions of each sector can be found in the relevant chapter.

The types of transaction

11.105 The second dimension is that of the type of transaction which relates to the particular account within which the transaction appears. These can be grouped broadly according to purpose, whether current, capital or financial.

The balance sheets

11.106 To complete the full set of accounts, the system includes balance sheets and a reconciliation of the changes that have brought about the change between the beginning and the end of the period.

11.107 In theory the net lending or borrowing from the capital account for each sector should equal the net borrowing or lending from the financial account. In practice because of errors and omissions in the accounts, a balance is rarely achieved and the difference is known as the *statistical discrepancy*. Across all accounts, when an input-output balance is available, these discrepancies must sum to zero. Consolidating the current and accumulation accounts would provide a balanced account which would look like many of the presentations of commercial accounts.

Balancing the quarterly sector accounts

11.108 On a quarterly basis the UK publishes sector current, capital and financial transactions accounts, and financial balance sheets, which show the distribution and redistribution of income between the sectors of the economy, additions to wealth, and the flows of funds from savers to borrowers. As these estimates are part of the integrated economic accounts data set that includes GDP (the current price income and expenditure components of GDP being data sources for the sector accounts), any imbalances in the sector accounts are part of the evidence considered in the GDP balancing process outlined below. However, the sector accounts themselves must also be subject to balancing.

11.109 When the sector and financial accounts are assembled data suppliers are required to deliver their source data on a specified date. The database of time series first balances each row of the matrix according to the agreed methodology, then aggregates a top to bottom account for each sector. Finally the sector statistical adjustment items, reflecting errors and omissions in the accounts, are calculated.

11.110 The sector statisticians, who each have responsibility for the sector balancing item and the key economic indicators for their sector, all then take part in the balancing process; a process which involves an iterative approach of data deliveries, assembly of the accounts and balancing meetings. The final adjustments to allocations of counterparts and unidentified transactions, or adjustments to sample survey results, are guided by targets for the size of sector statistical adjustment items and the requirement that key economic indicators tell a coherent story.

11.111 In the UK the Balance of Payments account is the same as the Rest of the World account in the sector accounts, so it is also included in the sector accounts balancing process.

Seasonal adjustment of the sector accounts

11.112 Quarterly current and capital accounts for each sector are seasonally adjusted, but there is no balanced seasonally adjusted matrix of financial transactions. Although net lending/borrowing may be seasonal, transactions in individual financial instruments are often not seasonal. This is because of the high level of volatility in financial transactions, and the high level of switches made between financial categories as investors and borrowers change their portfolio preferences. A discussion of the methods used to seasonally adjust the components is given in paragraphs 11.193–202 below.

Publication and processes

11.113 The ONS publishes quarterly and annual economic accounts estimates according to regular pre-released timetables on a variety of media. The quarterly estimates are published primarily in ONS First Releases, in electronic form via the ONS databank and in the UK Economic Accounts. The annual estimates are published mainly in the UK National Accounts (Blue Book) and in electronic form via ONS databank. The quarterly estimates and the techniques used to obtain the estimates, are described in paragraphs 11.119–11.132. The annual estimates, and the techniques used in the estimation process, are described in paragraphs 11.133–11.159.

11.114 For each quarter estimates are published in ONS First Releases and in electronic form as follows:

The *preliminary estimate of the growth in the volume of GDP* from the previous quarter at constant factor cost is published about three weeks after the end of the quarter to which it relates. It is based on the output estimate alone and contains very limited information on components.

Estimates of UK income, output and expenditure are published after about eight weeks. This includes a fuller analysis of the output component. There are also less detailed analysis by income and expenditure components.

A *full set of UK economic accounts* are published after twelve weeks. This includes GDP breakdowns by output, income and expenditure components, a full set of sectoral current, capital and financial accounts and the balance of payments.

The preliminary estimate of GDP (the 3-week estimate)

11.115 The preliminary estimate for gross domestic product provides estimates of the growth in the volume of GDP on the previous quarter at basic prices. This estimate is based on the estimate of the index of output of the production industries (the index of production) for the first two months of the quarter, and the retail sales estimates for the three months of the quarter, together with limited information on the output of the rest of the economy. Although at this stage estimates for most individual industry series are not sufficiently reliable for publication, the preliminary estimate provides a broad indication of the level of growth in quarterly GDP, which will become firmer at later stages in the process. This is mainly because of the absence from this measure of highly volatile and difficult to measure components such as inventories and company profits. Further, output components (for example, the index of production and retail sales) are available earlier than other components and tend to be revised less. Therefore the published quarterly estimates of GDP will match as closely as possible, the movements in the total of the output components.

The UK output, income and expenditure estimate (the 8-week estimate)

11.116 A single estimate of GDP with its income, output and expenditure components is produced approximately eight weeks after the end of each quarter, replacing and revising the preliminary estimate described above. The breakdown of GDP by categories of income and expenditure are published at current prices and output is published at constant basic prices. At this stage this quarterly GDP estimate is improved by the addition of, for example,: the index of production for the third month of the quarter, new information from inventories and capital expenditure inquiries, motor trades and customs data, improved data on the construction industry and a range of quarterly turnover inquiries in service industries.

11.117 As with the annual GDP estimate, the production of a single quarterly estimate requires the balancing of the information from the different approaches. However, because the detailed information which feeds into the I-O process is not available at this stage a different approach has to be taken.

The quarterly balancing of GDP

11.118 The aim of the quarterly balancing and adjustment process is to reduce inconsistencies in the accounts and to come to a firm view on movements in key aggregates. The published accounts should show all three approaches with similar movements and levels with credible explanations for movements in components. ONS judgement is that in general in the short term the output approach gives the best estimate of growth in GDP. The balancing process itself consists of three basic stages:

- scrutiny of the initial estimates

- judgemental adjustment to the estimates and,

- alignment adjustments.

These are looked at in turn below.

Scrutiny of the initial estimates

11.119 Each GDP component is the responsibility of a compiler who processes and validates the basic data. At fixed times in the quarterly cycle the compilers provide their best estimates on the basis of their source data and a balance showing the aggregates from the different GDP approaches is struck. When examined the resulting overall picture will typically show income and expenditure figures with a different profile from the output figures. Different levels of output may also emerge from each approach.

11.120 A period of validation and scrutiny then follows when compilers meet and test the plausibility of estimates and the coherence of the estimates across the accounts. Additional information collected by outside organisations is also used to provide an alternative to ONS's own statistics. This information may support existing estimates or provide a basis for adjustments. Vernon (1994) describes this process in full.

11.121 Supply side estimates - estimates of demand for commodities, based on components of output and international trade derived using the latest input-output framework - are also compiled and used to help identify inconsistencies in the accounts at low levels of aggregation. They provide alternative sources of estimates for some investment components, supplementing quarterly survey data, and help to reconcile estimates of output and expenditure. Lynch and Caplan (1992) describes this process.

Judgemental adjustments

11.122 After scrutiny of the initial estimates there may still be discrepancies between income, output and expenditure components. A decision is taken on the movement in GDP which will be published; In particular this decision takes into account movements in output components, but movements in the other components, and any other available information, are also considered.

11.123 Judgemental adjustments may then be made to component data so that aggregates match the movement in GDP. These adjustments are made to a variety of components and are within the error range of the components. They should also not remove the integrity of the individual component series.

Alignment adjustments and the calculation of GDP

11.124 After the scrutiny and adjustment process the movements in income and expenditure are unlikely to match those of output. The final balancing step is the incorporation of the mechanically calculated alignment adjustments which will sum to zero over a calender year. These adjustments smooth the quarterly paths of income and expenditure totals so that they match, as closely as possible, the movement in output without altering annual totals. In the expenditure analysis, the adjustments are allocated to changes in inventories and, within the income analysis, to the operating surplus of private non financial corporations, as these areas are considered to have the widest error margins.

11.125 The alignment adjustments are published in the quarterly economic accounts release described below and in *UK Economic Accounts*. More information about how the alignment adjustments are calculated is given in an article by Ted Snowdon in the November 1997 *Economic Trends*.

11.126 The results of the quarterly balancing process provide a solution to the discrepancies in the different measures of GDP by putting forward the most coherent estimates that arise from the integrated accounts.

The fully integrated quarterly accounts estimate (the 12-week estimate)

11.127 Some 12 weeks after the end of each quarter the ONS produces a full set of quarterly economic accounts, revising and expanding the information made available in the earlier estimate as well as revising estimates for earlier quarters in the current year.

11.128 By this stage in the estimation process the full final employment figures are usually available. These feed into both the income and to a lesser extent the output estimates of GDP. The quarterly profits inquiry has supplied its final balance and improved data is available for UK oil and gas production along with revised estimates for inventories and capital expenditure. In addition new data from the family expenditure survey, national food survey and the balance of payments are amongst some of those sources that become available for the first time at this stage.

11.129 The expenditure analysis is available at both current and constant prices. Output is in constant prices only and income at current prices. Current and accumulation accounts are also produced for each of the institutional sectors: non-financial corporations, financial corporations, general government, households and non profit institutions serving households. The accounts are fully integrated, but with a statistical discrepancy (previously known as a balancing item) shown for each sector account. This reflects the mismatch between the sector financial surplus or deficit and the identified borrowing or lending in the financial accounts.

11.130 As with the GDP estimate produced earlier in the quarter, the fully integrated quarterly accounts estimate is balanced using the process described above. However, in addition, because of the integrated nature of this estimate, any imbalance in the sector accounts has to be part of the evidence considered in this process. (The balancing of the sector accounts is detailed in paragraphs 11.108–11.112 above).

11.131 In the presentation of quarterly GDP and current and capital account estimates for sectors, the emphasis is placed on the seasonally adjusted series. Seasonal adjustment is done at component (or sub component) level and seasonally adjusted aggregates are calculated as the sum of seasonally adjusted components.) The seasonal adjustment of the economic accounts is discussed in more detail in paras 11.191–11.202 below.)

11.132 All of the quarterly accounts are fully integrated with the annual accounts. This means that as well as quarterly estimates always adding up to annual totals, for many series the annual totals are derived as the sum of the quarters. The annual accounts are discussed further in the paragraphs below.

Annual estimates

11.133 Annual estimates of the UK domestic and national product, income and expenditure are published each August in the ONS publication *The United Kingdom national accounts* (The Blue Book). This annual publication covers the latest 10-year span with summary tables providing extended cover of the preceding ten years on a consistent basis.

11.134 The series published in the Blue Book includes: GDP at current (market) prices, basic prices and factor cost; by type of expenditure, by industry and by sector of employment and type of income; GDP at constant prices by industry of output, and the GDP deflator; gross national income (GNI), gross national disposable income (GNDI), and real national disposable income (RNDI). UK summary accounts (the goods and services accounts, current accounts, accumulation accounts and capital accounts) are all published along with the relevant current, accumulation and capital accounts for the sectors and appropriate sub sectors of the economy (including the rest of the world).

11.135 *The UK Balance of Payments* (The Pink Book) includes data for the current account, the capital account, the financial account and the International Investment Position. The publication contains detailed tables showing data for the five main components of the current account; trade in goods, trade in services, compensation of employees, investment income and current transfers. The most important economic indicator published in the Pink Book is the current balance which shows the overall balance on the current account. The Pink Book also contains detailed methodological notes.

11.136 The annual estimates prepared for the Blue and Pink Books incorporate the results of the annual inquiries which become available in the first part of the year, although estimates for the latest year are still based largely on quarterly information. During the compilation process it is therefore likely that revisions will be necessary for several recent years and so estimates for the last 3 complete calendar years are reassessed. The process of reassessing these estimates involves the preparation of input-output tables. This input-output approach requires the amalgamation of all of the available information on inputs, outputs, income and expenditure and, because of the amount of information required, takes time to prepare. Similarly the production of the consolidated sector and financial accounts requires the preparation of top-to-bottom sector and sub sector accounts to identify discrepancies in the estimates relating to each sector. Thus, because of the thorough and detailed nature of this estimation process, estimates for earlier years are not revisited unless there are strong reasons for doing so.

GDP and the balancing of the annual accounts

11.137 As discussed above there are three different approaches to the estimation of current price GDP in the UK: the income approach, the expenditure approach and the output approach. In theory the three different approaches should produce the same result. However, the different approaches are based on different survey and administrative data sources, and each produces estimates which, like all statistical estimates, are subject to errors and omissions. A definitive GDP estimate can only emerge after a process of balancing and adjustment. ONS believes that the most reliable 'definitive' estimate of the current price level of GDP is derived using the annual input-output framework. Thus for the years when an input-output balance is available, GDP is set at the level derived from the balance. For subsequent periods this level is carried forward using movements in income expenditure and output totals. (This quarterly balancing process is described in paragraphs 11.119–11.132).

The input-output framework

11.138 The economic accounts are mainly concerned with the composition and value of goods and services entering into final demand and the incomes generated in the production processes. The UK supply and use balances, however, are concerned with the intermediate transactions which form inputs into these processes. The I-O analyses are constructed to show a balanced and complete picture of the flows of products in the economy and illustrate the relationships between producers and consumers of goods and services. In addition they show the interdependence between industries: what industries purchase from one another in order to produce their own output.

11.139 On an annual basis, I-O supply and use balances are used to achieve consistency in the economic accounts aggregates by linking the components of value added, output and final demand. Because the income, output and expenditure measures of GDP can all be calculated from the I-O supply and use balances, a single estimate of GDP can be derived by balancing the supply and demand for goods and services and reconciling them with the corresponding value added estimates. For the years 1989 to 1996 the balancing process has been used to set the level of current price GDP, and has disposed of the need for any statistical discrepancies.

11.140 The I-O process, which produces balances annually, has been speeded up considerably over the last few years and can now produce the first balance for a year just eighteen months after the end of that year. These full I-O supply and use balances, consistent with the national accounts Blue Book, are then published annually as separate publications with summary information provided in the Blue Book itself.

Some background on the structure of the balances

11.141 The I-O balance is based on a framework which incorporates estimates of industry outputs, inputs and value added. The balance is composed of two matrices: the 'Supply' matrix and the 'Combined use' matrix, each of which breaks down and balances one hundred and twenty-three different industries and products at purchasers' prices. Further details on the matrices are given in Chapter 13. A full description of the present methodology is given in the *UK Input-Output Balances Methodological Guide* (ONS, 1997). A new Guide will be produced showing the new ESA methodology. The following paragraphs draw mainly from the existing Guide.

Supply matrix

11.142 At a very aggregate level the supply matrix can be represented as follows:

	Sales by industry	**Imports of goods and services**	**Distributors' trading margins**	**Taxes *less* subsidies on products**
Sales by product				

11.143 The *make* matrix shows estimates of domestic industries' gross output, (gross sales adjusted for changes in inventories of work in progress and finished goods) compiled at basic prices. However, for the balancing process the estimates of gross domestic output are required at purchasers' prices, i.e. those actually paid by the purchasers to take delivery of the goods, excluding any deductible VAT.

11.144 To convert the estimates from output valued at basic prices to output at purchasers' prices requires the addition of:

- the value of imports of goods and services
- distributors' trading margins
- taxes (*less* subsidies) on products

Adding these series to the Make matrix gives estimates of the supply of products at purchasers' prices.

Combined use matrix

11.145 The Combined use matrix reveals the input structure of each industry in terms of domestic and imported goods and services. It shows the product composition of final demand and, for each industry, the intermediate purchases.

11.146 The product purchases are represented in the rows of the balance while purchases by industries, and final demands, are represented in the columns. At a very aggregate level the matrix can be considered in four parts as shown below.

	Industry purchase	*Final demands*
Product sales	**Shows purchases by industry to produce their output (i.e. intermediate consumption)**	**Shows final demand categories (e.g. final consumption) and the values of products going to these categories**
Primary inputs	**Shows the factor incomes of each industry, sales by final demand and taxes on production *less* subsidies (excluding product-specific taxes)**	

The body of the matrix, which represents product sales, is at purchasers' prices and so already reflects the product-specific taxes added to the supply matrix.

11.147 The I-O balance is effectively achieved when:

> For industries,
> *inputs* (from the Combined use matrix)
> *equal*
> *outputs* (from the Supply matrix).

> For products
> *supply* (from the Supply matrix)
> *equals*
> *demand* (from the Combined use matrix).

That is, when the data from the income, expenditure and output approaches used to fill the matrices all produce the same estimate of current price GDP.

The balancing process

11.148 The balancing process is carried out over a number of months and involves, the I-O team, and the compilers who feed data directly into the process from ONS surveys or through the economic accounts compilation process.

Initial estimates

11.149 Once the initial data estimates have been gathered estimates of the components of supply and demand for products are prepared, together with the estimates of industry outputs and inputs (thus value added). The resulting output-based estimates of current price value added are then compared with the expenditure and income measures and the checks and analysis which follow extend the validation checks which will already have been carried out on the initial data estimates. The investigations which follow often lead to the revision and redelivery of data.

11.150 In parallel with this work alternative estimates of value added for each of the one hundred and twenty-three industries are prepared using income-based data.

11.151 The coherence of these initial estimates is then assessed by:

- comparisons of value added for each industry using the income and output-based approaches, and

- comparisons of the components of supply and demand for each type of product (which effectively compare the output and expenditure approaches).

In addition a variety of time series, for example, growth rates and the ratio of value added to gross output, are compiled to aid the assessment.

11.152 At this stage the resulting income, output and expenditure aggregates will typically show different profiles over time.

Revised estimates

11.153 To obtain the revised estimates an iterative process begins to reconcile:

- the income and output-based estimates of industry value added and,

- the supply and demand for each product.

11.154 These estimates are scrutinised and validated and checked for their plausibility and coherence across all industries and products. Consistency and coherence over time are also important and the impact of revisions to earlier years and the quality of the relative data sources are also taken into account. In addition when necessary other sources, for example ONS survey data and company reports and accounts, are used to inform the investigation of particular areas. Discussions follow between the I-O team and data compilers and any issues are resolved.

Final estimates

11.155 As final estimates are received from data compilers the steps of assessment and scrutiny, comparison and reconciliation continue. For the time series under consideration the quality of source data, revisions performance and any specific estimation problems are taken into account. Any changes to estimates are agreed and the inconsistencies between supply and demand, and between output and income-based value added, are continually reduced. This process continues until convergence between the aggregate totals is achieved.

11.156 The single best estimate of GDP which emerges will reflect the relative merits of the output, income and expenditure estimates at the aggregate level. It will also have been assessed after consideration of the effect on current and constant price expenditure growth rates, the impact on the expenditure deflator, and the relation between current and constant price value added.

11.157 Once this GDP estimate has been fine tuned and agreed by all concerned the industry value added and the value added weights are fixed after a full reconciliation of the income-based components with the output-based estimate. Product supply and demand will still differ at this stage because of the lack of detailed source data on, for example, distributors trading margins and the allocation of other services provided by manufacturers, but adjustments are made until a balance is achieved. The Combined use matrix is then fully balanced by adjusting the intermediate purchases within the predetermined column and row totals.

11.158 This final step in the balancing process is to apply the r.A.s. method to the intermediate section of the Combined Use matrix. This process will adjust the intermediate purchases in line with pre-determined row and column totals, resulting in a fully balanced table. The term r.A.s. refers to an iterative mathematical process, where A is the coefficient form of the intermediate section of the Combined Use matrix. A is pre-multiplied by a diagonal matrix, with the vector r of replacement factors forming the diagonal, and post-multiplied by a diagonal matrix with the substitution vector s forming the diagonal. A single iteration applies the above process for each row and then for each column. After each iteration the replacement factors are changed appropriately and repeated until a desired balance has been achieved. The end result of this process is that supply equals demand for each product.

11.159 The end result is a full I-O balance where input equals output and supply equals demand for all one hundred and twenty-three product and industry groups.

Current prices, constant prices and index numbers

11.160 Over time changes in current price GDP show changes in the monetary value of the components of GDP and, as these changes in value can reflect changes in both price and volume, it difficult to establish how much of an increase in the series is due either to increased activity in the economy or to an increase in the price level. As a result, when looking at the real growth in the economy over time, it is useful to look at volume (or constant price) estimates of gross domestic product.

11.161 In constant price series, for all years the transactions are revalued to a constant price level using the average prices of a selected year, known as the base year. In most cases the revaluation (known as deflation) is carried out by using price indices - such as component series of the Retail prices index or Producer price index - to deflate current price series at a detailed level of disaggregation.

11.162 Some constant price series are expressed as index numbers in which the constant price series are simply scaled proportionately to a value of 100 in the base year. These constant price index numbers are volume indices. They are of the 'base weighted' or 'Laspeyres' form. (See Chapter 2.) Aggregate price indices are of the Paasche or current-weighted form. They are generally calculated indirectly by dividing the current price value by the corresponding constant price value and multiplying by 100. Examples are the GDP deflator and the consumers' expenditure deflator.

11.163 Value indices are calculated by scaling current price values proportionately to a value of 100 in the base year. By definition such a value index, if divided by the corresponding volume index and multiplied by 100 will give the corresponding price index.

11.164 In the UK economic accounts the expenditure approach is used to provide current price and volume measures of GDP. Because of the difficulties in accounting for changes in labour productivity it is not possible to obtain direct estimates of GDP at constant prices from the income data. However an approximate aggregate measure is calculated by deflating the current price estimates using the GDP deflator derived from the expenditure measure. The production approach has been traditionally used to produce a constant price measure only. However, an experimental current price production measure has been developed largely independent of the constant price production measure, which is used to allocate the industry breakdown of current price output for the latest year.

11.165 In the expenditure approach all of the expenditure components are expressed in terms of the average prices prevailing in the base year, and constant price GDP is aggregated from these. The output approach involves weighting together the detailed components, expressed in index number form, according to their relative importance in the base year. The choice of base year can be very important.

The base year and the need for rebasing

11.166 Volume measures of GDP are compiled over a set time period, with reference to a given base year. For example, from the 1998 Blue Book volume measures for the UK economic accounts are compiled using the base year 1995. In theory because the base year fixes the price structure used for comparison, it should be selected because it most closely reflects the price structure of the period of comparison. In practice this is not easy and the base year is simply moved forwards, or rebased, every five years in line with international recommendation, although there is a degree of flexibility in defining the 'set time period' to which this base refers.

11.167 The 1995 price structure has been used to compare the data from 1994 onwards. This means that for these years, the expenditure components of GDP will be calculated in terms of average 1995 prices, and the components of output will be indexed and weighted together using value added in 1995.

11.168 This rebasing is required because of the need to update the pricing structure being used. Over time comparisons of the volume series are complicated by changes in the relative prices of different goods and services and by qualitative changes in the goods and services themselves. As time passes some goods escalate in price more rapidly than others. Others change so much that they become, in effect, different goods or services from those produced previously under the same name. Because of these changes the relative prices of goods and services in the base year become less representative of the relative values put on them in the current period and the changes in measured volume will not be representative of recent growth.

11.169 However, rebasing does not mean that the whole series of constant price estimates (some going back as far as 1948) are recalculated using the relative weights of the new base year. This would mean imposing inappropriate weights on estimates for earlier periods: as already mentioned the base year is only used to compile estimates over a set time period. So currently, although constant price and volume estimates are expressed as 'at 1995 prices', for series prior to 1994 more appropriate pricing structures will have been used and for them 1995 is only being used as a reference year.

11.170 In order to link all of the constant price estimates a process called *chain linking* is used. Each series is divided into several blocks of years, and each block of years is associated with a base year and link year. These blocks are:

Period	Base year	Link year	
1948 to 1957	1958	1958	
1958 to 1962	1963	1963	
1963 to 1968	1963	1968	Output
1968 to 1972	1970	1973	Output
1963 to 1972	1970	1973	Expenditure and income
1973 to 1977	1975	1978	
1978 to 1983	1980	1983	
1983 to 1986	1985	1986	
1986 to 1994	1990	1994	
1994 to date	1995		

11.171 Within each of these blocks all constant price figures are calculated with reference to the same base year. In the link years, figures are calculated with reference to two consecutive base years, so that a linking factor may be obtained and the whole series, as published, may be shown with reference to the latest base year. By this process the whole period is rescaled to the same base year, but within each block the relative prices used to re-value to constant prices are those most appropriate to that period.

11.172 Reasonable comparisons can be made between the constant price values for any pair of years which fall within the same block. Otherwise comparisons between any two years which fall in different blocks give only a general indication of changes in the volume measured.

11.173 The choice of the link year is significant as it is important to find a period where relative prices are as stable as possible. For example, 1986 was chosen as the link year between 1985 and 1990 prices because relative prices stabilised in 1986 following the dramatic fall in oil prices in 1985. As a result it was thought more appropriate to measure growth in the period from 1983 to 1986 using 1985 prices and the period from 1986 onwards using 1990 prices.

Rebasing

11.174 Rebasing is carried out to ensure that :

- price structures remain representative
- the ratios assumed in compiling the production measure of GDP are corrected before they distort the estimates of the change in value added significantly
- the importance of industries as represented by value added is maintained.

The effects of rebasing

11.175 For most types of expenditure rebasing has the effect of reducing the estimates of growth slightly for periods after the link year. This reflects normal behaviour by consumers who will tend, on average, to increase their consumption of the goods and services whose unit prices have fallen or grown the least, in substitution for goods and services whose unit prices have grown more. Thus when rebasing takes place the weight given to the products whose consumption has been increased will be increased, and the weight of the products whose consumption has been reduced will also be reduced. A similar principle holds for the output components. At each rebasing an article is published in *Economic Trends* evaluating its effects, most recently in August 1998.

11.176 At the same time as the estimates are rebased, component series will be reviewed to improve coverage, sources and methods. These changes when taken along with the usual revisions to estimates normally have a bigger impact on the resulting series than rebasing itself.

The future of rebasing and annual chain linking

11.177 In the new ESA it is recommended that countries move to a chain-linked set of estimates of volume growth, with the weights used being updated annually rather than every five years. This is because changes in relative prices and weights can cause base-weighted indices to misrepresent current volume growth. As a result the UK are investigating how to introduce annual chain-linked estimates of growth into the accounts, in the near future.

11.178 The requirements for delivery of constant price data to Eurostat, which are comparable with those produced by other member states, are subject to regular discussion in Eurostat.

Accuracy and bias

11.179 The ONS strives to produce timely, consistent and coherent estimates of GDP which accurately represent productive activity in the economy. The basis of these estimates is strengthened by the interrelationship within the system, and the subsequent requirement that the many (and often independent) data sources are internally consistent. However, it remains very difficult to comment on the accuracy of GDP.

11.180 Estimates of GDP are built from numerous sources of information: business surveys, household and other social surveys, administrative information and survey data from the Inland Revenue and many other sources. Data are collected monthly, quarterly, annually or in some cases from ad hoc surveys. Some of the resulting estimates which feed into GDP will be firmly based whilst others may be weak.

11.181 Sampling errors can be calculated for estimates derived from random samples and the following table shows these for several of the ONS business inquiries that feed into GDP. They are conventionally quantified by means of the 'standard error'. The coefficient of variation expresses this as a percentage of the estimated value and indicates the relative reliability of the different components. A programme of work is currently underway which will lead to the publication of sampling errors for all major ONS business surveys.

	Standard error (£bn)	*Coefficient of variation (%)*
Annual Business Inquiry (1996)[1]		
Production:		
Total sales and work done	1.5	0.3
Total purchases	1.2	0.4
Gross value added excluding all taxes	0.4	0.3
Total net capital expenditure	0.2	0.9
Construction:		
Total sales and work done	1.4	1.6
Total purchases	1.2	2.2
Gross value added excluding all taxes	0.2	2.9
Total net capital expenditure	0.0	2.2
Overseas Trade in Services inquiry (1996)[2]		
Exports	0.3	1.9
Imports	0,1	1.4

	Standard error (£bn)	Coefficient of variation (%)
Quarterly Profits Inquiry [3,4] All industries	0.5	
Financial Assets and Liabilities Inquiry [3] All industries (total financial assets surveyed)	3.5	3.3

1   Production: SIC Division C–E. Construction: SIC Division F.
2   Production and services sectors *less* financial, travel and transport services, the public sector and services provided by Law Society members.
3   SIC Divisions C–O, excluding J.
4   Quarter-to-quarter changes rather than levels.

11.182   In addition to sampling errors, reliability is also affected by non-sampling errors such as limitations in coverage and measurement problems. Though there is limited information about non-sampling errors, it is likely that for some surveys non-sampling errors are the more important source of error. Data validation by survey statisticians, additional consistency checks performed by compilers and the inclusion of coverage adjustments where survey sources are known to have shortcomings reduce non-sampling error and improve the quality of the accounts.

11.183.   Even if the reliability of individual data sources were known, the complexity of the process by which GDP is estimated is such that it would be difficult to build up an overall estimate of reliability from the component series. The process of bringing together the three approaches to GDP into one measure, which uses detailed supply and demand balances, and brings in extra information about the reliability of the raw data and consistency with other sources (described in paragraphs 119–130), adds significantly to the reliability of the overall estimate of GDP, but this reliability cannot be measured scientifically. Our current approach to measuring reliability is to use the evidence from analyses of revisions to growth rates, outlined below.

Testing for bias in the initial estimates of GDP

11.184   ONS regularly monitors the revisions to growth rates of the components of GDP, in terms of income, output and expenditure, where the revision series analysed are the difference between the initial quarterly estimates and the corresponding 'final' estimates published three years later. The growth rates for each quarter are the percentage changes since the corresponding quarter a year ago.

11.185   In the analysis the growth rates of 23 components of GDP are tested for bias, or in statistical terminology to discover if the mean revision is significantly different from zero. The revision series are also tested for the effects of economic cycles; that is, whether the expansion and contraction phases of the economy cause any bias to emerge.

11.186   The results of the latest analysis published in *Economic Trends* (August 1996) relate to the data from 1982 to 1992 (the latest data which could be examined at the time as the analysis looks at revisions after three years). The main results were:

Only four of the 23 GDP components tested showed an indication of bias, but only over the full ten year period. These were the total expenditure components in constant prices, GDFCF in both constant and current prices and manufacturing output.

In the phase analysis which examined data over 13 years, three out of the 23 components showed significant bias during the expansion phases of the economy. Only the revisions to imports and to gross capital formation, both at constant prices, showed bias in the contraction phase.

11.187 The message here is that the figures do show some bias. However these results are less worrying than they may at first appear. Because the estimates examined only include data up to 1992, this latest analysis will not take full account of the recent improvements incorporated into the estimates of growth from 1989 onwards. Considerable effort has been made to improve the quality of the data used in the compilation of the economic accounts. In 1989 the then Central Statistical Office was reorganised to bring all of the main economic series feeding into the economic accounts under one management structure. Further improvements followed through the Chancellors' Initiatives launched in 1991 and 1992 which enlarged the size and frequency of many business surveys. Firm agreements have also been drawn up between the data suppliers and data compilers and these have helped improve the quality of the data, for instance in the more timely rectification of errors.

11.188 The ONS regularly looks for bias in initial estimates and has not up till now found reason to put in bias correction factors. Regular monitoring of the revisions to estimates continues, and is published in Economic Trends.

Sectoral balancing and the reliability of the estimates

11.189 Further assessment of the reliability of the consolidated economic and sector accounts can be gained by examination of the capital and financial accounts which should, in theory, show a balance between the net lending/borrowing in the capital account and financial account for each sector. However because of errors and omissions in the accounts, such a balance is rarely achieved. The resulting *statistical discrepancy items* required to equate these accounts are shown in the Blue Book.

11.190 These discrepancies provide a measure of reliability as they reflect errors and omissions in the accounts. Some components of the accounts, for example, estimates for general government, provide excellent coverage and are very reliable whilst others, for example life insurance and pension funds, are poorly covered. A detailed table, which looks specifically at the reliability of components of the sector financial accounts, is produced in Financial Statistics. However, because of the many sources of information that feed into the economic accounts it is not possible to generalise these 'reliability measures' to the aggregate GDP estimates.

Spurious accuracy and rounding to the nearest £1 million

11.191 One final point must also be made about the reliability of the statistics. In most of the published tables no attempt is made to round of estimates beyond the nearest £1 million. In some instances this shows figures which appear to have more precision than evidence warrants. The reasons for this presentation are as follows:

- rounded figures can distort differences over time or between items.

- some of the estimates in the tables are fairly precise, and if such an estimate is small, rounding would unnecessarily distort what it shows; yet if such series were not rounded to the nearest £1 million, the major aggregates of which they are components would appear precise even though other components were heavily rounded, and

- not rounding beyond the nearest £1 million aids users who prepare derived statistics by avoiding the accumulation of rounding errors which can occur when a number of rounded numbers are manipulated.

11.192 In presenting numbers to the nearest £1 million, the rounding is always such that the components add to the total of current prices, so that the accounts balance. In particular the quarterly estimates, both before and after seasonal adjustment, add up to the calender year totals. Also, the rounding is made consistent throughout the sector accounts.

Seasonal adjustment

What is seasonal adjustment?

11.193 Any analysis of GDP or its component parts will involve; an examination of the general pattern of the data, the long term movements within the series, and whether there were any unusual circumstance, such as strikes or bad weather, that may have affected the data. However this type of analysis is not easy using raw time series data because there may be short term effects associated with the time of year which will obscure other (possibly interesting) movements in the series. For example retail sales go up in December due to the effect of Christmas. The purpose of seasonal adjustment is to remove the variations associated with the time of year, ie the seasonal effects, and so allow the comparison of consecutive months or quarters.

11.194 A series is modelled in one of two form, either as an additive model:

$$Y = C + S + I$$

Or using a multiplicative model:

$$Y = C \times S \times I$$

Where C is the trend-cycle   -   the medium and long term movements in the series.

S is the seasonal component   -   which reflects the effect of climate and institutional events that repeat more or less regularly each year.

I is the irregular component   -   which represents unforeseen events of all kinds, including errors in the data.

Most economic series are decomposed using a multiplicative model.

11.195 Seasonal adjustment is the process of identifying and removing the seasonal component from a time series.

Seasonal adjustment and the economic accounts

11.196 In the presentation of quarterly GDP and current and capital account estimates for sectors, the emphasis is placed on the seasonally adjusted series. This seasonal adjustment is done at component or sub component level and in general the seasonally adjusted aggregates are calculated as the sum of the seasonally adjusted components. For example, the seasonally adjusted consumers expenditure total is calculated by summing over one hundred seasonal adjusted components such as fruit, fish and electricity.

11.197 The seasonal adjustment of the aggregates is carried out in this indirect way because it is believed to give a clearer interpretation of the changes in the final seasonally adjusted series: GDP is composed of many series which may exhibit different seasonal effects which are better identified (and therefore removed) at a disaggregated level. In addition, the component series and sub component series within the integrated economic accounts system, are important to users in their own right.

How do we seasonally adjust?

11.198 At the Office for National Statistics, seasonal adjustment is carried out using a package known as X11, or in its improved form, X11 ARIMA. X11 ARIMA is being phased in as the preferred method of seasonal adjustment for all series and is used for the majority of series that feed into the economic accounts. This seasonal adjustment is carried out each time new data becomes available (a process sometimes referred to as current updating or concurrent adjustment).

11.199 X11 ARIMA was developed by Statistics Canada as an extended and improved version of the US Census Bureau X-11 method. One of the main aims of the X11 ARIMA program is the identification and estimation of the seasonal component in order to produce a seasonally adjusted series. The program is based on moving averages using an iterative procedure to obtain successively better estimates of the seasonal component. The following steps are performed:

The series is modified using any user defined prior adjustments (permanent and/or temporary)

Trading day and Easter effects are identified and removed if significant

The programme fits an ARIMA (auto regressive integrated moving average) model to the series in order to extrapolate forward and backward, an extra year of data

The programme then uses a series of moving averages to decompose the time series into the three components (the trend cycle, the seasonal component and the irregular component). It does this in three iterations, getting successively better estimates of the three components. During these iterations extreme values are detected, removed and replaced.

A range of diagnostic statistics are produced, describing the final seasonal adjustment. These are used to asses how good the seasonal adjustment is and whether improvements can be made.

11.200 However, seasonally adjusting the components or sub components of the economic accounts is not simply a mechanical process. A good knowledge of the series, as well as an understanding of the X11ARIMA package, is crucial in the production of the seasonally adjusted series. The compilers of the accounts, having a very detailed knowledge of their series, are able to look for any likely problems in the process. For example:

Is the data collected in a way which may cause odd effects (eg a lag between the activity and when it is recorded)?

Has there been a change in the way the data is collected? This may cause trend or seasonal breaks.

Is the series likely to be affected by trading days or Easter effects?

Have there been any events which are likely to cause large outliers or breaks in the series? These maybe things like: the Gulf war; the budget moving from March to November; extreme weather conditions; or events which may only affect the individual series.

11.201 Thus combining the compilers detailed knowledge of the series with the technical expertise offered by the ONS Methods and Quality Division, individual series can be seasonally adjusted effectively.

Seasonal adjustment on the sector accounts

11.202 Quarterly current and capital accounts for each sector are seasonally adjusted, but there is no balanced seasonally adjusted matrix of financial transactions. Although net lending/borrowing may be seasonal, transactions in individual financial instruments are often not seasonal. This is because of the high level of volatility in financial transactions, and the high level of switches made between financial categories as investors and borrowers change their portfolio preferences.

The hidden economy

Adjustments made for the hidden economy in the UK economic accounts

11.203 Different parts of the economic accounts are affected by hidden activity to different degrees. When administrative data is used there is often a good reason for data to be 'hidden' from the collecting authority; for example, data on income from tax returns will be affected by people not declaring all their income, whereas there should be no reason not to declare activity to a confidential household surveys or business survey. On the other hand respondents tend to take administrative forms, such as tax returns, more seriously than surveys which are more likely to be affected by non-response, or poor quality responses (eg the FES records only about a half of alcohol consumption). Thus despite being more susceptible to hidden activity administrative sources may be more accurate than survey sources.

11.204 The economic accounts do not provide an explicit measurement of the hidden economy; hidden activity is captured well by some sources used by the accounts, less well by others.

11.205 There are a small number of adjustments which are explicitly made for the hidden economy within the UK economic accounts. There are also several adjustments made which are not explicitly aimed at measuring the hidden economy but partly adjust for hidden activity, for example income from farming is taken from a MAFF source rather than an Inland Revenue source, this is partially because the MAFF source captures income by workers earning less than the tax threshold, but it also captures hidden activity better than the IR source.

11.206 Unlike its predecessor, the 1995 ESA explicitly states that all economic activities regardless of their legal status should be included within the economic accounts. Thus, in principle the sale of illegal drugs, prostitution and the sale of stolen goods should all be included within the accounts.

11.207 The UK has been carrying out a project looking at available data sources for illegal activities; however, there are no immediate plans for including this activity within the accounts, due to the lack of accurate data and the general lack of a consensus on how illegal activities should be treated.

11.208 The UK is also reviewing at its current adjustments for hidden activity, in particular as part of the 'exhaustiveness' work programme to improve harmonisation of EU Member States' GNP estimates for the fourth resource.

Chapter 12

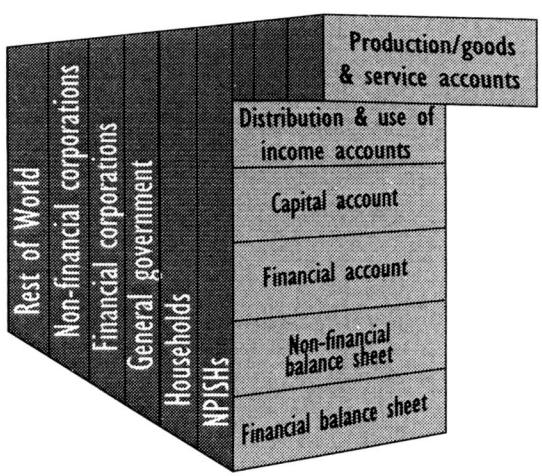

Production/goods & service accounts

Distribution & use of income accounts

Capital account

Financial account

Non-financial balance sheet

Financial balance sheet

Rest of World

Non-financial corporations

Financial corporations

General government

Households

NPISHs

Goods and services account

Chapter 12 Goods and services account

Introduction

12.01   This chapter describes the methodology for the estimation of the aggregate variables which are included in the goods and services account. This account, which is essentially a new feature of the presentation of the UK statistics, relates the resources which become available in the economy to the uses which are made of them. The information, which is presented in table 1.7.0 of the Blue Book (BB), is in respect of the whole economy. However, such tables can also be produced for individual groups of products, in which form they are more familiarly seen as part of the input-output (supply and use) tables. Although the presentation shown in the BB table is new, the variables have all appeared in the previous accounts as components of the expenditure and production measures of GDP, as well as within the framework of the supply and use tables.

12.02   The description of the methodology given here for compiling the estimates relates mostly to the aggregates of the variables, which is the form in which they appear in the goods and services account. Generally speaking, the totals are derived as the sum of the estimates for the main components, such as for individual industries or commodities. Where relevant, the description here makes some brief reference to the procedures for compiling component estimates, although this is covered in detail in other parts of the publication.

12.03   The goods and services account, as such, is compiled only at current prices, annually. However, some of the variables appearing in the account are also derived at constant prices and appear in other parts of the overall accounts. For this reason, the description of the estimation methodology included in this chapter will cover data at current and constant prices where appropriate. The text will include an outline description of each variable, together with some of the key features of coverage. Reference will also be made to the main changes compared with the previous System of National Accounts. Conceptual issues have been covered more fully in Chapter 3, and other aspects will be considered more fully in the chapters dealing specifically with the particular variables. A more detailed exposition of the input-output methods is contained in the ONS publication 'Input Output methodological guide' (ONS, 1997).

The (summary) goods and services account

12.04   The structure and composition of the goods and services account are shown below, for the whole economy, in summary form. For the full table, output and final consumption expenditure are disaggregated. In addition, it should be noted that the table shows final consumption expenditure (P3): for the whole economy this variable is the same as actual final consumption (P4), which is an alternative presentation (see Chapters 21, 22). Taxes less subsidies on products appear on the resources side of the account in order to provide a consistent valuation, essentially purchasers' prices, with the uses side.

Summary goods and services account for 1995
(Data to be included in 1998 edition)

Resources   £ million

Output (P1)
Taxes on products (D21)
less Subsidies on products (D31)
Imports of goods and services (P7)

TOTAL RESOURCES (P1+D21-D31+P7)

Uses

Intermediate consumption (P2)
Final consumption expenditure (P3)
Gross fixed capital formation (P51)
Changes in inventories (P52)
Acquisitions less disposals of valuables (P53)
Exports of goods and services (P6)

TOTAL USES (P2+P3+P51+P52+P6)

12.05 The estimates of the variables going into the account are made largely independently. The balance between resources and uses is achieved mainly through the mechanism of the input-output balancing process, as described in Chapter 13. The text will now consider the variables in turn. The order follows that of the table resources and uses - except that imports of goods and services are considered with the export figures towards the end of the chapter, and taxes and subsidies are covered last.

Output (P1)

12.06 Output is essentially the measure of productive activity taking place in a particular period. The concepts and definitions have been described in detail in Chapter 3, and other aspects are covered in Chapter 13 on the production account.

12.07 Output covers the production of both goods and services by businesses (including households) and the activities of government and the NPISH sector. Two important changes to the coverage of output reflect the new treatment of the illegal economy and Financial Intermediation Services Indirectly Measured (FISIM). However, these innovations are not ready for implementation.

12.08 As mentioned earlier, this section will largely describe the measurement of aggregate output at current and constant prices. In general, these aggregates will be determined by adding together separate estimates for different industries, which form the basis of the sampling arrangements for collecting information. Descriptions of how the detailed estimates of output are compiled by industry are given in paragraphs 13.32–13.47. The analysis by institutional sector is derived by breaking down the total as described in paragraphs 13.48–13.56.

The composition of output

12.09   Output, in aggregate (and also by industry) has three main components, which appear separately in the goods and services account. The description of the estimation of aggregate output is presented for each of them separately. The components are:

- Market output
- Output produced for own final use
- Other non-market output

12.10   Market output relates to goods and services sold at economically significant prices. The distinction between market and non-market output is discussed in Chapter 4. Some activities, such as the provision of education, health or social services, can be either market or non-market.

Market output

12.11   The first of the three components - market output - is by far the largest. Market output for the economic accounts is derived as the sum of:

- Sales of own production
- Changes in inventories of finished goods and work-in-progress
- Output not sold on the market.

Sales of own production

12.12   Information on sales of the production of goods and services is derived from various sources, but mainly statistical surveys directed at businesses. This provides a good source for estimates, since the economic accounts' requirements are broadly similar to the basis of commercial accounting. The aggregate figure is compiled essentially by summing estimates for different industries, which constitute the basis of the sampling process. However, the approach varies a little over the different industries, reflecting, in particular, the precise way in which output is defined for some of the distribution and financial services industries. These issues are covered in Chapter 13, which deals with the industry dimension of the estimates.

Sales of own production at current prices

12.13   The main sources of information on sales is provided by the annual inquiries into production, construction and the distribution and services trades carried out by the Office for National Statistics (ONS) (Surveys 25.1.3, 25.1.5, 25.1.8, 25.1.9 and 25.1.11). Other main contributions come from surveys conducted by the Ministry of Agriculture, Fisheries and Food (MAFF) for agriculture (farm business surveys), and by the Department of the Environment, Transport and the Regions for the construction industry. The information collected in all these inquiries is largely in the form required for the economic accounts, although some small modifications may be necessary to align definitions for the particular coverage required for the accounts. Sales figures need to be adjusted for changes in inventories of finished goods, work in progress and coverage of businesses. To adjust to basic prices, taxes net of subsidies on products need to be deducted. The ONS annual service trades inquiries do not, however, cover all industries. For those industries not being covered, and for certain other industries, including finance, insurance and real estate, estimates of output are based on income data, as described in Chapter 13.

Sales of own production at constant prices

12.14   Annual estimates of constant price market sales are derived, generally, by deflating the industry-based current price data, derived as above, by appropriate price series. The price information comes from a variety of sources, but principally the ONS's producer price index (including export prices) (25.2.7) and the retail price index (25.4.7). Other price data used come from ONS's surveys of certain prices in the service sector (25.2.7), and DETR's data on construction prices. Constant price data for agriculture are produced by MAFF. For all industries, where deflation is used, it is undertaken in as much detail as possible. Deflators for industries are derived by weighting together individual price series by the commodity composition of output. In a few industries, direct volume information is available, and this is used to extrapolate the level of output in the base year.

Changes in inventories of finished goods and work-in-progress

12.15   The increase in these inventories represents production which, in essence, has taken place but which has not yet been sold. The methodology used for measuring the contribution to output arising from changes in inventories is far from straightforward. The estimation of inventories for the economic accounts is discussed in detail in Chapter 15 on the capital account. The essence of the approach is to revalue the historic cost, book values (levels) of inventories, by industry, at end-year or end-quarter, in a way which approximates to the replacement cost (current purchaser's price) basis of recording required for the economic accounts. Most of the information on book values comes from ONS's quarterly surveys of inventories or the annual inquiries (25.1.3, 25.1.5, 25.1.7, 25.1.8, 25.1.9, 25.1.11 and 25.2.4), with the price information coming from the producer and retail price indices (25.2.7 and 25.4.7, respectively)

Output not sold on the market

12.16   There are three main categories for this third component of market output:

- Sales to establishments within the same enterprise group
- Payments to employees as income in kind
- Barter transactions

12.17   The last of these is not considered to be of any significant size in the UK, so no estimate is made for the accounts. For the first component, estimates of intra-establishment sales are recorded within the aggregate of sales included in the various statistical inquiries from which the figures of sales are derived. Thus, the only item which needs to be estimated is income in kind, for example company cars, and free or reduced price accommodation. Estimates are derived from a variety of sources, including Inland Revenue (25.4.3), the FES (25.4.5) and official surveys of labour costs (25.4.1). Some annual estimates are projections of periodic benchmark data. Few quarterly data are available; these are mostly derived by extrapolation and interpolation. Constant price estimates are compiled by deflating the value estimates, derived as above, by appropriate price series based on the RPI (25.4.7) and ad hoc sources.

Output produced for own final use

12.18   The second component - output produced for own final use - covers production by both households and corporations. Communal activity is also included. By definition, output of corporations for their own final use is confined to producing capital assets; that is, making machine tools or undertaking construction work for their own use. This information is collected, as for market sales above, in the ONS annual inquiries, and constant prices figures are derived by deflating the value estimates by appropriate price data. For households, the main items included are the services of owner-occupied dwellings, construction activity, payments to domestic staff and own production of agricultural produce. These are also included in household consumption expenditure, and their estimation is described in Chapter 22. Government also produces some products like software.

Other non-market output

12.19   The third component - other non-market output - covers output that is provided free or at prices which are not economically significant. Such production will be financed out of taxation or donations or other fees or the like. This kind of output arises in respect of the activity of the provision of many goods and services by the State, both central government and local government, and by non-profit bodies serving households (the NPISH sector). Non-market output is valued at the costs of production, that is the sum of the four components - intermediate consumption, compensation of employees, consumption of fixed capital, and other taxes on production less other subsidies on production. These four variables are described in general terms at various places in this publication. They will be covered only briefly here in the context of estimating aggregate non-market output.

12.20   With a few exceptions, as explained in Chapter 21, all general government output is classified as non-market. However, it should be remembered that allocation to sectors is made in principle at the level of the local kind-of-activity unit (KAU), and that therefore charity shops, for example, which sell output at market prices, are allocated to the non-financial corporations sector. The information used to measure non-market output for government and NPISH is essentially the same as that included for part of the estimates for final consumption expenditure for these sectors, as described in paragraphs 12.43–12.48. The estimation of non-market output at current prices can be seen as the sum of the four separate components of cost:

Intermediate consumption

12.21   Estimates of intermediate consumption for government - essentially the purchases of goods and services from other producers - for government are obtained from administrative sources, as a by-product of information needed for monitoring public expenditure. For central government the information, which is available quarterly, is on a cash basis rather than the accruals basis required for the economic accounts. Where the cash figures are not thought to represent the likely profile of the provision of the goods, for example for defence spending, the quarterly series are smoothed. For local government, the information on spending on goods and services relates to the financial year. The quarterly and calendar year figures are derived by interpolation, using what limited quarterly data are available on spending.

Compensation of employees

12.22   Figures of compensation of employees, covering wages and salaries and social security contributions, are obtained from central government and local government quarterly reports.

Consumption of fixed capital

12.23   Consumption of fixed capital represents the value of assets used up in the productive process, and is a cost of production for non-market output. Annual estimates are provided by the perpetual inventory model (PIM) methodology used for compiling the estimates of capital stock (see Chapter 16). Quarterly estimates are based on simple interpolation.

Other taxes on production less other subsidies on production

12.24   Estimates are based on a sector allocation of the appropriate taxes and subsidies.

Other non-market output at constant prices

12.25　Annual and quarterly estimates of non-market output at constant prices are derived from the extrapolation approach, in which base year estimates of output, obtained as above, are projected by indicator series deemed to represent the movement in output of the particular component. The indicators, for both annual and quarterly estimates, are based where possible on quarterly indicators of output, but otherwise on employment information (weighted together by pay levels), or deflated wages and salaries or value data, and also estimates of capital consumption at constant prices. By convention, where input measures are used productivity change has hitherto been assumed to be zero (apart from changes in the grade of labour) but for the 1998 Blue Book new indicators of productivity change for parts of government (notable health, education and social security) were introduced in respect of years from 1986 onwards.

Intermediate consumption (P2)

12.26　Intermediate consumption is the value of goods and services consumed in the process of production. Previously, this was generally known as 'inputs'. The concepts and definitions relating to intermediate consumption have been described in detail in chapter 3, including issues such as the boundary with GFCF and the consistency with output.

Changes from previous system

The definition of intermediate consumption contains a number of important changes compared with the previous system. The main changes are that mineral exploration, computer software and military equipment (other than weapons), previously included in intermediate consumption, are now classified as gross fixed capital formation. The measurement of financial intermediation services has also changed (*see* note on FISIM in Annex I to Chapter 20)

12.27　Reflecting issues of recording analogous to those affecting output, intermediate consumption in aggregate (and by industry) for the economic accounts is estimated as:

Purchases by producers
less the change in raw materials and supplies held in inventories

12.28　The following section describes how these two components are estimated. The description covers the estimation of the aggregate figures which are derived, largely, by adding together estimates for individual industries. The way in which the industry data are compiled is set out in paragraphs 13.71–13.85, and the breakdown of the total by institutional sector in paragraphs 13.86–13.93.

Purchases

12.29　Information on total purchases has, in the past, generally been available only from annual production inquiries (25.1.3), supplemented with detailed commodity information in periodic inquiries, again directed at production industries. More recently, information on total purchases has also been obtained in annual distribution and service inquiries (25.1.5, 21.1.7, 25.1.8, 25.1.9 and 25.1.11). No quarterly figures have been collected. Details of the estimation processes are given below.

Purchases at current prices

12.30 Estimates are made according to two main approaches. First, for production and most other industries, information on purchases is collected in the ONS annual inquiries. Use is also made of information collected in DETR's construction surveys and from MAFF. For the few industries where direct data are not available, and for certain other industries, including finance, insurance and real estate, estimates are derived as the difference between the output figures, derived as above, and estimates of value added, which have been made from the income side of the account.

Purchases at constant prices

12.31 Annual estimates of intermediate consumption at constant prices are derived, generally, by deflating the industry-based current price information, derived as above, by appropriate price information. The price data come from a variety of sources, but principally the ONS's producer price index (including import prices) (25.2.7), retail prices index (25.4.7), and certain service sector prices (25.2.7), and DETR's construction prices .

Change in inventories of raw materials and supplies

12.32 This component of intermediate consumption represents the contribution to inputs resulting from drawing on material and supplies held in stock. The problems of measurement are similar to those affecting the output estimate, as mentioned above. The methodology for the estimates of inventories is discussed in detail in chapter 15 on the capital account. Most of the information on book values, which forms the basis of the estimates, comes from ONS's quarterly surveys of inventories or the annual inquiries (25.1.3, 25.1.5, 25.1.7, 25.1.8, 25.1.9, 25.1.11 and 25.2.4), with the prices information coming from the producer and retail price indices (25.2.7 and 25.4.7, respectively).

Final consumption expenditure (P.3) /Actual final consumption (P.4)

12.33 These two variables are alternative presentations of the same aggregate. The relationship between the two is set out in Chapter 5 (5.75–5.85). Sources and methods are described in detail in other chapters. This section will focus on the first presentation, in which final consumption expenditure is defined as the sum of spending by the three constituent sectors -households, non-profit institutions serving households (the NPISH sector) and general government - on consumption (as opposed to investment) goods and services, since this corresponds to the way in which the statistics are collected.

Changes from former system of accounts

The aggregate of final consumption for these three sectors is, in principle, equivalent to the sum of consumers' expenditure and general government final consumption in the previous accounts. However, there are some important changes in the scope of final consumption. One main change is the intended identification of NPISH as a separate sector; previously, these were included within the personal sector. In respect of the consumption variables there are two important proposed additions to household consumption, which it has not been possible to implement yet; namely FISIM (see Chapters 2 and 20) and wider coverage of the hidden economy (see Chapter 2). FISIM also would have a small impact on the estimates for government and NPISHs. In contrast, amongst exclusions, domestic rates (but not council tax) and motor vehicle excise duty paid by households, previously both part of consumption, are now scored as taxes on income and wealth.

12.34   Descriptions of the derivation of the detailed figures of final consumption are set out elsewhere in this publication. The breakdown for households and the NPISH sector is given in Chapter 22. The corresponding breakdown for government (Classification of the functions of the government - COFOG) is in Chapter 21.

Household final consumption expenditure

12.35   Household final consumption expenditure represents, in the main, traditional consumer spending. It includes, however, imputed rent for the provision of owner-occupied housing services, income in kind and consumption of own production. Also included is household expenditure overseas, with spending by non-residents in the United Kingdom being excluded.

12.36   The household sector covers not only those living in traditional households, but also those people living in institutions, such as retirement homes, boarding houses or prisons.

12.37   The aggregate figures, both annual and quarterly, of household consumption are built up from the detailed estimates of spending on individual goods and services. However, as with other variables, aggregate and components are subject to possible modification in the balancing process. Before looking at the estimation methodology, it will be useful to give the main sources used for estimating household consumption. These have certain advantages and disadvantages, as discussed in Chapter 22. The sources are:

i)   sample surveys of spending by households and individuals
ii)   statistics of retail and other traders' turnover
iii)   other statistics of supply or sales of particular goods and services
iv)   administrative sources
v)   commodity flow analysis.

12.38   The broad approach to estimation is essentially to determine the best estimate for each good or service from the various sources which might be available. The final estimates are subject to the balancing process, as described in Chapter 11.

12.39   The following paragraphs describe how estimates of household consumption, in aggregate, are made for the accounts. Given that most of the estimates are made from short-period data, the annual and quarterly estimation are considered together within each of the sections dealing with current and constant prices. Fuller details of estimates for individual goods and services which form the basis of the aggregate are given in Chapter 22 on the household sector.

Current prices

12.40   For the estimates of household consumption at current prices, three continuous surveys of households or individuals, all providing quarterly information, are used. These are the ONS Family Expenditure Survey (FES) (25.4.5) and International Passenger Survey (IPS) (25.5.4), and the MAFF National Food Survey (NFS) (25.4.6). The FES and NFS are both surveys of households. The former collects information on the value of spending over the whole range of consumer goods and services. The NFS covers essentially food items, obtaining information on both the value and quantities of food purchases. The IPS is a survey of individuals, both resident and non-resident, conducted at ports, and seeking information on holiday and travel expenditure.

12.41 The main business surveys used in the estimation of household consumption are the ONS annual inquiries into the distribution and service trades (25.1.8 and 25.1.11), particularly retailing, and also the monthly retail sales inquiry (25.1.10). All these surveys collect information on the total value of sales. In addition, commodity information is collected regularly in the retailing inquiries, and occasionally in other surveys. Where relevant, statistical techniques can be used to make detailed commodity estimates from figures of total sales, and to deal with any of the other problems with these sources. Specific sources and administrative information embrace, for example, the privatised utilities and similar industries, government tax data, and certain trade association sources.

Constant prices

12.42 As with the current price figures, the aggregate estimate of household consumption at constant prices, for both the annual and quarterly accounts, is derived essentially by summing estimates for individual goods or services. The methodology for the detailed estimation is described in Chapter 22. The approach is a mixture of deflated values and direct or indirect volume estimates, essentially the revaluation of present quantity data by base year prices. The price information used for deflation is provided mostly by the ONS retail prices index (25.4.7), but certain other price information is used, particularly for services (25.2.7). Deflation is undertaken in as much detail as possible, consistent with maintaining an adequate level of accuracy in both the value and price data. The main areas where direct volume series are used relate to spending on housing services, drink and tobacco.

Government final consumption expenditure

12.43 Government final consumption expenditure covers spending, other than on capital goods, by both central and local government. Government purchases the non-market output of the government sector, which is produced from its intermediate consumption and value added. Spending therefore can be seen as the sum of compensation of employees and purchases of goods and services, and the consumption of fixed capital. Certain receipts for goods and services, which will appear in household consumption, are deducted from purchases. The derivation of the figures, which is summarised below, is described in more detail in Chapter 21.

Changes from former system of accounts

The definition of government consumption is largely the same as for general government final consumption in the previous statistics. Certain military expenditure is now classified to gross fixed capital formation and computer software also has changed treatment. Further information about government final consumption is contained in Chapter 21.

Current prices

12.44 Current price information for estimates of government consumption is obtained largely from administrative sources, as a by-product of the planning and monitoring of public expenditure. Information from central government, which is mainly on a cash rather than accruals basis, is available quarterly . However, the data for certain components of spending, such as defence, are not thought to reflect the pattern of provision of the goods, or the profile of output which has gone into their production. In such cases, the financial year spending is spread over the four quarters. For local government, figures on wages and salaries are available quarterly. However, information on purchases is collected only for financial years, and the calendar year and quarterly series are derived by interpolation, using what limited quarterly data are available.

Constant prices

12.45 Estimates at constant prices are obtained in two broad ways. In the first, estimates of spending on goods and services, made as above, are deflated by specially constructed price indices, based on producer prices (25.2.7), retail prices (25.4.7), and other information. Secondly, in respect of compensation of employees, the constant price series are mostly based on the movement in numbers of employees, usually with some analysis by grade. In some areas, the wages and salaries figures are deflated by indicators of pay rates. As discussed in the section on the measurement of non-market output, new indicators of productivity change have been introduced for health, education and social services from 1986.

Consumption expenditure of non-profit institutions serving households

12.46 The NPISH sector includes bodies such as universities, charities, churches, clubs, trade unions and friendly societies. As with government, final consumption covers compensation of employees and purchases of goods and services, less any receipts from sales of goods and services, which will mostly be in household consumption. An estimate of capital consumption is also included.

12.47 Given the very varied nature of the organisations making up this sector, and the particular definition of spending required for the accounts, the estimates are built up from a diverse range of sources. At current prices, detailed annual information is available from universities, trades unions and some other bodies. For charities, annual projections of an earlier survey were made up to 1996, when a new benchmark inquiry was carried out. This new information is largely projected as before, using information on charities' total income. Other annual information comes from a miscellany of sources. Quarterly estimates are interpolations of annual data.

12.48 At constant prices, the annual and quarterly estimates are obtained by deflating the value figures, derived as above, by appropriate price indices, largely based on retail prices (25.4.7). This includes specially-constructed prices for universities and charities, but more general indices elsewhere.

Gross Capital Formation (P.5)

12.49 Gross capital formation is made up of three components:

Gross fixed capital formation
Changes in inventories
Acquisitions less disposals of valuables.

These components are very different in nature and magnitude.

Gross fixed capital formation (P.51)

12.50 Gross fixed capital formation (GFCF) relates principally to investment in tangible fixed assets such as plant and machinery, transport equipment, dwellings and other buildings and structures. However, it also includes investment in intangible fixed assets, improvements to land and also the costs associated with the transfer of assets. The investment relates to assets which are used repeatedly in the production process for more than one year. Coverage is described in detail in Chapter 6 (6.6 and 6.15).

> **Change from former system of accounts**
>
> Previously in the UK accounts GFCF had been described as GDFCF (Gross domestic fixed capital formation). The dropping of the term 'domestic' is of itself no significance as regards the concept and coverage of this variable. However, the new accounts extend the range of items to be regarded as capital assets. The main additions, which were previously included in intermediate consumption, are intangible fixed assets. However, also included in GFCF, all within tangible assets, are cultivated assets, such as livestock, which are used repeatedly to produce goods and services; expenditure on military structures and equipment (other than weapons), previously mostly in government final consumption; and historic monuments.

12.51   International guidelines for the goods and services account suggest that GFCF should be broken down between:

- net acquisitions of tangible fixed assets;
- net acquisitions of intangible fixed assets;
- major improvements to non-produced non-financial assets;
- cost of ownership transfer on non-produced non-financial assets.

Net acquisitions of tangible and intangible fixed assets should be split into acquisitions of new assets, acquisitions of existing assets and disposals of existing assets. However, as explained in Chapter 15, the provision of this detail is not practicable and so only the total series is shown.

12.52   The approach to estimating GFCF, in aggregate, is described below. The estimation approaches used for analyses of the aggregate are included in Chapter 15. The industry analysis, from which total GFCF is derived, is covered there. The asset and sector data are largely breakdowns of the total.

12.53   Generally speaking, the current price estimates of GFCF can be made in two broad ways. There is, first, the conventional approach of collecting information from businesses in statistical surveys or from the public sector. The alternative is to use the commodity flow approach, in which estimates of supply (output plus imports) of products are allocated to the components of use - here, specifically, GFCF. The advantages and disadvantages of these two approaches are discussed in Chapter 15.

12.54   In the UK, estimates of GFCF use both the direct and model-based approaches. Given the statistical requirements, and the relative advantages and disadvantages of the two methods, there is rather more emphasis on the survey approach. The commodity flow approach is probably better at estimating the short-term movement in investment, rather than the level. The approach plays a particular role in estimating GFCF for certain assets, such as private sector dwellings, ships and aircraft and facilitating the consistency of the accounts generally.

Current prices

12.55   There are four main sources of direct information for deriving annual and quarterly estimates of GFCF. These are, for most assets:

- ONS annual surveys of businesses, covering production (25.1.3), construction (25.1.5), distribution (25.1.8, 9) and service industries (25.1.11), and quarterly capital expenditure inquiries across industry (25.2.2);
- public sector returns for both central and local government;

- ONS surveys of public corporations (25.2.3);
- for private sector dwellings the DETR survey of construction output, together with information on improvements from the FES (25.4.5);
- banking statistics and building society surveys (25.3).

As mentioned above, the directly-collected information is supplemented with estimates from the commodity flow approach. One particular use of this approach is to provide an improved quarterly profile where it is thought recording practices - for example cash instead of accruals - might not provide a sensible quarterly series.

Constant prices

12.56   The estimation of GFCF at constant prices - essentially by deflating the value figures - is one of the more difficult parts of the compilation of the economic accounts. This arises because the variable nature of capital goods, many of which will be unique, leads to severe difficulties in establishing appropriate price series. This is a particular problem for the deflation of building work. Deflation of GFCF generally is based in large part on the information collected for the producer price index, and including, importantly, prices of imported goods (25.2.7). The weights for the industry-based price indices are provided by the detailed asset analysis estimated for the supply and use tables. Further details of the deflation process are given in Chapter 15.

Changes in inventories (P52)

12.57   Inventories in the economic accounts relate to goods which are held by producers prior to further processing or sale. Inventories also includes work-in-progress. Information about inventories, in appropriate form and valuation, is relevant to the expenditure, production and income measures of GDP. The main components of inventories - the asset breakdown - are:

- materials and supplies;
- work-in-progress;
- finished goods
- goods for re-sale.

12.58   This section considers, in broad terms, how changes in inventories are calculated for the goods and services account, where they may also be regarded as a component of expenditure on GDP. Some discussion has already been included about the estimation of inventories for output and intermediate consumption. A full description of the calculation, which is made for each of the four asset groups, is contained in Chapter 15. Chapter 15 also covers the estimation of the analyses by industry, by asset and by institutional sector.

12.59   The broad principle of the approach to estimate changes in inventories is to revalue the historic cost, book values (levels) of inventories, by industry, at end-period, collected in the ONS inquiries (25.1.3, 25.1.5, 25.1.7–9, 25.1.11 and 25.2.4), in a way which approximates to the replacement cost basis of recording required for the economic accounts. The price data used for revaluation are based on indices of prices of the types of goods which are stocked, but making some allowance for the length of time such goods are held in stock before being used. The price information comes mostly from the producer and retail price inquiries (25.2.7 and 25.4.7).

12.60   In addition to its role in the goods and services account and for expenditure on GDP, as mentioned earlier, the inventories calculation provides the basis for the inventories components for output and intermediate consumption in the production measure. In addition, for the income measure, the source information collected on operating surplus will include the effect of holding gains. The estimate of holding gains, made as part of the inventories calculation, needs to be deducted from the profits data collected to provide the proper basis of the figure of operating surplus required for the accounts.

Changes from former system of accounts

In the previous accounts, inventories were known as stocks, and changes in inventories as stockbuilding. In the new SNA, the definition of inventories has been widened to include certain items which were previously not included. The new coverage now embraces:

- work-in-progress of cultivated assets and livestock for slaughter.
- work in progress in certain service industries, such as computer software and films, and
- Government non-strategic stocks, as well as strategic stocks hitherto included.

Acquisitions less disposals of valuables (P.53)

12.61 Valuables are goods which are not used in the production process, and which are generally held as a store of value. They include precious stones and metals, antiques and art objects. The presentation in the accounts, bringing together acquisitions less disposals of valuables within gross capital formation, is new. There are problems with both the concept of valuables and the derivation of any estimates. Valuables can enter the UK economy in two ways. First, they may achieve 'valuables' status as time evolves (*see* 8.7). Secondly, they may enter (or leave) through overseas trade. Measurement of the former is appropriate to the 'other changes in assets' account. The figure for the goods and services accounts, therefore, relates together with dealers' margins on transactions in valuables to the overseas trade element, using information supplied by HM Customs and Excise (25.5.3). Deflation to constant prices is undertaken using information from the RPI (25.4.7). Under derogation from the EU the United Kingdom will exclude gold transactions by monetary financial institutions.

Exports of goods and services (P.7) and imports of goods and services (P.6)

12.62 Exports and imports of goods and services relate to transactions between residents and non-residents. The figures, in addition to their inclusion in the goods and services account and in expenditure-based measure of GDP, are also major components of the balance of payments accounts. Definitional and conceptual issues, identifying some important inclusions and exclusions, are set out in Chapter 24. This section will describe the estimation of the aggregates, which are built up by adding together estimates for separate components. Given the similarity in measurement as between exports and imports it will be sensible to describe the estimation methodology for (i) exports and imports of goods and (ii) exports and imports of services.

Exports and imports of goods

12.63 The way in which exports and imports of goods are defined is given in Chapter 3. It is worth mentioning the key features here as they will be relevant to the estimation of the aggregate for the goods and services account. The general principle obtaining is that exports and imports of goods occur in the accounts when ownership changes between residents and non-residents (even though goods may not move across frontiers). There are however some exceptions to this principle. Given that the annual and quarterly figures are both derived from monthly trade statistics, the summary of the methodology presented here will be combined.

Current prices

12.64   The annual and quarterly estimates of exports and imports of goods are based on monthly information collected from a dual system of reporting for the overseas trade statistics. For trade with the European Union, since the beginning of 1993 information has been collected by Customs using an EU-wide system called Intrastat (25.5.3), based on the administration of Value Added Tax (VAT). For trade with countries outside the EU, the estimates are based on information provided on Customs declarations (25.5.3). Prior to the introduction of Intrastat, trade with EU countries was also based on Customs documents.

12.65   Some information about the two systems is given in Chapter 25 on data sources. The main feature of the two approaches is that, while the Customs declarations source covers all traders, the Intrastat system obtains information from the largest operators who cover all but 3% of the value of intra-EU trade. The figures include estimates for this residual part of trade.

12.66   The basic information provided under both systems relates, broadly, to goods crossing the UK boundary, either inwards or outwards. Valuation for the trade statistics is also made at the UK boundary. For exports, this is termed 'free on board' (fob); for imports, the valuation will also include the cost of insurance and freight from the boundary of the exporting country. The valuation for imports is denoted by cif.

12.67   The basic trade statistics information needs to be adjusted to accord with the balance of payments coverage and valuation principles of the economic accounts. These are described in detail in Chapter 24. In brief, there are two main coverage adjustments. The first is to exclude goods which appear in the trade statistics, but for which ownership does not change, such as contractor's plant and goods for minor processing or repair. As a corollary, it is necessary to include in the accounts certain goods where ownership does change, but which do not enter the trade statistics, such as ships and other goods purchased by non-residents in international waters. In respect of valuation, imports are adjusted to a fob basis by subtracting an estimate of the insurance and freight costs which are included in the trade statistics valuation.

Constant prices

12.68   The monthly current price data outlined above provide the basis for the derivation of the annual and quarterly constant price estimates of exports and imports of goods. The approach used is essentially deflation of the monthly-based value information by appropriate price indices. Increasingly, use is being made of information on export and import prices collected in ONS statistical inquiries (25.2.7). Where specific import and export price data are not available, deflation is based on the use of domestic producer price indices adjusted to reflect factors such as exchange rate movements or the ratio of value to weight from the information within the monthly trade statistics. Deflation is undertaken in as much detail as possible, but taking account of the accuracy of the data.

Exports and imports of services

12.69   The definitions of exports and imports of services have been set out in Chapter 3. Of particular interest is the way in which consistency is ensured between the f.o.b. valuation for goods and the estimates for transport services. It should also be noted that estimates of household spending abroad and non-residents' spending in the UK appear, respectively, in the imports and exports of services, thus complementing their treatment in household final consumption.

Current prices

12.70   Annual and quarterly estimates of exports and imports of services, embracing a wide range of activities by government, business and finance, and transport and travel, are derived from a variety of sources. Increasingly, estimates are being based on ONS quarterly surveys of activity by firms (25.5.2). A particular difficulty with these surveys is the need to identify on the statistical register, which is used as a basis for the surveys, those firms which are involved in trade in services. Further important sources include the International Passenger Survey (25.5.4) for travel and the Bank of England for financial services. This information is supported by data from various Government sources, such as DETR and HM Treasury, and by external surveys such as those carried out by the Chamber of Shipping and the Civil Aviation Authority. Most of the information is available quarterly. Where quarterly data are not available, estimates are made by extrapolation and interpolation.

Constant prices

12.71   Estimates of exports and imports of services at constant prices are derived mostly by deflating the value figures, obtained as described above, by appropriate price indices. Direct volume estimates are available in a few cases only. The price information takes various forms, including relevant information from the RPI (25.4.7), or specific data on for example freight rates. Information is also used on prices in various overseas countries, which are adjusted by the appropriate foreign exchange rates.

Taxes on products (D.21) and subsidies on products (D.31)

12.72   Given the role these variables have in the economic accounts, it is sensible to consider them together. The concept of taxes and subsidies - taxes essentially increase the price of a product and subsidies reduce it - has been explained in Chapter 5. In the goods and services account, output, and hence resources, are valued at basic prices. Basic prices (*see* Chapter 4, paragraphs 4.50–4.56) include only taxes and subsidies on production, such as non-domestic rates, and not taxes and subsidies on products, for example VAT, which are regarded as being borne by consumers rather than producers. However, uses are valued at purchaser's (or market) prices, that is including, essentially, all taxes and subsidies. Thus, to balance the valuation used for resources and uses in the goods and services account, it is necessary to add to the resources side of the account the figures for taxes less subsidies on products.

Taxes on products (D.21)

12.73   These are taxes payable per unit of good or service being produced. The main UK taxes on products are value added tax, and taxes on drink, tobacco, hydrocarbon oil and betting, and stamp duties. Further details are given in Chapter 21 on the government sector.

Current prices

12.74   Estimates of taxes on products are derived on a quarterly basis, mostly from government administrative returns. They are accounted for on a cash basis. As such, they may not always reflect the profile of spending to which they should be related. This is particularly true of Value Added Tax (VAT). VAT accruals are based on returns by VAT registered traders to Customs and Excise showing their VAT liable transactions and VAT accruing during the period. The returns are on staggered three month periods, interpolated to provide calendar quarter figures. For other taxes where it is necessary to approximate more closely to the timing of the liability, the estimates which go into the accounts are the cash receipts, lagged by one month. Where this is done, corresponding 'accounts receivable' and 'accounts payable' are included in the financial accounts to preserve consistency with the cash basis used there.

Constant prices

12.75   At constant prices, estimates are made in a number of ways. The main approaches are (i) multiplying quantity estimates of consumption by the rate of tax applying in the base year, and (ii) for ad valorem taxes, the base year ratio of tax to spending is simply applied to the present constant price estimates of spending.

Subsidies on products (D.31)

12.76   These are subsidies payable per unit of good or service produced, with the aim of reducing the purchase price to the consumer. The main subsidies in the UK, which are paid by both central and local government, relate to agriculture, transport and housing. Further details are given in Chapter 21 on the government sector.

Current prices

12.77   Annual estimates of subsidies are derived from administrative and various other sources. As with taxes, the information is provided on a cash rather than an accruals basis. The annual and quarterly figures of subsidies for public corporations are derived from their accounts, on the assumption that they will better reflect the timing of when the subsidy accrues. Again as with taxes, accounts receivable and payable are included in the financial accounts. Elsewhere, quarterly estimates are largely the annual figure spread uniformly through the year.

Constant prices

12.78   Estimates at constant prices are derived mainly by multiplying quantity data for the period in question by the corresponding subsidy in the base year.

Reconciliation with expenditure and production measures

12.79   It was mentioned earlier that all the variables appearing in the goods and services account are components of the expenditure and production measures of GDP. The simple reconciliation, given below.

GDP at market prices

Production measure

	Output (P1)
less	Intermediate consumption (P2)
plus	Taxes on products (D21)
less	Subsidies on products (D31)
equals	GDP at market prices

Expenditure measure

	Final consumption expenditure (P3)
plus	Gross fixed capital formation (P51)
plus	Changes in inventories (P52)
plus	Acquisition less disposals of valuables (P53)
plus	Exports of goods and services (P6)
less	Imports of goods and services (P7)
equals	GDP at market prices

12.80   It should be noted that actual final consumption (P4) could have been used instead of final consumption expenditure (P3). Further, as with the goods and services account, the total GDP aggregate is presented at market prices. In the new system of accounts, the focus is more on this valuation than on factor cost, in which all taxes are deducted and all subsidies are added in.

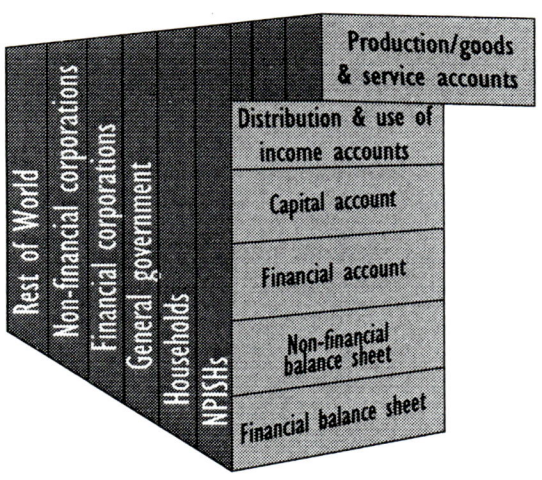

Production account
and other production-
based statistics

Chapter 13 Production account and other production-based statistics

Introduction

13.1   This chapter describes, in three sub-chapters, the methodology for the estimation of the variables for (i) the production account, (ii) the compilation of the supply and use tables, and (iii) the production-based estimate of constant price GDP. The description covers the whole economy estimates, but in particular the component series, including the industry variables, from which the aggregate is usually derived, and, where relevant, the institutional sector analysis.

13.2   There is some overlap between the first two sections, in that the output and intermediate consumption variables of the production account are also included, in more disaggregated form, in the supply and use tables. This will be explained further in the text.

Changes in terminology

Previously in the UK national accounts statistics, output will have generally been referred to as 'gross output', to distinguish it from 'net output'; the latter was defined as (gross) output less inputs. In the new SNA, the term:

Gross output is replaced by *output*;

net output is replaced by *gross value added*;

inputs is replaced by *intermediate consumption*; and

output measure of GDP is replaced by *production measure of GDP*.

'*Gross*' means that the measure is before deducting consumption of fixed capital (depreciation).
To summarise the above, the old relationship of;
Net output = gross output minus inputs
is now superseded by:
Gross value added = output minus intermediate consumption.

The production account

13.3   The production account, which is very much a new feature of the presentation of the UK statistics, relates output, as a resource, and intermediate consumption, as a use, to establish, as the difference, gross value added. If consumption of fixed capital is further subtracted, value added is measured on the net basis. Conceptual issues relating to the production boundary and to the variables included in the production account have been discussed in Chapter 4.

13.4   The production account for the whole economy for 1995 is shown below.

		£ million
Uses	**Resources**	
Intermediate consumption	Output	
	Market output	
	Output for own final use	
	Other non-market output	
	Taxes less subsidies on products	
Balance:		
Gross domestic product		
less Consumption of fixed capital		
Net domestic product		

13.5   It should be noted that, in respect of the whole economy, and as shown above, taxes less subsidies on products are also included as a resource. This enables a balance to be struck for Gross (or Net) Domestic Product at market prices. Production accounts are also compiled for individual institutional sectors or industries, with the balance – value added – showing the sector's or industry's contribution to GDP or NDP. However, there is an important difference in presentation in that, in the sector and industry accounts, taxes less subsidies on products are not included. In other words, the sum of value added by individual industries or institutional sectors would differ from GDP at market prices by the aggregate of taxes less subsidies on products. A similar adjustment is needed for financial services indirectly measured (FISIM). (*See* Annex 1 to Chapter 20.)

13.6   With the exception of consumption of fixed capital, the methodology for estimating the aggregates of the variables of the production account has been set out, in summary form, in the previous chapter, which also included some key features of the variables. This sub-chapter will consider some of these aspects more fully, but with the main emphasis on the derivation of the estimates of the production account variables by industry and by institutional sector. This will be done for each of the variables of the account separately. The industry analysis will relate essentially to the main 17 groups of NACE Rev 1. However, the text will also embrace the detail required for the 31 standard groups, which are formed by further disaggregation of manufacturing and of mining and quarrying. The analysis by institutional sector will cover the seven main sectors. The description relates to annual current price information.

Output (P1)

13.7   Output is essentially the measure of productive activity – of goods or services – taking place in a particular period. Output covers the normal production of both goods and services by businesses, and the activities of government, households and the NPISH sector within the production boundary. The concepts and definitions have been described in detail in Chapter 4.

Changes from the previous system

There are three significant changes to the coverage and measurement of output in the new system.

- the first is the inclusion of growth of cultivated products, such as crops and livestock. Previously, such output was recorded only at the moment of the harvest of the crops or the slaughter of the animal. Now, like any other production which extends over a number of periods, the activity is recorded in output as work-in-progress.
- the second change, which is discussed in Annex 1 to Chapter 20, relates to Financial Intermediation Services Indirectly Measured (FISIM). This is not yet implemented.
- the third is the extension of the production boundary to cover (in principle) illegal activities, such as drug dealing and prostitution. Needless to say, such activities are almost impossible to evaluate accurately, but it is intended to make some allowance explicitly for certain of these but because of the inherent difficulties they have not yet been incorporated. (*See* Chapter 2).

In addition, some work in progress has been identified and measured in service industries.

The composition of output

13.8   Output, in aggregate, and also by industry, is built up as the sum of three main components, which appear separately in the production account for the whole economy and, as appropriate, for the seven institutional sectors for which the full range of accounts is compiled. The components are:

- Market output – that is, output sold at prices which are economically significant. This will cover the traditional output of firms, embracing both goods and services.

- Output produced for own final use – that is, output produced by a business for its own use, and certain activities of households

- Other non-market output – that is, output at prices which are not economically significant. This will cover almost all government and NPISH sector output.

13.9   A price which is economically significant is one which affects the level of both supply or demand for a particular good or service. More specifically, such a price is defined to exist where more than 50 per cent of the production costs of output are covered by sales. For certain market activities, for example retailing and financial intermediation, output is defined in a particular way, as explained in the section below on the industry analysis.

13.10   Some activities can produce either market or non-market output. The main examples relate to the provision of education, health or social services. In principle, where producers engage in both market and non-market production, separate producer units should be established, and estimates made for all the main variables for the accounts for the separate units. Where this is not possible, producers are divided into market and non-market according to which output predominates.

Market output

13.11 The first of the three components – market output – is by far the largest, accounting for around 90% of total output. The typical production of firms, covering both goods and services, would be classified as market output. Where productive activity takes place, output arises in a number of forms. The main outcome of the production process will be that the goods or services produced in a given period will be sold to another producer or to a final buyer in that period. However, goods which are produced may not be sold immediately, but will be put into inventories (or stocks). A further possibility is that, where the production process extends over more than one reporting period, there is no finished product, as such. Here, output is represented by the change in work-in-progress which has occurred during the period of measurement. While these outputs comprise the bulk of market output, also included is certain economic activity the output of which is not sold on the market.

13.12 Thus, it is convenient to see market output for the national accounts defined as the sum of:

- Sales of own production;

- Changes in inventories of finished goods and work-in-progress; and

- Output not sold on the market.

13.13 This way of looking at output is useful in its own right. However, just as importantly, this formulation largely reflects the way in which, in practice, output is recorded and estimated for the national accounts. The reason for this approach is that, most generally, it is not realistically possible to measure output directly. However, the separate estimates of sales and inventories, which are made as part of commercial accounting practices, provide a means of estimation.

13.14 The inventories component is particularly important for industries such as aerospace and shipbuilding. In respect of the production of major capital assets, output is recorded by work in progress in the particular period. This will often take the form of a progress payment made by the purchaser to the producer. Alternatively, output may be estimated by spreading the total value of the asset over the period of production in proportion to the costs incurred in each period. When the asset is completed and sold, the sale is recorded in that period, but the accumulated value of work-in-progress recorded in the latest period and previous periods is unwound by a single negative entry in the period of sale. Thus, the value of output recorded over the period of production is the value of the sale, but spread over the whole production span.

13.15 Estimates of market output also embrace those parts of what may be more commonly non-market activities, such as education and health, which are sold on the market. Market output for the national accounts is valued at basic prices. The way in which each of these components is estimated for the accounts is now described.

Sales of own production

13.16 As mentioned in the previous chapter, the principal data sources for estimating this main component of output are the ONS annual inquiries into production (*see* section 25.1.3), construction (25.1.5), and the distribution and services trades (25.1.8–12). Other contributions come from surveys conducted by the Ministry of Agriculture, Fisheries and Food (MAFF) for agriculture, and by the Department of the Environment, Transport and the Regions (DETR) for the construction industry. The ONS service trades inquiries do not, however, cover all industries. For these and certain other 'market' industries, alternative estimation methods are used, as will be described below.

Changes in inventories of finished goods and work-in-progress

13.17   This form of output represents production which, in essence, has taken place but which has not yet been sold. The item will have been made and transferred to inventory as a finished good; alternatively, it still remains to be completed at the end of the reporting period. In extremis, for work on a major capital project, all output in a given period may consist of work-in-progress, with no sales. An exception is work-in-progress for structures to be used by the builder. These are treated as gross fixed capital formation (GFCF).

13.18   The measurement of the contribution to output arising from changes in inventories is far from straightforward. This is a reflection of the particular way in which inventories need to be valued for the national accounts. This requires a valuation based on 'replacement cost', which relates to the value at the time the good is used in the production process. The valuation issue is discussed in the chapter on inventories (Chapter 15).

13.19   Clearly, to attempt to value all transactions at every stage of the production process in this way is hardly a realistic proposition. Hence, in practice, for their commercial accounts most firms value items for stock essentially on the basis of the 'historic cost', that is the cost at the time of purchase. The two valuations will not, most generally, be the same. However, if the historic cost basis were used in the national accounts, part of the change in the level of inventories would reflect the effect of price changes, rather than a genuine 'volume' change. With rising prices, unless allowance is made, the effect would be to understate the cost of inputs to the productive process and to overstate the contribution to output.

13.20   The methodology used for the estimation of the inventories component of output for the production account is fairly complicated. It is discussed in detail in Chapter 15 on the capital account, as part of the general methodology for inventories in the accounts. In principle, it involves revaluing the historic cost, book values (levels) of inventories, by industry, at end-year or end-quarter, in a way which approximates to the replacement cost basis of recording required for the national accounts. The contribution to output arising from inventories is then derived as the difference in the revalued stock levels at the beginning and end of the particular period. Estimates are made initially at constant prices and then reflated, by the average prices obtaining in the period, to current prices. The approach is known as the adjustment for holding gains (or stock appreciation). Most of the information on book values comes from ONS annual industrial surveys, mentioned above, and quarterly stocks surveys (25.2.4), with the prices information needed for revaluation coming from the producer and retail price indices (25.2.7 and 25.4.7, respectively).

Output not sold on the market

13.21   The three main components under this head are:

- Sales to establishments within the same enterprise group

- Payments to employees as income in kind

- Barter transactions

13.22   The last of these – essentially the exchange of the provision of goods or services without the use of cash – is not considered to be of any significant size in the United Kingdom, so no estimate is made in the accounts. For the first component, estimates of intra-establishment sales are recorded within the aggregate of sales included in the various statistical inquiries from which the figures of sales, above, are derived. Thus, the only item which needs to be estimated is income in kind. This component, which is covered in more detail in Chapter 14, relates to receipt, by an employee, of goods and services instead of money. The main examples of income in kind in the UK are the provision of company cars, of rent-free or reduced rent accommodation, and meal vouchers. Estimates are derived from a variety of sources, including Inland Revenue (25.4.3), the FES (25.4.5) and official surveys of labour costs. Some annual estimates are projections of periodic benchmark data. It should be noted that the figures of income in kind are included not only in output, but also in compensation of employees in the income-based measure of GDP, and in household consumption for the expenditure-based estimate.

Output produced for own final use

13.23 The second component – output produced for own final use – covers production by households, government and businesses. Communal activity is also included. For households, the main items included are the services of owner-occupied dwellings; construction activity, such as the building of a new dwelling or major improvements to existing dwellings; payments to domestic staff, and own production of agricultural produce. However, any payments for work done by members of the household for the benefit of the household are excluded. Information is derived from a variety of sources, as is explained as part of the description of the estimation of the details of household consumption, in Chapter 22.

13.24 For government this consists of the production of fixed assets for its own use, in recent years mainly capitalised computer software.

13.25 In respect of firms, the principal own-account output will relate to making machine tools, or undertaking construction work, such as building or improvements to factories. This information, previously included with sales, is now being collected separately in the ONS main annual statistical inquiries mentioned in paragraph 13.17.

13.26 Own-account production is valued at the basic price for similar market activities. If this is not possible, valuation is based on costs. The recording of this form of output has its counterpart in final consumption or GFCF in the expenditure measure of GDP, and also in compensation of employees in the income measure.

Other non-market output

13.27 The third component – other non-market output – covers output that is provided free or at prices which are not economically significant. Such production will be financed out of taxation or donations or other fees or the like. The main example of this kind of output arises in respect of the activity of the provision of goods and services by the State, both central and local government, such as the defence of the realm, general public administration and services such as health and education. Non profit institutions serving households (the NPISH sector) also produce non-market output.

13.28 The borderline between producers of market and non-market output is not always easy to draw. Even where it is, information on the separate activities is not always available in the required form. Partly reflecting this, in the UK statistics, almost all producers within both central and local government are classified as non-market producers. The exceptions are, for central government – the Export Credits Guarantee Department, and for local government – the housing revenue accounts and trading services.

13.29 In the absence of a meaningful price for valuation, non-market output is, by international convention, valued at the costs incurred in the production of the goods and services. This is defined as the sum of intermediate consumption, compensation of employees, consumption of fixed capital, and other taxes on production less other subsidies on production. Where non-market producers engage in market activities, receipts from such activities are deducted from total costs to arrive at non-market output.

13.30 The information used to measure non-market output for government and the NPISH sector is described in Chapters 12, 21 and 23.

The detailed analyses of output

13.31 Chapter 12, on the goods and services account, described how the aggregate estimates of output, by the three main components, were made. The main sources and approach have been outlined in the preceding paragraphs. The following sections describe the analysis of output by industry and by institutional sector.

The analysis by industry

13.32 This section describes how output is estimated for the 17 broad industry groups of NACE. A more disaggregated analysis – for 123 industries – is made for the input-output analysis, as described later in this chapter (paragraph 13.109 *et seq*). The description here relates to annual current price data. Constant price estimates of output by industry do not appear in the accounts as such. However, such estimates may be used in the derivation of constant price value added by industry (*see* paragraphs 13.141 *et seq*). Although output is made up of the three components – market output, own-produced output, and other non-market output – the combined figure only is required for the industry analysis.

13.33 The analysis of output by industry largely reflects the way in which the aggregate estimate is derived. There are though some important definitional and measurement differences for the different industries, which are discussed below. The description of the methodology gives some indication of the importance of the various institutional sectors, within the industry group. This is brought together in the later section which looks at the analysis by institutional sector.

13.34 Classification of output to an industry is, in general, based on the principal activity of the unit reporting the expenditure, and as recorded on the Inter Departmental Business Register (IDBR) (*see* Chapter 25), and used in the statistical inquiries. For the production industries, the reporting unit is often an individual factory, which can largely provide the full range of statistical information required for the national accounts. For the distributive and service industries, the industrial classification is generally based upon legal units which are individual companies in the corporate sector, sole proprietorships, partnerships etc.

13.35 Given the nature of the three components of output, much of the description included below will relate to the estimation of market output, that is the contribution from sales and changes in inventories. Figures, by industry, for both these variables are based on the common source of ONS annual industry inquiries (25.1), so, where appropriate, the description below applies to both. For the other component of market output – income in kind – the allocation by industry is provided by the Inland Revenue. For government, administrative records are used for industry allocation while that for the NPISH sector is direct from constituent bodies.

Agriculture, hunting and forestry

Fishing

13.36 Estimates of output for agriculture are mostly compiled by MAFF as part of the derivation of the accounts for the national farm. Some small addition is made for non-farm agriculture enterprises, As mentioned earlier, estimates include as output, the growth in cultivated assets, such as livestock and also growing crops. This output is equivalent to work in progress. The figures also go into inventories or GFCF in the expenditure measure of GDP, and the method of estimation is described in the chapter on the capital account (Chapter 15). The figures of output also include own-account production by households in the industry. This, again, is a component of expenditure GDP – it is part of household consumption – and the method of estimation is explained there (Chapter 22). Estimates for forestry – now embracing growing trees – are supplied by the Forestry Commission.

Mining and quarrying

Manufacturing

Electricity, gas and water supply

13.37   Estimates for these three industry groups are derived from information on sales and inventories collected in ONS annual inquiries to the production industries (25.1.3). This source also provides the basis for the further sub-division to move from 17 industry groups to 31 groups. This involves splitting mining and quarrying between 'energy producing materials' and 'other', and splitting the industry aggregate for manufacturing into 14 separate detailed manufacturing industries. Output of major capital goods is usually measured by progress payments made by the owner to the producer over the whole period of production and not at the time of completion of the production. Such payments appear in GFCF in the expenditure measure. Alternatively, and for work on own-account, the total estimated value of the output is spread over the quarters of production in proportion to the costs in these periods. This expenditure is also recorded as part of GFCF. Most of output for these industries is attributable to non-financial corporations, but there are small elements for households.

Construction

13.38   Annual estimates of construction output are based on a number of essentially alternative sources. The reason for this is to try to cater, as best as possible, for certain problems of measurement for this industry, for example in respect of activity in the hidden economy. The sources are the ONS annual inquiry (25.1.5), DETR's construction surveys, and periodic English House Conditions Survey, and also Inland Revenue information on self-employment incomes (25.4.2). The output of the construction industry does not include own-account capital formation by businesses and households, although this is included in the construction product. The estimation for the latter is described in the chapter on household consumption (Chapter 22). The estimation process also makes use of figures assembled for a supply/demand balance for the industry, using estimates from the various sources described above.

Wholesale and retail trade; repair of motor vehicles, motorcycles and personal and household goods

13.39   In respect of wholesale and retail industries, output is defined, not as the value of sales, but as the margin on these sales, that is the difference between sales and the purchases of goods for sale (with allowance for change in inventories, including any losses for theft or damage). The estimates of margins are derived from information on sales and purchases, and inventories, as collected in the ONS annual inquiries to the wholesaling, retailing and motor trades industries (25.1.7–10).

Hotels and restaurants

13.40   Estimates are based on information on sale and inventories from the ONS annual catering inquiry (25.1.11).

Transport, storage and communication

13.41   Company reports are the only source for the air transport industry, but are also used to adjust the inquiry figures for other industries.

Financial intermediation

13.42   The definition of output of financial intermediation raises a number of issues of concept and measurement. The estimates of output for the industries in this group are based not on direct measurement from an inquiry, but on information going into the income measure of GDP. This involves, first, establishing an estimate of gross value added for each of the main industries within the group, derived as the sum of compensation of employees and operating surplus, and then applying an estimated ratio of output to gross value added to the GVA figures. (*See* the general notes on FISIM and insurance in Chapter 20 and its annexes.)

Real estate, renting and business activities

13.43   Annual estimates of output are derived for the separate industries within this group. The approach used for renting and real estate follows the income-based approach for financial intermediation, described above. Also included under this head is the output of the services of owner-occupied dwellings, the estimation of which is described in the chapter on the household sector (Chapter 22). For the other business service activities in this group, estimates of output are based mostly on information on turnover/receipts from the ONS annual inquiries (25.1.11, 12).

Public administration and defence; compulsory social security

13.44   All units in this industry are non-market producers. Most output is non-market output. Such output is estimated as the sum of intermediate consumption, compensation of employees, consumption of fixed capital and other taxes less subsidies on production. Information is obtained, mostly, from government administrative sources. Consumption of fixed capital is estimated from the perpetual inventory method (PIM) approach for deriving figures of capital stock (see Chapter 16).

Education

Health and social work

13.45   Output of these two industry groups will include both market and non-market output. As far as possible, attempts are made to distinguish between market and non-market producers and between their market and non-market output. Information from market producers is derived from the turnover/receipts data collected in the ONS annual inquiries (25.1.11), other than for education, which uses the income-based approach, described above for financial intermediation. Estimates for non-market producers are from the public expenditure planning and monitoring system, and are derived as for public administration.

Other community, social and personal service activities

13.46   These industries will also include market and non-market activities. For the market producers, estimates of output are based largely on information on turnover/receipts from the ONS annual service sector inquiries (25.1.11). For the non-market component, figures are derived as for public administration.

Private households with employed persons

13.47   The estimates are derived from the income approach. Output is by international convention deemed to equal the remuneration of domestic staff.

The analysis by institutional sector

13.48   This section describes the estimation of output for the seven main institutional sectors. Given the increased emphasis on the analysis by institutional sector, some of the key features for the sectors are also described.

13.49 Estimation is a mixture of direct measurement and, for the private sector, allocation of industry-based totals. First, the data for central and local government and for public non-financial corporations, are collected from separate surveys and so are readily identified. For the private sector, estimates for financial corporations are derived from the appropriate industry classification used in the surveys, and information for the NPISH sector is also readily identified. For other private sector investment, a breakdown is made between private non-financial companies and households, largely based on the sector allocation appearing on the statistical register of businesses (IDBR). It should be borne in mind that the long-run figures on a sector basis will be influenced by changes in the composition of the sectors, for example as a result of privatisations.

Taxes on products (D21) and Subsidies on products (D31)

13.50 Given the role these variables have in the national accounts, it is sensible to consider them together. The concept of taxes on production and subsidies – taxes essentially increase the price of a product and subsidies reduce it – has been explained in Chapter 3. In addition, the role of taxes and subsidies generally in the valuation of the national aggregates, such as GDP, is discussed in Chapter 12. This section will consider the main features as they relate to the production account.

13.51 Figures on taxes and subsidies on products are included in the production account for the whole economy, but not for similar accounts for institutional sectors or industries. The estimates of output appearing in the production account include figures for other taxes and subsidies on production, for example national non-domestic rates. However, they do not embrace taxes and subsidies on products, such as VAT, which are regarded as being borne by consumers. Thus the inclusion of taxes less subsidies on products is necessary to enable GDP at market prices to be derived.

Taxes on products (D21)

13.52 These are taxes payable per unit of good or service being produced. The main UK taxes on products are value added tax, and taxes on drink, tobacco, hydrocarbon oil and betting, and stamp duties. Some information was given in Chapter 12 on the goods and services account, and further details are included in Chapter 21 on the government sector. Estimates are derived quarterly, mainly from Customs and Excise records. They are essentially accounted for on a cash basis. Adjustments are made to some taxes to put them closer to the accruals basis. Where this is done, corresponding 'accounts receivable' and 'accounts payable' are included in the financial accounts to preserve consistency with the cash basis used there.

Subsidies on products (D31)

13.53 These are subsidies payable per unit of good or service produced, with the aim of reducing the purchase price to the consumer. The main subsidies in the UK, which are paid by both central and local government, relate to agriculture, transport and housing. Further details are given in Chapter 21 on the government sector. Estimates are derived from the public expenditure monitoring system and the accounts of public corporations. As with taxes, some adjustments are made to an accruals basis, with corresponding accruals adjustments in the financial accounts.

The detailed analysis of taxes and subsidies on products

13.54 As mentioned above, unlike the other variables in the production account, the account for the institutional sectors does not include taxes less subsidies on products. The aggregate estimates are, however, broken down by industry, for input-output purposes, as described in paragraphs 13.105–13.107.

Intermediate consumption (P2)

13.55 Intermediate consumption is the value of goods and services consumed – transformed or used up – in the process of production. Previously, intermediate consumption was generally known as 'inputs'. The concepts and definitions relating to intermediate consumption have been described in detail in Chapter 4. The description has covered some important inclusions and exclusions, including the boundary with gross capital formation (particularly the treatment of maintenance and repairs) and with compensation of employees.

13.56 A further consideration with the definition of intermediate consumption is to ensure consistency with the measure of output. For example, goods and services purchased from other establishments in the same enterprise are included in intermediate consumption, while those produced and consumed by the same establishment in a particular accounting period are excluded. It should also be noted that rental payments for use of fixed assets are included as part of intermediate consumption. However, where assets are owned, no imputed rental is included.

13.57 Intermediate consumption is valued at purchasers' prices for the national accounts. It is recorded and valued at the time the good or service is used in the production process – essentially the replacement cost principle (see Chapter 15) – which is not usually the time of purchase. The principles of measurement of intermediate consumption are analogous to those discussed above for output. Similarly, on the practical side, it is clearly not readily possible to record and value information on the use of goods in the production process in this way. In practice, producers will record the purchases of goods and services intended for use in production and also the change in their stockholding of such goods. Thus, intermediate consumption in aggregate (and by industry) for the national accounts is estimated as:

Purchases *less* the change in raw materials and supplies held in inventories

13.58 These variables are now considered separately.

Changes from former system of accounts

The definition of intermediate consumption contains a number of important changes in coverage compared with the previous system. The main changes are that mineral exploration; computer software; entertainment, literary and artistic originals, and military equipment (other than on weapons), all previously included in intermediate consumption, are now to be excluded and classified as gross fixed capital formation (GFCF).

Purchases

13.59 The data sources for estimates of purchases are ONS annual production, construction and distribution and service (25.1) inquiries, together with information from DETR's construction surveys and from MAFF. For those industries not covered in these inquiries, as for output, estimates of intermediate consumption are derived from alternative, income-based sources, as explained below.

Change in inventories of raw materials and supplies

13.60 This component of intermediate consumption represents the contribution to inputs resulting from drawing on material and supplies held in stock. It is also a component, along with inventories of finished goods and work-in-progress, which is included in the expenditure measure of GDP. The measurement issues here parallel those affecting the contribution to output of inventories of finished goods and work-in-progress. The key features are repeated here largely for convenience, with a fuller explanation given in Chapter 15.

13.61 The methodology used involves revaluing the historic cost, book values (levels) of inventories of raw materials and supplies, by industry, from ONS industrial surveys (25.1, 25.2), at end-year or end-quarter, in a way which approximates to the replacement cost basis of recording required for the national accounts. The contribution to intermediate consumption arising from inventories is then derived as the difference in the revalued stock levels at the beginning and end of the particular period. Estimates are made initially at constant prices and then reflated, by the average prices obtaining in the period, to current prices. Price information comes from the producer and retail price indices (25.2.7 and 25.4.7, respectively).

The detailed analyses of intermediate consumption

13.62 The previous chapter described how the aggregate estimates of intermediate consumption were made. The main sources and approach have been outlined in the preceding paragraphs. The following sections describe the analysis of intermediate consumption by industry and by institutional sector.

The analysis by industry

13.63 This section describes how intermediate consumption is estimated for the 17 broad industry groups of NACE. A more disaggregated analysis – for 123 industries – is made for the input-output analysis, as explained later in this chapter. The description here relates to annual, current price data. Constant price estimates of intermediate consumption, by industry, do not appear in the accounts, as such, other than in supply and use tables. However, such estimates may be used in the derivation of constant price value added by industry.

13.64 The analysis of intermediate consumption by industry largely reflects the way in which the aggregate estimate is derived, and is very similar to what is done for the output variable. The description of the methodology gives some indication of the importance of the various institutional sectors. This is brought together in the later section which looks at the analysis by institutional sector.

13.65 Classification of intermediate consumption to an industry is, in general, based on the principal activity of the unit reporting the expenditure, and as recorded on the IDBR, and used in the statistical inquiries. For the production industries, the reporting unit is often an individual factory, which can provide the full range of statistical information needed for the accounts. For the distributive and service industries, the industrial classification is generally based upon legal units which are individual companies in the corporate sector, sole proprietorships, partnerships etc. Note that the consumption of financial intermediation services (FISIM) is not allocated to industries, but is recorded as the intermediate consumption of a notional industry.

Agriculture, hunting and forestry

Fishing

13.66 As with output, estimates of intermediate consumption for agriculture are mostly compiled by MAFF, as part of the derivation of the accounts for the national farm. Some small allowance is made for non-farm agricultural enterprises, and also for own-account production by households. Estimates for forestry are supplied by the Forestry Commission, and for fishing by MAFF.

Mining and quarrying

Manufacturing

Electricity, gas and water supply

13.67 Estimates of intermediate consumption for these three broad industry groups are derived from information on purchases and material and supplies inventories, collected in ONS annual inquiries to the production industries (25.1). This source also provides the basis for more detailed analysis into 31 groups.

Construction

13.68 As with output, annual estimates of intermediate consumption for construction are based on various sources, largely reflecting certain problems of measurement, for example in respect of the effect of activity in the hidden economy. The main sources are the ONS annual inquiry (25.1.5) and DETR's construction surveys. In addition, the estimates make use of data assembled for a supply/demand balance for the industry.

Wholesale and retail trade; repair of motor vehicles, motorcycles and personal and household goods

13.69 Estimates of intermediate consumption are derived from information on purchases and on stocks of materials and supplies, collected in ONS annual inquiries to the wholesaling, retailing and motor trades industries (25.1.7–10). Purchases of goods for resale are not part of intermediate consumption. These have already been deducted in arriving at the definition of output.

Hotels and restaurants

13.70 Figures are based on information on purchases and on material and supplies stocks obtained in the ONS annual catering inquiry (25.1.11).

Transport, storage and communication

13.71 Estimates of intermediate consumption are based on information on purchases obtained in ONS annual inquiries to certain industries within the group, and, for the remainder, on various *ad hoc* sources, including company reports. As with output, the coverage of the figures for road transport is less firmly based than for other industries, reflecting problems in identifying and measuring only that road haulage activity undertaken by transport industries.

Financial intermediation

13.72 Figures for intermediate consumption are derived as the difference between the output estimate and the estimate for gross value added, based on income data. This is equivalent to applying an estimated ratio of intermediate consumption to value added to the latter figure.

Real estate, renting and business activities

13.73 Estimates are derived for the separate industries within this group. For renting and real estate, the method follows the income-based approach for financial intermediation, described above. Intermediate consumption of the services of owner-occupied dwellings is as described in the chapter on the household sector (Chapter 22). For the other business service activities in this group, estimates of intermediate consumption are based on purchases information from the ONS annual service sector inquiries.

Public administration and defence; compulsory social security

13.74 As described in the estimation of output, intermediate consumption is a component part of the estimation of non-market output for this group. The information is obtained from public expenditure sources.

Education

Health and social work

13.75　These industries can be market and non-market. For the former, other than for education, estimates of intermediate consumption are derived from purchases data obtained in ONS annual inquiries (25.1.11). For education, estimates are based on the income approach, as mentioned above for financial intermediation. Estimates for non-market producers are derived from public expenditure monitoring sources, and are compiled as explained above for public administration.

Other community, social and personal service activities

13.76　Again, these industries include both market and non-market activities. For the market component, estimates of intermediate consumption are based largely on information on purchases from ONS annual service sector inquiries (25.1.11). For the non-market producers, figures are derived as for public administration.

Private households with employed persons

13 77　This industry has no intermediate consumption. Output equals the compensation of employed staff.

The analysis by institutional sector

13.78　The estimation of intermediate consumption for the seven main institutional sectors largely follows the methodology used for output, as described above.

Consumption of fixed capital (K1)

13.79　Consumption of fixed capital (previously capital consumption) in a given period represents the amount of fixed assets used up as a result of wear and tear and foreseeable obsolescence, including provision for insurable losses. The relevance of the estimate of capital consumption in the accounts is to adjust figures from the 'gross' basis of recording to the 'net' basis.

13.80　Provision for consumption of capital is an important feature of commercial accounting, where it is termed depreciation. However, the estimates used for the national accounts differ in two important respects from those obtaining in commercial accounts. The differences are first that the national accounts figures are based on current replacement cost (equivalent to current purchasers' prices), rather than historic cost, and secondly, they generally use a longer length of life over which the asset is depreciated.

13.81　The figures going into the accounts are derived as part of the estimation process for the measuring the stock of capital. This uses a model-based approach – the Perpetual Inventory Method (PIM) – rather than direct measurement. The methodology uses data on GFCF and prices, together with assumptions on the length of life of the various assets and their retirement distribution and, by convention, a straight line method of depreciation. It is described in Chapter 16 on the non-financial balance sheets. However, it will be useful to outline here the main principles of measurement.

13.82 Within the model, the estimates of capital consumption (and capital stock) are made for largely homogeneous groups of assets. For each group, there will be available a long-run series of annual current price figures of GFCF. The first step is to revalue the current price data to a constant price base using appropriate price series. Then, using the assumptions about expected length of life of assets in each group, the constant price data are depreciated, on a straight line basis, over the assumed life. Thus, an element of depreciation is imputed for each asset group for each year of its assumed life. The next step is to aggregate, for each year, across the range of all assets covered, to provide an annual series of consumption of fixed capital at constant prices. Finally, to derive estimates at current (replacement) cost, which is the main basis used in the accounts, the series of annual constant price estimates for each asset group are reflated by the appropriate price indices and then summed across the assets for each year.

The detailed analyses of consumption of fixed capital

13.83 The model generates consistent analyses by industry, type of asset and institutional sector, based essentially on the classification of the input data.

Supply and use tables

General descriptions

13.84 This section describes the supply and use tables in the UK national accounts. The main focus will be on the sources of data used to compile the estimates. However, a brief explanation will also be given of the structure and nature of the supply and use tables. A fuller description of various aspects of the I-O work is contained in the ONS *Input-Output Methodological Guide* and in the associated material referred to therein.

13.85 The supply and use tables provide a detailed analysis, by product, of the production and use of goods and services in the economy, and also the income generated by that activity. The analysis described here is based on two main tables or matrices, namely a Supply table and a Combined use table, as shown in very summary form below.

13.86 The Supply table (Table 13.1) shows the total availability (supply) of individual products (goods and services), by industry, for use in the economy from both domestic production and imports. The information on domestic production, by industry, (section A) is at basic prices. Imports (section B), covering both goods and services, are valued on the free on board (f.o.b.) basis. The table also includes distributors margins (C) and taxes less subsidies on products (D). These, together, convert the basic price valuation of production and imports to purchasers' prices for total supply. It should be noted that a distinction is made between industries and products, though the same classification is used for both. Producing units are classified to industries according to which products they make. If they produce more than one product they are classified according to whichever accounts for the greatest part of their output.

Table 13.1: Supply matrix

Sales by industry group									
Sales by product	1	2	3	4	Domestic product output	Imports of goods and services	Distrib utors' trade margins	Taxes less subsidies on products	Total supply of products
1	A	A	A	A		B	C	D	B1
2	A	A	A	A		B	C	D	B2
3	A	A	A	A		B	C	D	B3
4	A	A	A	A		B	C	D	B4
Total output	A1	A2	A3	A4	Y		Nil		Z

13.87 The Combined use table (Table 13.2) has two dimensions. Across the rows the table shows how the supply of individual products, by industry, derived as in Table 13.1, goes to the various components of demand - first to intermediate consumption (E) and then to the final demand categories of consumption by households (F), the NPISH sector (G) and government (H), and to GFCF (I), inventories (J) and exports (K). Also shown is the allocation of other taxes (other than on products) less subsidies on production (L) to intermediate demand, and the breakdown of value added (essentially compensation of employees and operating surplus) (M) by industry. The columns for the individual industries show the make up of total inputs, as represented by the separate products purchased for intermediate consumption (E) and value added (M). The Combined use matrix is equivalent to a detailed production account for the whole economy. The data in the table are valued at purchasers' prices.

Table 13.2

Purchases of product	Purchases by industry group					Purchases by final demand categories							Total demand for products
	1	2	3	4	Total intermediate demand	House holds	NPISH	GG	GFCF	Invent ories	Exports	Total final demand	
1	E	E	E	E		F	G	H	I	J	K		B1
2	E	E	E	E		F	G	H	I	J	K		B2
3	E	E	E	E		F	G	H	I	J	K		B3
4	E	E	E	E		F	G	H	I	J	K		B4
Total inter mediate purchases													Z
Other taxes less subsidies on prod uction	L	L	L	L									
Gross value added	M	M	M	M									
Total inputs	A1	A2	A3	A4	Z								

13.88   As can be seen from the two tables, for each industry total output (A1 to A4 in table 13.1) is equal to total input (A1 to A4 in table 13.2), and for each product, total supply (B1 to B4 in table 13.1) is equal to total demand (B1 to B4 in table 13.2). Further, with total output (at basic prices) denoted by Y, total intermediate consumption by Z, and taxes *less* subsidies on products by X, then GDP at market prices is defined as Y+X-Z.

Changes from former system of accounts

Largely reflecting the introduction of the new system of national accounts, a number of changes have been made to the presentation of supply and use tables. The previous series sales by final demand is no longer shown. The part of it relating to sales of existing goods has now been moved to the rows for the relevant products. The rest of it – sales by final demand, e.g. by government - are now shown as output rather than being netted off final demand. A second change is that the 'Taxes *less* subsidies' row of the previous supply and use tables now relates solely to producers; no components are appropriate to final demand.

Uses of input-output tables

13.89   Input-output tables can provide a basis for economic analysis, for example to evaluate the impact on individual industry groups of fiscal or economic policy decisions. They can also be used as a framework for reconciling and balancing the various expenditure, income and production estimates assembled for the accounts, including the determination of the level of GDP. The way in which the accounts are balanced has been described in Chapter 11. ONS also uses the I-O data to provide the value-added weights used in compiling the constant price production-based measure of GDP, and also weights for combining detailed price data, for example for the producer price index, both in its own right and for deflating value data (on output, intermediate consumption, GFCF or inventories) to constant prices.

Form of supply and use tables

13.90   At present supply and use tables in the UK are compiled annually, at current prices. Estimates are available about 18 months after the end of the year to which they relate. Thus, the mid-year Blue Book for year (t) will include the supply and use tables for year (t-2). The analysis is undertaken for 123 separate industry/product groups. This degree of detail is presented in separate publications which give the main supply and use tables, together with a number of supplementary tables. For the Blue Book the information published is the Combined use matrix analysed by 11 industry/ product groups. These broad estimates will have been based on data assembled at the full 123 group detail. Use is also made of the structure of the accounts for compiling the national accounts estimates for year (t-1), and also for the quarterly accounts.

Some key developments

13.91   It is worth mentioning that a number of key developments are currently in hand to extend the scope and usefulness of supply and use tables. These include (i) improved use of the I-O framework for deriving estimates for the latest year (t-1), (ii) the production of annual tables at constant prices, and (iii) the compilation, for a much more aggregated industry/product split, of quarterly tables at both current and constant prices.

Estimation methodology

13.92   The methodologies used for compiling the aggregates of all the main variables appearing in the supply and use tables, and also the industry analysis of output and intermediate consumption, have been described elsewhere in this publication. Detailed documentation is not yet available on the derivation of the estimates, by product, of industry output and of final demand. This will be published in 1999 in the *Input-Output Methodological Guide*.

Estimation of constant price GDP from production statistics

General description

13.93   Estimates of GDP can be derived in three ways – from expenditure, income or production statistics. The figures from the three sources are used within the balancing process to determine the optimal estimates of GDP, at both current and constant prices, annually and quarterly. This section describes the derivation of GDP at constant prices from production statistics, that is, in principle, based on the estimates of output and intermediate consumption by industry. Given that the GDP estimates are essentially compiled on a quarterly basis, the description focuses on the quarterly figures. However, reference is made to where firmer annual data are used to improve the accuracy of the quarterly series.

13.94   The estimates of GDP, based on production statistics, are derived by summing the data for individual industries. As mentioned earlier in this chapter, in respect of current price data, an industry's contribution to gross domestic product is its value added, that is the difference between output and intermediate consumption. The same applies at constant prices. Thus, in principle, total value added at constant prices could be derived by aggregating the separate industry components, themselves derived as the difference between output and intermediate consumption, each valued at constant prices. This approach to estimation is known as double deflation. In practice the method is somewhat difficult to apply, since it requires a considerable amount of detailed value and price information, by product.

The base year

13.95   The derivation of estimates at constant prices requires the determination of a base year to which the data relate. What this means is that the estimates are derived according to the pattern of prices obtaining in the base year, and different base years, with different patterns, will give rise to different constant price estimates. One particular feature concerns the frequency with which base years are updated. Generally, this has been every five years. This is largely adequate when relative prices do not change much. An alternative is to update the base year every year – this is known as annual chain linking. The UK accounts, as described in the section on the principles of measuring volume in Chapters 2 and 11, currently use a five-yearly chaining.

The form of the index series

13.96   Also as explained in Chapters 2 and 11, the index series used is a Laspeyres index; that is a base weighted index of volume.

Approach to estimation

13.97   In the absence of reliable information which might be used for deriving direct estimates of quarterly constant price value added, by industry, by the theoretically correct process of double deflation, the general approach adopted in the UK to estimating production-based GDP is to extrapolate the base year estimate of value added by 'indicators' which are assumed to reflect the quarterly movement in constant price value added. In theory, the optimum approach would be to project the estimates for output and intermediate consumption by separate indicators, deriving value added as the difference between the projected estimates. This has some similarities with the double deflation approach, but is again generally not possible due to data limitations. As a consequence, the methodology used is to project the figure of base year gross value added (that is, the difference between output and intermediate consumption), without implying any estimation for either output or intermediate consumption. This emphasis on the movement in the indicator series is important, since while the series may not provide sensible estimates of the *level* of value added, the assumption is that they should, in the short term, provide a reasonable estimate of *change*. The indicators are of course volume measures themselves, either in quantity form, or derived by deflating value data. Estimates are made for detailed components of individual industries, based on the projection approach, which are then combined to derive broad industry and various aggregate figures.

Measurement issues

Deflated value or quantity indicators

13.98   As just mentioned, indicators can take the form of deflated values or actual quantity data. The choice of what to use will largely be determined by data availability. However, it is worth observing that, other things being equal, it is preferable to use deflated values rather than quantity data. There are three reasons for this. The first reflects the fact that the information used for compiling the indicator series is based on sample data. The value and price information will tend to move in a narrower range than might the quantity figures. The second reason is that allowance for quality change (*see* paragraphs 13.145–148) is less difficult through price series than for quantity series. Thirdly, quantity indicators tend to relate to the principal products of industries, and may not adequately reflect secondary output.

The indicator series

13.99   The various indicator series which might be used, together with their main advantages and disadvantages, are given below.

   (a) *Output*
        The most frequently used indicator is output, in some form or other. The series may be deflated values, normally using an output price index, or direct quantity measures. This indicator is of main benefit where the ratio of value added to output (at constant prices) is close to unity or likely to be stable.

   (b) *Input (materials)*
        The use of material input, again deflated values or quantities, is likely to be preferable to an output indicator for those industries producing non-standard or complex goods, such as capital goods industries and the construction industry.

(c) *Input (employment)*

An indicator based on employment, say numbers or, better, hours worked, will provide an indicator for the labour component going into the productive process. Ideally, where used, some allowance should be made for productivity (*see* paragraph 13.156).

(d) *Input (materials and employment)*

This indicator is essentially a combination of (b) and (c) above. Materials may be regarded as a crude proxy for hours worked, if otherwise not available. The relative weighting for the two components may be based on their importance in the productive process or on some other means.

(e) *Input (employment and capital consumption)*

In this combined indicator, capital consumption may be seen as a rough proxy for productivity. The weights for the two components may be based on the relative importance of wages and salaries, and operating surplus.

13.100 There is one other approach (not currently used in the United Kingdom) which might be considered, based on direct estimation rather than indicators. It involves deflating an estimate of value added for the industry group by an appropriate price index. It may be possible to estimate current price value added from the income side, with deflation normally undertaken using an output price index. This approach is equivalent to deflating output and intermediate consumption by the same price index. This is often referred to as single deflation.

13.101 The choice of indicator for projecting base year value added depends on a number of factors. The main consideration is obviously that it should be a good proxy for the movement in constant price value added. Put another way, the issue is the stability in the ratio of the indicator to the series for constant price value added. Where an output indicator is used, stability is required in the output to value added ratio, at constant prices. Variability in the ratio will arise for various reasons, such as changes in the product mix, in the type and source of inputs, and in the production process itself, including contracting out intermediate processes and technological change. The effect of the various changes is likely to be greater for current price estimates than for their constant price counterparts. Other things being equal, it will also be greater for quarterly estimates, where output might be markedly seasonal, than for annual figures. This ratio will also move differently for the different kinds of indicators. On the whole, the use of indicator series should not be a problem for periods of up to a year. However, it is essential to align quarterly estimates to firmer annual estimates, and also to take into account the relationship between quarterly and annual data in making estimates for the latest period. There are two further points about the selection of indicators. The first is that, in many cases, the choice is to a large extent governed by data availability. Secondly, where there is a choice of indicators, it is important to assess the intrinsic accuracy of the particular estimates which might be made. In such cases, the extrapolated estimate might best be based on some weighted combination of the indicators which are available.

13.102 In the UK, largely reflecting appropriateness and availability, the most commonly used indicator is gross output. The employment indicator is used mainly for the public sector. The table below shows the importance of the different types of indicator used to estimate quarterly constant price GDP.

Table 13.3

Type of indicator	1990 weights at mid-1997
Double deflation	3
Output	
Quantity (goods)	6
Quantity (services)	10
Deflated values (goods)	20
Deflated values (services)	41
Input	
Materials	1
Employment	18
Other	1

13.103 For GDP in aggregate, around 450 separate indicator series are used, each representing the movement in constant price value added at the particular level of detail. (The series used in the 1990 based index are described in an ONS Methodological Paper, No 5 - *Gross domestic product: Output methodological guide*.) The aggregated estimates – for GDP and the broad industry groups – are derived by combining the estimates for the individual components. In practice, the component series are derived in index number form and combined using the value added weights of the base year.

13.104 The appropriateness of the indicators as proxies for estimating movements in constant price value added is under continual review. The review covers such issues as their plausibility for determining movements for other than the short-term, the relationship between annual and quarterly data, and whether any modifications need to be made to the estimates on account of possible *ad hoc* changes in the particular ratio of the indicator to value added.

Problems of measurement

13.105 There are various problems which arise with the use of indicators to measure movements in value added at constant prices. Where output indicators are used, information for the production industries – either physical quantities of goods produced or deflated sales (adjusted for stock changes) – generally provides a ready and reasonable basis for estimation. However, the choice of indicators for the service industries is much less easy. For some service industries, measures of output do suggest themselves fairly readily. For example, movements in constant price value added for retailing may be estimated using information on deflated retail sales, while, for transport, passenger-kilometres and tonne-kilometres are seen as sensible indicators. However, the output of many business and professional services, and above all of financial institutions (such as banks or building societies), is difficult to define, let alone measure. The solutions adopted by the United Kingdom are shown in the description of the industries below.

13.106 There are also problems with the measurement at constant prices of the non-market output of general government such as public administration, defence, public education and health. These services are usually provided free or at subsidised prices, often with no easily identifiable unit of output or comparable service being sold on the market. Over the years, consideration has been given to a range of series which might form suitable indicators for measuring output of the public sector. These have included, for example, numbers of, or time spent on, medical consultations within health, or numbers of pupil hours within education. However, up until recently the comprehensive set of output indicators for estimating constant price value added for the public authorities was far from comprehensive.

13.107 Increasing use of performance measurement within government has made more information available. With effect from the 1998 Blue Book ONS has introduced the use of direct measures of government output in the areas of health, education and social security. The output measures have been taken back to 1986 in the published figures.

13.108 In other areas of government non-market output, following the traditional approach and paralleling the way in which output is measured at current prices, the movements in the constant price figures are based on information on inputs of labour and capital. The indicators for the former, which is by far the greater weight, are based on numbers employed. Capital inputs are represented by estimates of constant price consumption of fixed capital.

13.109 The use of employment (or other input) indicators, for the government sector and also elsewhere, raises the key issue of whether any allowance should be made for changes in productivity. It is necessary to consider separately the position for the non-market and market sectors of the economy. In respect of non-market activity, largely reflecting problems of estimation, by convention, no allowance has hitherto been made for productivity changes. However, as noted above, the 1998 Blue Book introduced new direct measures of the major areas of government output, thereby incorporating the effects of changes in productivity. Where employment is still used, and for market sector activities, an estimate of productivity change is incorporated in the figures. The point should be made that, while employment was once used fairly widely as an indicator series over the service industries, it has been progressively replaced by indicators based on deflated turnover. Where appropriate, estimates of productivity change are derived from those industries where employment is not used as an indicator. The industries where productivity adjustments were made were divided into high, medium and low productivity groups, based on their 'technological' characteristics. The adjustments made did not exceed a prescribed amount. This approach, while far from perfect, was considered better than assuming zero productivity. It should be noted that adjustments for productivity are also necessary where price indices based on wages and salaries are used for deflating output. Finally, estimates of output per head, can also be used to provide some broad check on the appropriateness of the series for constant price value added series derived by way of the indicator approach.

Alignment with annual data

13.110 The constant price production-based estimate of GDP is compiled as a quarterly series, with the emphasis on change, rather than levels. However, levels are also of importance and need to be properly measured. In practice, in the context of total GDP, this is achieved largely through the reconciliation of the production, expenditure and income estimates at current prices. However, in addition a key part of the production-based constant price approach is to ensure that satisfactory estimates are available, for the separate industry groups, through the indicator approach. At the moment, for some industries, this is achieved by the use of firmer or more appropriate annual indicators, which, when available, replace the quarterly data. In addition, modifications are made, from time to time, to the estimates for the separate industries, based on the current price industry estimates.

Quality changes

13.111 The issue of how to deal with quality changes in goods and services affects all parts of the constant price accounts. For some services, quality is difficult to define and no less easy to measure. It will be useful to mention briefly here some of the key points as they affect the measurement of value added, by industry, at constant prices. To begin with, two general points might be made. First, a change in the quality of a good or service should be regarded as a change in volume rather than in price. Secondly, the identification and measurement of quality change is better undertaken by way of price series than quantity series.

13.112 Where indicators of physical quantities, such as tonnes of steel or passenger kilometres, are used, little can be done to allow for changes in quality. In the latter case, for example, a whole range of factors such as frequency of train service, convenience and comfort are all associated with the quality of the transportation service being provided. However, to try to make some allowance for these factors in determining output indicators is extremely difficult. In practice, the best that can be attempted, generally, is to build up the estimates from as detailed a level as is possible. In the example chosen, separate indicators are used for season ticket and other ticket holders.

13.113 Where value series are used in the determination of indicators, it is often possible to make some allowance for quality in the price series which are used for deflation to constant prices. This allowance is made, for example, by seeking information from contributors about changes in specification of products. An alternative is to use statistical techniques such as hedonic price estimation, in which quality is regarded as a function of technical characteristics. For production industries, the producer price indices, which are used extensively for deflating sales or output, incorporate some allowance for quality change. For the service industries, although similar allowance is made in the retail prices indices, used for deflation for retailing and certain service industries, the problems of identification and measurement are altogether more complex.

13.114 The retailing industry provides a good example of some of these difficulties. Here, the quality element of service is represented by aspects such as convenience, range of goods stocked and sold, their display, personal or self-service. While the retail prices index can make some allowance for changes in quality of goods, it does not seek to allow for changes in the kinds of service mentioned above. The problems with allowing for quality changes are widespread over the accounts. Further, they are not peculiar to the UK, affecting all countries. Any improvement is likely to be extremely costly and time-consuming. One point to note is that the problems will affect all three measures of GDP.

Hidden economy

13.115 The accounts cover some aspects of the hidden economy (*see* paragraph 13.38 for example). Because of the obvious difficulties of measurement, the ESA proposals on illegal activity have not yet been implemented.

Data Sources

Derivation of the weights

13.116 The estimation of aggregate GDP involves combining index numbers of the movement in activity for individual indicators by weights which reflect the value added in the base year appropriate to the particular indicator. This section describes how the weights are derived.

13.117 The weights are derived at two levels. The first relates to the (123) industries used in the input-output analysis. The way in which these estimates are assembled has been described in the section on input-output earlier in this chapter. In brief, two approaches are used. For most industries, value added is obtained as the difference between output and intermediate consumption, using information from ONS business inquiries (25.1). However, for some industries, estimates are derived from the income side, using Inland Revenue information on wages and salaries and operating surplus (*see* above).

13.118 The second level for the weights relates to the further disaggregation of the information for the 123 industry groups, as described above. The more detailed estimation is usefully considered separately for index of production industries and others. In respect of the former, weights are derived for about 250 indicators of the 87 industries analysed for the supply and use tables. The main source of the further disaggregation is direct estimates of value added from the more detailed analysis of the ONS annual business inquiries, as defined above. For some other industries, the further disaggregation is based on output data. Elsewhere, for example, for the motor vehicle industry, the breakdown is based on sales information from the ONS product inquiries (25.1.7), while for oil and natural gas, the greater detail is provided by DTI, based on relevant information on inputs and outputs supplied by the companies.

13.119 For the 39 main industries outside the index of production, weights are derived for about 250 separate indicator series. The main source for this further disaggregation is information on compensation of employees and operating surplus from Inland Revenue statistics, which is used to sub-divide the main industry figure. Where data on operating surplus are not available, the allocation is based on compensation of employees, either from IR sources or from estimates based on employment and average earnings. Information from the business inquiries has been used in some industries, for example gross margin within retailing.

The derivation of the indicators by industry

13.120 The following section describes in broad terms the indicators used for the estimation of value added at constant prices. The text focuses on the quarterly indicators, but important differences in the annual approach are also described. A full list of the indicators used will be published in due course. In some cases the definitive indicator is a quarterly (or monthly-based) series. In others separate quarterly and annual indicators are used. A third possibility is that only annual series exist, in which case the quarterly estimates are obtained by interpolation and extrapolation. In all three cases, the quarterly data are aligned to the firmer annual estimates. It should also be noted that the quarterly estimates include, as appropriate, any adjustment which might be deemed necessary to allow for possible biases, as shown by the relationship to the annual figures.

13.121 The concept of value added is being measured on an industry not a commodity basis. Where sufficient information is available, the ancillary activities of a company are treated as separate units and classified to the appropriate industry. However, in many cases this is not possible and residual, say, distribution or transport or service activities, might be included within a manufacturing firm, or *vice versa*. As with other parts of the accounts, this will impact on the 'purity' of the analysis by industry, but the effect on the accuracy of the overall estimate is likely to be *de minimis*.

13.122 Finally the description of the industry estimates relates essentially to the methodology for estimating value added on the basis of indicators rather than direct measurement, which is used only for the agriculture and electricity industries.

The analysis by industry

Agriculture, hunting and forestry

13.123 The large amount of price and quantity data available for agriculture makes it possible to use the double deflation method for compiling annual estimates for this industry. Double deflation is essential in this area, since the relationship between output and intermediate consumption could well be affected by the vagaries of the weather, and single deflation or a single indicator approach would be inadequate. A large number of items of output and of intermediate consumption are separately distinguished, each carrying its own base-period weight. A three-year average (of the years 1994, 1995 and 1996, for the 1995-based estimates) is used for the purposes of revaluation, in order to reduce the effect of fluctuations arising from weather conditions and other factors. As mentioned in the section on current price output, the industry now includes, as part of its output, work in progress relating to livestock and growing crops. For items in year round production, such as eggs and milk, quarterly indicators of output are used to project base-year value added.

13.124 A particular feature of the estimation process is the need to accommodate the availability of products and the variability in prices over different periods of the year. The first issue is resolved by treating similar products – for example new and old potatoes – as different products in the estimation process. For prices, where data relate just to say a single quarter, then, in constructing indices, the 'base year' is defined solely in terms of that quarter.

13.125 For forestry, the quarterly estimates are largely interpolations of the annual data which are based on the area of land planted and numbers of trees felled.

Fishing

13.126 Estimates are based on the quarterly weight of fish (by species) landed from UK registered vessels, weighted together by the value of landings in the base year.

Mining and quarrying

Manufacturing

Electricity, gas and water supply

13.127 Given the commonality of sources and approach, the estimation methodology for these three industry groups is considered together. The annual and quarterly estimates are derived from information collected and assembled for the monthly index of production (IoP), an economic indicator of value in its own right as well as part of GDP. With the exception of electricity, where double deflation is used, estimates of constant price value added are based on the movements in various indicator series of output.

13.128 The basic information for the IoP comes from ONS monthly and quarterly surveys (25.1), together with certain information collected by other departments and trade associations. Increasingly, information has been obtained monthly, with the use of outside sources declining. The information collected on sales or production is mostly collected on a value basis, though volume data are collected for items such as oil, gas and coal, and cars and commercial vehicles. The value figures for each detailed industry are deflated using appropriately weighted price information from the producer price index (25.2.7).

13.129 Where the value data collected are based on sales or deliveries, as discussed in the section on current price output, the figures need to be adjusted for the change in inventories of finished goods and work-in-progress in order to provide the appropriate measure of output. The ONS monthly and quarterly inventory inquiries (25.2.4) provide the basis for these adjustments. One other general adjustment is made to the estimates. This is to allow for different numbers of working days in each month. Thus, the IoP essentially measures average output per day, so enabling direct comparisons to be made for different months.

13.130 For the electricity industry, the movement in constant price value added can be affected appreciably by both output and intermediate consumption. For the former, this will reflect the effect of different tariffs for industrial, commercial or domestic users, while for the latter, different types of fuel are used for generating electricity. Thus, as neither output nor input, alone, provides a suitable indicator for estimating constant price value added, double deflation is used.

Construction

13.131 The construction industry indicators are essentially output-based, and use information on work done in the industry collected quarterly by the DETR in Great Britain and by the Northern Ireland departments. However, reflecting various problems of measurement for the industry, particularly the effect of the hidden economy, use is also made of information from the ONS annual construction inquiry (25.1.5) and Inland Revenue data on self-employment incomes (25.4.2). The value estimates derived from the above sources are deflated by price indices of construction activity collected by DETR.

Wholesale and retail trade; repair of motor vehicles, motorcycles and personal and household goods

Hotels and restaurants

13.132 For all industries under these heads, there are separate quarterly and annual indicators. The latter, which are mostly based on deflated turnover from the more firmly-based annual inquiries (25.1.11), replace the quarterly series, when available. For retailing, the quarterly indicators are based on the deflated value of sales from the monthly retail sales inquiry (25.1.10). The indicators relate to different types of business, such as butchers or jewellers, and the sales values are deflated by using appropriately weighted retail price series. For wholesaling, the various quarterly indicators used are based on IoP and retail sales inquiry data, and turnover data collected quarterly or monthly. Elsewhere, for the short-period estimates, deflated VAT turnover is used for hotels and restaurants, and vehicle registrations and petroleum deliveries for the motor trades. Deflation, generally, uses producer price (25.2.7) and retail price (25.4.7) information, as appropriate.

Transport, storage and communication

13.133 The indicators for this industry group, which are again largely output series, are almost wholly based on quarterly data. For rail, the indicators relate to passenger-kilometres and, for freight, tonne-kilometres. For passenger transport, road haulage and storage the indicators are mainly deflated turnover. Deflation uses retail prices information and corporate services prices. For transport support services, information relates to activity at ports and airports, and also includes certain volume data on exports and imports. Within the communication industry, a wide range of mainly volume indicators is used for postal services, while for telecommunications information is based on data provided by British Telecommunications, supplemented by turnover figures, appropriately deflated, collected in ONS quarterly inquiries (25.1.11).

Financial intermediation

13.134 This industry covers banks, building societies, other financial institutions, insurance companies and auxiliary activities to finance. The measurement of the movement in constant price value added for these industries raises a number of problems. First, for banks and building societies, this largely reflects the way these organisations operate, and the fact that the product traded is essentially money. While direct fees are charged for a range of services provided, income related to other services is derived from the difference between interest received on loans and that paid on deposits. In respect of this second group of services, the interest margin is treated as the charge for sales of services by the financial intermediation industry.

13.135 The counterpart to these charges is treated as purchases by a notional industry which consequently has a negative value added (*see* 'Adjustment for financial services' below).

13.136 However, ESA95 seeks to make allowance for the implicit costs incurred by depositors and lenders, and to treat such costs as payment for services, for either final consumption (by households, NPISH sector and general government) or intermediate consumption by actual industries or sectors. This is undertaken through the concept of the estimation for 'Financial intermediation services indirectly measured' (FISIM), which is discussed in Annex 1 to Chapter 20. The methods by which this might be implemented have not yet been agreed.

13.137 The output of banks is measured by: employment data (with allowance for productivity); number of bank credit and debit clearings; outstanding domestic and overseas sterling and foreign currency loans and deposits deflated by the all items retail prices index (excluding mortgage interest), using the US dollar exchange rate to allow for exchange rate movements. For building societies output is indicated by: employment data (with allowance for productivity); annual number of advances outstanding (interpolated and extrapolated quarterly); quarterly number of advances made and total liabilities deflated.

13.138 For insurance, where the general treatment in the accounts has been changed (*see* Annex 2 to Chapter 20), the indicators are household expenditure on life assurance at constant prices and, for non-life insurance, deflated premiums, annually, with quarterly estimates based on interpolation and extrapolation. For the other financial institutions and for auxiliary services, various indicators are used, including deflated hire purchase debt outstanding, numbers of stockbroking transactions and, again, household expenditure on life assurance, at constant prices.

Adjustment for financial services

13.139 Since the interest flows contributing to the weight for financial intermediation as described above do not contribute to GDP in total an adjustment is necessary to subtract them. The 'adjustment for financial services' is equivalent to a notional industry which purchases the output concerned. Since this notional industry has no output, its value added is a negative value equal to those purchases. The indicators are the same as those used above, so that movements in GDP as measured in this way are the same as they would be if value added were measured in the conventional way throughout. Nevertheless, as the annex to Chapter 20 points out, there is final consumption expenditure on such services which should be allowed to affect GDP.

Real estate, renting and business activities

13.140 A diverse range of mainly quarterly indicators is used for estimating value added for the separate industries within this group of industries. The largest single component relates to output of services of owner-occupied dwellings, where the indicator used is based on the constant price figures included in household consumption, with some deduction for maintenance costs (*see* Chapter 22). The indicator series is the movement in the housing stock, with some adjustment for quality changes. The indicator for real estate is mostly deflated income from renting of property, available annually and interpolated quarterly. For business services, the indicators are largely deflated turnover data for individual industries from ONS monthly and quarterly service sector inquiries (25.1). Various deflators are used, including industry-specific price information collected by ONS, and estimates based on movements in wages and salaries, incorporating some allowance for the effect of productivity change. For some industries, ONS employment data (25.4.1) are used, with some adjustment for productivity.

Public administration and defence; compulsory social security

13.141 This industry group covers central and local government administration and defence. The output of social security is indicated by the number of claimants. Otherwise the indicators used are based mainly on employment or deflated wages and salaries, from administrative sources, with a small contribution from capital consumption at constant prices. The same broad approach is used for annual and quarterly estimates, although the annual figures generally use more detail, for example disaggregating the employment data by grade or rank of staff. By convention no explicit allowance is made for productivity change.

Education

Health and social work

13.142 These two industries embrace both market and non-market output. Indicators for non-market output are based on a range of output indicators including pupil-years for education and for health the Department of Health's cost-weighted activity index, numbers of visits to general practitioners, numbers of prescriptions dispensed, numbers of NHS eye tests and numbers of courses of dental treatment given under the NHS. For social work the indicators are based on the numbers receiving different types of care. For market activity the estimates use employment data, but including some allowance for productivity change.

Other community, social and personal service activities

13.143 The indicators used here are all quarterly. In most of the broad industry groups there is a mix of market and non-market activities, for example in respect of sanitary services and museums and art galleries. For the local government component, the indicators are based on employment data, with a small contribution coming from capital consumption at constant prices. For the market sector element, for some industries the indicators are based on employment data, with some allowance for productivity changes included. In respect of other industries, indicators for cinemas, theatres and related entertainment, and laundries and hairdressing, are all based on deflated turnover – from either ONS inquiries (25.1.11 and 25.1.12) or VAT data - using best estimates of activity-specific prices, while sport and recreation uses the movement in household expenditure at constant prices.

Private households with employed persons

13.144 The indicator for this very small component is based on household expenditure on private domestic services at constant prices (*see* Chapter 22).

Chapter 14

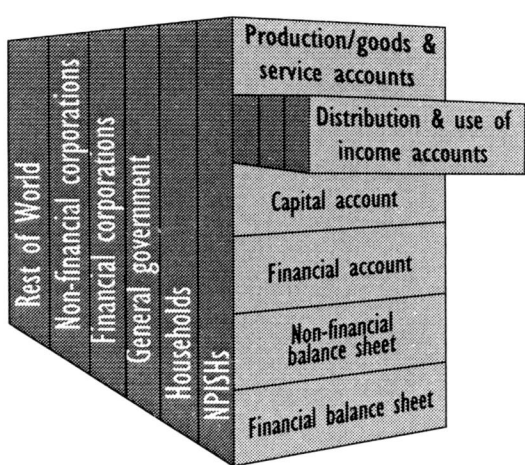

Distribution and
use of
income accounts

Chapter 14 Distribution and use of income accounts

Introduction

14.1   This is one of a series of chapters which focus on one type of account in the new system. This chapter describes the methodology for the estimation of the variables in the distribution and use of income accounts. These are part of the current account within the economic accounts and were previously called the income and expenditure accounts. The conceptual basis of this account is described in Chapter 5. These accounts take the gross or net domestic product (B.1) from the production account (see Chapter 13) and show how this is used or distributed. What is left at the end of the distribution and use of income accounts is the undistributed income – this is the gross or net saving of each sector, which is carried forward to the capital account. Capital transactions are regarded as redistribution of wealth rather than income in the ESA and are recorded within the capital account (Chapter 15).

14.2   The distribution and use of income accounts exist for all the main institutional sectors. To obtain the disposable income and savings of each sector we need to take account of transfers in and out of the sector. The accounts are not consolidated, so that in the whole economy account, transfers such as social contributions and benefits appear in both uses and resources.

14.3   These accounts describe the distribution and redistribution of income and its use in the form of final consumption. The distribution and use of income are analysed in four stages, each of which is presented as a separate account:

- the generation of income account
- the allocation of primary income account
- the secondary distribution of income account
- the use of disposable income account

For some purposes the use of disposable income account may be replaced by:

- the redistribution of income in kind account
- use of adjusted disposable income account

14.4   This set of accounts, together with the production account, form the current accounts of the integrated system of economic accounts. The variables covered in this chapter are shown in Table 14.1. For convenience some of the detailed disaggregation in this account has not been included for certain variables.

14.5   It is important to note that primary incomes form part of the input-output balancing process described in Chapters 11 and 13. The sources described in this chapter are the initial inputs to that process.

Table 14.1: The distribution and use of income accounts

Resources	Uses
Generation of income account	
Gross value added (gross domestic product) (B.1g)	Compensation of employees (D.1)
	Taxes on production and imports (D.2)
	Subsidies, received (D.3)
	Balances:
	Operating surplus, gross (B.2g)
	Mixed income, gross (B.3g)
	Operating surplus, net of capital consumption (B.2n)
	Mixed income, net (B.3n)
Allocation of primary income account	
Operating surplus, gross (B.2g)	Property income, paid (D.4)
Mixed income, gross (B.3g)	Balance of primary incomes (national income), gross (B.5g)
Compensation of employees (D.1)	
Taxes on production and imports, received (D.2)	
less Total subsidies, paid (D.3)	
Property income, received (D.4)	
Secondary distribution of income account	
Balance of primary incomes (national income), gross (B.5g)	Current taxes on income, wealth (D.5)
Current taxes on income, wealth (D.5)	Social contributions (D.61)
Social contributions (D.61)	Social benefits other than social transfers in kind (D.62)
Social benefits other than social transfers in kind (D.62)	Other current transfers (D.7)
Other current transfers (D.7)	Balance: disposable income, gross (B.6g)
Use of disposable income account	
Disposable income, gross (B.6g)	Final consumption expenditure (P.3)
Adjustment for change in net equity of households in pension funds (D.8)	Adjustment for change in net equity of households in pension funds (D.8)
	Gross saving (B.8g)

Generation of income account

14.6   This is the first of the Distribution and Use of Income accounts. It shows the sectors, sub-sectors and industries which are the source, rather than the destination, of income. It shows the derivation of the 'profit' arising from production, called the operating surplus (or mixed income in the case of unincorporated businesses in the households sector).

14.7   This account analyses the degree to which value added covers the compensation of employees (their wages and salaries etc.) and other taxes less subsidies on production. So it gives a figure for the operating surplus of the total UK economy: the surplus (or deficit) on production activities before distributions such as interest, rent and income tax charges have been considered. Hence the operating surplus is the income which units obtain from their own use of the production facilities.

14.8   Note that taxes on production and imports are shown as a *use* by producing sectors in this account but not as a *resource* of government. This is because they do not relate to productive activity by government, and cannot therefore contribute to its operating surplus. They become a *resource* of government in the allocation of primary income account which follows.

Allocation of primary income account

14.9   This account is the second of the four in the distribution and use of income group. It shows the resident units and institutional sectors as recipients, rather than producers, of primary income. It demonstrates the extent to which operating surpluses are distributed (for example by dividends) to the owners of the enterprises. Also recorded in this account is the property income received by an owner of a financial asset in return for providing funds to, or putting a tangible non-produced asset at the disposal of, another unit. The receipt by government of taxes on production *less* subsidies is shown in *resources*.

14.10   This account and the subsequent accounts can be calculated only for the institutional sectors, because they deal with resident units as recipients of primary income rather than as producers. The conceptual basis for this account is described in Chapter 5.

Secondary distribution of income account

14.11   This account is the third in this group. It describes how the balance of primary income for each institutional sector is allocated by redistribution through such current transfers as taxes on income, social contributions and benefits and others. It excludes social transfers in kind.

14.12   Balancing item of this account is disposable income (B.6g) which reflects current transactions and explicitly excludes capital transfers, real holding gains and losses and the consequences of events such as natural disasters.

14.13   Social contributions are shown in both the uses and resources part of this account: as uses for households; and as resources for the sectors responsible for the management of social insurance. When social contributions are payable by employers for their employees, they are also included:

- in compensation of employees (in the uses part of the employers' generation of income account, as they are part of wage costs); and in
- compensation of employees, in the resources section of the households' allocation of primary income account, since they form part of the remuneration of households.

Use of disposable income account

14.14   This illustrates how income is consumed and saved: the remaining balance at the end of this account is the saving.

14.15   This the last of the distribution and use of income accounts. In the system for recording economic accounts, only the government, households and NPISH sectors have final consumption. This account illustrates how their disposable income is split between final consumption and saving. In addition, for households and pension funds, there is an adjustment item in the account which corresponds to the way that transactions between households and pension funds are recorded. (This adjustment is D.8: Adjustment for the changes in the net equity of households in pension funds' reserves.)

14.16   The balancing item for this account, and thus for this whole group of distribution and use of income accounts, is gross saving (B.8g).

Use of Inland Revenue data

14.17 The major source for the annual data on the generation of incomes is the Inland Revenue, which is the department responsible for collecting most taxes on income and wealth in the United Kingdom. The current tax system is described in the Annex to this chapter. Inland Revenue data have always been widely used in UK national income statistics. The Inland Revenue have developed the statistical analysis of tax records so that a substantial amount of detail can be obtained from them and they can be used for the UK economic accounts. Tax data is used as the source for the annual estimates of profits of private non-financial corporations, of most mixed incomes (income from self employment) and of total wages and salaries. About two-thirds of the annual estimates of UK income come from tax data.

14.18 Although the quality of the data is extremely good and detailed by its very nature there are deficiencies from a economic accounting viewpoint. There are delays because of the process of assessment and appeal; there are incomes below the taxable level which need to be estimated; there can be difficulty in assigning taxes to the period in which they accrue (as distinct from when they are paid); and there is evasion. These issues are dealt with below.

Delays

14.19 Because there is some delay in providing good quality estimates from the Inland Revenue, other methods are used to provide initial estimates.

Evasion adjustment

14.20 An estimate is made to allow for certain factor incomes not being declared to the Inland Revenue and to bring the income measure of GDP into line with the expenditure measure in the longer term. The expenditure measure of GDP is largely derived from a wide range of household and business surveys designed for statistical purposes and from government accounting data. Thus the expenditure measure of GDP should include some (though probably not all) of the expenditure which comes from the incomes not declared to the Inland Revenue. In fact the expenditure measure of GDP is, almost always, higher than the unadjusted total of factor incomes.

14.21 An allowance called the *net allowance for evasion* is made. It is based on the trend of the '*initial residual difference*' between the two measures of GDP. This difference comprises:

- the under-reporting of factor incomes;
- other errors of estimation, for example, those arising from the use of samples.

14.22 The estimates for evasion only attempt to allow for the under-reporting errors. Errors of the other type are dealt with in the balancing process described in Chapter 11; they can be either positive or negative. An article in the February 1980 issue of *Economic Trends* gives a more detailed description of the evasion adjustments.

14.23 The part of the hidden economy which is concealed from the expenditure estimate of GDP as well as the income measure is not estimated; neither are estimates made for any transfer incomes on which tax may be evaded, since these do not (even in concept) enter into gross domestic product. So the evasion adjustment does not claim to measure either the total size of the hidden economy or the total of incomes on which tax is evaded.

14.24 Each year the net allowances for evasion are updated due to revisions to, and later movements in, the initial residual difference. As a percentage of GDP the total allowance currently made rises from two per cent in the early 1970s to three per cent in the mid-1970s before falling to one and a half per cent in 1981 and to one and a quarter from 1982. The evasion adjustment increased to one and a half per cent in 1992.

14.25 The adjustment is mainly allocated to mixed incomes, but small adjustments are also made to compensation of employees and to the operating surplus of corporations.

Allocation of taxation to ESA categories

14.26 Lists of the UK taxes allocated to each ESA category are shown in Chapter 21.

Treatment of amounts due but not paid

14.27 Values in the accounts are shown on an accrued basis; that is to say incomes are recorded at the same period as the corresponding production, or when they become due. The difference between this and the cash flow in the period is reflected in *Other accounts payable/receivable* in the financial account and balance sheet. Though it must be noted that estimates for trade credit are very weak.

The published series

Gross value added (gross domestic product), B.1g

14.28 This is the resource in the generation of income account, brought forward from the production account. It represents the sum of all the values added by the constituent resident units engaged in production, before allowing for consumption of fixed capital. Net value added (net domestic product) is the same measure after deducting the consumption of fixed capital.

14.29 This variable and its derivation is fully described in Chapter 13 : Production account.

Compensation of employees (D.1)

14.30 This is the total remuneration in cash or in kind, payable by an enterprise to an employee in return for work done by the latter during the accounting period. It has two components:

- wages and salaries payable in cash or in kind (D.11)
- employers' social contributions (D.12).

The first of these is the most significant, accounting for 88% of the total. The conceptual basis for this item is described in Chapter 5. Compensation of employees appears in the accounts for all sectors of the UK economy.

Wages and salaries (D.11)

14.31 These include the following:

- basic wages and salaries payable at regular intervals;
- enhanced rates of pay for overtime, night work, hazardous circumstances, etc.;
- cost of living allowances, local allowances and expatriation allowances;
- bonuses paid on productivity or profits, Christmas bonuses, allowances or transport to and from work, holiday pay for official or annual holidays; commissions, tips and directors' fees paid to employees; and other similar bonuses.

14.32 Also included here are wages and salaries in kind such as meals and drinks, price reductions in subsidised canteens, vehicles provided for the personal use of employees, the provision of sports, recreation or holiday facilities for employees and their families, the provision of work place creches and other similar benefits.

14.33 In principle, wages and salaries exclude amounts which employers continue to pay to their employees temporarily in the case of sickness, maternity, redundancy etc. – these are classified as unfunded employee social benefits (D.623) with the same amounts being shown under employers' imputed social contributions (D.122). Similarly, other unfunded employee social benefits are excluded. However, in practice it is difficult to make the distinction comprehensively.

14.34 Estimates of wages and salaries for those within the PAYE system are derived from a one per cent sample of the tax deduction documents. The documents are sent by the Inland Revenue under statutory authority to the Department of Social Security, who note the details of social security contributions on their central computer records at Newcastle-upon-Tyne. A separate computer file based on 1 per cent of these records is compiled for statistical analysis including details of pay and tax by industry, area, etc. The estimates for pay obtained for the whole PAYE population are obtained by multiplying the one per cent sample estimates by appropriate grossing factors. These have been obtained by comparing the employees' National Insurance contributions totals obtained from the one per cent sample with the total employees' National Insurance contributions recorded. Allowance is then made for information received more than twelve months after the year to which it applies. The total number of tax deduction documents exceeds 30 million each year and the sample of 300,000 records is sufficient to estimate total wages and salaries with a standard error of about $^{1}/_{4}$ per cent.

14.35 Tax records relate to tax years which run from 6 April to 5 April. Interpolation of calendar year estimates is carried out using Office for National Statistics figures of numbers of employees in employment and average earnings. The quarterly path is also constructed using monthly and quarterly surveys of numbers in employment and the quarterly survey of average earnings (25.4.1). The pay in cash and in kind of the armed forces is based on returns from the Ministry of Defence.

14.36 Some supplementary estimates are made to cover income partially covered, or not at all covered, by the PAYE system.

- Estimates of incomes below the threshold at which the PAYE system becomes operative are based on data from the Family Expenditure Survey (FES) on the numbers concerned and their average annual earnings.
- The FES is also used in place of PAYE estimates for domestic service which are recognised as being incomplete.
- Earnings of employees of farming enterprises are measured by data compiled by the Ministry of Agriculture, Fisheries and Food.
- Estimates of the wages and salaries of HM Forces are compiled from GEMS returns produced quarterly by the services on the same basis as the figures published subsequently in the Appropriation Accounts presented to Parliament.
- Evasion adjustments are added to the initial estimate of income to allow for income earned but not declared to the Inland Revenue (further information can be found in paras 14.20–25 above).

14.37 Various adjustments are made to cover tips, the earnings of juveniles, fringe benefits such as company cars, profit sharing schemes, and income-in-kind. For the latter, estimates are included to cover: coal provided free or at concessionary rates to miners and other colliery employees; food provided for merchant seamen and fishermen; the imputed rent of housing provided free by employers; the net cost to employers of meal vouchers and meals provided in canteens; the net cost to employers of food and accommodation provided for staff in catering establishments other than canteens.

14.38 The wage and salary figures include bonus and other ad hoc payments in the quarter in which they are paid. Such payments might be relevant to activity which has taken place in previous periods. As such, it might be thought that they should be put on to an accruals basis by spreading them over the earlier periods. This is arguable in concept, and not fully possible in practice. Any re-allocation would also require modifications to other series such as household consumption, operating surplus and taxes on income, for example, to ensure consistency both within and over the three measures of GDP. The basis for some of these adjustments, particularly to household consumption, would be extremely hazardous, to say the least. Thus, the quarterly series of wages and salaries will reflect the effects of 'lumpy' payments. However, the figures for operating surplus and taxes on income are adjusted, if necessary, to be as consistent as possible with these figures.

Employers' social contributions (D.12)

14.39 These represent the value of the social contributions incurred by employers in order to ensure their employees are entitled to social benefits. They are treated as supplements to wages and salaries. They comprise payments by the employers to insurers and the cost of benefits paid directly to employees or former employees (unfunded schemes). These employers' contributions are treated as a component of the compensation of employees, who are deemed to pay them to the insurers (in the case of funded schemes) or back to the employers (in the case of unfunded schemes). The payment of benefit is recorded in D.62 in both cases.

14.40 Estimates of employees' and employers' contributions to funded pension schemes are derived from data collected by the Office for National Statistics. That part of employers' contributions which refers to the National Insurance Fund, the National Health Service and the Redundancy Fund is estimated quarterly from data prepared by the Government Actuary's Department. As with employees' contributions, separate estimates are made for employers' contributions to superannuation schemes, the estimates being obtained by interpolation and projection of annual figures. Contributions to public sector unfunded schemes come from government accounts and are equal to the accruing superannuation liability contributions paid by government departments. For unfunded schemes, mainly the teachers' and National Health Service employees' schemes where the government constructs a notional fund, the contributions recorded are the actual amounts paid by employers.

Taxes on production and imports (D.2)

14.41 These are taxes levied by general government or by institutions of the European Union relating to the production and import of goods and services, the employment of labour, the ownership or use of land, buildings or other assets used in production. Such taxes must be paid whether or not profits are made by the unit. Taxes on the personal use of vehicles etc. by households are recorded under current taxes on income, wealth, etc. The economic accounts break these taxes down into: taxes on products (D.21) and other taxes on production (D.29). These estimates come from the accounts of government and are of good quality.

Taxes on products (D.21)

14.42 These are taxes that are payable per unit of a good or service produced or transacted.

14.43 The single most important element of this category is value added tax (VAT), part of which is deemed to be paid to the UK central government and part direct to institutions of the EU. UK taxes on products are listed in Chapter 21. The conceptual definition and coverage of this heading is considered in more detail in Chapter 5.

Other taxes on production (D.29)

14.44   These are those taxes on the production of goods and services or trade in them, which are not proportional in any way to output or sales. In the United Kingdom they are collected by central government, with the exception of local non-domestic rates which were a tax on the rental value of property and were levied by local authorities up to 31 March 1990 (1989 in Scotland). From April 1989 in Scotland and from April 1990 in England and Wales rates were abolished and replaced (as far as businesses were concerned) by national non-domestic rates (NNDR) These are a tax set by central government on the rental value of non-domestic property, the proceeds of which are distributed to local authorities. NNDR is collected by certain local authorities (collecting authorities) as agents of central government. Local authority rates are still collected in Northern Ireland.

14.45   The main taxes in this group are NNDR and motor vehicle duties paid by businesses. The full list is shown in Chapter 21. The sources of data on taxes on production and imports are as follows:

- data on receipts by HM Customs and Excise, as published in their annual report and in *Financial Statistics*;
- other central government tax receipts as published in either the Consolidated Fund Accounts or other accounts of government departments;
- data on national non-domestic rates as recorded in local authorities' statutory annual returns to the Department of the Environment, Transport and the Regions (DETR) and summarised in *Local Government Financial Statistics* for England and similar publications for Scotland, Wales and Northern Ireland.

14.46   Taxes are essentially accounted for on a cash basis. For the economic accounts, adjustments may be made to the figures to put them closer to the accruals basis. Where this is done, corresponding 'accounts receivable' and 'payable' are included in the financial accounts to preserve consistency with the cash basis used there.

14.47   These estimates are from government accounts sources and are thus of good quality.

Subsidies (D.3)

14.48   Subsidies are current unrequited payments that the UK general government, or institutions of the European Union, make to resident producers. The object of such payments is to influence their production levels or the payment level of the factors of production. Subsidies are divided into: subsidies on products (D.31) and other subsidies on production (D.39). The main source of data is the system of public expenditure planning and monitoring. These estimates are of good quality. Subsidies are recorded as far as possible on an accruals basis.

Balance: Operating surplus (B.2) / Mixed income (B.3)

Operating surplus (B.2)

14.49   This forms part of the balance of the generation of income account. It represents the income which the units obtain from their own use of their production facilities. That is, value added *less* compensation of employees *less* taxes on production payable *plus* subsidies received. It is the last balancing item in the economic account which can be calculated for both the industries and the institutional sectors and sub-sectors. This item includes the surplus from the own account production of accommodation services by owner-occupier households.

14.50   However, for unincorporated enterprises in the households sector, the balancing item of the generation of income account is called 'mixed income'.

14.51 Despite its position as a balancing item in the accounts, the operating surplus is initially estimated directly from Inland Revenue statistics of private corporations' profits and returns from public corporations, and the results modified as a result of input output balancing described in Chapters 11 and 13.

Relation to commercial accounting profits

Commercial accounting practice and economic accounting differ in certain respects, one of which is shown by the calculation of operating surplus. This figure for corporations within the economic accounts differs significantly from what users of corporate accounts would expect. The main reason for this is that whilst commercial accounting practice accounts net interest received as profit, economic accounting standards state that interest flows must be treated as redistribution of primary income: that is, as property income, not profit. On this basis financial corporations typically show negative profits – offset by large interest receipts. However, there is a special treatment accorded to such financial corporations in the economic accounts - see the annex to Chapter 20. Economic accounts also distinguish between profit and holding gains (inventory appreciation) so that profit is calculated on a replacement cost basis. Not all companies do this.

Mixed income (B.3)

14.52 This is the term for the generation of income account's balancing item for unincorporated enterprises owned by members of households (either individually or collectively) in which the owners, or other members of their households, may work without receiving a wage or salary. In the former system of UK economic accounts, this was known as income from self-employment. Any payment for work by the owner or members of their family cannot be distinguished from the owner's profits as entrepreneur. All unincorporated enterprises owned by households that are not quasi-corporations are said to fall in this category, except owner occupiers in their capacity as producers of housing services for their own final consumption, included in the operating surplus (B.2), and households employing paid domestic staff, an activity that generates no surplus. This is described in detail in Chapter 13.

14.53 Despite its position as a balancing item in the accounts, mixed income is initially estimated directly from Inland Revenue statistics of self employment income, and the results modified as a result of input output balancing described in Chapters 11 and 13.

Property income (D.4)

14.54 This is the income receivable by the owner of a financial asset or a tangible non-produced asset in return for providing funds to, or putting the tangible non-produced assets at the disposal of, another institutional unit. Property income is split into six categories:

- interest (D.41)
- distributed income of corporations (D.42)
- reinvested earnings on direct foreign investment (D.43)
- property income attributed to insurance policy holders (D.44)
- rent (D.45)

less
- an adjustment for financial intermediation services indirectly measured (FISIM, P.119)

Of these, interest is the most significant.

14.55 Interest is the amount that a debtor becomes liable to pay to a creditor, over a given period of time without reducing the amount of principal outstanding, under the terms of the financial instrument agreed between them.

Changes from former system of accounts

The ESA95 recommends the use of accrued interest, while the previous United Kingdom system combined cash flows where available with estimates based on balance sheet figures and interest rates.

14.56   Dividends and other distributive income are a form of property income to which the shareholders become entitled to as a result of their ownership of corporations. A special case are the distributions by quasi corporations to their owners, where there is one payment sector and only one recipient sector.

14.57   These figures come from the dividends and interest matrix (DIM) system developed within the ONS. The DIM is a framework for estimating dividend and interest flows to and from each sector of the economy. It is constructed on the standard classification of financial instruments (a term covering all types of financial assets and liabilities – bank loans, corporation shares, government bonds, etc.). For most instruments, payments of interest or dividends are made by a single sector and the amount payable in each period is known. However, the amounts receivable by each sector are frequently not known. The only firm information is that, for any given instrument, the sum of income receipts by all sectors must be equal to the sum of payments by all sectors. The DIM framework allocates the total payable between the receiving sectors using figures on the sectors' holdings of the instrument.

14.58   This framework thus provides logical links within the economic accounts between the estimates of each sector's transactions in different financial instruments, of the level of their assets and liabilities in these instruments and of their receipts and payments of dividends and interest on the instruments. It also has the advantage of being a general methodology for producing such estimates which is common to all sectors.

14.59   The estimates of these amounts are derived by allocating the total among the receiving sectors according to their average holdings in the period. Their average holdings in the period are derived from their balance sheet figures (see Chapter 18). A similar approach is adopted for assets for which there is a single receiving sector but the amounts paid by the various paying sectors are not known.

14.60   Where more detailed or separate information exists for certain sectors, which gives reliable figures for individual flows to or from individual sectors, these data are used for those sectors and the balance sheets are then used to allocate the remaining amounts between the other sectors.

14.61   If the total flow of interest for a particular instrument is not known, the balance sheet figures are used to estimate the total interest flows using known or estimated interest rates.

14.62   Thus these figures and those for each sector's transactions in and holdings of financial assets and liabilities are logically linked together, coherent, consistent and complete.

14.63   For several types of assets, the holders of them may be in the same sector as the issuers. For example, banks borrow funds from other financial corporations; a corporation may hold some share in another corporation as part of its portfolio of shares. Within the figures calculated by the dividends and interest matrix the aim is to include intra-sector transactions (as both payments and receipts) wherever possible. Intra-sector holdings are generally included in the unconsolidated balance sheet estimates. It therefore follows that similar coverage of income flows is obtained where these balance sheets are used to allocate income between sectors. Consistency of coverage is necessary if, for example, income data and balance sheets are being used to check that the implied rates of return are sensible and consistent with other information.

14.64   In the system of economic accounts, interest (D.41) is shown:

- among resources and uses in the allocation of primary income account of the sectors (this is different from most business accounting, where interest paid is usually shown as a fixed charge similar to other costs of production in the operating account);
- among resources and uses in the external account of primary incomes and current transfers.

14.65   Dividends (D.42) are shown:

- among uses in the allocation of primary income account of the non-financial and financial corporations' sectors;
- among resources in the allocation of primary income account of the shareholders' sectors;
- among uses and resources in the external account of primary income and current transfers.

14.66   More detail on the dividends and interest matrix (DIM) can be found in the article 'Sector allocation of dividend and interest flows - a new framework' published in the October 1992 edition of *Economic Trends*.

Adjustment for Financial Intermediation Services Indirectly Measured (FISIM, P.119)

14.67   FISIM is an indirect measure of the value of the services for which financial intermediaries do not charge explicitly. The total value of FISIM is measured in the system as the total property income receivable by financial intermediaries other than insurance corporations and pension funds *less* their total interest payable. For this purpose, property income receivable should exclude any from the investment of their own funds.

14.68   FISIM is not at present allocated to user sectors, so the entries shown for interest are those for actual interest payable and receivable. The adjustment is made to deduct the value of FISIM from interest resources of financial corporations. An equivalent offsetting adjustment is attributed to the nominal sector deemed to be consuming FISIM. To simplify the presentation of the accounts, the nominal sector is only shown separately in the 'not sectorised' column of the matrix tables in Blue Book section 1.8. The counterpart to the adjustment to the financial corporations sector is included in the aggregate account for the total UK economy. (*See* Annex 1 to Chapter 20.)

Reinvested earnings on direct foreign investment (D.43)

14.69   This is defined as equalling the operating surplus of the direct foreign investment enterprise *plus* any property incomes or current transfers receivable *less* any property income or current transfers payable (including actual remittances to foreign direct investors and any current taxes payable on the income, wealth, etc. of the direct foreign investment enterprise).

14.70   A direct foreign investment enterprise is one in which a foreign investor owns 10 per cent or more of the equity. The foreign shareholder's share of the surplus defined in the previous paragraph is deemed to be distributed to it (in this account) and re-invested by it (in the financial account).

14.71   These figures may be either positive or negative on either side of the account. Resources are net earnings from outward direct investment by UK investors; uses are net earnings on inward direct investment in the UK by the rest of the world. For the UK, both resources and uses are positive, but there is some variation between specific sectors and industries.

14.72   Figures for this come from the ONS and Bank of England surveys of overseas direct investment (25.5.1).

14.73 These earnings are recorded when they are earned. They appear:

- as uses and resources in the allocation of primary income account of the sectors;
- similarly, as uses and resources in the external account of primary incomes and current transfers.

Property income attributed to insurance policy holders (D.44)

14.74 This is the primary income received from the investment of insurance technical reserves (*see* Chapter 5). It is shown as a use of the insurance sector and a resource of the policy holding sectors.

14.75 Figures for property income on these reserves come from surveys run by the ONS to insurance corporations and pension funds (25.3.3). A similar set of ONS income and expenditure surveys are completed by a sample of pension funds. These data are compared with the results of the DIM system for interest and dividends which, in turn, is based on annual balance sheet and quarterly transactions in financial assets inquiries to these institutions (see Chapter 25 for more information on these inquiries).

14.76 The division of this property income between the policy holder sectors takes into account information on the breakdown of premiums paid by product, the timing of premiums (many households now pay their annual premiums over a number of months, by direct debit), and the level of reserves held by type of business. Information from the Association of British Insurers, which represents about [98%] of UK insurers, supplements the data from the ONS surveys of insurance enterprises for this purpose.

Rent (D.45)

14.77 This property income is payable to the owners of land and sub-soil assets in return for allowing others to exploit the assets in question. Such assets include the ownership of inland waters and rivers; rent is payable to the owners for the right to exploit the waters for recreational or other purposes. In the former UK system *Rent* included rental on buildings, now in principle a constituent of output and intermediate or final consumption. It is not possible to separate the two elements in many cases and component series are classified to one or the other according to a judgment of which is the predominant element. In practice, the main constituents of rent for the UK are for agricultural land; and royalties in respect of permits to explore for oil and gas.

14.78 Rents are recorded in the period when payable and are shown:

- as resources and uses in the allocation of primary income account of sectors;
- among resources and uses in the external account of primary incomes and current transfers.

14.79 The total rent income over the whole UK economy is estimated from:

- government receipts of rent on land;
- government receipts of royalties for oil and gas exploration;
- information from MAFF on agricultural rents.

Balance of primary incomes/national income (B5)

14.80 This is the balancing item of the allocation of primary income account: the total value of the primary incomes receivable, *less* the total of the primary incomes payable.

Current taxes on income, wealth etc. (D.5)

14.81 Most of these consist of taxes on the incomes of households or profits of corporations; and of taxes on wealth that are payable regularly every tax period (as distinct from capital taxes levied infrequently). This category is broken down simply into taxes on income (D.51) and other current taxes (D.59). It covers all compulsory, unrequited payments (in cash or kind) levied periodically by general government and the rest of the world on the income and wealth of institutional units.

14.82 Information from the Inland Revenue provides most of the data needed here.

Taxes on income (D.51)

14.83 These consist of taxes on incomes, profits and capital gains. They include taxes on individual income (through the PAYE system) including the income of owners of unincorporated enterprises; and corporation tax. They are assessed on the actual or presumed incomes of individuals, households, NPISH or corporations.

14.84 These data come from the Inland Revenue. For individuals the PAYE system (see para 14.35 above) is supplemented by annual tax returns for higher rate taxpayers, and data of tax paid on deposits with monetary financial institutions (MFIs). Discussion of the Inland Revenue data used in the economic accounts is also given in the Annex to this chapter.

Other current taxes (D.59)

14.85 These include:

- taxes that are payable periodically on assets, e.g. ownership or use of land;
- Council tax, and its predecessors the Community Charge and domestic rates;
- payments by households of vehicle excise duty, or for fishing licences.

A full list of the UK taxes currently included in these categories is shown in Chapter 21. A conceptual discussion of them can be found in Chapter 5.

14.86 These taxes are recorded:

- as uses of the sectors in which the taxpayers are classified in the secondary distribution of income account;
- as resources of general government in the secondary distribution of income account;
- as uses and resources in the external account of primary incomes and current transfers.

14.87 These estimates are obtained from a variety of sources. Taxes on capital come from the Inland Revenue (see above). The figures for motor vehicle duties (motor vehicle excise duty and driving licence fees) come from Department of Transport records. Figures for the council tax, community charge (payable by adults in England and Wales from 1990 to 1993 and in Scotland from 1989 to 1993) and local domestic rates came from the DETR and corresponding departments for Scotland, Wales and Northern Ireland.

Social contributions and benefits (D.6)

14.88 Social benefits are transfers to households (in cash or in kind), intended to relieve them from the financial burden of a number of risks or needs. They are either made through collectively organised schemes, or outside such schemes by government units and NPISH. They may be financed by social contributions (social insurance benefits) or non-contributory (social welfare benefits). The secondary distribution of income account includes both social contributions (D.61) and social benefits other than social transfers in kind (D.62).

Social contributions (D.61)

14.89   These are actual or imputed payments to social insurance schemes to make provision for social insurance benefits to be paid. They are available in the following detail:

- Actual social contributions (D.611);
- Employers' actual social contributions (D.6111);
- Employees' social contributions (D.6112);
- Social contributions by self- and un-employed persons (D6.113);
- Imputed social contributions (D.612).

14.90   In the United Kingdom these comprise contributions to the national insurance scheme, the national health service, the Redundancy Fund, and the Maternity Pay Fund and premiums associated with the minimum state pension scheme. They include contributions in respect of Statutory Sick Pay (whereby employers pay their employees sickness benefit in return for a deduction from the contributions they pay). The main source for this information is the Inland Revenue, which collects the contributions on behalf of the Department of Social Security (DSS), the Department for Education and Employment (DfEE) and the Government Actuary's Department (GAD).

14.91   Under the PAYE system, when an employer deducts income tax from the wages and salaries of his employees he also deducts National Insurance contributions (NIC) and pays them to the Inland Revenue for forwarding to the Department of Social Security (DSS). The Inland Revenue collects most National Insurance Contributions on behalf of the Department of Social Security. The information collected basically covers all employees in employment who are liable to tax and/or NICs. The NIC threshold is currently (1998) £61 per week. (As employers may submit computerised returns to the Inland Revenue, derived from their payroll systems, in practice some employees falling below the NIC threshold are included. In total about 37 million documents are received.) Information from a one per cent sample of DSS National Insurance records is extracted annually. For this sample of individual records, additional information about pay, tax, pensions, industrial/business classification, employers' contributions and category of contributions is extracted from end-of-year Inland Revenue PAYE documents. These data are transferred to a statistical file which is analysed to provide information on employees' NICs by industry, region, etc. This database or project is known as the National Income Statistics Project.

14.92   The social contributions shown in the central government accounts are the estimated amounts deducted by employers. The difference between these deductions and central government receipts is shown as in accounts receivable and payable in the financial account.

Social benefits other than social transfers in kind (D.62)

14.93   These comprise social security benefits in cash, private funded social benefits in cash or kind, unfunded employee social benefits and social assistance benefits in cash. They are intended to provide for the needs that arise from certain events or circumstances: for example, sickness, unemployment, retirement, housing, education. They may be social insurance benefits or social assistance benefits. The constituent benefits are listed in Chapter 21. The social transfers in kind that are excluded are those paid by government as individual consumption (see Chapter 21).

14.94   These estimates are derived from Department of Social Security and GEMS records.

Other current transfers (D.7)

14.95   Current transfers are all transfers that are not transfers of capital.

14.96 Five sub-categories are shown in the Blue Book tables. These are:

- net non-life insurance premiums;
- non-life insurance claims;
- current transfers within general government;
- current international co-operation;
- miscellaneous current transfers.

They are described separately below.

Net non-life insurance premiums (D.71)

14.97 These relate to insurance policies taken out by enterprises or individual households. The policies taken out by individual households are those taken out on their own initiative and for their own benefit, independently of their employers or government and outside any social insurance schemes. These premiums comprise both the actual premiums payable by policy holders during the accounting period and the premium supplements imputed from the property income attributed to insurance policy holders. The total of these has to cover the payments of service charges to the insurance corporations as well as the payments of the insurance itself. After deducting the service charges from total non-life insurance premiums, the remainder is described as net non-life insurance premiums. Only the net non-life insurance premiums constitute current transfers. The service charges represent purchases of services and so are shown as intermediate or final consumption, as appropriate.

14.98 These estimates are derived from the ONS survey of insurance companies (25.3.3).

Net non-life insurance claims (D.72)

14.99 These are the actual amounts payable in settlement of claims that become due during the current accounting period. The settlement of a claim is regarded as a current transfer to a claimant. (This heading does not include payments to households in the form of social insurance benefits.)

14.100 These estimates are derived from insurance inquiry returns to the ONS (25.3.3).

Current transfers within general government (D.73)

14.101 In the United Kingdom these comprise current transfers from central to local government. They include both specific grants (that is, those financing a specified expenditure) and general grants. (They do not include transfers of funds committed to finance gross fixed capital formation; such transfers are regarded as capital.) The transfer of national non-domestic rates resources from central to local government is included here.

14.102 These estimates are obtained from public expenditure monitoring records.

Current international co-operation (D.74)

14.103 Resources of the United Kingdom are predominantly current transfers in cash or in kind between the government and institutions of the European Union. Uses comprise foreign aid and social security benefits paid abroad. More detail is given in Chapter 21. These estimates come from government accounting records and are of good quality.

Miscellaneous current transfers (D.75)

14.104 This is the residual category where current transfers not specified elsewhere are recorded. They may take place between resident institutional units or between residents and non-resident units.

14.105 The main component of this is grants paid by central government to higher and further education institutions. It also includes UK contributions to the EU budget based on the so-called '4th resource' (derived from the value of the UK's ESA79-based gross national product at market prices). These data come from government sources and are of good quality.

Disposable income (B.6)

14.106 This is the balancing item in the secondary distribution of income account. It represents the balance of primary incomes for a sector or institutional unit *plus* all current transfers receivable (except social transfers in kind) *less* all current transfers payable (except social transfers in kind).

Redistribution of income in kind account

14.107 The function of this account is described in detail in Chapters 5 and 21. It is an alternative treatment which treats all consumption expenditure by general government and NPISH for the benefit of individual households as social transfers in kind, and derives an adjusted disposable income for households which is available for actual consumption and saving; and for general government which is available for collective consumption on behalf of the community as a whole.

Social transfers in kind (D.63)

14.108 By convention the final consumption expenditure by general government which is treated as individual, and thus as social transfers in kind, is defined in terms of COFOG, the international classification of the functions of government. It comprises expenditure on education, health, social security and welfare and part of housing, recreational and cultural activities, and transport. It excludes from all of these expenditure on general administration, regulation and research.

Final consumption (P.3)

14.109 This is described in detail in paragraphs 12.42–60. Actual final consumption of households comprises their own consumption expenditure plus the individual consumption expenditure of general government and NPISH. Actual final consumption of general government comprises its own collective consumption expenditure. NPISH have no actual final consumption.

Adjustment for the change in net equity of households in pension funds (D.8)

14.110 This is estimated by the ONS from pension funds accounts. (*See* Annex 2 to Chapter 20.)

Annex 1

The UK income tax system

14a.1 Individuals may receive many different types of income but not all kinds are taxable. Examples of the kinds of income on which tax may be payable are pay (including income in kind), profits from a business, occupational pensions, state retirement pension and widow's pension, unemployment benefit, interest from building societies and banks, dividends on shares, and income from property.

14a.2 Income tax is charged under a series of Schedules and Classes, for example :

Schedule A	-	rent (other than from furnished lettings) and other receipts from land and buildings;
Schedule C	-	interest on certain British Government securities and securities of foreign overnments and public corporations and local government;
Schedule D Case I	-	profits of trade;
Schedule E	-	income from offices, employments and pensions.

14a.3 Individuals' taxable income is calculated by added together all their sources of income liable to tax and then deducting any allowances and reliefs to which they are entitled. The tax is calculated by applying the appropriate rates of income tax.

14a.4 The way in which an individual's income tax liabilities are met depends upon the types of income they have and on their total tax liability. Tax may be collected by a direct payment from the individual or by deduction at source by a third party such as an employer, or a monetary financial institution where an account is held. For most employees who receive income under Schedule E, the tax due is calculated by the employer and the appropriate amount deducted before the employee paid. This is the Pay-As-You-Earn system and is discussed in greater detail below.

14a.5 Since 6 April 1991 interest paid to investors by monetary financial institutions (MFIs) is credited net of basic rate tax, unless the investor has registered to receive the interest without deduction of tax. Thus basic rate taxpayers' liability of the income tax has been met in full and there is usually no need for them to make a return to the Inland Revenue in respect of this income. For higher rate taxpayers, where an additional amount of tax is due on the investment income, the additional tax is assessed and collected separately and they usually have to complete a tax return.

14a.6 Individuals may also receive investment income from corporations in the form of dividends and be liable to income tax on this income. Other forms of income may also give rise to tax liabilities and the amount due is usually collected directly from the individual after details have been supplied in a tax return. More detailed tax statistics than those published by the ONS can be found in the Statistical and Economics Office of the Inland Revenue's publication 'Inland Revenue Statistics' which is published each year.

14a.7 Corporation tax is chargeable on the profits made by private and public corporations and unincorporated associations. The profits estimates used for the economic accounts relate to corporations' Schedule D cases I and II earnings only; i.e. earnings from the profits of trade or the profits of trades or professions or vocations not dealt with under any other schedule. Other cases of income are excluded. Inland Revenue statisticians maintain a standing file based on the data they are passed from the corporation tax assessments work. The file consists of a sample which contains the 35,000 largest companies (which contribute about 85 per cent of these profits) and 1 in 10 of the remainder: some 150,000 corporations in all. The basic information obtained for each sample company is its Schedule D Case I profit or loss in each accounting year. If provisional figures are not available for a corporation, a figure is imputed on the basis of similar cases and figures on the amount of tax paid by the corporation in previous years.

14a.8 Although trading losses made in earlier or later years can be used to offset a trading profit in the current year, these offsets are not reflected in the sample members' figures which are used for the economic accounts estimates.

Pay-As-You-Earn

14a.9 The main source of data used to measure wages and salaries is tax records. Much the greater part of taxes on employment income are collected in the United Kingdom by a system known as 'Pay-As-You-Earn' (PAYE). All (approximately) 1.3 million UK employers have a statutory obligation to operate a PAYE system to deduct tax, and social security contributions, at source from wages, salaries and occupational pensions of the 22 million taxpaying employees and pensioners. About 90 per cent of all UK individuals liable to income tax are chargeable under schedule E and almost all (about 98 per cent) of income tax chargeable under schedule E is collected under the PAYE system. In 1994–95 tax under schedule E yielded about three quarters of gross income tax receipts. After the end of each tax year details of pay and tax contributions for each employee are sent to the Inland Revenue by employers. The Inland Revenue code the industry of employment using the Standard Industrial Classification, 1992.

14a.10 Other forms of income such as income from abroad or from letting accommodation may also give rise to a tax liability. The amount due is usually collected directly from the individual after details have been supplied in a tax return.

Annex 2

Value added tax (VAT)

14b.1   Customs and Excise are responsible for collecting Value Added Tax (VAT). It is charged on a wide range of home-produced goods and services. There are currently (1998) four rates:

- standard (currently 17.5 per cent);
- on fuel and power (5 per cent);
- insurance premium tax (4 per cent);
- zero rate.

There are also a number of exceptions.

14b.2   VAT is assessed by, and collected from, those registered businesses which produce, sell or otherwise provide taxable goods and services. It is charged at every stage in the process of production, manufacture and distribution. Each enterprise pays VAT on the goods and services it purchases for its business. This is known as input tax and as a general rule may be reclaimed from Customs and Excise if it relates to a taxable output. When enterprises supply taxable goods and services they are obliged to charge VAT on their value. This is termed output tax and is normally passed on to the consumer in the selling price.

14b.3   Generally, every three months a business has to collate all its output tax and reclaimable input tax and make a return to Customs and Excise. Special accounting schemes apply to retailers. Where a business regularly claims tax back (where input tax is more than output tax), returns may be made on a monthly basis. Since Autumn 1992, the largest 1,600 enterprises which regularly pay VAT (those which pay more than £2million VAT a year) have been required to pay tax monthly on account against quarterly returns. Customs and Excise receive some 7 million VAT returns each year.

14b.4   A trader with a turnover of £49,000 or more a year in taxable goods and services is legally required to be registered with Customs and Excise for VAT purposes. Traders with lower turnovers may seek voluntary registration. Currently approximately 1.6 million businesses are registered for VAT of which about 0.3 million are ones which regularly claim this tax back.

14b.5   A number of services are exempt from VAT. Exemptions include most banking services, health and postal charges. Businesses dealing with wholly exempt supplies cannot reclaim any input tax. However, Customs and Excise have devised special rules on input tax recovery for the 10,000 traders who are involved in a mixture of exempt and taxable supplies.

14b.6   Single group registration for large corporation groups is allowed. This avoids the need to charge VAT on transactions between the corporations in the VAT group.

Chapter 15

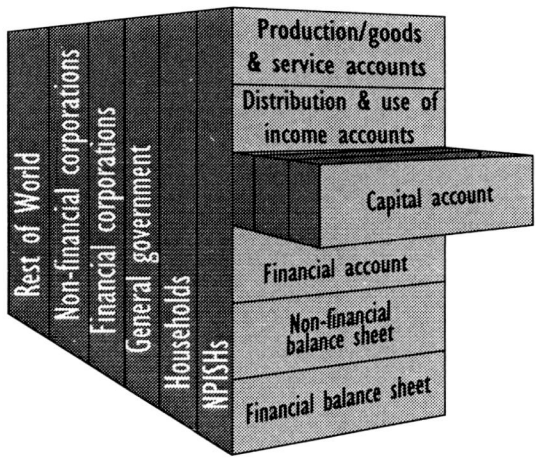

The capital account

Chapter 15 The capital account

Introduction

15.1   This chapter describes the methodology for the estimation of the variables which go to make up the capital account. In many cases it describes analyses made to construct the accounts which are not published. It also includes a section on consumption of fixed capital. (A description of the perpetual inventory methodology for the estimates of capital stock, closely linked to capital consumption, is given in Chapter 16.) A summary description of the estimation process for the *aggregates* of the main variables of the account and also the key features of the variables have been included in Chapter 12 on the goods and services account. This chapter expands on these issues, including the changes in coverage compared with the previous system of accounts, and explains the compilation process for the component series, that is the breakdown by industry, asset or institutional sector, as appropriate. The analysis of the variables by institutional sector provides the basis for the derivation of the capital account for each of the sectors, as part of the full range of sector accounts discussed in the individual sector chapters.

15.2   The capital account, as presented in the form for the whole economy, is a new feature of the UK statistics. The account shows how investment in non-financial assets (essentially gross capital formation) is financed out of saving and capital transfers. Saving, it will be recalled, is the balance emerging from the production and the distribution and use of income accounts. The capital account strikes a balance – 'borrowing' if investment exceeds saving and net capital transfers and 'lending' if investment is less than saving and transfers. The description of the account for the whole economy is readily extended to the accounts for individual institutional sectors. Although the presentation of the whole economy account is new, most of the variables would have appeared in previous Blue Books in the sector summary table, while capital accounts, in a slightly different format, were also compiled for individual sectors.

The (summary) capital account

15.3   The capital account is presented in two parts. As background to this chapter, the variables making up the two sub-accounts are shown below, in summary form. In some accounts published in the Blue Book, GFCF and capital transfer are disaggregated below the level shown here. The table gives figures for 1995.

The (summary) capital account for 1995

£million

Change in net worth due to saving and capital transfers account (III.1.1)

Saving, gross (B.8g)

plus Capital transfers receivable (D.9)
less Capital transfers payable (D.9)

Change in net worth due to saving and capital transfers
    (B.10.1g= B.8g +D.9(balance))

Acquisition of non-financial assets account (III.1.2)

Change in net worth due to saving and capital
transfers (B.10.1g)

Gross fixed capital formation (GFCF) (P.51)
Change in inventories (P.52)
Acquisitions *less* disposals of valuables (P.53)
Acquisitions *less* disposals of non-produced non-financial assets (K.2)
Net lending (+)/borrowing(-)
    (B.9=B.10.1-P.51-P.52-P.53-K.2)

15.4   Thus, in the upper table, the accounts show that saving (B.8g) – the balance between national disposable income and final consumption expenditure from the production and distribution and use of income accounts – is reduced or increased by the balance of capital transfers (D.9) to provide an amount available for financing investment (in both non-financial and financial assets). Then, in the lower table, total investment in non-financial assets is the sum of P.51, P.52, P.53 and K2. Thus, if investment is lower than the amount available for investment, the balance will be positive and is regarded as lending (if negative the balance is borrowing). It represents the resources available to the rest of the world. Where the capital accounts relate to the individual institutional sectors, the net lending/borrowing of a particular sector represents the amounts available for lending to or borrowing from other sectors. The value of net lending/net borrowing is the same irrespective of whether the accounts are shown before or after deduction of K.1, fixed capital consumption (provided a consistent approach is adopted throughout). The variables are explained in more detail below.

15.5   The lending or borrowing figure arising in the capital account is identical, in principle, to the balance which emerges from the transactions in financial assets (see Chapter 17). However, because of measurement problems in the accounts generally, in practice these will not be the same. Part of the balancing process for the economic accounts statistics for years before the latest one shown (i.e. for years t-1 and earlier) involves assessing and modifying the component variables so that the estimates of net lending/borrowing made from the current and capital accounts, and from the financial accounts, are the same at the level of the whole economy, and reasonably close to each other at the sector level. The balancing process is described in Chapter 11.

The analyses and some key features of the capital account

15.6   Most of the variables included above have been described, in some form or other, earlier in the publication. For example, the three variables – GFCF, changes in inventories and acquisition less disposals of valuables – have all been covered, in broad terms, in Chapter 12. As mentioned in the introduction, this chapter will describe the methodology for the estimation of the *detailed* components for the capital account variables required for the account itself and also for other parts of the economic accounts. This will cover the *asset* analysis included in the capital account and also an alternative asset breakdown, as well as the analysis by *industry* and by *institutional sector*. The description of the detailed analyses will be in respect of each of the variables separately. The industry analysis will relate essentially to the 17 main standard groups of NACE 1. However, the text will also embrace the detail required for the 31 groups which are formed by further disaggregation of manufacturing and of mining and quarrying. The analysis by institutional sector will cover the seven main sectors:

- Private non-financial corporations (S.11002/3)
- Public non-financial corporations (S.11001)
- Financial corporations (S.12)
- Central government (S.1311)
- Local government (S.1313)
- Households (S.14)
- Non-profit institutions serving households (NPISH) (S.15)

The type of assets covered will be explained for each variable. The estimation methodology will be described for each variable, as appropriate, for annual and quarterly data, at current and constant prices. However, the capital account as such is compiled only at current prices, annually and quarterly.

15.7   Before looking at the nature of and estimates for individual variables, three points should be made. First, as mentioned above, the capital account is concerned only with transactions in non-financial assets. The distinction between non-financial and financial assets is largely self-evident. Some discussion of this has been given in Chapters 6 and 7. Secondly, this chapter includes a section on consumption of fixed capital, which represents the value of fixed assets used up in the production process. The deduction of a figure of capital consumption transforms the basis of recording from a gross to a net basis. It is included here because both the net and gross presentations are valid, provided that both saving and fixed capital formation are dealt with consistently. Finally, it should be remembered that in making the changes in coverage required by ESA95 (for example the inclusion of computer software), it has not always been possible to collect data retrospectively. In order to produce a continuous consistent historical series, some assumptions have had to be made, and as a consequence the quality of the data for earlier periods is poorer than for the recent past.

15.8   The variables of the capital account are covered in this chapter in the order presented in the two sub-accounts above. The derivation of consumption of fixed capital is given at the end of the chapter. It is closely linked with the perpetual inventory methodology for deriving estimates of the capital stock, which is outlined in Chapter 16.

15.9   The description of the capital account will be presented in terms of the account for the whole economy. This is readily extended to the accounts for individual institutional sectors, which represent the main analysis of the account. However, as will be explained later, the whole economy account should not be seen as the sum of the separate sector accounts. Rather, the estimation process seeks to establish figures for the whole economy, and then to ensure that the sector data are consistent with these aggregates.

Change in net worth due to saving and capital transfers account (III. 1.1)

15.10   The balancing item on this account shows the amount available for the net acquisition of financial and non-financial assets. This amount is equivalent to saving plus net capital transfers receivable. The variables making up the account are considered below.

Changes in liabilities and net worth

Saving, gross (B.8g)

15.11   This is the balance of resources and uses which derives from the production and distribution and use of income accounts. For the whole economy, it is the balance of national disposable income over final consumption expenditure. The Blue Book gives measures both gross and net of fixed capital consumption.

Capital transfers (D.9)

15.12   The definitional and conceptual issues relating to capital transfers, including the distinction between capital and current transfers, were set out in Chapter 6. The features of capital transfers, including details of the kind of transfers made and also the allocation of payments and receipts by institutional sector, are given in the sectoral chapters, in particular Chapter 21 on the government sector, which is the source or recipient of most capital transfers. The text here will summarise only the main issues.

15.13   The capital account shows separate figures for receipts and payments for three broad categories, as below. In respect of the whole economy, the balance of receipts and payments will reflect any transfers received from or paid to overseas. For individual institutional sectors, the accounts will also show the flows between sectors. Given that most capital transfers are made by central government and to a lesser extent, local authorities, these sectors are taken as the source of information on receipts as well as payments. In the description given below, the 'receivable' and 'payable' components are not shown separately, but are discussed under the type of transfer.

Capital taxes (D.91)

15.14   Capital taxes cover taxes levied on assets or the transfer of assets, principally death duties (including inheritance tax). Taxes on capital gains, which were included in capital taxes in the previous accounts, are now included in current taxes on income and wealth (D.5). Capital taxes are paid to central government, mostly by households, but companies also make payments. Information, on a quarterly basis, is derived from Inland Revenue sources, as part of the administration of the tax system.

Investment grants (D.92)

15.15   These cover grants made by government to residents or non-residents to finance all or part of the costs of acquiring fixed assets. In the UK, the main type of investment grants are central and local government grants for housebuilding and improvements and grants to universities, but European Union investment grants to UK residents are also included. Estimates are made, on a quarterly basis, largely from information available as part of the arrangements for monitoring public expenditure.

Other capital transfers (D.99)

15.16   Other capital transfers cover a variety of transactions, including central government grants to local government and cancellation of debts, particularly by central government (*see* Chapter 21). Note that grants from government to public corporations are treated as equity investment if they come from a part of government that has a claim on the corporation, particularly if they relate to the cancellation of debt in the context of privatisation. Estimates are again made largely from information required for the monitoring of public expenditure.

The analysis by institutional sector

15.17   This section describes the allocation of the figures of receipts and payments of capital transfers to each of the seven main institutional sectors, as shown below. Given that the bulk of payments emanate from central government, this also provides the main source for information on the recipient sector. It should be borne in mind that the long-run figures on a sector basis will be influenced by changes in the composition of the sectors, for example as a result of privatisations. The main transactions for the individual sectors are shown below.

Private non-financial corporations

15.18   Receipts by these producers relate mainly to investment grants; payments are mostly taxes and contributions to capital formation by local government.

Public non-financial corporations

15.19   This sector embraces the public corporations. The series is much smaller than in the pre 1998 Blue Books, because many capital transfers from government have now been classified as equity investments.

Financial corporations

15.20   There are no receipts of capital transfers and only small payments of taxes.

Central government

15.21   Central government makes major capital transfers to private and public non-financial corporations, local government and NPISH, as detailed for the recipient sectors. Receipts are of capital taxes.

Local government

15.22   Local authorities receive capital transfers from central government to cover a variety of service spending, particularly on housing, and transport. They also receive transfers from private non-financial corporations. Grants are made to households for housing improvements.

Households

15.23   The household sector's main receipt is from local authorities for housing improvements. Payments relate to taxes, mainly death duties.

NPISH sector

15.24   Receipts by the NPISH sector are mostly central government grants to housing associations and to universities.

Change in net worth due to saving and capital transfers (B.10.1)

15.25   This is the sum of saving and receipts less payments of capital transfers. It represents the amount available for the acquisition of financial and non-financial assets.

Acquisition of non-financial assets account (III.1.2)

15.26   The second sub-account shows how saving and net capital transfers are used in the acquisition of non-financial assets, such as gross fixed capital formation, or additions to inventories or valuables, or of non-produced assets. The description of the variables of the account relates to the structure of the account, as summarised earlier in this chapter.

Changes in liabilities and net worth

Change in net worth due to saving and capital transfers (B.10.1)

15.27   This is the balance from the previous account.

Changes in assets

15.28   Included on the assets side of the table are three separate components which together make up gross capital formation (P.5). They are three very different variables, in nature and magnitude, and will be considered separately. The components are:

- Gross fixed capital formation (P.51);
- Changes in inventories (P.52);
- Acquisition less disposals of valuables (P.53).

15.29   Although the analyses will mostly show the components separately, occasionally they will relate to the aggregate of the three variables – gross capital formation. It is important that this aggregate is not confused with the main component of gross *fixed* capital formation.

Gross fixed capital formation (GFCF) (P.51)

15.30   GFCF relates essentially to investment in tangible and intangible fixed assets and improvements to land. These are defined in detail in Chapters 3 and 6. GFCF constitutes additions to the nation's wealth. Investment is in assets which are used repeatedly in the production process for more than a year. The way in which *aggregate* estimates of GFCF are compiled has been outlined in Chapter 12 on the goods and services account. Amongst other things, the chapter also mentioned the main changes in the scope of GFCF, compared with the previous system, and some key features of the variables. In addition, the chapter mentioned the two main approaches for estimating GFCF, that is from expenditure data or the commodity flow model. This section will expand on some of the more important of these points and also describe the estimation process for the analyses of GFCF.

Changes from previous system

In the new accounts, the scope of GFCF has been widened compared with ESA79 in a number of ways. It now includes mineral exploration, computer software, and entertainment, literary and artistic originals, the first of which would previously have been included in intermediate consumption, while the last was outside the production boundary. These items compise the category of intangible fixed assets.

Also now included, as part of tangible fixed assets, are cultivated assets, such as trees or livestock, which are used repeatedly to produce goods, and historic monuments. Further, expenditure on military structures and equipment (other than weapons), previously in government intermediate consumption, is now also part of tangible fixed assets. The new system seeks (with only partial success) to reclassify transactions in land to K.2.

An attempt has been made in the new ESA to try to clarify the borderline between GFCF and intermediate consumption in respect of small tools. By convention, expenditure on these items should be classified to GFCF only if it exceeds around £400 at 1995 prices.

Gross Fixed Capital Formation in the previous accounts was termed Gross Domestic Fixed Capital Formation. The dropping of the word 'domestic' is of no significance.

Tangible and intangible fixed assets

15.31 A distinction is made between investment in tangible and intangible fixed assets. The former are made up of the conventional assets, such as plant and machinery, transport equipment, dwellings and other buildings and structures, and also cultivated assets. Intangible assets, which represent most of the expansion in scope of GFCF, have been defined in the box above. Further details are given in the section on assets, later in this chapter.

Produced and non-produced assets

15.32 A further key distinction is between produced and non-produced assets. The former cover the conventional fixed assets, as mentioned in the previous paragraph. Non-produced assets are essentially assets which are needed for further production, but have not themselves been produced. The main example is land. Improvements to such assets, for example land reclamation, forest clearance, drainage, and flood prevention, are recorded as GFCF. Acquisition of non-produced assets is classified to K.2 (*see* below).

Acquisitions and disposals

15.33 It should be noted that information on GFCF relates to both acquisitions and disposals of assets. This basis of recording is crucial since many assets, particularly buildings, have long service lives and will change ownership several times during the course of their productive life before they are scrapped. Recording both sides of these transactions means that only the transfer costs of the deal (see below) are included in aggregate GFCF. Reflecting the net approach, for a given producer, GFCF can be negative.

New and existing assets

15.34 Although not shown in the summary table above, a distinction is made in ESA95 between investment in new assets and in existing assets. As just mentioned, many assets will change ownership several times during their life. An existing asset is one which has already been acquired by at least one owner and whose value will already have been included in GFCF. At the present time it is not possible to make this distinction in the UK accounts. The 'acquisition minus disposal' basis of recording avoids double counting investment in existing assets.

15.35 Within the expenditure measure of GDP, the purchase of an existing asset may be recorded in a different variable within the expenditure measure from the sale. For example, if an existing asset, e.g. a car, is sold by a producer to a household, the producer, say in the non-financial corporation sector, records the sale as negative GFCF, while the household records the purchase as final consumption in the household sector. Similarly, an export of an existing asset will be treated as negative GFCF and a positive export.

Major improvements and repairs

15.36 The figures of GFCF should include improvements to fixed assets, such as major renovations or modifications. The borderline between such improvements, on the one hand, and repairs (which are part of intermediate consumption), on the other, is not easy to draw (*see* Chapter 4, paragraphs 4.33 to 4.38). The estimates for the economic accounts have to rely very much on the recording practices adopted by businesses for distinguishing between these two components.

Land

15.37 In the new accounts, acquisitions and disposals of land should, in theory, be identified separately from the buildings on the land. Transactions in land are to be recorded, not within GFCF, but as acquisitions and disposals of non-produced non-financial assets (K.2). In general, the separation of land and buildings is not considered feasible for the UK accounts, so the value of land underlying structures is included with the total value of the asset as going into GFCF. However, separate identification is possible for agricultural land which undergoes a change of use, and this makes up the main component going into acquisitions and disposals of land and other non-produced tangible assets (K.21, not published separately). In the accounts, the purchase and selling price of land is the same. The transfer costs of the transaction are recorded separately as part of GFCF (see section on transfer costs below).

Transfer costs

15.38 Largely reflecting the problems with identification of land, explained above, the transfer cost component within GFCF relates to both land and buildings, and not just land, as in the ESA (see Chapter 6). This therefore remains the same as previous practice (see paragraph 15.66).

Financial leasing

15.39 The treatment of investment made under a financial lease, which is new in ESA 95 but which the UK adopted in 1991, involves classifying assets acquired on financial leases to the industry of the user rather than to that of the owner. The acquisition of assets under finance leases are recorded as fixed capital formation by the user sector, financed by an imputed loan. Rental payments are split into interest and principal repayments.

Progress payments

15.40 Where production of a capital asset extends over a number of quarters, GFCF is generally recorded on the basis of progress payments made in each period by the owner to the producer, and not when the asset is finally completed. As an alternative to recording progress payments, the total estimated value of the asset is spread over the period of production in proportion to the costs arising in each quarter. Where production is speculative and there is no contract, the production is recorded in inventories, first as work-in-progress and then as finished goods, becoming GFCF if and when it is sold or used by the producer. One consequence of this is that the value of assets will be recorded in advance of their being used productively.

GFCF by non-residents

15.41   In the accounts, by convention non-residents are not allowed to own immovable fixed assets such as buildings or land. A notional resident unit is created which is deemed to purchase the asset. The purchase of the asset is matched by a financial transaction with the non-resident unit.

Valuation

15.42   Valuation of GFCF is mostly on the basis of purchasers' prices. However, special arrangements apply for progress payments and, for example, own account construction and certain intangible assets. Details are given in Chapter 6 (Table 6.4).

Estimation methodology

15.43   As mentioned in Chapter 12, there are two broad approaches for estimating GFCF. These are (i) the expenditure (or demand) approach, in which information is collected from firms in statistical surveys or, for the public sector, through administrative returns, and (ii) the commodity flow approach, which derives estimates from information on the supply of fixed assets. It will be useful to consider the advantages and disadvantages of these two methods.

15.44   The survey approach will be largely based on the practices obtaining in commercial accounting; that is, recording actual fixed investment or progress payments made in the period. In the main, this represents a good source of data, although it may not always be precisely on the basis required for the accounts. The main problem is likely to be the borderline with intermediate consumption, and in particular the timing and the treatment of repairs and maintenance, and possibly certain minor items of expenditure such as small tools. However, an important advantage of the survey approach is that the information will be largely consistent with the estimates of operating surplus appearing elsewhere in the accounts, and also with the data going into the financial accounts. The approach also provides the basis for the detailed industry and institutional sector analysis, which is made on the basis of the relevant classification of the reporting unit. The main disadvantage of this methodology, though, is that only a fairly broad analysis by type of asset is possible.

15.45   The commodity flow approach, in a general sense, is a model-based methodology in which estimates are made of the demand for products from figures of supply. Thus supply, as measured by production and imports, will go to demand in the form of intermediate consumption of producers, or final consumption by households, the NPISH sector, and government, or inventories or exports, or here, specifically, to GFCF. As a product-based method, the approach permits a much more disaggregated identification of capital goods, and hence provides a more detailed asset breakdown. This is of great benefit for the compilation of input-output tables, and also for deflation, either in detail, directly, or by providing the weights for combining price indices for deflating the aggregated industry figures, as collected in the statistical inquiries.

15.46   The improved asset information is one advantage the model approach has over the survey-based estimates. The commodity flow approach does not provide information about the industry or sector of purchase, nor about acquisitions and disposals, or the split between new and existing assets. Such disaggregations cannot be constructed without other information. A further obvious difficulty is that of determining that part of the total supply of a product which should be allocated to GFCF (as opposed to the other components of demand). In practice, amongst the various components of demand, estimates are usually available on exports, while allocation to the other variables is normally based on the pattern of demand contained in the most recent input-output analysis. Reflecting data availability and other factors, the allocation procedure will be much less robust for quarterly than for annual estimates. Further, allocation will be easier for some products than for others, with particular difficulties for goods which can go to all types of demand, or where the inventories component, which is not measured all that accurately in aggregate, let alone by product, is likely to be an important part of demand. One final problem is the need to ensure that the valuation of the variables is consistent across the whole commodity flow model. This is particularly relevant to the need to move, for each product, from the basic price valuation of production through transport costs, margins and taxes to the purchasers' price valuation of demand.

15.47　In practice, in the UK, estimates of GFCF use both the survey and model-based approaches. However, given the statistical requirements for particular kinds of data, and the relative advantages and disadvantages of the two methods, there is much more emphasis on the survey approach. The commodity flow approach is probably better at estimating the short-term movement in investment, rather than the level. The approach plays a particular role in estimating GFCF for certain mostly single-type, large-scale assets, such as private sector dwellings, ships and aircraft.

15.48　One further point about the estimates might be made at this stage. This is that, in the annual inquiries, although information is requested for the calendar year, contributors are allowed to make returns which relate to their own accounting year. Such reporting years cover periods of twelve months ending at some time between 6 April in the year to which the inquiry relates and 5 April in the following year. The resulting aggregates are therefore the sum of information relating to a variety of twelve month periods. As far as possible given the often erratic nature of the quarterly figures of GFCF, estimates for manufacturing are adjusted to a calendar year basis. These calendar year data, supplemented by any commodity flow results, are used to produce the definitive annual series to which quarterly series are benchmarked.

The detailed analyses of gross fixed capital formation

15.49　Chapter 12 on the goods and services account described how the aggregate estimates of GFCF were made, covering both annual and quarterly figures, at current and constant prices. As mentioned there, the estimates are built up largely from quarterly and annual information collected from private sector and public sector sources. For the former, statistical surveys, principally by ONS (25.1, 25.2), but embracing also information collected by DETR and MAFF, provide data on investment in the various assets, other than dwellings, by each industry. These data are supplemented by estimates of the private sector's total GFCF in dwellings, again largely based on DoE information, and of transfer costs on land and buildings, from Inland Revenue sources. Figures of public sector investment, for all assets, are based on quarterly returns from central government, local authorities and public corporations. Generally speaking, all information from quarterly sources is revised in the light of firmer annual data, where available. In addition to the survey and related information, estimates also use data from the commodity flow approach.

15.50　At constant prices, estimates of GFCF are obtained by deflating the value figures by appropriate price data, largely from ONS's producer price inquiry (para 25.2.7). Deflation raises some difficult problems which are discussed further below in paragraphs 15.111–15.118.

15.51　The following sections describe the analyses of GFCF by asset, industry, and institutional sector. Although the accounts are published on a sector basis and the basis of the estimates (for the private sector) is, essentially, the industry, it will be sensible to consider first the analysis by asset. This is principally because it is necessary to identify the various types of asset in the descriptions of the other breakdowns.

The analysis by asset

15.52   There are two broad analyses of GFCF by asset to be considered. These are set out in the table below. The first analysis is the breakdown recommended in the ESA95 for inclusion in the capital account; the second is the more traditional analysis, by type of asset, as used in previous Blue Books. The following depreciation retains the separate identification of 'plant and machinery' and 'vehicles, ships and aircraft', though in the ESA95 these are combined as 'machinery and equipment'. Both analyses use the same three broad headings. The distinction between them relates to the analysis within the two categories of tangible assets and intangible assets.

Asset analyses of GFCF

Recommended for capital account | **By type**

1. Acquisitions less disposal of tangible fixed assets
of which
    acquisitions of new assets   dwellings
    acquisitions of existing assets   other buildings and structures
    disposals of existing assets   plant and machinery
                              vehicles, ships and aircraft

    cultivated assets

2. Acquisitions less disposals of intangible fixed assets
of which
    acquisitions of new assets   mineral exploration
    acquisitions of existing assets   computer software
    disposals of existing assets   entertainment, literary and artistic
                              originals
                              other

3. Additions to the value of non-produced non-financial assets
of which
    major improvements to assets   major improvements to assets
    cost of ownership transfer of   cost of ownership transfer of
    non-produced non-financial assets   non-produced non-financial assets

---

15.53   In the capital account, the ESA95-recommended analysis is essentially between acquisitions and disposals, and new and existing assets. These distinctions have been outlined above. The identification of acquisitions and disposals is an integral part of the collection process for much of the information on GFCF. However, there is no information readily available on acquisitions and disposals in respect of dwellings; nor is a distinction made in the general collection arrangements as between new and existing assets. For these reasons the information on GFCF recorded in the capital account are only on a net basis, that is for acquisitions *less* disposals. The collection of data in this way also restricts the availability of more detailed analysis.

15.54 Information on GFCF collected in the ONS quarterly and annual statistical inquiries includes, as appropriate, the different types of tangible fixed asset, shown above, as well as providing the information for the estimates of intangible fixed assets and improvements to non-produced assets. It will be useful now to give some further information on the composition of the various asset groups.

Tangible fixed assets

15.55 *Dwellings* These include houses, bungalows, flats and maisonettes, and also houseboats. All expenditure on the construction of new dwellings, including self-build, and improvements to existing ones is included. Architects' and quantity surveyors' fees are also treated as part of capital formation. The GFCF figures include, as far as is possible, items of equipment which are an integral part of the completed dwelling, such as boilers and fitted kitchen equipment, but moveable equipment installed by tenants and owner-occupiers is included in household consumption.

15.56 *Other buildings and structures* This was termed 'Other new buildings and works' in the previous system. Although the asset coverage is roughly the same, certain GFCF, which will have been recorded as part of new buildings and works, will now appear as improvements to non-produced assets (see below), while certain military expenditure, previously part of central government final consumption, will now be classified here. The category includes buildings, other than dwellings, and most civil engineering and construction work. Both new construction, and extensions and improvements to existing buildings and works, are included. Machinery and equipment forming an integral part of buildings and works, for example lifts, heating and ventilating plant, is in general included. The distinction between buildings and plant is sometimes difficult to draw; in such cases the normal commercial accounting practice for distinguishing between them has to be followed. Railway track and gas and water mains are included under this heading, but electricity and telephone lines and cables are classified as plant and machinery. Architects' and surveyors' fees are included, but not fees and costs incurred in transferring the ownership of land and buildings, which are covered in transfer costs (see paragraph 15.38).

15.57 *Plant and machinery* As mentioned in paragraph 15.52, this group is one of the two sub-divisions of the ESA95 group 'machinery and equipment'. The coverage is broadly the same as previously, but also now included is certain military expenditure (other than weapons) previously classified to government consumption. By and large, the heading includes the traditional plant and machinery, as identified in commercial accounts. As mentioned above, plant and machinery forming an integral part of buildings and works, for example lifts, heating and ventilating plant, is in general included in other buildings and structures. Electricity and telephone lines and cables and ducts are included as part of plant and machinery. The distinction between plant and machinery and vehicles is not always clear. In practice, assets such as agricultural tractors, road-making vehicles, and vehicles used within warehouses and railway goods depots, for example, are all classified to plant and machinery.

15.58 *Vehicles, ships and aircraft* The distinction between vehicles and plant and machinery has been mentioned in the above paragraph. Vehicles embraces all vehicles intended mainly for use on public roads. Cars purchased by businesses are included. However, cars bought by households are part of final consumption expenditure and are excluded from GFCF. Also included is railway rolling stock. Ships covers all vessels owned by UK companies and registered as UK vessels on the General Register of Shipping and Seamen. Hence UK owned ships transferred from overseas to United Kingdom registration count as new capital formation and conversely ships re-registered abroad are deducted. Similarly, all civil aircraft owned by UK registered companies are included irrespective of their location. Expenditure borne by the government on the development of prototype civil aircraft is excluded together with all single-use military aircraft.

15.59　*Cultivated assets* This category now includes investment in cultivated assets. This covers livestock and trees which are used repeatedly and continuously for more than one year to produce other goods and services. The methodology for estimation is described in the industry analysis for agriculture and forestry.

Intangible fixed assets

15.60　Intangible fixed assets represents a new category of GFCF, embracing the three groups shown below. Previously, such expenditure should have been recorded in intermediate consumption. However, some is likely to have been recorded in GFCF.

15.61　*Mineral exploration* This expenditure covers the costs of drilling and related activities such as surveys. It is included in GFCF whether or not the exploration is successful. This expenditure was treated as intermediate consumption in the previous statistics. Estimates of this expenditure are readily available from the companies undertaking the activity (see paragraph 15.74).

15.62　*Computer software* This covers computer software, such as the purchase or development of databases, which is used in the productive process for more than one year. Given that much computer software is likely to be included indistinguishably with its computer hardware, there has long been some uncertainty on how this expenditure has been treated in the statistical returns made for the economic accounts. Where software can be identified separately, many companies are already capitalising such expenditures in their accounts so it is already being included in the estimates of GFCF.

15.63　The series for computer software identified in asset analyses, therefore, represents only that software which is not already being recorded as part of hardware. In other words it is not a comprehensive series for software. The problem is with the allocation by type of asset – as between plant and machinery and intangible fixed assets – not the aggregate of GFCF, which is not affected. Given that most industries may undertake investment in computer software, the estimation methodology is mentioned here, rather than in the industry section. The estimates are based on information collected in the ONS quarterly capital expenditure inquiries (para 25.2.2).

15.64　*Entertainment, literary and artistic originals* This heading embraces original films, sound recordings, manuscripts, tapes etc on which musical and drama performances, TV and radio programmes, and literary and artistic output are recorded. Defining and estimating for this new variable has proved extremely problematic. Having pursued these issues for some time, the United Kingdom has adopted a largely pragmatic approach to estimation. For the moment, the figures included in the accounts relate only to the TV and radio, publishing and music industries. These are thought to be the main industries with these kind of expenditures, while major collection difficulties were foreseen in extending the coverage more widely. On the conceptual side, the determinant of spending does not relate only to the physical original itself, but also to the copyright attached to it. A description of the methodology for estimation is given in the industry analysis, under 'other community, social and personal services'.

Additions to the value of non-produced non-financial assets

15.65　*Major improvements to assets* This component covers essentially improvements to land such as reclamation, clearance to enable production for the first time, drainage and flood prevention. Site clearance of land previously used in production for further development is not included here, but is part of GFCF in dwellings or other buildings and structures. This component is new in the sense of now being identified separately in GFCF. In the previous accounts, the expenditure will have been recorded under the dwellings or building and structures heading.

15.66 *Cost of ownership transfer of non-produced non-financial assets* Transfer costs cover stamp duties, legal fees, dealers' margins, agents' commissions and other costs incurred in connection with the transfer of ownership of land and buildings. These costs are treated as part of GFCF. ESA95 recommends that transfer costs under this head should related solely to non-produced assets, that is essentially to land, with the costs for other transactions being included with the asset itself. However, given the difficulty and uncertain sense of endeavouring to separate the land and building components, the estimate recorded in the accounts follows previous practice and relates to both buildings and land. The estimates of transfer costs are based on information from Inland Revenue and the Land Registry, and certain other sources including DETR.

The analysis by industry

15.67 The analysis of GFCF by industry largely reflects the way in which the aggregate estimate, at least for the private sector, is put together. This major component is added to estimates for the public sector and also for dwellings and transfer costs, to obtain total GFCF. In the new system, the industry analysis is based on the NACE classification. The description of the methodology includes, for most industries, some indication of the importance of the various institutional sectors. This is brought together in the later section which looks at the analysis by institutional sector. Estimates on an industry basis at constant prices are described in paragraphs 15.111 to 15.118.

15.68 Classification of GFCF to an industry is in general based on the principal activity of the unit reporting the expenditure. For the private sector, for the production industries, the reporting unit is often an individual factory, which can provide the full range of statistical information required for the economic accounts. For the distributive and service industries, the industrial classification is generally based upon legal units which are individual companies in the corporate sector, sole proprietorships, partnerships etc.

15.69 The following are the main industrial sectors separately identified.

GFCF by industrial	
A + B	Agriculture, forestry and fishing
C	Mining and quarrying
D	Manufacturing
E	Electricity, gas and water supply
F	Construction
G + H	Wholesale and retail trade, hotels and restaurants, etc
I	Transport, storage and communication
J + K	Financial intermediation and business services
L	Public administration and defence
M + N	Education and health
O	Other community, social and personal service activities

Agriculture, hunting and forestry; fishing

15.70 GFCF for *agriculture*, which now includes estimates for cultivated assets, is derived from a variety of mostly MAFF sources. The principal source is the annual Farm Business Survey which provides information for the main types of asset. Quarterly estimates are derived by interpolation. Within the asset analysis, other buildings and structures comprise farm and horticultural buildings. Expenditure on drainage, irrigation and other land improvements and operations are included in improvements to non-produced assets. A new category is included for cultivated assets. The estimates included in the accounts cover only dairy cattle and breeding stock; GFCF in other cultivated assets is thought to be negligible. The estimates for livestock are based on the difference between values of stock, at the end and beginning of the year, derived from numbers of animals multiplied by their estimated average value.

15.71 Estimates for *forestry*, which now includes cultivated assets, are based largely on the annual report and accounts of the Forestry Commission. The main item of expenditure is the establishment of forests (which includes the cost of draining the ground, clearance, planting, care of young trees and associated administrative costs). These expenditures are classified to improvements to non-produced assets. There are also expenditures on buildings and structures, mainly roads, and small amounts on vehicles. The Forestry Commission information provides direct estimates for public non-financial corporations. Estimates for the private sector are based on the statistics of the area of private forests planted and an assumed per hectare valuation as for the public sector.

15.72 Gross fixed capital formation in *fishing* vessels is estimated by the Department of Trade and Industry, who monitor, for such vessels, additions to and deletions from the Department of Transport's General Register of Shipping and Seamen. Very rough estimates are made for the small amount of fixed capital formation in other machinery and equipment and in buildings and structures.

Mining and quarrying

15.73 Two separate industry groups are covered under this heading 'Extraction of mineral oil and natural gas' and 'Other mining and quarrying'. The GFCF estimates for both industries now include the cost of any exploration, whether successful or not. Previously, this expenditure had been included in intermediate consumption.

15.74 In respect of *extraction of mineral oil and natural gas*, estimates of GFCF, including exploration costs, are based on information collected in the DTI quarterly inquiry into the oil and natural gas industry. The estimates cover fixed capital formation by exploration licensees, production licensees, operators appointed by production licensees and specialised contractors selling services to the industry. The fixed capital formation of contractors who may operate in the North Sea or other oil fields is included only where the contractors are registered UK companies. Where operators work as part of a consortium, expenditures are allocated in proportion to their shares.

15.75 Detailed information by appropriate asset type is collected, within which platform modules and mobile drilling rigs, for example, are classified to machinery and equipment. All other categories are classified to other building and structures. The estimates of GFCF are essentially on an invoice basis and will include progress payments.

15.76 For *other mining and quarrying*, estimates of GFCF, including exploration costs, in the coal industry are derived from information collected in ONS quarterly capital expenditure inquiries (25.2.2). The estimates are aligned to firmer information collected in annual inquiries (25.1.3). Mine shafts are classified to other buildings and structures.

15.77 Most of the GFCF in mining and quarrying is included in private non-financial corporations' investment, with a small element for public non-financial corporations.

Manufacturing

Electricity, gas and water supply

15.78　These two groups of industries have common sources for the estimates of GFCF and are considered together. The main source is the ONS quarterly capital expenditure inquiries (para 25.2.2), which are benchmarked on the firmer information from the annual inquiries to production industries (para 25.1.3). Some particular features for individual industries are described below.

Manufacturing

15.79　Information on GFCF is also available for the fourteen sub-industries within manufacturing which make up the more detailed whole economy analysis (by 31 industries). It should be noted, though, that, in respect of the latest year, the quarterly basis of the initial annual estimates does not provide information of sufficient accuracy to permit the full industrial analysis to be made; this has to await the availability of the firmer annual data. Almost all investment is by private non-financial corporations. The estimates also cover GFCF by public sector establishments, based on the sources and uses of funds statements from the relevant public corporations.

Electricity, gas and water supply

15.80　Since the previous version of Sources and Methods, the supply activities of these three industries have been privatised. New data sources, in the form of the ONS annual and quarterly inquiries mentioned above, have been introduced.

15.81　For electricity, expenditure on power stations is broken down into the two groups 'other buildings and structures' and 'plant and machinery'. Transmission lines, whether overhead or underground, and transformers and switching equipment, are classified to plant and machinery, as are major plant spares for power stations and the initial expenditure on nuclear fuel elements for nuclear power stations. Fixed capital formation by the nuclear fuel industry is included with the electricity industry. For gas, gas mains and services are classified to other buildings and structures, while gas holders and meters appear under plant and machinery. Finally for the water industry, estimates are available on water resources (e.g. reservoirs, water treatment etc) and water supply (e.g. mains). A rough apportionment of the expenditure is made between other buildings and structures, machinery and equipment, and improvements to non-produced assets. The whole of GFCF in these industries is now allocated to private non-financial corporations.

Construction

15.82　Estimates are based on the ONS quarterly capital expenditure inquiries (para 25.2.2), which are benchmarked on the firmer information from the annual inquiry (para 25.1.5). Fixed capital formation by building and civil engineering departments of central government, local authorities and public corporations is excluded. Most investment is by the private non-financial corporations sector. The estimates also take into account, particularly for this industry, the balance between resources and use.

Wholesale and retail trade; repair of motor vehicles, motorcycles and personal and household goods

Hotels and restaurants

15.83　These two groups of industries use common sources, namely the ONS quarterly capital expenditure inquiries (para 25.2.2), which are benchmarked to firmer information collected in the annual inquiries for wholesaling, retailing, motor trades and catering (para 25.1.7–1.9). Particular features for the individual industries are mentioned below.

15.84 The figures for *wholesaling* include the wholesale distribution of motor vehicles. For *retailing*, GFCF in respect of TV hire is excluded, appearing within 'business services'. For *catering*, given the relatively large number of small units within this industry, estimates of GFCF are less precise than for other industries.

Transport, storage and communication

15.85 Estimates of GFCF for this industry group are derived from a variety of sources. The methodology will be described separately for each of the main component industries.

Railways

15.86 This covers expenditure on fixed capital formation by British Rail and the private rail operators. Capital formation on urban railway networks, for example, London Regional Transport, is included in 'Other inland transport'. Estimates are obtained from the ONS quarterly capital expenditure inquiries (para 25.2.2). Annual figures are the sum of the quarterly estimates. The permanent way, including overhead power lines and signalling equipment, is classified to other buildings and structures. The sector allocation is to private and public non-financial corporations.

Other inland transport

15.87 This industry covers private and public sector passenger transport and the road haulage industry. Information on private sector GFCF is collected in the ONS quarterly capital expenditure inquiry (para 25.2.2). The information is revised in the light of the firmer information from the annual inquiries. For the public sector, which includes, for example, London Regional Transport (buses and trains), the Passenger Transport Executives and the Northern Ireland Transport Holding Company, information is obtained from quarterly sources and uses of funds returns and later from their annual reports. Self-drive vehicle hire is covered within 'business services'. Again, the sector figures relate largely to private and public non-financial corporations.

Sea transport

15.88 GFCF in ships is estimated by DETR, using information from a quarterly inquiry into shipowners' capital expenditure and international trade credit and certain other sources.

15.89 The estimates are constructed on a mixed deliveries and progress payments basis. The figures for GFCF and work in progress on ships are computed as:

	(a)	deliveries of new and second hand ships from abroad;
minus	(b)	sales of second hand ships abroad;
plus	(c)	progress payments on new ships being constructed in the UK *less* sales of ships to UK breakers;
plus	(d)	additions and alterations to UK ships carried out in the UK or abroad.

Deliveries of ships from abroad and sales abroad are identified by changes in the Shipping Register and valued at the cost shown for the vessel in the shipowners' inquiry or, if not available, according to a standard set of valuation rules. Progress payments on ships being constructed domestically are also obtained from the shipowners' inquiry, which gives, for individual ships, information about amounts charged to capital account. Complete coverage is ensured by also monitoring new orders received by British Shipbuilders. The DETR inquiry is the source of information on additions and alterations to UK ships carried out in the UK or abroad and on GFCF in other buildings and structures, and on containers, which are classified as machinery and equipment.

Air transport

15.90 Estimates of GFCF are derived from the Civil Aviation Authority quarterly inquiry into the balance of payments and capital transactions of UK airlines. Among other things, this approach ensures that imported aircraft appear in the capital formation statistics at the same time as in the balance of payments.

Other transport services

15.91 This group of industries covers various supporting services to the transport industry, such as the British Airports Authority, the Civil Aviation Authority, and the British Waterways Board. Also included is the fixed capital formation of central and local government in airports, harbours, docks and canals. The information is obtained from the ONS quarterly capital expenditure inquiry (25.2.2), and is benchmarked on the firmer data from the annual inquiries.

Communication

15.92 Estimates of GFCF are based on information collected quarterly, by ONS, from the small number of operators in the postal and telecommunications industries. The annual estimates are the sums of the quarterly figures. Telephone cables and ducts are included under machinery and equipment.

Financial intermediation

15.93 Estimates of GFCF are derived from three main sources, all quarterly. These are Bank of England inquiries directed at banks, building societies and certain other financial institutions (25.3), ONS inquiries of insurance companies (25.3) and the ONS capital expenditure inquiries (25.2.2). Most of the investment under this head appears in the financial corporations sector. Investment made under financial leases is classified to the industry of the user.

Real estate, renting and business activities

15.94 A wide variety of industries are covered under this head, including computer services, and legal, accountancy, advertising and business and management services. Estimates are based on information collected in the ONS quarterly capital expenditure inquiry (25.2.2). The figures are aligned to the firmer information from the annual inquiries.

Dwellings

15.95 Estimates of GFCF in dwellings, which includes investment in new dwellings and expenditures on improvements to existing dwellings, is derived from private and public sector sources. For the private sector, figures of new investment are derived from the DETR quarterly survey of construction output, together with estimates for the comparatively small self-build component; figures of improvements are also based on the DETR survey, supplemented with information from the same Department's periodic English House Conditions Survey and ONS's annual Family Expenditure Survey (para 25.4.5). Production statistics are used as the basis for the estimates of new investment because it is difficult to determine how much of the expenditure on new dwellings relates to the dwellings themselves excluding land.

15.96 The information from the DETR's survey of construction output is used to estimate private housing. However, it is necessary to make a number of adjustments to the survey data for use in the economic accounts. These adjustments are, first, three additions for components not covered in the survey:

- the value of output in Northern Ireland;
- the value of housing output of housing associations, which are classified to the public sector in the DETR statistics and to the NPISH sector in the economic accounts; and
- professional fees incurred in relation to private sector housebuilding work.

There are also two adjustments needed in relation to the inventories of dwellings. These relate to:

- work in progress by the construction industry on uncompleted dwellings for the private sector; and
- builders' inventories of completed dwellings.

Reductions in these inventories add to investment; increases have a negative effect.

15.97　For improvements to dwellings, the use of the various sources for the estimates reflects the need to try to allow, as far as possible, for likely understatement in expenditure associated with the hidden economy.

15.98　Almost all private sector expenditure on dwellings is allocated to the household sector, with a small residual element attributed to private non-financial corporations.

15.99　Public sector GFCF in new dwellings and improvements is mainly by local authorities, both acting as housing authorities and in providing accommodation for certain staff such as police. Estimates are derived from the local authorities' quarterly capital payment and capital outturn returns, part of the public expenditure monitoring system. Also included is fixed investment in dwellings by New Town Development Corporations, and by certain other public corporations and by central government, in new married quarters for the Forces and in dwellings built for civilian employees such as prison staff.

Public administration and defence; compulsory social security

15.100　This includes all fixed capital formation by central and local government in relation to the general administration of the State, and to services provided collectively to the community. It does not include GFCF on specific, 'individual' services such as health or education which are included within those specific industries. Fixed capital formation by central government includes expenditure on the Crown Estates, the Houses of Parliament, civil defence, employment services, social security offices and law courts. Estimates under defence now include expenditure on military structures and equipment, previously classified as final consumption. Local government services classified here include the police and fire services, town and country planning together with other local authority capital formation not allocated to specific services. The estimates also include a small amount of capital formation by new town development corporations. The estimates also cover capital expenditure on roads and road improvements, and on street lighting and car parks. Purchases of road-making equipment by local authorities are included, but purchases by civil engineering contractors engaged in road works are included in the fixed capital formation of the construction industry.

15.101　Information for central government is based on the government accounting system. Quarterly estimates are derived from the analysis of public expenditure, while firmer annual figures are subsequently provided in the Appropriation Accounts. For local authorities, estimates are based on the quarterly capital payment and outturn returns, again updated by firmer (financial) year information. Estimates are made net of any capital receipts.

Education

15.102　Estimates of GFCF cover both public and private sector education. For the former, the local authority sector includes primary and secondary schools. Central government covers approved schools and remand homes for children. Universities, though financed partly by central government grants, are classified to the NPISH sector, as are further education establishments.

15.103 The information source for the estimates of central and local government GFCF is the same as for public administration (see above). Estimates of GFCF by universities are based on quarterly returns of grant aided expenditure made by the Higher Education Funding Council for England. GFCF for private schools is estimated from an annual survey of its membership by the Independent Schools Information Service. As ISIS does not cover all independent schools, the estimates are grossed up using pupil numbers. Quarterly estimates are derived by interpolation and extrapolation.

Health and social work

15.104 Estimates of GFCF cover both public and private sector health and social services. The information source for the estimates for the NHS is, again, the government accounting system, while local authority information for public health and social work is again based on the capital payment and outturn returns. For the private sector, estimates are based on information collected in ONS annual inquiries (para 25.1.11), supplemented with certain other inquiry data, and, for doctors and dentists, information from a sample of Inland Revenue Schedule D assessments. Where necessary, quarterly estimates are derived by interpolation and extrapolation.

Other community, social and personal service activities

15.105 This group covers a wide range of predominantly private sector service industries, including sewage and refuse disposal, recreational, cultural and sporting activities, laundries, hairdressers, professional and business associations, and trade unions. The public sector component includes, for central government, research establishments in the civil and defence fields, and for local authorities, residential support for the elderly, children in care and the mentally and physically handicapped, municipal theatres, concert halls, libraries, museums and art galleries, sports and leisure centres. Also included is the British Broadcasting Corporation.

15.106 Information for the private sector is largely based on the ONS quarterly capital expenditure inquiries (para 25.2.2), aligned to firmer data from the annual inquiries (para 25.1.11). For the public sector, the information sources are the government accounting system, and the local authority capital payments and outturn returns, as above.

15.107 Included for the first time is investment in entertainment, literary and artistic originals. For the moment, as mentioned in paragraph 15.64, this has been defined to embrace TV, radio, publishing and music industries. Estimates are based on information from BBC annual reports and various other sources.

The analysis by institutional sector

15.108 The allocation of GFCF to sector is based on the sector using the asset. In respect of central and local government, the estimates flow directly from the public expenditure monitoring system. Information for public non-financial corporations is also readily available. For the private sector, estimates for financial corporations are derived from the appropriate industry classification used in the surveys, while estimates for the NPISH sector are also available directly. For other private sector investment, a breakdown is made between private non-financial companies and households, largely based on the sector allocation appearing on the statistical register of businesses.

15.109 For dwellings, the investment is almost entirely by the household sector. For transfer costs the residual is the household sector.

15.110 It should be borne in mind that the long-run figures of GFCF on a sector basis will be influenced by changes in the composition of the sectors, for example as a result of privatisations. Further, the figures are net of any receipts from sales, which has some impact on the series for government.

Estimates of GFCF at constant prices

15.111 Estimates of GFCF at constant prices are derived essentially by deflating the value figures by relevant producer price indices or other price data. Deflation, which is based on price series for the different types of asset identified above, is undertaken for the separate industries within the private sector, for the component sectors of the public sector, and for dwellings and transfer costs.

15.112 The estimation of GFCF at constant prices constitutes one of the more difficult parts of the whole economic accounts. There are two main reasons for this. The first reflects the variable and often unique nature of capital goods, which makes it very difficult, in some instances, to establish sensible deflators. The second problem arises from the limited analysis by type of asset. As a result, even if a satisfactory measure of price movement exists for a particular asset, it may not always be possible to identify the same asset in the GFCF figures collected in the inquiries. In practice, the price indices available have to be aggregated to match the asset analysis of expenditure which is collected. One particular feature of the deflation is to ensure that import prices are included in the estimation of the deflators as well as UK producer prices.

15.113 An aggregate deflator is compiled for each of the following tangible asset types:

- *military equipment* (other than weapons) is deflated by specially constructed price indices based on producer prices;
- *other machinery and equipment*, for which price indices for particular capital goods are weighted together to form deflators for purchasing industry groups. The weights are based on information on the pattern of purchases assembled for a special survey of producers of capital goods, by industry in the base year;
- *dwellings,* for which estimates at constant prices are derived from housing costs;
- *other buildings and structures* (see below);

15.114 The problem of the often unique nature of assets is particularly acute when attempting to revalue expenditure on building work. The approach adopted is to use information on tender prices from samples of contracts collected by the Department of the Environment. In doing so, two adjustments are made to the price data, first in order to reflect the time taken between tender stage and the construction work, and secondly to take account of 'cost of variation of price' clauses which are a feature of many major construction contracts. The price information is linked back to the proper basis of the base year.

15.115 There is one further point relevant to the deflation process which should be mentioned. The producer price indices used to compile deflators for the capital formation statistics measure price changes for goods at delivery stage. While this is satisfactory for most of GFCF, which is recorded at the time goods are paid for, there are potential problems for those major capital assets which are valued by progress payments. Where this occurs, payments will lag the work done. Thus in order to match better the payments and the prices, it is necessary to lag the price data, overall by one quarter (except where there are stage payments). This is done for the price indices used for some components of machinery and equipment, and for buildings and structures other than dwellings.

15.116 *For intangible assets,* deflation is undertaken in a variety of ways. For mineral exploration a specially constructed price index is used, based on costs of exploration.

15.117 *Non-produced assets* For transfer costs, deflation is based on price in the base year projected by volume measures.

15.118 Estimates at constant prices by industry are derived largely by deflating the asset breakdown within each industry. The make up of the deflators for the individual assets has been covered above.

Changes in inventories (P.52)

15.119 Inventories in the economic accounts relate to goods which are held by producers prior to further processing or sale, or which will enter intermediate consumption or be sold without further processing. Inventories also include work-in-progress. In the previous accounts, inventories were known as stocks, and changes in inventories as stockbuilding.

15.120 As well as appearing in the capital account, where essentially they may be regarded as a component of the expenditure measure of GDP, changes in inventories are also part of output within the production measure. A summary description of the role of inventories in the accounts has been given in Chapter 12. This also included a brief description of the complex methodology required to estimate the contribution of inventories to the accounts. The calculation will be explained in detail in this section. Before looking at this, it will be useful to set out some of the key features of inventories.

Changes from former system of accounts

In the new ESA, the definition of inventories has been widened. The coverage now embraces in principle any work-in-progress in cultivated assets, such as livestock, trees and growing crops, and in certain service industries, such as computer software and films, and all inventories of government market producers not just strategic stocks as hitherto.

Main inclusions and exclusions

15.121 Within the broad definition of inventories as goods which enter the productive process, as mentioned above, inventories now also embrace all types of 'growing' natural resource managed by a producer unit. Two main exclusions from inventories are:

• intangible assets, such as the stock of 'goodwill' for a business, which appear elsewhere in the capital account, and
• inventories held by final consumers for future consumption.

15.122 In the latter case, the economic activity involved in producing these goods will already have been measured by other parts of final expenditure. Examples of such goods would be the stocks of food or of consumer durables owned by the household sector; stocks of road mending material held by local authorities; stocks of medicine in hospitals, or the stocks of ammunition and weapons for the armed forces held by central government.

Cultivated assets

15.123 Work in progress for cultivated assets relates to plants, trees and livestock which either have a single use, that is they produce an output only once, such as crops, or are used repeatedly to produce regular output, for example breeding cattle or fruit trees. In the latter case, if the activity is on own-account, it is classified, not to inventories, but to GFCF. Although cultivated assets is a new category, certain estimates relating to breeding cattle were made for stockbuilding in the previous accounts. These were essentially annual, the quarterly figures being interpolations and extrapolations. In the new accounts, amongst other things, attempts are made to derive improved quarterly estimates of work in progress for growing crops. This is described in the section on the industry analysis of inventories.

Services

15.124 Work in progress in this other new area relates to services which take a long time to produce, such as legal, design, accounting and research. Measurement issues are raised in the section on the industry estimates.

Asset structure

15.125 The main components of inventories are much the same as before. The one change is that goods for resale now form a separate category. The broad asset breakdown is:

(a) *Materials and supplies* – goods to be used in the process of production such as fuels, raw materials, semi-finished products, packaging material and office supplies;

(b) *Work-in-progress* – output produced by an enterprise that is not yet finished, in particular for major capital assets and also agriculture;

(c) *Finished goods* – goods produced by an enterprise now in final form to be sold to other parts of demand;

(d) *Goods for resale* – goods acquired by, say, retailers for resale, without further processing.

Inventories held abroad

15.126 As a consequence of the definition of residence these are regarded as being held by a quasi-corporate direct investment enterprise in the rest of the world. They are not included with inventories.

Progress payments

15.127 Some goods take a long time to manufacture. In this case, the future purchaser of these goods will generally make *progress payments* to the manufacturer in the course of production, rather than paying for the item in full on completion or delivery. When this occurs, part or all of the inventories component of work in progress is paid for prior to the final delivery. These payments take place in industries like aerospace, shipbuilding or heavy engineering. Fixed capital formation (and also output) includes progress payments which have been made as part of the expenditure by the purchasing industry. Work-in-progress of the producer industry is reduced commensurately. In this way, the expenditure and production measures of GDP will be more consistent with the income measure, which will include the wages and salaries associated with the work done.

Inventories in the economic accounts

15.128 The estimates of inventories which go into the capital account, and also the expenditure and production measures of GDP, are essentially the changes in the levels (or book values) of inventories between the beginning and end of the period of interest. However, for economic accounts purposes, it is necessary that inventories are valued at the time they enter the production process, for example when the raw material is used or the finished good is sold. This form of valuation and recording is clearly far from practical. Firms cannot realistically revalue each item of stock as and when it is used or produced. (For convenience, the text will generally refer just to the former.) In practice, a number of different approaches are used, as considered below.

15.129 In statistical returns, as in commercial accounts, firms will record the levels of inventories at the beginning and end of the accounting periods, based on 'historic costs', that is, the price at the time of purchase of the good, rather than at the time of its use - the replacement cost basis. What this means, therefore, is that, in a period of rising prices, the change in the value of the level of inventories will include an element which is due purely to prices. This element is called the 'holding gain' (previously 'stock appreciation'), and it needs to be excluded from the estimate of change in the recorded levels of inventories to leave the pure 'volume' increase required for the economic accounts. If inventories could be measured and valued at the time of their use in the production process – the replacement cost basis – holding gains would not affect the change in their levels which could be used directly in the accounts.

Valuation of inventories

15.130 In commercial accounts, inventories, other than long-term work-in-progress, are valued at the lower of cost and net realisable value; the former is what they cost, the latter what they could be sold for at the time of valuation. For work-in-progress, valuation is essentially at cost, plus estimated profit less any foreseeable loss. There are several different ways of calculating the historic cost of any inventories on hand at the end of an accounting period. These include:

(a) *First in first out (FIFO)* This system assumes that materials acquired at the earliest date are first drawn from store, so that inventories will consist of the most recently acquired items.

(b) *Last in first out (LIFO)* This convention uses the opposite assumption to FIFO. It assumes that materials drawn from store will consist of the most recently acquired items, so that inventories relate to the items first purchased.

(c) *Average cost* Inventories are valued at some average price, based on existing value and volume of holdings. The average is recalculated periodically, for example when new goods are received. Any subsequent purchase is then valued at that price until the average is recalculated.

(d) *Standard cost* Under a standard cost system every element entering into production is given a unit value, based on recent costs or current costs or expected future costs. Once determined, the value of inventories can be obtained by multiplying the quantity of each commodity in stock by its standard cost. This standard is usually maintained for a year. This approach is not, however, used in preparing commercial accounts.

(e) *Unit cost* A company may know the exact cost of acquiring inventories by recording the date and price at which each item was bought. This approach requires much detail and is rare.

(f) *Base stock* This method assigns a fixed unit value to each item in stock up to a certain number of units and any excess over this number is valued by some other method.

15.131 Of the above, in line with accounting practices, the FIFO and standard cost approaches are by far the most widely used. The other feature of valuation is the use of the lower of historic cost or net realisable value. This is relevant when the market prices of items in stock fall below their historic cost or when the book values are written down, for instance because of slow-moving, obsolete or deteriorating goods. In commercial accounting, the effect in both cases is to treat the reduction as a trading loss. However, in the economic accounts, just as the effect of increases in prices need to be eliminated, then so do reductions. The reductions are equivalent to negative holding gains.

Holding gains and the relationship between inventories and operating surplus

15.132 It is useful here to consider the relationship between the measurement of changes in inventories and operating surplus (profits) within the accounts. As mentioned earlier, the difference in the change in the book value of inventories measured at replacement cost and historic cost, is called the holding gain (stock appreciation). The holding gain, which is part of operating surplus in the historic cost estimate included for commercial accounts, is excluded from the estimate of operating surplus which goes into the income measure of GDP.

15.133 Trading enterprises measure operating surplus as the total revenue earned during a period less all the costs involved in making the items which produced that revenue. This measure is not the same as the total revenue earned less total costs incurred during a period. Some items sold during a period incorporate costs incurred in earlier periods. Similarly some costs incurred during the current period will only be reimbursed by revenue earned in later periods. However, in order to measure operating surplus according to the first definition above, it is not necessary to keep track of all the costs incurred in producing each item that is sold. The approach which can be used is to calculate profit in a particular period as

the revenue earned from sales during a period
less the costs associated with those sales,

where the cost of sales equals:

the opening value of inventories
plus the costs incurred during the period
less the closing value of inventories.

15.134 The profit may be calculated in this way because the costs incurred during the period, together with the opening book value of inventories, measure all the costs that could have been involved in making the items generating the revenue earned in that period. However some of these costs will eventually be reimbursed by revenue earned in later periods; these costs are measured by the closing level of inventories. Deducting these costs from the cost of sales, in effect cancelling the part of costs that are relevant to future periods, leaves the costs that are directly associated with the revenue earned during the period.

15.135 The paragraph above defined operating surplus as the difference between the revenue from selling an item and the cost of making it. The cost involves purchasing inputs to the productive process, for example buying raw materials or paying labour for work done. These costs are incurred before the revenue from the sale, and at a time of general inflation the price of these inputs will change between the time when they are bought and the time the item is finally sold. Measured in this way, and for a given selling price, the profit will depend on the change in the prices of these inputs in the period between the final sale and the time the associated costs were incurred.

15.136 In the economic accounts, profits are measured by valuing the cost of sales as if these costs were incurred in the current period. The difference between the two approaches is the holding gains element, which needs to be deducted from operating surplus, as conventionally measured, in calculating the income measure of GDP. The procedure described above is often referred to as a way of calculating a value for operating surplus that maintains the physical quantity of inventories. It confines operating surplus to the factor incomes derived from productive activity and removes any element of capital gain. It might be noted that the holding gain is conceptually identical to the 'cost of sales adjustment', which is one of the adjustments used in current cost accounting in commercial accounts in order to transform the operating surplus from an historic cost to a current cost basis.

Inventories: methods of calculation

Inventories calculated on the FIFO convention

15.137 This section describes the derivation of the estimates of the change in inventories and of holding gains required for the economic accounts. The estimates – quarterly and annual – are based on information obtained from the ONS statistical inquiries (25.2.4) or other sources. The information is essentially the book values (historic cost) of inventories, generally but not always broken down by type of inventories (as in paragraph 15.137), which is collected across industries. The description given here is based on the methods used for manufacturing industries. The methods used for other industries are similar.

15.138 The broad purpose of the calculation is to revalue the book values of inventories at the beginning and end of the period to what approximates to a replacement cost method of valuation. In doing so, the effect of price changes is removed, and inventory building can be determined as the difference in the revalued levels of inventories.

15.139 The calculation is undertaken for each industry and for each of four broad types of inventory independently. The method of estimation is based on two sets of data and two assumptions. The initial data are:

- the book values of inventories at the beginning and end of the period, generally a quarter;

- producer price indices (PPIs) as at end months.

The first stage is to calculate two price indices:

- weighting the PPIs according to the composition of the inventories, thus representing replacement cost at the end of the month;

- lagging this to allow for the pattern of purchases over time.

It is to produce the second index that requires the two assumptions. They are:

- that firms use either the standard costs basis to value their inventories, or the lower of FIFO based costs or net realisable value; and

- the average length of time goods are held in inventory before use or sale.

Table A: Calculation of book value inventories price index lags

	Stock Volume	Average Monthly Purchases	Amount of stock assumed purchased in month:		
			n	n-1	n-2
Industry 1	75	40	40	35	
Industry 2	50	65	50		
Industry 3	60	35	35	25	
Industry 4	45	50	45		
Industry 5	30	10	10	10	10
Total	260		180	70	10
Lagging pattern			0.692	0.269	0.039

Table A shows how the lagging factors would be derived for a typical group of industries to derive a book value price index, on the FIFO assumption.

15.140 Having determined the book value price index, it is now possible to revalue inventory levels in a way which approximates to the replacement cost basis of valuation. The principle of the calculation, which can be applied with suitable modification to each of the four asset groups, is given in the worked example in Table B below. The period of estimation is assumed to be a quarter. It will be assumed that inventories are valued according to the FIFO principle. The book value of inventories (A) at the end of the quarter represents the value of goods in inventories purchased over the previous months and valued at the (historic cost) price of purchase. The current price series for inventories, representing the prices at which stocks will have been purchased and their commodity composition, is given at (B).

15.141 The first step in the calculation is to revalue the book values to constant prices (usually the base year of the accounts). The example will assume that goods in stock at the end of the period will, have been purchased in equal amounts over the previous three months. Hence, the appropriate book value price index for revaluing the level of inventories to constant prices will, usually, be the *average* of the prices in the previous three months. Thus, the price index for revaluing end-December inventories to the appropriate replacement cost basis is the average price of the three months October, November and December. However, for the end-September level, the price index to be used is **not** the average of July, August and September, (127), but the index for September alone (126). The reason for this is that the valuation is based on the lower of cost or net realisable value. As the realisable price of inventories at the end of the September figure (126) is lower than that of the cost (127), then the former figure is the one to be used.

15.142 Applying the relevant prices to the book values gives constant price figures as at (C). The difference between these two figures gives the change in inventories (stockbuilding) required for the economic accounts at base year prices. This figure is now converted to a current price basis by inflating by the average price of inventories over the quarter, to give the estimate of the change in inventories at current replacement cost. Finally, the holding gain may be derived as the difference between the change in the value of the level of inventories as recorded and as at replacement cost.

Table B: Estimation of holding gains for inventories

Basic data

(A) Book values: end September 1000 (bs)
                   end December   1600 (bd)

(B) Inventories price index (base year = 100)

July	128	(pj)
August	127	(pa)
September	126	(ps)
October	130	(po)
November	131	(pn)
December	135	(pd)

Holding gains calculation

Inventories at constant prices

end September 794   ks=bs/ps
end December   1212 kd=bd/(ave[po+pn+pd])

Change in inventories at constant prices

418   kc=kd - ks

Change in inventories at current replacement cost

552   cc=kc x(av[po+pn+pd])

Holding gain

48   hg= (bd - bs) - cc

Figure 15.1 Calculating the change in inventories and holding gains

First in First Out (FIFO) industries input data

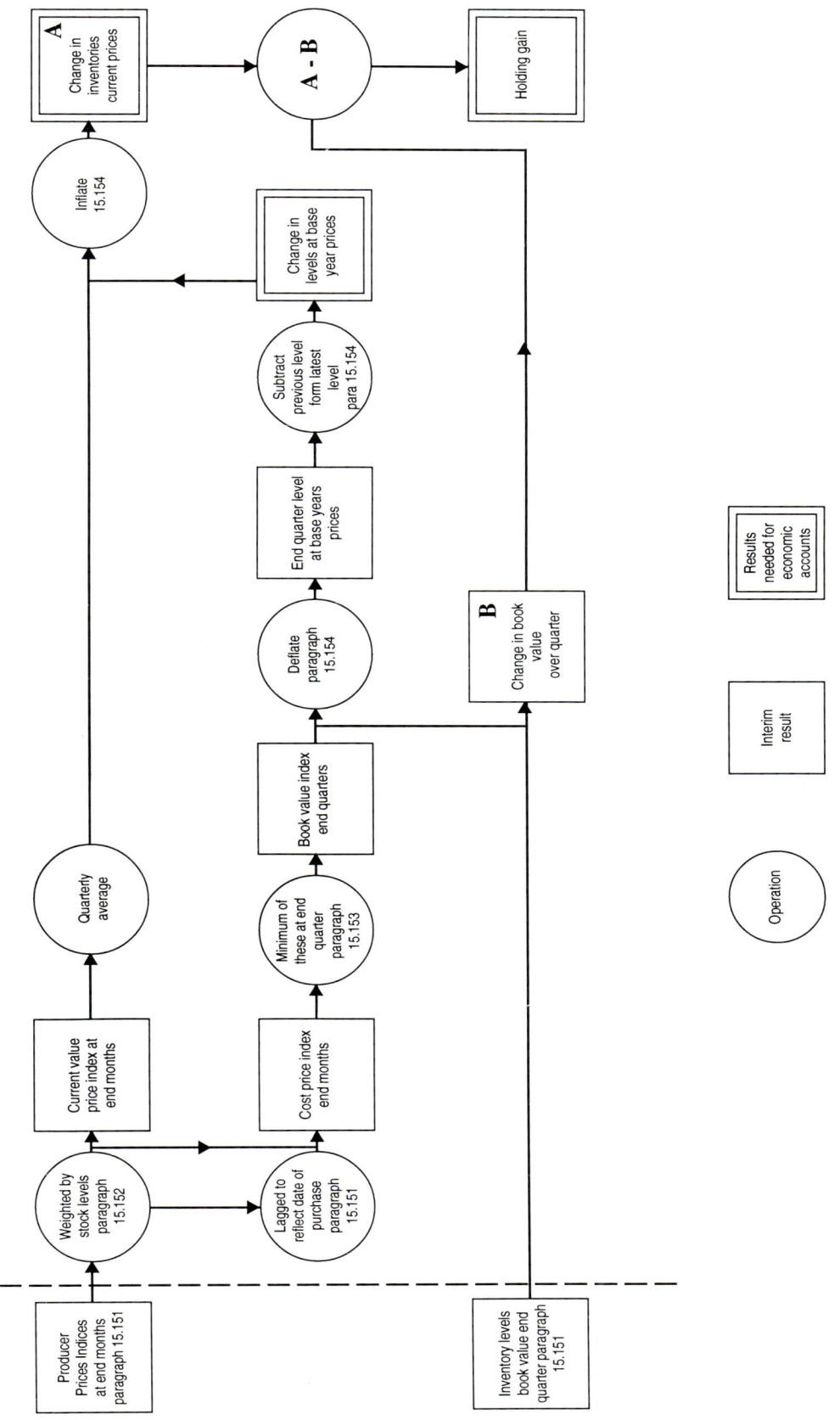

Inventories calculated by the standard cost convention

15.143 This section looks at the above calculation where inventories are valued by standard costs. Under this system, inventories are valued by some fixed value per unit of stock. This unit value may be based on recent costs, current costs, or expected future costs. Where standard costs are used, contributors to statistical inquiries are asked to give the value of their inventories at the beginning and end of each quarter at the same standard cost. If the standard has changed, the statistical return asks for the opening stock at both the old and new standard. The change in the value of inventories, when measured at the same basis of cost, represents a purely physical change in stocks at that standard cost. This can be revalued to give stockbuilding at current or constant prices. The standard costs represent the prices obtaining at a particular moment. These prices are equivalent to the current price inventories index. Therefore, inventories valued at standard cost can be revalued by the current price inventories stocks price index, though this has to be lagged or led to make it refer to the same moment as the standard cost. On average, and across all manufacturing industry, studies have shown that standard costs at any moment refer to the costs prevailing about two and a half months before that time. Therefore, the assumption is made that the current price inventories index for the last month of any quarter measures the standard costs used for completing the statistical return about inventories for the next quarter.

15.144 The standard cost stock levels are deflated to constant prices using this lagged current price inventories deflator. The method used is similar to those used for FIFO stocks shown by the example in Table B. The difference between the constant price stock levels is the change in inventories at constant prices. This is revalued, at the average price of the latest period, to give the change at current prices. The constant price levels are also revalued to give notional book value levels and changes consistent with a FIFO valuation of inventories. This allows holding gains to be estimated for this component of inventories, for inclusion in the aggregate figure of holding gains.

The inventories price indices

15.145 The current price inventories indices are key components of the estimation process. The series are based on ONS's producer price data for individual commodities (para 25.19). These separate price series need to be weighted together to give the aggregate price index for the stocks of any particular industry. These are base weighted, or Laspeyres, price indices. The weights are based on the commodity composition of inventories. In the absence of any direct information, the commodity pattern is estimated for each type of stock (materials and supplies, work in progress, finished goods and goods for resale). The next few paragraphs describe how this is done and how the estimates of the book value price indices are derived.

Price indices for materials and supplies

15.146 The industry price indices used for inventories of materials and supplies are compiled by weighting together relevant producer price data by weights which represent the commodity composition of these inventories. The commodity breakdown cannot be estimated directly. Instead, it is based on information on purchases, as collected in ONS's purchases inquiry, utilising the assumption that the breakdown of inventories held by an industry is the same as the pattern of purchases by that industry. (This is only strictly true if all purchases are held in stock for the same time.) In addition, allowance is made for the fact that some types of purchases are not stocked, for example electricity and gas. The calculation is made at the 'activity' industry level, that is about 220 industries at the 4 digit level within the classification of manufacturing industry.

15.147 However, information on the book value of inventories is compiled for only 32 industry groups in manufacturing which broadly represent the SIC classes. Therefore the 220 indices for each activity level industry are aggregated to match the 32 group level industries. This is done by weighting together the activity level price indices by the volume of stock held by each industry at the end of the year in question. This could not be measured directly because the book value of inventories relates to items bought at different times and prices. The volume of stock was therefore estimated by multiplying the average monthly volume of purchases during the year by the average stockholding period for each activity level industry. The resulting aggregate inventories price indices are the current price inventories price indices referred to above. They resemble any other aggregate producer price index; they measure the changes in the current prices of the goods that comprise an industry's stock.

15.148 The book value price indices, which are appropriate to the pattern of stockholding at a given point of time, can then be determined by lagging the current price inventories price indices, as appropriate, as described above. These can then be used to revalue the levels of inventories, as explained earlier.

Price indices for work-in-progress

15.149 The revaluation of work-in-progress presents a special problem, since these inventories are not traded and are not priced. Therefore it is not possible to determine a price series directly, as for materials and supplies. Furthermore, as will be seen later, the price indices appropriate to the book values cannot be derived simply by lagging the current price inventories indices. For work in progress, the current price indices and the book value price indices must be compiled separately.

15.150 As a basis for this, it is assumed that two factors – materials and supplies, and labour and overheads – contribute to the costs of work in progress. These costs are incurred during the various stages of the build up of work-in-progress. The cost of work-in-progress can therefore be divided between the contribution of materials costs and of labour/overhead costs. Information from the ONS annual production inquiry shows that, for most industries, costs are split roughly equally between materials and labour/overheads. For a small number of industries different splits are necessary. The exceptions are covered in the description of the industry estimates given below.

15.151 In respect of material costs, it is not possible to use the price indices for materials and supplies (derived as above) for revaluation, since the non-stockable items (electricity, gas and water), which were excluded from the price indices, clearly form part of the costs of work-in-progress. For this reason, the material and supplies price indices used in the calculation need to include these costs. The indices used are based on sector input price indices.

15.152 There is no direct measure of the element of labour and overheads costs. However, the current price index of inventories of finished goods can be thought of as made up of a materials component and a labour/overheads component. This latter component can then be estimated as the difference between the finished goods price index and the materials price index, both of which are measured. This series for the labour/overheads component costs of finished goods is thought to form an adequate estimate of the labour/overheads component of work-in-progress in an earlier period. This rests on the simple fact that goods move through and out of work-in-progress and into finished goods (and eventually into sales). Costs being incurred on work-in-progress in the current month will influence the costs of items entering finished goods inventories in a future month. The exact timing of the relationship will depend on the length of the production process (that is, the build-up period for work in progress).

15.153 Thus, the price indices to deflate work-in-progress could be compiled using a combination of the inventories deflators for materials and supplies and for finished goods. However, lags and leads need to be applied to these deflators in order to establish the appropriate book value price index. The size of the lags is determined by the average time stocks of materials and supplies were held before they were used. The estimate of lead was based on the average time an item was part of work-in-progress. In making the estimates, it was found that the weights for the materials and supplies index had a negligible effect on the book value price indices of all industries, except the mineral oil processing industry. In other words, mineral oil processing apart, the book value price indices are estimated as a weighted combination of the indices for finished goods alone. Specifically, the book value price index is estimated as the weighted combination of the finished goods prices for the next four months. Similarly, the current price indices, which are used to revalue the constant price stockbuilding figures to current prices, are composed of:

a) the current price materials and supplies inventories indices (inclusive of gas, electricity and water) with lags for periods of up to 14 months and with weights (both positive and negative) that add to zero;

b) the current price finished goods inventories indices, with leads for periods up to four months ahead and with weights that add to one.

15.154 For most industries a 50:50 weighting of the materials and supplies, and the finished goods price indices is adequate for deriving the current price inventories index. However, for some industries, a different weighting was required. This is explained below for particular industries.

15.155 The current and book value price indices, derived as above, can the be used to revalue the book values of work-in-progress, and determine the contribution to changes in inventories, as described earlier.

Price indices for inventories of finished goods

15.156 The methods used to construct the price indices for finished goods inventories are similar to those used for stocks of materials and supplies, and again use producer price indices. As with materials and supplies, no direct information on the composition of the stockholding of finished goods is available for weighting together price series. The commodity composition is therefore estimated using information on product sales collected in the ONS quarterly sales inquiries (25.1.4). To do this it was assumed that the value of the inventories of a finished good was directly proportional to the sales of the good. Put another way, all finished goods were held in stock for the same length of time. On this assumption, the ONS's sector output price indices could be taken to represent the inventories price indices for the 220 manufacturing industries at the activity level. These were then weighted together by the volume of stock to give a current price index for each of the 32 industry groups.

15.157 The book value price index is derived from the current price index by adjusting the latter index according to the pattern of lags derived from the stockholding periods for the inventories held by the 220 activity level industries. These periods were calculated in the same way as for materials and supplies, but using sales from the annual production inquiry rather than purchases. Estimates of change in inventories and holding gains are then derived, as explained earlier, using the appropriate current price and book value price data.

The detailed analyses of inventories

15.158 The description above, together with that in Chapter 12, has set out how the estimates of inventories, largely in aggregate, are made. In the main, the sources used are ONS quarterly and annual inquiries, and producer and other price information for revaluation purposes. The following sections now describe analyses of inventories by asset, industry, and institutional sector (not expected to be published in the Blue Book 1998).

The analysis by asset

15.159 Information by asset is collected in most of the ONS's various statistical inquiries which collect inventories data (25.2.4). For some inquiries, where only a total figure for inventories is collected, assumptions are made about the asset breakdown. The estimates of the contribution to the change in inventories for each of the four types of inventory has been described in the above paragraphs. As an integral part of the calculation, the estimation methodology also provides the required information at constant prices.

The analysis by industry

15.160 This section describes the estimation of the change in inventories for individual industry groups. One of the main aspects to be discussed will be the way in which the book value price index appropriate to work in progress is derived for certain industries within manufacturing. Also included will be an indication of the sector allocation of inventories. This information is brought together in the later section on the split of inventories by institutional sector.

15.161 As with the other 'business' variables, classification of inventories to a private sector industry is in general based on the principal activity of the unit reporting the expenditure. For the production industries, the reporting unit is often an individual factory, which can provide the full range of statistical information required for the economic accounts. For the distributive and service industries, the industrial classification is generally based upon legal units which are individual companies in the corporate sector, sole proprietorships, partnerships etc. For public administration the classification reflects the coverage of the general government in the sector accounts.

15.162 The contribution of inventories to the economy overall, and for the individual industries, is the change in levels (stockbuilding) at replacement or base year prices. Information is published on both levels of inventories and the change between periods, the latter at both current and constant prices, and also broken down by type of inventory.

15.163 The table below gives estimates of the book values of inventories in 1995 for the main industry groups.

Book value of inventories by industrial sector	
A+B	Agriculture, hunting and forestry; fishing
C	Mining and quarrying
D	Manufacturing
E	Electricity, gas and water supply
F	Construction
G+H	Wholesale and retail trade; repair of motor vehicles, motorcycles and personal and household goods; hotels and restaurants
L	Public administration and defence
I+J+K	Other industries

Agriculture, hunting and forestry; fishing

15.164 Estimates of change in inventories and holding gains for agriculture, based on the levels of farm stocks, re made by MAFF as part of the estimation of the annual income and expenditure of the farming industry. The estimates now include more appropriate quarterly estimates of work in progress in livestock and growing crops. Estimates for broad type of asset are made as follows:

(a) *Material and supplies* Quantities of input stocks of fertilizers and foodstuffs are estimated on the basis of usage and purchases data.

(b) *Work-in-progress* Quantities of work in progress for non-breeding livestock are measured from the December ten per cent sample 'census' of the agricultural industry.

(c) *Finished goods* Output stocks of wheat, barley and oats are measured directly; estimates of the stocks of potatoes, apples and pears are based on production and usage data.

15.165 The value of all these types of inventories are estimated by calculating the cost of production for each commodity in stock. Each commodity is considered to have been produced by a fixed bundle of inputs (e.g. labour, fuel etc.) over time. The value of the inventory is then the sum of the value of the inputs, where each input is valued according to the price at the time it was used. This calculation is modified by an efficiency factor which measures the way changes in the weather etc alters the relation between inputs and outputs. These data are annual. The quarterly data are interpolated. For growing crops, quarterly figures of work in progress are based on spreading the total value of the crop over the quarters of production in proportion to the cost incurred in each quarter. The estimates of inventories are allocated between the household and private non-financial corporations sectors. The ratio of farming company income to total farming income (both from Inland Revenue) determines the split between these sectors.

15.166 Estimates of inventories for forestry – essentially growing trees – are based largely on Forestry Commission data, adjusted for private forests.

Mining and quarrying

15.167 Two separate industry groups are identified under this head: 'Extraction of mineral oil and natural gas' and 'Other mining and quarrying'. For the former, estimates are derived from quarterly information provided to DTI(Energy) by the oil companies. The estimates are mainly based on quantity information, inflated by appropriate market prices. For other mining and quarrying, information is based on the ONS quarterly inventories inquiry (para 25.2.4), aligned to firmer information from the annual inquiry (para 25.3). Book values are deflated by appropriate producer price indices. For both industry groups, inventories are attributed to the private or public non-financial corporations sectors.

Manufacturing; Electricity, gas and water supply

15.168 These groups of industries have common sources for the inventories estimates, and are considered together. The main source is the ONS quarterly inventories inquiry (para 25.2.4), which is benchmarked on the firmer information from the annual inquiries of production (para 25.1). Some particular features for individual industries are described below.

<u>Manufacturing industry</u>

15.169 Within manufacturing, estimates of changes in inventories and holding gains are made for 32 separate industry headings. There are certain problems with the derivation of the price indices used to deflate book values of work in progress. These arise for industries where work in progress is a particular feature of the industry, and the general approach for estimating the price series does not apply. It will be recalled that the book value price indices are based on separate price indices for materials and supplies and for finished goods. The current price inventories indices need to be lagged and led, and also weighted together by the relative importance of these two components in overall costs. The working of this process for particular industries is described below.

15.170 For some industries the equal split of costs between materials and labour/overheads was not appropriate, so different weightings had to be used. The industries affected were mineral oil processing, iron and steel, chemicals, food, distilling, other drink, data processing, motor vehicles and brewing. Of these the position on mineral oil processing and distilling is considered further below.

15.171 For the distilling industry, the position on inventories is heavily influenced by whisky stocks which are left for several years to mature. Consultation with the industry suggested an average build-up period of $9^1/_2$ years (38 quarters). It also emerged that the most significant contribution to overheads in this industry arises from warehouse rental charges. To take account of this, a rent index, relating to industrial premises in Scotland, is used as part of the labour/overhead costs. The current price work in progress cost index is calculated by weighting together materials and labour/overheads costs deflators in the proportion of three to one, based on earlier information from the annual production inquiry. The book value index is evaluated by lagging the current price index over a 38 quarter period.

15.172 In respect of the shipbuilding industry, two issues need to be mentioned. First, in order to avoid double-counting, work in progress is ignored, on the assumption that the expenditure is recorded in GFCF. The second problem is that, as no output price indices are available, it is not possible to derive deflators for work in progress and finished goods. The approach adopted weights together, equally, a price index for materials and an estimated labour costs index. The latter is based on information on earnings, their relationship to labour costs, and productivity movements.

15.173 Similar problems arise for the aerospace industry. Here, adjustments are made to inventories to avoid double-counting with GFCF. Secondly, for revaluation, the work in progress and finished goods deflators are estimated directly using the aerospace industry 'combined costs' indices with a lagged materials component to allow for the materials stockholding period.

15.174 By sector, the main allocation is to private non-financial corporations.

Electricity, gas and water supply

15.175 Estimates use a range of price series, distinguishing between fuel and non-fuel items, for revaluing book values for these industries. In respect of electricity and gas, non-fuel stocks do not include domestic appliances in showrooms, which are included in retail stocks. All inventories are currently allocated to the private non-financial corporations sector.

Construction

15.176 Estimates of changes in inventories and holding gains are based on the ONS quarterly inventories inquiry (para 25.2.4), which is benchmarked on the firmer information from the annual inquiry into construction (para 25.1.5). Up to a few years ago, the estimates were based on a variety of DETR and other sources, including construction activity other than on dwellings, and, for dwellings, stocks of uncompleted dwellings and of unsold completed dwellings. Construction inventories are allocated between the household and private non-financial corporations sectors. The basis of the split is an estimate of the output of the self employed compared with that of contractors, which is largely based on ONS employment data.

Wholesale and retail trade; repair of motor vehicles, motorcycles and personal and household goods

15.177 Inventories held by these industries are almost entirely goods for resale. Estimates are based on the ONS quarterly inventories inquiry to retailing, wholesaling and the motor trades (para 25.2.4). The quarterly estimates are benchmarked on the firmer information from the corresponding annual inquiries (paras 25.1.7, 25.1.8 and 25.1.9). The price information for revaluing the book values uses both producer (including import) and retail prices. Figures relate mostly to private non-financial corporations, and also to households.

Hotels and restaurants

15.178 Annual estimates are based on the ONS catering inquiry (para 25.1.11). Quarterly estimates are interpolated. The price indices used to revalue book values again use both producer (including import) and retail prices.

Transport, storage and communication

15.179 Changes in inventories in these industries are likely to be comparatively small. For transport, limited annual information, covering rail and bus transport, is obtained from company reports. Quarterly figures are interpolations. For communication, certain quarterly figures are collected. Book values are revalued by appropriate producer prices. The figures relate almost entirely to the private non-financial corporations sector, but with an element for public non-financial corporations.

Financial intermediation

15.180 There are two main components under this head. First, estimates of silver bullion stocks held by dealers, which are derived from quarterly information on the weight of silver stocks held by UK banks provided by the Bank of England, and using the London Spot Market price for silver for appropriate valuation. Secondly, other metals trading, where estimates are based on quarterly information on the volume of stocks of a number of separate metals provided by the London Metal Exchange, and using LME price data for appropriate valuation.

Real estate, renting and business activities

15.181 Hitherto, the effect of any stockholding by this industry group was considered de minimis, and no estimates were made. In the new system of accounts, estimates are being made for work in progress in relation to activities such as legal, design, accounting and research services. The estimates are based on information collected in ONS annual service sector inquiries (para 25.1.11). Quarterly estimates are interpolated.

Education

Health and social work

Other community, social and personal service activities

15.182 The effect of any stockholding by these industry groups is thought to be very small, and no estimates are made.

Public administration and defence; compulsory social security

15.183 Estimates now include all inventories held by government market producers, not just strategic stocks, as previously. Information is derived, quarterly, as part of the arrangements for monitoring public expenditure. Inventories do not include those of central government bodies engaged in manufacturing activities, which are covered in that industry. Further, for such bodies, holding gains are not relevant to the concept of trading surplus, and are therefore not calculated. Strategic stocks, although not part of the activity of a trading enterprise, are often eventually resold and are not consumed in the course of the activities of central government. In respect of the Intervention Board for Agricultural Produce, estimates are based on monthly information provided by the Board on physical quantities held in stock of various commodities, together with relevant prices for revaluation. Holding gains are not calculated.

The analysis by institutional sector

15.184 The allocation follows the same groupings as for GFCF.

Acquisitions less disposals of valuables (P.53)

15.185 Valuables as defined here are essentially goods which are not used in the production process, and which are generally held as a store of value. They include, in principle, precious stones and metals, antiques and art objects. The presentation in the accounts, bringing together all valuables within gross capital formation, is new. However, it is likely that some valuables would have been included in the previous accounts, in final consumption or gross capital formation.

15.186 There are problems with both the concept of the variable and the derivation of any estimates. Valuables can enter the UK economy in two ways. First, they may achieve 'valuables' status as time evolves,for example, as might occur for an antique or a painting. Secondly, they may enter (or leave) through overseas trade. Measurement of the former type of valuables is extremely difficult and there is no attempt to measure it. The figures for the UK accounts will relate solely to the net value of dealers' margins on transactions (which are recorded in this item along with the value of the goods) plus overseas trade.

15.187 Estimates are based on information supplied by HM Customs and Excise.

Estimates at constant prices

15.188 Estimates of the aggregate, at constant prices, are made by deflating the separate export and import figures, using deflators based on information provided by Customs and Excise.

Acquisitions less disposals of non-produced, non-financial assets (K.2)

15.189 These comprise non-produced, non-financial assets, mainly land, used in the production process, and also intangible assets. They are divided into two groups, as described below.

Acquisitions less disposals of land and other tangible non-produced assets (K.21)

15.190 This is defined as covering transactions in land and in subsoil assets, such as deposits of coal, oil and gas, located on land or under the sea. The former is meant to cover land separately from the building on the land. However, as discussed earlier, it is not readily possible to make this separation in the estimates for the UK accounts. Hence, the figures recorded in the capital account for this item will not embrace the land component of the combined buildings and land transaction. This information will be included, as acquisitions and disposals, in the estimates of GFCF on dwellings and on other buildings and structures. The coverage for this heading will be restricted to agricultural land and land which undergoes a change of use. This practice does not affect the aggregate of gross capital formation, only the way the information is recorded, by asset.

Acquisitions less disposals of intangible non-produced assets (K.22)

15.191 The source of the data on land is the Inland Revenue survey of Property Transactions, based on a sample of returns to valuation offices where stamp duty is collected. The sector analysis is derived from the same survey.

15.192 Most transactions in patents, leases and other transferable contracts take place within sectors, but transactions with the Rest of the world are derived from ONS surveys. No information is yet available on purchases and sales of goodwill, but again, they are thought to be mainly intra sector transactions or transactions with the Rest of the world.

Consumption of fixed capital (K.1)

15.193 Consumption of fixed capital in a given period represents the amount of fixed assets used up in the course of production as a result of wear and tear and foreseeable obsolescence, including provision for insurable losses. The relevance of the estimates in the accounts is to adjust figures, for the whole economy or for the various institutional sectors, from the 'gross' basis of recording to the 'net' basis. Provision for consumption of capital is an important feature of commercial accounting, where it is termed depreciation. However, the estimates used for the economic accounts differ in two important respects from those obtaining in commercial accounts. The differences are, first, depreciation for commercial accounts is most likely to be based on the historic costs of the assets, rather than on the replacement cost, and secondly, the period over which assets are depreciated is likely to be shorter than that relevant to the economic accounts estimates.

15.194 As explained in Chapter 4 on concepts, consumption of financial capital is calculated for all fixed assets (other than valuables) which themselves have been produced as part of the productive process. Capital consumption applies to both tangible and intangible fixed assets. It does not apply to land, but it does apply to improvements to land and to transfer costs associated with its sale and purchase, and to other transfer costs.

15.195 Capital consumption is relevant only to 'normal' wear and tear and planned obsolescence, and also predictable losses due to fires or storms, for example. However, losses resulting from major events such as wars or natural disasters, or sudden technological change, are not treated as capital consumption, but are recorded in the 'other changes in the volume of assets' account. The United Kingdom is not publishing this account initially. The coverage of capital consumption in the new accounts has been extended, in parallel with the changes to the scope of GFCF, as described in the box accompanying paragraph 15.30. In addition, it now embraces structures such as dams and roads, which for the previous accounts were assumed to have infinite lives and were not depreciated.

15.196 The figures of capital consumption are derived as part of the estimation process for the measuring the stock of capital. This uses a model-based approach – the Perpetual Inventory Method (PIM) – rather than direct measurement (see paragraph 16.17).

Chapter 16

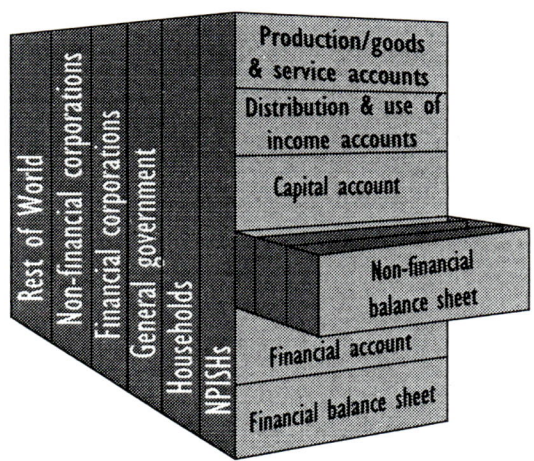

- Rest of World
- Non-financial corporations
- Financial corporations
- General government
- Households
- NPISHs

- Production/goods & service accounts
- Distribution & use of income accounts
- Capital account
- Non-financial balance sheet
- Financial account
- Financial balance sheet

Non-financial balance sheets

Chapter 16 Non-financial balance sheets

16.1   The information required for non-financial balance sheets is only partly available at present. The available data derives from the perpetual inventory model (PIM) referred to in the previous chapter. The remainder of this chapter describes the construction of this model.

Perpetual Inventory Model

16.2   The PIM uses data on purchases and sales of fixed assets and assumptions about their lives, to estimate the value of capital stock in existence at a particular time, and the value of such stock considered to be consumed (consumption of fixed capital) by the production process in any period. The methodology is carried out for a small number of large homogeneous groups of products. It uses the basic data on Gross fixed capital formation (GFCF) and prices relating to this investment, which go into the national accounts estimates, together with three key assumptions – on the length of life of the various assets, their retirement distribution and the method of depreciation.

Asset lives

16.3   The most important assumption made for the PIM calculation concerns the length of life of the groups of assets used in the estimation process. Assets have a finite life, ranging from say two to three years for computers to decades for roads. The length of asset life relevant to the estimation process relates to planned obsolescence, including allowance for accidental damage. It is not easy to estimate average lengths of life, since the assessment requires some form of forecasting, and past experience may not be a good guide. This particular issue was discussed in greater length in the previous version of Sources and Methods. The main points are summarised below.

16.4   Their long period of use requires estimates of lengths of life for assets, such as buildings or structures which were created in the early part of this century, and even before. At the same time, it is, of course, necessary to have long-run information on the levels of GFCF for such assets. The estimates of lengths of life for asset which were produced so long ago were made by unofficial compilers, almost entirely from external sources. Subsequently, over the years, other estimates were made based on information on depreciation allowances, from Inland Revenue, or on fire insurance from insurance companies. In the early 1980s, more up-to-date estimates of asset lives were obtained from manufacturing companies, and these were incorporated in the estimation process. More recently, in 1993, as part of a study by the National Institute of Economic and Social Research for the ONS, some further direct information on lengths of life was obtained. The study also made some more general assessment of the estimates of life lengths being used in the model. Despite this work and the associated improvements, the problems of estimating the likely lengths of life of assets remain as a major area of uncertainty in the PIM approach.

Retirement distribution

16.5   Within the various groups of assets used in the calculation, individual assets will clearly have varying lengths of life, and this feature needs to be incorporated in the model. It is possible to think in terms of different distributions of life lengths – such as uniform or normal – around the average. The NIESR study showed that the estimates of capital stock (and hence capital consumption) were not all that sensitive to the retirement distribution used. The present methodology incorporates the uniform distribution, that is, assets are assumed to retire with a uniform distribution either side of the average length of life for the group.

Method of depreciation

16.6   It is also necessary to establish how assets are to be depreciated over the estimated life-span. There are two broad ways of doing this. First, in the straight line method, the asset is assumed to depreciate by a constant value each year. Thus an asset with a 20 year life costing £20 million will be depreciated by £1 million in each year of its expected life. In the second approach, a constant *proportion* of the value of the asset at the start of each year is depreciated in that year. Thus, in the example given, the proportion might be taken as 10%, and depreciation in each year is then 10% of the depreciated value of the asset at the end of the previous year.

16.7   It is generally considered that the straight line approach is more appropriate to buildings and structures, and the constant proportion approach to plant and machinery. However, in part reflecting simplicity and part wider convention, the straight line method is used throughout the calculation in the UK. In practice, the estimates are not all that sensitive to the type of distribution used; the estimates of life lengths are much more important.

The PIM estimates of consumption of fixed capital

16.8   Against this background, the methodology used for the PIM estimates is now outlined. The calculation is undertaken, on an annual basis, for broad groups of largely homogeneous assets. The groups used are governed very much by the way the information on GFCF is collected, and therefore relate to the types of asset described in paragraphs 15.52–15.66. Ideally, the homogeneity of the groups should also reflect assets with largely similar lengths of life., for example by separating out computers and similar equipment which have a short length of life. For each group, the starting point of the PIM calculation is the long-run series of annual current price GFCF figures. Given that some assets have lives in excess of 100 years, it is in principle necessary to have estimates of GFCF stretching back before the present century. The existing, main dataset for the PIM calculations goes back to 1947. Before this time, estimates of GFCF were derived from a variety of mostly external sources by external practitioners.

16.9   The first step in the calculation is to revalue the long-run current price figures of GFCF for each group to the prices of a common base year. For simplicity and consistency the base year is that of the rest of the constant price dates in the national accounts. The price information ('P') is largely the same as would have been used in determining the estimates of constant price GFCF for the annual accounts. As with the value figures, much of the price data prior to 1947 are very much estimates, often in the earlier years being based on movements in wage rate and costs of materials. The next step is to depreciate the constant price values for each group of assets over the expected life of the assets, using the straight line method of depreciation. In doing this, as mentioned above, a uniform distribution of retirements around the average is assumed. Thus estimates of capital consumption at constant prices are established for each group of assets for each year ('KPCC'). These can then be summed, for each year, over the various asset groups to yield annual estimates of capital consumption at constant prices. Finally, estimates are obtained at current prices by inflating the constant price estimates for each group (KPCC) by the appropriate price index numbers (P), and again summing across the asset groups for each year.

The analysis by industry, by asset and by institutional sector

16.10 Disaggregated estimates of consumption of fixed capital are required for various parts of the accounts, in particular for the main institutional sectors for which the capital account is compiled. The analyses are derived as an integral part of the PIM calculation, using the relevant breakdown appropriate to the analysis of GFCF.

The PIM estimates of capital stock

16.11 The PIM methodology also provides the basis for the estimates of capital stock. This section describes briefly what the estimates are and how they are used, and explains how they are derived from the PIM model.

16.12 The gross capital stock is a measure of the value of the stock of assets which are surviving from past investment. The valuation used is current replacement (purchasers') price. Capital stock can also be measured on a net basis, by subtracting from the gross figure an estimate of the accumulated capital consumption of the surviving assets.

16.13 The estimates serve a number of purposes. These include, principally, (i) providing a basis for forecasting future levels of GFCF, (ii) assessing productive potential, and (iii) as a basis for calculating rates of return.

16.14 The description given above for estimating capital consumption is readily adapted to the estimation of capital stock. The initial step, as with the capital consumption estimates, is to revalue the long-run current price figures of GFCF for each asset group to a common base, using appropriate price indices ('P'). The retirement distributions are then applied to each set of annual data, to give, for each asset group, the constant price value of estimated surviving stock in each year ('KPS'). These data are then summed across the asset groups for each year, to yield the annual estimates of constant price gross capital stock. To derive current price estimates of gross capital stock, the constant price estimates (KPS) for each asset group are reflated to current price indices using the relevant deflators (P).

16.15 The estimates of net capital stock are obtained by subtracting the accumulated capital consumption of the surviving assets, estimated as described above, from the gross figure.

PIM and direct measurement

16.16 As evident from the description of the use of the PIM approach, the accuracy of the estimates relies crucially on the assumptions about lengths of asset life, including the ability to allow for the effect of economic cycles and technological change, and about depreciation. The methodology is very much an empirical one; it has not been possible to check the estimates against reality. Evidence has suggested that the figures of capital stock probably overstate the actual values.

16.17 In the last few years, the ONS has been considering whether the estimates of capital stock can be made by direct measurement, rather than by modelling. One obvious advantage of such an approach is that the estimates would relate to the actual assets surviving, rather than the model's estimate of survivals. Following a small pilot survey into the feasibility of direct measurement, it seems that, with increasing computerisation of asset registers, information is available to enable direct estimates to be made. However, there would obviously be increased costs, both for firms and for the ONS, in such an approach. The position remains under review.

Chapter 17

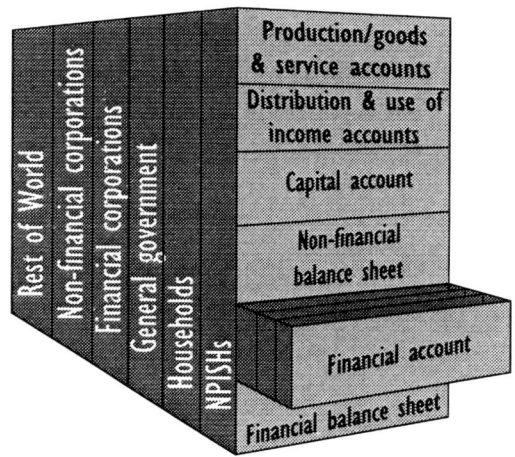

The financial account

Chapter 17 The financial account

General description

17.1   The financial account shows the acquisition and disposals of financial assets and liabilities. These transactions in financial assets and liabilities are between different institutional units within the UK economy and with the rest of the world. Thus this account records the movements in the financial assets and liabilities that form net lending or borrowing.

17.2   Financial transactions measure the net acquisition of financial assets, or the net incurrence of liabilities for each type of financial instrument. Some sectors are net lenders while others are net borrowers. The UK government during the 1990s is an example of the latter. Institutional units with surplus funds make them available for others to use.

17.3   All financial transactions involve a simultaneous creation or liquidation of a financial asset and the counterpart liability, or a change in ownership of a financial asset, or an assumption of a liability. A financial asset is always balanced by a liability of another unit (except in the cases of monetary gold and special drawing rights). For example, a loan is an asset of the lender (or holder) and a liability of the borrower; a share is a liability of the company issuing it and an asset of the holder. The financial account shows both net transactions between sectors (transactions between institutional units within a sector are consolidated) and also on a non-consolidated basis. The ESA 1995 legislation requires member states to prepare the financial account and the balance sheet on both a consolidated and a non-consolidated basis.

17.4   A financial transaction between two institutional units increases net lending/net borrowing by one institutional unit and, by the same amount, decreases net lending/borrowing by the other. So this account shows how the net borrowing of a deficit sector is financed by a reduction in its assets and/or by it incurring liabilities. Sectors with surplus resources make them available to others by acquiring assets (e.g. by purchasing government bonds, or money market instruments such as commercial paper issued by corporations etc.) or by reducing their own liabilities.

17.5   Financial transactions always have counterpart transactions in the system: these may be other financial transactions or non-financial transactions. Most transactions involving the ownership of goods or assets, or obtaining services, have some counterpart entry in the financial account. For example, buying a new car may involve trading in one's existing car (which is not a financial asset) and will also usually involve a payment in cash, or cheque (a reduction in financial assets) or taking out a loan (an increase in financial liabilities). There are also many transactions which are recorded entirely in the financial account: one financial asset being exchanged for another, or a liability such as a loan being repaid with an asset. At the same time new financial assets may be created by an institutional unit incurring liabilities, for example a corporation may issue commercial paper. Transactions entirely within the financial account affect the distribution of the portfolio of assets and liabilities and may change their total amounts but they do not alter the difference between total financial assets and liabilities. Chapter 7 describes in detail the four entries needed in the national accounts to cover a single transaction fully.

17.6   In the system, the financial account is the final account in the full sequence of accounts that record transactions between institutional units. It is the only account which does not have a balancing item that is carried forward. In the system the balance of the financial account is identical with (but of opposite sign to) the balancing item of the capital account. However, in practice a discrepancy will usually be found between them because they are calculated using different statistical data. These are given the term of statistical discrepancies in the UK accounts.

Uses of the financial account

17.7   This account shows how funds flow from sectors that are net savers to sectors that are net borrowers, usually via financial intermediaries. Thus it records the distribution and redistribution of financial assets and liabilities among the sectors of the economy.

17.8   The financial account shows how each sector's financial transactions are related to its saving and capital expenditure and how financial transactions generally are related to one another within and between sectors. So it provides a way of examining the financial effects of economic policy and assistance for decisions regarding future policy. It can be used to investigate factors influencing the holdings of, and transactions in, different types of financial instrument: in particular changes in interest rates. Corporations and others also use it to understand the financial environment they operate in, for example, by helping them to assess competition in the market for funds.

17.9   For financial institutions the financial account shows the large amounts of money which are channelled through them as financial intermediaries. The scale of this makes it important to be aware of changes in their sources of funds and in the use of those funds. The transactions of the financial institutions affect the liquidity, current and capital expenditure of other sectors, and the financing of the public sector net cash requirement.

17.10   The account is a matrix of transactions. It provides a discipline which requires the constituent statistics to be defined and compiled consistently. The total transactions in one instrument must agree: it is not possible to have the government issuing £x billion of gilts (government bonds) and all other sectors (including the rest of the world) recording purchases of a different amount.

17.11   In forecasting, past relationships between economic and financial variables are discerned and extrapolated into the future. In the forecasting process the advantage of this accounting matrix is that it keeps the various components of the forecast in touch with each other. It compels the forecaster to consider explicitly all the implications of a change. For example, if the public sector net cash requirement (PSNCR) were to be increased, the borrowing for this must be made from the other sectors. Hence the forecast must allow for the impact of this on the supply of funds to other potential borrowers. These other borrowers will have to find funds elsewhere (perhaps at a higher rate of interest) or cut back on their own activities.

17.12   The relationship of the financial account and key economic indicators is considered in the section 'Publication and key economic indicators' later in this chapter (paragraphs 17.51–17.57).

Concepts

17.13   This section considers the issues of valuation, timing, currency translation and consolidation.

Valuation

17.14   In the financial account the aim is to record transactions at the transaction value. So the value recorded in £ sterling should be the value at which the assets/liabilities involved are created, liquidated, exchanged or assumed, on the basis on commercial considerations only. Since each sale of an asset is balanced by a purchase the total value of sales for a certain type of asset should equal the total value of purchases. In an ideal world (for statisticians!) every transaction would be accurately recorded by both parties simultaneously and at the same valuation, identifying the sector of the other party involved. Collection and collation of these records would yield a perfect financial account. However, this is not possible and we have to use a partial system of data collection and recording in which there are known to be errors.

17.15   Where flows in foreign currency are denominated in foreign currencies the results are not as reliable as those denominated in sterling. This is because they depend upon eliminating the effects of variations in exchange rates. The preferred approach is that foreign currency be translated into sterling at the rate prevailing at the time of the transaction; but in practice average rates over a longer period are often used. A separate section below covers currency translation in more detail. Assumptions have to be made about the mix of currency holdings and about the average rates at which the transactions took place.

17.16   The transaction value in the financial account does not include service charges, fees or commissions etc. These are recorded as payments for services. In many cases the accounts recognise charges implicit in the spread between buying and selling prices. These also are excluded and the transaction measured at mid-market price. Likewise taxes on financial transactions are also excluded from this account; they are treated as taxes on services within taxes on products. However, it may be difficult for one party to a transaction, made by a third party on their behalf, to state the true cost - for example after selling shares for them their broker will have sent them the proceeds net of dealing costs such as brokers commissions, stamp duty. Respondents to ONS surveys of financial transactions are thus asked to show sales of assets net of dealing costs, and conversely to include them within the costs of assets acquired. An aggregate level adjustment is subsequently made to these figures to ensure they are on the basis required. Respondents are thus able to supply the good quality data they have available. Further details on specific instruments are given below.

Timing

17.17   In Chapter 2 the quadruple entry principle of accounting was described. For any transaction there are (at least) two parties. Each party's account must show two balancing entries. As a simple example, if a household purchases goods for £100 from a non-financial corporation, the household sector account would show the purchase in consumption expenditure balance by a reduction in its cash or bank deposits. The counterpart sector, non-financial corporations, would show a sale balanced by an increase in cash or bank deposits. In practice it is possible that the parties to any transaction may record it in their accounts in different time periods. With the application of accruals accounting it is likely that the household sector's recorded purchases may not be timed consistently with its recorded change in currency or deposits. To balance the national accounts it is necessary to record all sides to the transaction at the same point in time. When the counterpart of a financial transaction is a non-financial one, both are recorded at the time the non-financial transaction occurs. For example, when goods or services are sold, an 'account receivable/payable' is created in the financial account, (a form of passive trade credit), which is recorded when the entries are made in the non-financial account. This financial transaction is unwound when payment is actually made.

17.18   In principle, transactions in financial instruments are deemed to take place when ownership changes, rather than when payment is made. However, many loans, for example, come into existence only when the amount is paid. For other instruments settlement is so quick as to virtually be instantaneous; transactions in government securities are settled the following day.

17.19   Implementing this principle where one financial instrument is exchanged for another, gives rise to three possibilities (assuming all the necessary information is available):

- both financial transactions are transactions in means of payments (means of payment consist of currency and transferable deposits, monetary gold and special drawing rights): they are recorded at the time the first payment is made;

- only one of the two financial transactions is a transaction in means of payment: they are recorded at the time payment is made;

- neither of the two financial transactions is a transaction in means of payment: they are recorded at the time the first financial transaction takes place.

17.20   Timing consistency between sectors is affected by measurement problems. Where the sources for the parties to the transaction differ they are unlikely to be completely consistent. For many entries in the financial account inconsistencies between sectors are avoided by estimating both ends of a transaction from the same source. For example, deposits with building societies are estimated from building society data for both the non-monetary financial institutions and the household plus NPISH sectors.

17.21   Timing differences can lead to distortions in flows over a shorter (e.g. monthly) accounting period but are less important in the longer account periods as published here. They are likely to be small compared with total flows during a quarter or year. This is not always the case for the net flows (acquisitions net of realisations) measured in the financial account. However, timing differences do tend to even out over successive periods and care is taken to ensure that large specific transactions (such as arise when there is a major privatisation) are recorded consistently.

Currency translation

17.22   When flows in foreign currency are denominated in foreign currencies they have to be expressed in terms of domestic currency (i.e. sterling) before they can be used in the UK national accounts. In order to do this assumptions have to be made about the mix of currency holdings and about the average rates at which the transactions took place. The preferred approach is that currencies are translated into sterling using the rate which applied at the time of the transaction. However, this is frequently impractical and the calculations are based on an average rate. This renders these figures less reliable than those flows denominated in sterling.

17.23   In contrast, the ONS inquiry to Securities Dealers gives the following instruction about foreign currency transactions:

[these] can be translated into sterling using one or a combination of the following methods: at middle-market closing rates on the day of the transaction; or, using sterling equivalents as given on contracts; or, by combining amounts in the same currency and using average rates of exchange for the quarter.

17.24   However, non-financial corporations, whose turnover in financial assets and liabilities is relatively small, such as contributors to the ONS's quarterly Financial Assets and Liabilities Survey (FALS), are asked only for end-period levels and told that

foreign currency should be translated to sterling at the exchange rate on the dates to which the return relates.

They are also asked to give details of the sums and foreign currency involved, together with any other rates used, if applicable. The estimation of transactions in sterling requires assumptions about the pattern of transactions over the period and movements in the exchange rates.

17.25   One reason for the difference between these two sets of instructions is that the units to which these two different forms are sent are of different types. The securities dealers handle many client institutions' transactions each day, and have transactions amounting to hundreds of billion pounds each quarter. The transactions of the non-financial corporations receiving the FALS forms will generally be much smaller and are less likely to be occurring on a daily basis. So to minimise the reporting burden the FALS form requests only balance sheet data at start and end of the period. The securities dealers' form recognises the business they transact and the fact that they have their own computer and other systems which may record foreign currency transactions in just one of several different ways.

17.26   For some sectors their transactions in foreign assets are derived from balance sheet returns, made by them in sterling, once a year. For the balance sheets the foreign currency values had been translated at the rate on that day into sterling. These are translated back into local currency, and calculations are made of the implicit net transactions since the previous balance sheet, yielding a figure which then is translated back into sterling at average exchange rates over the period for the financial account. This process is made less reliable by the fact that some institutions may report their end year balance sheets for a date other than 31 December.

17.27   There are many different foreign currencies involved in daily transactions of UK financial institutions in the world's main securities markets. There is thus scope for discrepancies arising in the national accounts of all the parties involved, because of the fluctuations each minute in exchange rates and the size and quantities of the flows involved. The ONS instructions to responding corporations have to take into account the reporting or form-filling burden which is placed on the respondents and balance this against what is required for the national accounts. Where necessary, further details may be requested from respondents if, for example, the sums reported are particularly large or look out of line with expectations or if there are difficulties when the whole set of data for the UK financial accounts are produced. (See the sections on assembly and balancing processes in paragraphs 17.41–17.50 below).

Reinvestment of accrued interest

17.28   In principle, interest accruing but not paid out is treated as being reinvested in the relevant financial instrument (see Chapter 7, 7.31) in principle this applies to interest on deposits (F.22, F.29), Securities other than shares (F.3), and Loans (F.4). In practice it applies only to certain instruments, where both accruals and cash flows statistics are available. Accruals data are derived from the DIM matrix (*see* Chapter 14, 14.62–14.74). Cash flows are taken from general government accounts.

Consolidation

17.29   The ESA 1995 specifically requires member states to produce both consolidated and non-consolidated versions of their financial account (and balance sheet) tables. While full consolidation would require all flows within a specified sector or sub-sector to be eliminated, non-consolidated accounts are less easy to define. Loans (for example) between financial institutions which are not in the same sub-sector should not be consolidated for the financial accounts.

17.30   Transactions in assets should be shown separately from transactions in liabilities for each sub-sector and sector.

17.31   Under the previous system there were some inconsistencies in the approaches to consolidation in the financial account for different sectors of the UK economy. For example, banks' deposits were shown gross (i.e. one bank's loan to another would be shown as such) while loans between building societies were netted out or consolidated.

Classification of financial instruments

17.32   The ESA 95 specifies a classification of financial instruments (assets and liabilities) which is required by law from the United Kingdom and all other member states. This comprises seven categories:

F.1	Monetary gold and special drawing rights
F.2	Currency and deposits
F.3	Securities other than shares
F.4	Loans
F.5	Shares and other equity
F.6	Insurance technical reserves
F.7	Other accounts receivable/payable

17.33   This classification is based primarily on the liquidity and the legal characteristics of the financial assets. This is discussed in more detail in Chapter 7.

17.34   In the UK national accounts a more detailed approach is adopted to meet the needs of users of the accounts and reflect the importance of the UK financial system. It also reflects further detail allowed for in the ESA 95 and the likely requirements of the European Central Bank for administering the European Monetary Union (EMU).

17.35   For the UK the full list of instruments being used is shown in Table 17.1. A distinction between short-term and long-term assets and liabilities, outlined in the ESA 95, is not regarded as particularly meaningful in the UK context, but groupings of instruments can be made to approximate to a long/short-term split. For example, long-term loans are obtained by aggregation of outward direct investment loans, loans secured on dwellings, financial leasing loans and other loans by UK residents. A detailed description of the derivation of each financial instrument from data sources is given later in this chapter.

Table 17.1: Classification of financial instruments

F.1 Monetary gold and SDRs	
Monetary gold	F.11
Special drawing rights	F.12
F.2 Currency and deposits	
Currency	F.21
Transferable deposits	F.22
Deposits with UK MFI's	F.22
Sterling	F.2211
Foreign currency	F.2212
Deposits with the Rest of the World MFI's	F.229
Other deposits (National savings and tax instruments)	F.29
F.3 Securities other than shares	
Short term - money market instrument (MMI's)	F.331
MMI's issued by UK general government	F.3311
UK local authority bills	
MMI's issued by UK MFI's	F.3315
MMI's issued by other UK residents	F.3316
MMI's issued by Rest of the World	F.3319
Long term-bonds	F.332
Bonds issued by UK general government	F.3321
UK local authority bonds	F.3322
Bonds issued by UK MFI's	F.3325
Bonds issued by other UK residents	F.3326
Bonds issued by Rest of the World	F.3329
Financial derivatives	F.34
F.4 Loans	
Short term loans	F.41
Loans by UK MFI's (excluding loans secured on dwellings and finance leasing)	F.411
Loans by Rest of the World MFI's	F.419
Long term loans	F.42
Direct investment loans	F.421
Loans secured on dwellings	F.422
Finance leasing loans	F.423
Other loans by UK residents	F.424
Other loans by the Rest of the World	F.429
F.5 Shares and other equity	
Shares and other equity excluding mutual fund shares	F.51
Quoted UK shares	F.514
Unquoted UK shares (including reinvested earnings on inward direct investment)	F.515
Other UK equity (including direct investment in property)	F.516
Shares and other equity issued by the rest of the world	F.519
Mutual fund shares	F.52
UK mutual fund shares	F.521
Rest of the World mutual fund shares	F.529
F.6 Insurance technical reserves	
Net equity of households in life insurance reserves and pension fund reserves	F.61
Prepayment of insurance premiums and reserves for outstanding claims	F.62
F.7 Other accounts receivable/payable	
Trade credits and advances	F.71
Other accounts receivable/payable except trade credits and advances	F.79

Data sources

17.36   The main data sources for the financial account are ONS inquiries, data collected by the Bank of England and the Building Societies Commission and administrative data from the Treasury. The ONS and the Bank of England work together closely to produce the financial account because of the importance of monetary financial institutions both in their own right within the account and as an important source of information for counterpart sectors' holdings of instruments such as bank deposits. The inquiries are described in 25.2 and 25.3.

17.37   Data of the total purchases and sales (flows) of different financial instruments are only available for certain sectors, notably those where the ONS carries out quarterly inquiries into financial transactions. These inquiries generally collect acquisitions and realisations separately for the longer-term types of instrument such as shares and bonds, but holdings (balance sheet) data for short-term instruments such as currency, deposits and money market instruments (e.g. bills). The institutions from which these gross data for transactions are obtained comprise insurance corporations and pension funds, UK mutual funds (investment trusts and unit trusts), and certain other non-monetary financial institutions.

17.38   For other sectors of the economy and the rest of the world the financial transactions are compiled from other sources. These include the Bank of England's capital issues database (25.2.10) and the London Stock Exchange (for figures on capital issues). Issues of government stock ('gilts') and Treasury bills come from central government sources.

17.39   The remaining figures for other sectors and the rest of the world are derived from changes in financial balance sheet data, that is, from changes in financial assets held and liabilities outstanding between the opening balance sheet and the closing balance sheet, revalued as appropriate to allow for price and exchange rate movements. For example the Bank of England conducts monthly and quarterly balance sheet inquiries to all large and many smaller banks. The Building Societies' Commission similarly has a monthly balance sheet inquiry which covers all building societies. Financial transactions of monetary financial institutions are derived from these two sets of data from all such institutions. The Treasury also provides data from administrative sources which are used in the financial account.

17.40   For one or more sectors there may be no data for particular instruments – for example, because data collection is not regarded as practicable (e.g. from households) and/or because periodic benchmark data show that they hold and transact only very small amounts of the instrument concerned. Some surveys run by the ONS collect greater detail from corporations annually than they do each quarter. This is done to limit the form-filling burden that is placed upon the corporations sampled and reflects the fact that they may only collate detailed information themselves once a year (for their annual accounts). Other surveys are made of only the largest corporations in a sector (or industry group) each quarter but once a year the forms may be sent to a larger number, including smaller businesses.

Assembling the financial account

17.41   This section introduces the financial account database and covers the use of counterparts and residuals to obtain the complete matrix of data.

17.42   The accounts are assembled using a database of time series that builds up from data sources to the matrix of transactions by sector that make up the financial accounts. For each time series the database records where it come from, which cell in the sector accounts matrix it belongs to and how it is aggregated with other series to produce the cell totals. The information is held in greater detail than is required for the 1995 ESA, largely depending on the detail available in the source data. In some areas, however, the data obtained from the source each quarter may be less detailed than that required within the timetable for first publication or for ESA. In some cases there are no direct data sources and the derivation of the data in these circumstances is described below.

17.43   It is primarily the household and NPISH sectors where there are no direct data sources. However, depending on the instrument involved, this problem affects the rest of the world sector and certain others. There are two techniques used to obtain data in these circumstances:

      (i) counterparts; and
      (ii) residuals.

These both make use of the principle that each acquisition of an asset must have a matching disposal of an asset or a take up of a liability.

17.44   The supplementary detailed information held in the database is used to assess counterpart transactions and derive residuals. For example, in most tradable securities held in domestic currency the household sector is the residual but, for most tradable securities held in foreign currency the rest of the world sector is the residual. Each instrument in the matrix is balanced individually across all sectors as well as the transactions of each sector being reviewed for the sector accounts.

Counterparts

17.45   Some entries in the financial accounts are estimated from information supplied by the other party – the counterparty – to the transaction. This method is particularly used for loans and deposits. For example, all sectors' holdings of banks deposits are estimated using counterpart information from banks, which provide a full sector analysis of their deposit liabilities. In some instances, information from a particular sector may be collected only once a year rather than quarterly; or there may be no regular source of information from it – as with the household sector.

Residuals

17.46   The other way to estimate data for which there is no information is to treat its cell in the matrix as the residual for that particular financial instrument. This is based on the fact that the total of transactions in assets must equal the total of transactions in liabilities for any instrument. The observed data will not identify all the transactions and so the residual sector is allocated the unidentified transactions. Under the old system the residual was usually allocated to a single sector – although it was sometimes apportioned between a number of sectors. The residual sector was often the personal sector. When the ESA separation of the household and NPISH sectors is implemented it will be necessary to consider afresh the allocation of residuals between these two sectors.

The balancing process

17.47   Data suppliers who feed data directly into the financial account database are in the ONS and the Bank of England. They deliver source data on specified dates to the central system, at several points each quarter. The database of time series first balances each category of the matrix, according to agreed methodologies, and then aggregates to produce a top-to-bottom account for each sector.

Since there are errors and omissions in the estimates for some or all cells (instruments and sub-instruments) for each sector, a non-zero statistical discrepancy appears for each sector. This is the difference between net lending/borrowing as measured by the current and capital accounts and net lending/borrowing as measured by the financial transactions account.

17.48 For each sector there is a nominated statistician who is responsible for the coherence of the top-to-bottom account for that sector, and each has a target for the size of the statistical adjustment item. These targets are set according to the quality of the data sources. These statisticians also each have responsibilities for one or more individual instruments. They meet to assess the plausibility of the standard procedures for the latest quarters and to agree adjustments to the allocation of the counterparts and residuals, or to the sample survey results, or to suggest that further investigations need to take place in respect of certain cells' data (which may involve contacting survey respondents).

17.49 The balancing process also looks at the effect of any proposed adjustments on key economic indicators within the accounts, and whether there is a coherent economic story when placed together with other data.

17.50 After the meeting of statisticians to discuss the data set there is a redelivery of source data for the changes agreed at the meeting. Further changes may be made, in an iterative process at the end of which a set of accounts is produced which meets these targets and is fully consistent, coherent and plausible. These are then published. The UK is one of the world-leaders in both the timeliness and the quality of its production of a coherent set of accounts.

Publication and Key Economic Indicators

17.51 Key indicators from the financial and sector accounts, such as household saving, are published in a First Release about 12 weeks after the end of the reference period. At the same time full sector accounts are made available electronically on the databank. Full paper publication follows in UK Economic Accounts, which includes a commentary linking together the economic stories for the different sectors.

17.52 Two important economic aggregates could be identified in the financial account under the previous UK system of accounts. These were the public sector net cash requirement (PSNCR) and the money supply (M4). These are not immediately available from the new ESA tables because they are cash measures.

Public sector net cash requirement

17.53 The PSNCR indicates the extent to which the public sector (central government, local government and public corporations) borrows from other sectors of the economy and the rest of the world to finance the balance of expenditure and receipts arising from its various activities. It is measured on a cash basis whereas the ESA 1995 based accounts are calculated on an accruals basis. This means that the PSNCR can no longer be obtained simply by aggregating items in the financial account for the three sub-sectors of the public sector. Instead it is calculated and published separately. The intention is to publish a reconciliation with the ESA financial account in due course.

Money Supply (M4)

17.54 Money supply is a balance sheet concept; changes in money supply may be related to the financial account. There is no single correct definition of money. The narrowest definition is M0 which represents notes and coins in circulation outside the Bank of England and also includes bankers' operational deposits at the Bank. A broader definition of money is M4. M4 comprises holdings by the UK private non-monetary sectors (i.e. the M4 'private sector') of notes and coin, together with their sterling deposits at banks and building societies in the UK (including certificates of deposit and other paper of not more than 5 years maturity issued by banks and building societies) and, from December 1995, liabilities of UK banks and building societies arising from repos.

17.55   M4 is recorded as the sterling part of F.21 and F.2211; Paper of 1–5 years is part of F.3315.

17.56   The main counterparts to changes in M4 are the PSNCR *less* public sector debt sales to the private sector and external and foreign currency finance of the public sector and M4 lending to the private sector.

Quality and reliability of the data

17.57   The quality and reliability of the data vary. They depend upon factors such as:

- the frequency of collection of the data (quarterly or annual or less infrequent);
- whether they come from a full census (e.g. most Bank of England figures for banks) or from a sample (most ONS surveys are samples);
- sampling error and bias;
- whether financial transactions are collected directly (as on the ONS surveys of non-monetary financial institutions) or derived from balance sheet data (e.g. banks' and building societies' data);
- the quality of the data: are respondents asked to supply data that they themselves use? If not, do they recognise the importance of them for the UK as a whole?;
- whether there is a statutory obligation for the data to be supplied;
- the timeliness of data receipt;
- whether the sector is dominated by a few very large institutional units (e.g. the top eight or so insurance corporations write at least half of all insurance business in the UK).

The financial instruments

17.58   The remainder of this chapter considers each financial instrument in turn. For each instrument there is a description and then a discussion the sources of the data and the methodology used for each sector. The order followed is that specified by ESA95 and will be used in the Blue Book UK National Accounts 1998.

F.1   Monetary gold and special drawing rights (SDRs)

17.59   These consist of assets held by the government in the Exchange Equalisation Account (EEA). The EEA is the central government account at the Bank of England in which transactions in the official reserves are recorded. These reserves comprise the UK official holdings of gold, convertible currencies, Special Drawing Rights (SDRs) and the UK reserve position in the IMF. From July 1979 they also include European Currency Units (ECUs) acquired from swaps with the EMI. The EEA is managed by HM Treasury. All earnings on the official reserves and transactions with the IMF pass through the EEA.

17.60   These financial assets are the only ones for which there are no counterpart liabilities in the system. Transactions in them always involve changes in the ownership of financial assets of the UK or the rest of the world.

17.61 Monetary gold is gold held as a component of foreign reserves by monetary authorities, or by others who are subject to the effective control of the authorities. In the UK only gold owned by the government is recorded in the reserves.

17.62 Under the ESA95 Regulation, the UK has a derogation for monetary gold under which it will continue (as in the previous UK system) to record the gold held by monetary financial institutions (MFIs) as a financial asset rather than as a valuable. However, gold held by MFIs is recorded in Currency and deposits (F.2) as foreign currency; it is not recorded as monetary gold.

17.63 Special Drawing Rights (SDRs) are international reserve assets created by the International Monetary Fund (IMF) and allocated to its members to supplement existing reserves. This category records all transactions in SDRs. SDRs are not considered liabilities of the IMF since IMF members to whom SDRs are allocated do not have an actual (unconditional) liability to repay their allocation of them. SDRs are held only by official holders. The IMF issues a quota of SDRs to its members that can be swapped for reserve currencies. A reserve currency is one that other countries are willing to hold in their official reserves.

17.64 The market valuation for SDR transactions is provided by the Bank of England. This is a change from the conventional valuation used in the previous system of accounts.

17.65 The UK's reserve position in the IMF is not part of this financial instrument, F.1, but is within F.229. It is a deposit with no maturity and equals the difference between the UK quota and the sterling holding by the IMF. The reserve position will fluctuate depending on other countries' drawings or repayments of sterling to the IMF.

17.66 There are two sectors which transact monetary gold and SDRs: central government and the rest of the world. The flow which represents the net change in F.1 is supplied quarterly by the Bank of England. Both gold and SDRs are recorded by the Bank of England within the Exchange Equalisation Account. Since the source of the transactions data is official records they are considered to be of high quality.

F.2 Currency and deposits

17.67 This category consists of all transactions in currency and deposits. Thus it includes currency in circulation and all types of deposits in both domestic (i.e. sterling) and foreign currency. Gold held by UK monetary financial institutions (MFIs) is also recorded here (rather than as a valuable), continuing the treatment under the previous ESA. The category has three major components:

- F.21 Currency
- F.22 Transferable deposits
- F.29 Other deposits

F.21 Currency

17.68 This covers domestic (i.e. sterling) and foreign currency notes and coins, and gold held by UK MFIs.

17.69 Sterling currency notes are a liability of the central bank, while coin is regarded as a liability of central government. Coin is issued by the Royal Mint, but Bank of England notes are issued by the Issue Department of the Bank of England, which is part of the central bank sector. The Bank of England holds securities (mainly gilts) as backing for the note issue. This is a change from the former treatment, in which the note issue was a liability of central government and the Issue Department was treated as part of central government. The flows data for central government relate to net issues of coins less holdings of notes and coin by central government for its own use. The figures for the monetary financial institutions sector relate to issues of notes by the Issue Department (liabilities) and changes in holdings of sterling notes and coin (assets) by the sector (including Scottish and Northern Ireland bank notes).

17.70   The figures for public corporations' transactions in sterling notes and coins are based on working balances held by the Post Office. The transactions of private non-financial corporations are derived as part of the residual which is shared with the household and NPISH sectors. Holdings by the local government sector are assumed to be negligible. The data for building societies come from the Building Societies Commission (25.3.2). The data for other financial intermediaries come from the Bank of England and those for insurance corporations are based on a benchmark survey of their holdings, derived from a statutory DTI inquiry. For the rest of the world sector the data are estimated from figures of transactions obtained from the International Passenger Survey (25.5.4), of which certain proportions are assumed to be in sterling notes and coins.

17.71   A special adjustment allows for the fact that the amount of sterling notes and coin held by corporations is high on Thursdays (due to payment of weekly wages in cash) and the amount held by households is high on Fridays (reflecting withdrawals to cover weekend spending). Further adjustments are made to reflect the build up of households' notes and coins over the Christmas and Easter periods.

17.72   Data for foreign currency notes and coin are crudely estimated from the travel data collected in the ONS International Passenger Survey (25.5.4). Transactions by monetary financial institutions (MFIs) are derived from banking statistics (25.3.1), which also include transactions in gold (*see* paragraph 17.64). They are currently attributed to the private non-financial corporations sector.

F.22   Transferable deposits with monetary financial institutions (MFIs)

17.73   These are deposits which are immediately convertible into currency or which are transferable by cheque, banker's order etc., without any kind of significant restriction or penalty. This category comprises:

- Sterling deposits with UK MFIs (F.221);
- Foreign currency deposits with UK MFIs (F.222);
- Deposits with other (i.e. rest of the world) monetary financial institutions;

F.221 Transferable deposits with UK MFIs

17.74   UK MFIs which can accept deposits currently comprise banks and building societies. These are considered separately in the following paragraphs.

Domestic and foreign currency UK bank deposits (F.2211, F.2212)

17.75   These comprise all forms of deposit held with institutions authorised by the Bank of England or entitled to accept deposits in the UK. These include UK incorporated institutions and UK branches of European authorised institutions (briefly referred to as 'banks'). Deposits with UK financial corporations other than banks ceased at the end of 1981 when, under the terms of the Banking Act 1979, institutions not in the banking sector were no longer permitted to accept deposits. These figure are thus not separately identified in the tables.

17.76   The figures include:

(a)   deposits with the Bank of England Banking Department, other than from the monetary sector (intra-sector deposits);

(b)   deposits from rest of the world offices of banks;

(c)   receipts under Save-As-You-Earn (SAYE) and sharesave contracts, on which banks pay interest net of tax;

(d)   liabilities under sale and repurchase agreements of British Government securities and other paper.

17.77   The figures may be divided between domestic (i.e. sterling) and foreign currency. Both sight and time deposits are included within each.

17.78   Deposits evidenced by the issue of negotiable paper (certificates of deposit, commercial paper, promissory notes, bills and other paper) are generally omitted from this sub-instrument and included in money market instruments F.331. The figures also exclude the institutions' liabilities in the form of shares and other securities, reserves and other internal accounts.

17.79   The figures are compiled by the Bank of England from returns provided by the institutions (25.3.1). The increase in banks' liabilities to other sectors is measured as the increase in other sector's deposits, as reported in the returns, adjusted by transit and suspense items and the interbank reporting difference.

Transit and suspense accounts, and the interbank difference

There are two adjustments to the data collected from banks. They relate to:

(a)   the inclusion of the "inter-bank differences" (both sterling and foreign currency); and

(b)   sterling and foreign currency transit and suspense items

The inter-bank difference arises because in practice total reported liabilities to other banks do not match the reported claims on other banks. Research indicates that banks misclassify their customers as between UK monetary and non-monetary financial institutions (NMFIs) and non-residents. The difference is assumed to relate to F.221, transferable deposits with MFIs. The depositing sector is considered to be mainly NMFIs for sterling deposits and the Rest of the world sector for foreign currency.

Sixty per cent of sterling and 100 per cent of foreign currency items in transit are assumed to relate to bank deposits and are deducted from that series. The counterpart of the adjustments is assigned to non-financial corporations (sterling) and the Rest of the World sector (foreign currency).

The remaining forty per cent of the net change in sterling transit items is added to F.411, Short term loans by UK MFIs. The counterpart borrowing sector is assumed to be non-financial corporations.

17.80   Figures for counterpart sectors are almost entirely obtained from the analysis provided in the banking sector returns. However, data for insurance corporations and pension funds come from ONS surveys of a sample of these institutions (25.3.3).

Domestic and foreign currency deposits with building societies (F.2213)

17.81   The figures comprise shares, deposits, REPOs and (from 1969) receipts under Save-As-You-Earn contracts on which building societies pay interest net of tax (i.e. 'retail' deposits) plus (from 1983) time deposits on which the societies pay interest gross of tax (i.e. 'wholesale' deposits). The retail deposits include interest credited to depositors' accounts.

17.82   Deposits evidenced by the issue of negotiable papers (certificates of deposit, commercial paper and other paper) are omitted from this sub-instrument and included in money market instruments F.331.

17.83   The figures do not include building society negotiable bonds. These are assumed to be entirely taken up by the monetary sector and included in bank lending: sterling (F.411).

17.84   The figures are based on a system of returns from the societies, which are operated by the Building Societies Commission (25.3.2). Sector transactions in deposits are derived from the societies' monthly balance sheet returns.

F.229   Deposits with rest of the world MFIs

17.85   This consists of transactions in transferable deposits, both domestic (i.e. sterling) and foreign currency, held by rest of the world monetary financial institutions (MFIs). From July 1979, European Currency Units (ECUs) are included here. ECUs were acquired from swaps with the European Monetary Co-operation Fund (EMCF) before 1 January 1994 and with the European Central Bank (ECB) afterwards. The UK's reserve position in the IMF appears not in this item but in other deposits, F.29.

17.86   Assets with banks abroad are derived from the international banking statistics of countries in the Bank for International Settlements (BIS) reporting area and cover the UK non-bank private sector's deposits with banks in the rest of the world. These data come from the BIS via the Bank of England. Some assets of general government and public corporations may also be included but are not separately distinguishable. The financial flows are estimated from changes in the levels, adjusted for exchange rate movements. They omit, as far as possible, the effects of any discontinuities in the levels series, but no attempt has been made to allow for any deficiencies in the financial flows prior to 1984. Trends in the data have been informed by trends in rest of the world bank borrowing and deposits derived from the ONS Financial Assets and Liabilities Survey. Transactions by UK MFIs are derived from inquiries carried out by the Bank of England. Estimates of securities dealers' transactions with banks abroad are derived from their asset levels as reported in an ONS survey. Estimates for earlier years, back to 1986, are based on information from published annual accounts.

F.29   Other deposits

17.87   This category covers national savings, tax instruments, the UK reserve position in the IMF and in the European Monetary Co-operation Fund (EMCF) before 1 January 1994 and with the European Monetary Institute (EMI) subsequently. It also covers the financing of central government from the EC No. 1 Account (*see* Chapter 21) and deposits by public non-financial corporations with the National Loans Fund.

National savings comprise the following:

Deposits

17.88   Changes in outstanding deposits, including estimated accrued interest, with

- the ordinary departments of the National Savings Bank (NSB), the NSB investment account from January 1981, and
- Trustee Savings Banks up to third quarter 1979;

Certificates and bonds

17.89   Net sales or repayments of income bonds, deposit bonds, premium savings bonds, British savings bonds (removed from sale on 31 December 1979), national savings stamps, gift tokens, national savings certificates, SAYE contracts and yearly plan agreements, plus increases due to accrued interest and index-linking.

17.90   Most national savings instruments are treated as financial claims by the household sector on the general government sector, but National Savings Bank investment accounts, income bonds and deposit bonds have in the past also been taken up by other sectors, in particular non-financial corporations, other financial intermediaries and financial auxiliaries sectors. Figures for public corporations are included from the second quarter of 1982 which reflect changes in the working balance held by the Post Office on account of its national savings business.

Tax instruments

17.91   Tax instruments are also included in this instrument. They are instruments issued to tax payers in exchange for advance payments to meet future tax liabilities. They comprise certificates of tax deposit and, in earlier periods, tax reserve certificates, and tax deposit accounts. Transactions not identified by sector are entered under private non-financial corporations.

UK reserve position in IMF, ECB

17.92   Part of the UK official reserves are included here. The reserve position in the IMF is treated as a deposit with no maturity and equals the difference between the UK quota and the sterling holding by the IMF and ECB. The reserve position will fluctuate depending upon other countries' drawing or repaying sterling to the IMF. The breakdown of the UK official reserves is available as a balance sheet figure and the transactions are calculated as the change in these levels by the Bank of England each quarter. The figures for this come from Government administered sources and are of high quality.

F.3 Securities other than shares

17.93   The main sub-division here is between money market instruments (F.331) and bonds and preference shares (F.332). The final sub-division is for financial derivatives (F.34).

F.331 Short term securities: money market instruments (MMIs)

17.94   In the UK financial account money market instruments are taken to represent short-term securities. They are given in further detail, according to the issuing sector, as follows:

- F.3311 UK central government
- F.3312 UK local authority bills
- F.3315 UK monetary financial institutions (MFIs)
- F.3316 Other UK residents
- F.3319 Rest of the world

F.3311   Short term money market instruments issued by UK general government

17.95   These comprise both domestic (sterling) and foreign currencies (ECUs etc.) Treasury bills issued by the UK general government sector (S.13).

17.96   The Treasury bills covered exclude those held by the National Debt Commissioners, the Exchange Equalisation Account and central government departments.

17.97   Bills held by the Bank of England as the sterling counterpart of foreign currency deposits arising from central bank assistance are also excluded.

17.98   Net sales by central government constitute the increase in Treasury bills outstanding. This information comes from the Bank of England each month.

17.99   For periods up to the end of 1986 the private non-financial corporations sector was the residual for this instrument. For the years 1987 to 1990 inclusive the transactions of this sector were assumed to be zero. From 1991 onwards the changes in holdings recorded in the Financial Assets and Liability survey (25.2.5) are used to estimate transactions by large corporations. Information for public corporations comes from the Bank of England.

17.100   Information is available from the Bank of England for the transactions of banks; building societies (the Building Societies Commission supply aggregate data from their quarterly balance sheet inquiry). Data for insurance corporations and pension funds are derived from quarterly ONS inquiries which collect holdings data (25.3). Total transactions by other financial intermediaries and financial auxiliaries are calculated as the instrument residual after the rest of the world part has been taken out; some data is available however from ONS and Bank of England inquiries. The actual amount of the residual allocated to this sector is the difference between the two.

17.101   The transactions of local government are derived from quarterly balance sheet inquiries on borrowing and lending carried out by the DoE, the Welsh Office and the Scottish Office and from Northern Ireland.

17.102   Holdings by the household and NPISH sectors are known to be small. Their transactions are thus assumed to be zero.

17.103   Rest of the world residents' transactions relate to Central Monetary Institutions (CMIs) and non-CMIs transactions and are provided by the Bank of England, based on information on banks' custody holdings on their behalf. Total rest of the world sector holdings are now the sum of these, plus an amount worth fifty per cent of the residual, to allow for the under reporting of non-CMI transactions.

17.104 For *local authority bills (local government) short term money market instruments* the total figure is taken from returns made quarterly by local authorities in the UK. The figures are sectorised using a combination of local authority and counterparty sector sources. This is described in more detail below.

17.105 Figures for local authority short term money market instruments are taken from returns made by local authorities to the Department of the Environment, Scottish Office, Welsh Office and from central government accounts. Borrowing from public corporations is also that shown by local authorities, supplemented by certain additional transactions identified by these corporations. Similarly data for transactions of such local authority debt by other domestic sectors is derived from a combination of local authority data and information from the appropriate sector (for example, the ONS quarterly inquiries of insurance corporations; returns made to the Bank of England by banks). The other financial intermediaries and financial auxiliaries sectors are the residual ones. Information about rest of the world activity in local authority money market instruments comes from a variety of sources, including the Bank of England. It is derived from changes in levels and by residual, where investment by the rest of the world is derived from total issues (net of redemptions) *less* total net identified purchases.

F.3315 Short term money market instruments issued by UK monetary financial institutions (MFIs)

17.106 This comprises UK corporate sterling and foreign currency commercial paper issued by MFIs, UK bank and building society Certificates of Deposit (CDs) (in sterling and foreign currency) together with promissory notes, bills and other short-term negotiable paper issued by banks in both sterling and foreign currency and acceptances of commercial bills by banks.

17.107 For non-financial corporations, reported data (for certificates of deposit only) is compiled by the Bank of England, derived from results of the ONS quarterly Financial Assets and Liabilities Survey (25.2.5). In addition, part of the residual (ten per cent for sterling; two per cent for foreign currency certificates of deposit and commercial paper) is attributed to private non-financial corporations.

17.108 Data for issues of banks' money market instruments are available from the first quarter of 1984. They, plus banks' transactions, are derived by the Bank of England from monthly balance sheet returns provided by institutions classified to the sector (25.3.1). Data for building societies' transactions in banks' short-term money market instruments are derived from a monthly balance sheet inquiry carried out by Building Societies Commission (25.3.2). For non-monetary financial institutions, hard data for the transactions (acquisitions and realisations) in non-equity securities each quarter are available from the ONS insurance corporations, pension funds and securities dealers inquiries (25.3.3). For CDs only end quarter levels are collected so the transactions are derived from them. Information is also collected on the unit trust and investment trust forms.

17.109 Local government transactions in banks' short-term money market instruments are derived from the DoE, Welsh Office and Scottish Office and quarterly Borrowing and Investment inquiry to local authorities and from Northern Ireland. This inquiry collects balance sheet data from which the transactions are derived.

17.110 Household and NPISH sectors' holdings of banks' money market instruments are known to be very small. These sectors are allocated a total of 2 and 8 per cent respectively of the residual level of holdings for sterling certificates of deposit and commercial paper and one per cent of the residual holdings of foreign currency versions of these two types of short-term money market instruments. Transactions are derived from the holdings data.

17.111 Reported holdings of banks' sterling instruments by the rest of the world sector come from a monthly Bank of England balance sheet form. In addition, forty per cent of the residual assets of certificates of deposit (CDs) are allocated to this sector, along with twenty per cent of the residual for medium term notes and other short term paper (MTNs/OSTP). For foreign currency instruments, the rest of the world sector is allocated holdings of ninety two per cent of the residual for CDs and seventy per cent for MTNs/OSTP. Transactions figures are derived from the balance sheet data for the rest of the world.

17.112 Where the residual difference between issues by banks and reported transactions by other sectors has to be allocated across counterparty sectors, an agreement between the ONS and the Bank of England determines the sectoral allocations. Attributed data are allocated as various percentages of the residual (for sterling) or total (for foreign currency) holdings of the different types of short-term money market instruments.

17.113 Banks' liabilities also include commercial bills which have been accepted by banks. The Bank of England collect data on them. Bills rediscounted with the Bank of England's Issue Department are shown as an asset of the central bank. Information on other sectors' assets are obtained from inquiry sources. The lending associated with these acceptances is included under bank lending, F.41.

17.114 Building society issues and holdings of building society short-term money market instruments are reported by the societies on a monthly form to the Building Societies Commission (25.3.2).

17.115 For private non-financial corporations, transactions in building societies' certificates of deposit are derived by the Bank of England from the results of ONS's Financial Assets and Liabilities survey (25.2.5). In addition a portion of the residuals for certificates of deposit and commercial paper are allocated to this sector.

17.116 Banks' transactions in building society short-term money market instruments are derived from a monthly balance sheet inquiry carried out by the Bank of England (25.3.1). Data for non-monetary financial institutions are derived from the ONS quarterly inquiries to insurance corporations, pension funds and securities dealers, which collect end quarter holdings (25.3.3). A portion of the residual transactions are also allocated to this sub-sector, reflecting the data for types of institutions within this group that are not surveyed.

17.117 Central government, local government and public corporations are assumed not to hold short term money market instruments issued by building societies, and thus have no transactions in them.

17.118 Households and NPISH are allocated a small portion of the residual for certificates of deposit and commercial paper.

17.119 Data on the rest of the world's holdings of building societies' short term money market instruments are collected by the Building Societies Commission (25.3.2) who then pass them on to the Bank of England. The transactions are derived from them. An allocation of the residual is made to this sector.

17.120 An agreement between the ONS and the Bank of England determines the sectoral allocation of the residual difference between issues by banks and building societies and reported assets by the other sectors. The residual allocations are done separately for certificates of deposit; commercial paper; medium term notes and other short-term paper.

17.121 The quality of the liability data for UK MFIs issues of short term money market instruments is generally good, as it is based on comprehensive surveys of MFIs. Data for the transactions in these instruments as assets by MFIs are of good quality for the same reasons. The figures for other sectors are less reliable since they include a portion of the residual.

F.3316 Short term money market instruments issued by other UK residents

17.122 This comprises the following: UK corporate commercial paper issued by non-MFIs (both domestic - sterling - and foreign currency) and UK local authority bills.

17.123 Data on issues of sterling commercial paper by non-monetary financial corporations are provided by the Bank of England based on returns to the Bank by commercial paper issuers. Asset holdings are based on inquiry sources with the residual being divided among private non-financial corporations, non-MFIs, households, NPISH and the Rest of the World. The data for transactions are derived from these balance sheet figures.

17.124 Euroclear data from the Bank of International Settlement (BIS) are the source for issues of foreign currency commercial paper on the Euro-markets. The liabilities are divided between private non-financial corporations and non-monetary financial institutions. Issues of foreign currency commercial paper elsewhere are obtained from returns made by the main issuers, supplemented by estimates; all are private non-financial corporations. Again, the figures of transactions in the financial account are derived from these balance sheet data.

17.125 Bank assets in the form of money market instruments (bills, commercial paper, other short-term paper) issued by UK corporations are reported on one or more of the Bank of England forms. Building Societies' holdings of money market instruments of UK corporations are reported on returns made by the societies to the Building Societies Commission. The aggregate data are then passed to the Bank of England. Insurance corporations and pension funds provide data to the ONS each quarter on their holdings of these assets. The transactions data are derived from these holdings data each quarter.

17.126 Information about rest of the world purchases and sales of UK corporations' money market instruments comes from a variety of sources, including the Bank of England. It is derived from changes in levels and by residual, where investment by the rest of the world is derived from total issues (net of redemptions) less total net identified purchases.

17.127 Data quality for issues of sterling commercial paper is good. MFI asset figures are of good quality because they are based upon comprehensive surveys of these institutions. The figures for other sectors are less reliable since they are derived from holdings data where a portion of the residual has been allocated to them. The foreign currency data are generally not as good (except for MFIs).

F.3319 Short term money market instruments issued by the rest of the world

17.128 This instrument includes sterling commercial paper issued by rest of the world residents. The rest of the world liability figures are obtained as the residual, being the total of the figures for the other sectors which hold these financial instruments as assets.

17.129 The data are derived from a variety of sources such as the ONS Securities Dealers' inquiries (25.3.3), the ONS Financial Assets and Liabilities Survey (25.2.5) and the Bank of England. In addition, data are being collected by the ONS from insurance corporations and pension funds from 1998 (starting with data for 1997) (25.3.3).

F.332 Long term securities other than shares: bonds

17.130 Long-term bonds are securities by which one party (the issuer) is bound to pay money to another, over the course of more than one year. Before the bond reaches maturity (at which point it must be repaid) the holder will usually receive interest payments, which are commonly made twice yearly.

F.3321 Long term bonds issued by UK central government

17.131 Long term bonds issued by UK central government are commonly referred to as British Government Securities (BGS) or gilts. They are the main form of longer term government borrowing and may be of the conventional bond type or index-linked. They are marketable securities denominated in sterling or foreign currency. British Government foreign currency notes and bonds are included, together with government issues abroad.

17.132 UK central government bonds may be conventional or index-linked. Conventional bonds pay interest to the holder in the form of a fixed coupon, usually every six months until maturity. Index-linked bonds pay out interest linked to the annual rate of increase in the retail prices index (RPI) - for example the interest per £100 gilt may be the rate of increase in the RPI plus 2.5%. Most gilts have a fixed redemption date, but some are undated; that is, they have no fixed maturity date.

17.133 The figures for central government sales comprise total cash issues less redemptions (including the purchases by government sinking funds) and less net purchases by the National Debt Commissioners. Data are from official sources and therefore of high quality.

17.134 Figures for private non-financial corporations' transactions in gilts are grossed-up sample estimates based on changes in holdings (at book value) reported by large corporations in the ONS Financial Assets and Liabilities Survey (25.2.5), and previously the Survey of Company Liquidity. Data for public corporations' transactions in gilts related only to Girobank before privatisation, and were provided by the Bank of England.

17.135 The figures for banks are derived from balance sheet returns made to the Bank of England (25.3.1). Data for building societies come from monthly returns, showing both transactions and holdings, made to the Building Societies' Commission (25.3.2). Other financial corporations make quarterly returns of transactions to the ONS, supplemented by annual balance sheets (25.3.3).

17.136 Local government bodies report purchases and sales of gilts to the Department of the Environment (DoE) each quarter.

17.137 Data for the household and NPISH sector comprise the residual difference between net sales of central government and identified transactions of other sectors. A survey of charities in 1996 indicated that they had balance sheet holdings of gilts of about £3.4 billion: however, there is no information about the size of transactions in these holdings and so any split of the residual between households and NPISH would be an assumed one.

17.138 The figures for the transactions of the rest of the world sector are based on custody holdings of sterling bonds held by non residents reported by banks, information from the Stock Register, occasional surveys and evidence from balancing the accounts. The ONS is able to estimate the total value of foreign currency bonds issued by the UK government but is unable to measure exactly how much is purchased by non-residents. To balance the accounts it is assumed that any investment in foreign currency bonds that cannot be directly attributed to UK domestic sectors must therefore be investment by the rest of the world. Adjustments are made when necessary to convert the figures of external liabilities from a nominal to a cash transactions basis.

17.139 Issues and redemptions of British government foreign currency bonds include an issue in 1991 of ECU 2,500 million HM Government 10 year bonds, issues in 1992 of DM 5,000 million and US $3,000 million HM Government 5 and 10 year bonds respectively; and issues commencing 1992 of 3 year HM Government ECU Treasury notes. Data for these issues are from official sources and thus of high quality. The holders of these bonds are UK banks, insurance corporations and pension funds, other financial institutions and the rest of the world sector. Counterparty data are obtained from Bank of England and the ONS inquiries; the rest of the world is the residual sector for this particular sub-instrument within long term bonds issued by UK central government.

F.3322 Long term bonds issued by UK local authorities

17.140 Long-term bonds issued by local government are known simply as 'Local Authority bonds'. Figures for issues of such bonds come from returns made by local government to the DoE.

17.141 Data for central government transactions in local government bonds come from the Treasury. Public corporations' transactions in local government bonds come from the Bank of England. Data from DoE, Welsh Office and Scottish Office returns from local government provide the estimates for private non-financial corporations' holdings, from which transactions are derived. Households and NPISH are jointly the residual sector for these bonds.

F.3325, F.3326 Long-term bonds issued by UK MFIs and other UK residents

17.142 These comprise bonds issued by private non-financial corporations and UK financial corporations and others as well as UK corporate quoted and unquoted preference shares.

17.143 They include Eurobonds, medium term notes, debentures and loan stock and preference shares issued by UK private sector corporations. They also include issues by non-profit making bodies such as housing associations and universities.

17.144 Estimates are compiled from a variety of sources including the London Stock Exchange, Bank of England and ONS statistical surveys.

Liabilities

17.145 Estimates of issues are based on the prices at which the bonds etc. are offered on the market. For most sectors, issues of Eurobonds, medium term notes and other debt securities listed on the London Stock Exchange are obtained directly from the Stock Exchange. Estimates for other market issues, including issues of preference shares listed on the Stock Exchange, are compiled by the Bank of England. All figures are net of redemptions. However, figures for building societies are obtained from statistical inquiries carried out by the Building Societies Commission. Building Societies issue permanent interest bearing shares (PIBS), Eurobonds, medium-term notes, domestic debentures and loan stock.

17.146 The figures for non-market issues are based on data from the University of Nottingham's Centre for Management Buy-Out Research and returns by financial corporations. Investment by foreign parent companies in the loan capital of UK companies in which they have a direct investment stake is also included, based on information from the ONS and Bank of England inquiries. The reliability of the bonds and preference share data as a whole is affected by the rough estimates for some non-market issues.

Assets

17.147 Transactions in long term bonds by private non-financial companies include their estimated expenditure on preference shares and convertible loan stock when acquiring independent corporations. They also include their portfolio transactions in long term bonds (estimated from the ONS Financial Assets and Liabilities Survey (25.2.5)). Figures for transactions by public corporations are taken from their annual accounts.

17.148 Data for banks, building societies and other financial corporations are net transactions taken from inquiries run by the Bank of England, the Building Societies Commission and the ONS (25.3). Those for other financial corporations are, from 1968, adjusted for estimated costs incurred in transactions and are based on quarterly returns of cash transactions.

17.149 Government sector transactions are taken from inquiries to local authorities and the accounts of central government. They include central government sales of debentures in privatised corporations.

17.150 In the absence of any known sources of information, transactions by the household sector in bonds and preference shares are estimated as a constant £10 million a quarter with NPISH being allocated a constant £40 million per quarter.

17.151 Rest of the world direct investment transactions in preference shares and loan capital are estimated from the ONS statistics of cross-border acquisitions and mergers and foreign direct investment inquiries. They include injections of loan capital (including preference shares) into UK subsidiaries and the value of preference shares acquired during take-overs of independent corporations. Estimates for the rest of the world sector's portfolio transactions in long term bonds and preference shares are derived as a residual and are obtained as the difference between total UK capital issues and the aggregate of the transactions of all other sectors. Any errors or omissions in the transactions for other sectors (which could be large) are reflected in the residual figure.

F.3329 Long term bonds issued by rest of the world

17.152 These comprise bonds issued by foreign governments, municipal authorities and corporations as well as rest of the world corporate and unquoted preference shares insofar as they are traded with United Kingdom residents. Similarly, Eurobonds, medium-term notes, debentures and loan stock and preference shares issued by rest of the world private and public sector corporations are recorded here.

17.153 Figures for 1963 to 1979, for all sectors, were based partly on exchange control returns and partly on a Bank of England inquiry.

17.154 Transactions by private non-financial corporations include their investment in long-term bonds. Their figures are derived from the ONS Financial Assets and Liabilities Survey (25.2.5), which collects balance sheet information.

17.155 Data for banks, building societies and other financial corporations are quarterly figures of net transactions. They are taken from inquiries run by the Bank of England, the Building Societies Commission and the ONS (25.3). Adjustments are made to the reported figures for other financial institutions to remove the commission charges and other local costs included in the gross acquisition and sales which are not appropriate to the financial account. New questions have been added to the ONS inquiries to insurance corporations, pension funds, unit and investment trusts to collect the additional information now required for ESA 95 and the fifth edition of the Balance of Payments manual (BPM5). These ask whether the bonds held were issued by EC or non-EC residents. It is expected that the new inquiry data will be available from 1998 and will cover 1997 onwards.

17.156 Data for the UK government's holdings in the official reserves are supplied by the Bank of England. The quarterly transactions are derived from them.

17.157 Transactions by the household and NPISH sectors in rest of the world bonds are based on Inland Revenue data but are subject to a wide margin of error.

17.158 The rest of the world figures consist of total UK direct investment in the rest of the world in share capital plus portfolio investment. This liability figure is obtained as the residual, being the total of the figures for the other sectors (which hold these bonds as assets).

F.4 Loans

17.159 These are split into short-term loans (F.41), normally with an initial term to maturity of one year or less, and long-term loans (F.42).

F.41 Short term loans

17.160 These are normally loans with an original maturity of one year of less. Information on these loans is provided in the UK financial account at a detailed level. This is: sterling loans by UK MFIs (F.4111); foreign currency loans by UK MFIs (F.4112); loans by rest of the world MFIs (F.419). All loans by other sectors are assumed to be long-term.

F.411 Short term loans by UK MFIs

17.161 These comprise UK banks' and building societies' portion of short term loans by UK MFIs in sterling (F.4111) and foreign currency (F.4112). For ease of presentation, the discussion first covers UK banks' sterling and foreign currency loans and secondly UK building societies' sterling and foreign currency loans.

UK bank short term loans

17.162 The figures comprise lending by banks through advances; loans; the counterpart of banks' liabilities in commercial bills and some other paper; and also claims under sale and repurchase agreements of British Government Securities and other paper. They exclude investment in securities. Such loans are made in both sterling (F.4111) and foreign currencies (F.4112). Loans in categories recognised as long-term are excluded here and shown under various sub-headings within long-term loans (F.42). However, the lending included within this item will include some unspecified loans of more than one year original term.

17.163 The figures are compiled by the Bank of England from returns provided by banks. The counterpart sector analysis of loans and advances is obtained from the returns, subject to adjustment to allow for items in the course of clearance. *(See* box on transit items in Chapter 17, paragraph 17.81.)

Building society short term loans

17.164 This covers the building societies' part of loans by UK MFIs in sterling (F.4111) and foreign currency (F.4112). This sub-instrument comprises advances, net of repayments, of loans not secured on dwellings. The greater part of such lending is to households, but loans to financial and non-financial corporations, including the societies' own subsidiaries, also feature. The figures, including all sector transactions, are drawn from returns submitted by the societies to the Building Societies Commission. The quality of these data is good.

F.419 Loans by rest of the world monetary financial institutions (MFIs)

17.165 These comprise loans in both sterling and foreign currency by MFIs in the rest of the world.

17.166 For UK non-financial corporations data are received (quarterly in arrears) from the Bank for International Settlements [and the European Investment Bank]. The estimates are also influenced by data from the ONS Financial Assets and Liabilities Survey (FALS) (25.2.5). The FALS inquiry is run both quarterly and annually and collects information on the assets and liabilities of private non-financial corporations.

17.167 The Bank of England supply information for borrowing by UK MFIs (25.3.1). The other source of data for such borrowing by the financial sector is the quarterly ONS inquiry to security dealers which records their borrowing from rest of the world banks (25.3.3).

17.168 Estimates of UK public sector borrowing from commercial banks in the BIS reporting area are derived from official records and deducted from the BIS data to give estimates for total UK private sector borrowing, which is assumed to relate entirely to borrowing by non-financial corporations.

F.42 Long term loans

17.169 The loans recorded here comprise both outward and inward direct investment loans (F.4211 and F.4212), loans secured on dwellings (F.422), finance leasing (F.423),other loans (F.424) and Other loans by the rest of the world (F.429).

F.421 Direct investment loans

17.170 This is split into outward direct investment loans (F.4211), which are a financial asset of the UK, and inward direct investment loans (F.4212) which are a liability of the UK.

17.171 A direct investment in a corporation means the investor has a significant and lasting influence on the operations of the corporation (normally represented by a holding of 10% or more voting shares in the corporation). This is reflected in the direct investment inquiry forms sent out by the ONS to corporations with effect from the end of 1996 (annual inquiries) and the first quarter of 1997 (quarterly inquiries). The pre-ESA 95 threshold for defining direct investment was 20 per cent. From the investigations made prior to the change of definition it appears that there will be little impact on the direct investment loan figures reported to, and published by, the ONS as a result of it.

F.4211 Outward direct investment loans

17.172 Such loans are an asset of the UK. Loans provided by UK corporations to their branches, subsidiaries or associated corporations in the rest of the world are classified as outward direct investment loans. Investment may be for the purpose of acquiring fixed assets or inventories, but that is regarded as a financial transaction in the same way as additions to, or payments of, working capital, other loans and trade credit. Transactions in real estate are not included here but in F.51, Shares and other equity.

17.173 The outward direct investment inquiries run by the ONS (25.5.1) are sent to all corporations in the oil, insurance and other financial intermediaries sectors. The remaining private corporations are sampled by value of assets: coverage is 99 per cent by value of assets, which represents 85 per cent of the population (of corporations) by number.

17.174 Estimates for outward direct investment loans by MFIs are derived from quarterly and annual inquiries carried out by the Bank of England.

17.175 Public corporations, households and NPISH are not surveyed and are deemed to have no loans between UK parents and overseas subsidiaries.

F.4212 Inward direct investment loans

17.176 These figures comprise inward investment in the UK private sector (other than transactions in UK corporation securities which are entered under F.3 or F.5). They are a liability of the UK. However, identified inward transactions in real estate are included in F.51 'Shares and other equity'.

17.177 Inward direct investment loans data include figures for direct investment which are obtained from the ONS inward direct investment inquiries (25.5.1). These inquiries to private non-MFI corporations have a complete coverage of the oil, insurance and OFI sectors. The remaining corporations are sampled by value of assets: coverage is 92 per cent by assets (which represents 85 per cent of the population by number). Public corporations are not recipients of inward direct investment.

17.178 The figures for banks are obtained from banking sector returns made to the Bank of England (25.3.1). Building societies are not covered since, being owned by their members, they cannot be foreign owned.

17.179 Households and NPISH do not have loans between UK subsidiaries and foreign parents.

F.422 Loans secured on dwellings

17.180 Loans secured on dwellings comprise loans, net of repayments, made by UK banks (F.4221), building societies (F.4222) and other lenders (F.4229), mainly for house purchase and home improvement. It is assumed that these loans are granted exclusively to the household sector to non-profit institutions serving households, such as housing associations.

17.181 Loans secured on dwellings are obtained by sector as follows.

- The figures for banks and building societies are obtained separately from returns made to the Bank of England and the Building Societies Commission (25.3.1, 25.3.2).
- For insurance corporations and pension funds the figures are compiled by the ONS from quarterly returns made by samples of these institutions (25.3.3).
- Data for other financial intermediaries' loans secured on dwellings relate mainly to lending by centralised mortgage lenders and the acquisition of mortgage portfolios by loan securitisation subsidiaries of banks and other lenders. Loans made by the Trustee Savings Banks (to end-1981) are also included here (25.3.3).
- Central government loans are those made by central government under the Housing Act, 1961 and the Housing (Scotland) Act, 1962 to private housing associations; loans made by the Housing Corporation to housing societies and loans by new town corporations. The data are obtained from official records;
- Mortgage advances for house purchase and improvements, including loans for the purchase of council houses by tenants and loans to housing associations form local authority loans secured on dwellings. Quarterly figures are obtained from returns made to the Department of the Environment (England) and the Welsh Office (Wales). Figures for Scotland are interpolated from financial year totals. (There are no figures for Northern Ireland.)

F.423 Finance leasing loans

17.182 The figures show the amounts lent under finance leases by banks and other financial intermediaries, primarily specialist finance leasing subsidiaries of banks. They are calculated as the value of assets leased out under new leases, net of repayments made through the estimated capital element of rental payments by lessees. The total lending of the lessor sectors is obtained from returns made by banks to the Bank of England, from surveys of other financial intermediaries carried out by the ONS and by reference to information collected by the Finance and Leasing Association from its members.

17.183 The greater part of the borrowing under this instrument is attributable to non-financial corporations, after allowing for leasing to financial corporations and general government.

17.184 The coverage of the figures is limited to leases which are financial in substance. Other forms of lease are operating leases (including hire and rental). These are not regarded as generating transactions in financial instruments.

F.424 Other loans by UK residents

17.185 This comprises other loans by: non-financial corporations, including credit extended by retailers; non-MFIs, including insurance corporations and pension funds; central government; public corporations; local authorities; NPISH. Such loans made by these sectors, and the data sources for them, are considered separately in the following paragraphs.

Other loans by non-financial corporations (excluding credit extended by retailers)

17.186 These are picked up as counterparts of loans taken out by what are mainly non-monetary financial corporations.

Credit extended by retailers

17.187 The figures relate to hire purchase agreements and to certain other forms of credit including sales on budget accounts, credit sale agreements and personal loans repayable by instalments. Credit advanced by retailers in the form of trading checks exchangeable only in their own shops is included, but credit advanced by check traders as such is not covered. The figures also exclude monthly accounts and sales on bank credit cards such as Access and Visa (included in bank lending: F.4111) as well as (probably large) amounts of unpaid bills (included in trade credit, F.7, insofar as they are identified) and credit extended by other types of retailer.

17.188 Estimates are derived from the figures of new credit extended and repayments, adjusted to exclude charges, taken from the ONS Retail Sales Inquiry (25.1.10). It is assumed that all credit advanced by retailers it to the household sector and thus the figures represent the debt of individual consumers.

17.189 The figures for the public sector represent credit given by public corporations through gas and electricity showrooms (no longer applicable since privatisation).

Other loans by non-monetary financial intermediaries

17.190 This item includes instalment credit, other loans and advances by finance houses and other consumer credit companies, other than those included in the banks' sector, (excluding loans secured on dwellings) and loans made to their parent organisation by private sector pension funds. Also included are loans made by insurance corporations and by special finance agencies. Loans made by public corporation pension funds to their parent companies are included. Similarly, loans made by local government pension funds to their parent authorities are included. Counterpart figures for groups of institutions are allocated to households, NPISH, local government or non-financial corporations as appropriate.

17.191 Data come from ONS surveys to such financial institutions (25.3.3), some of which collect a split by counterparty sector (for example, the ONS quarterly survey of the financial transactions of pension funds collects such loans split into: loans to parent organisation; loans to financial institutions; loans to financial corporations; loans to other than financial organisations; loans to other UK residents; and loans to rest of the world).

Other loans by public sector

17.192 These figures relate to net lending by central government to the following sectors:

(a) building societies - under the House Purchase and Housing Act, 1959;
(b) non-financial corporations - including, from September 1972, shipbuilding credits re-financed from voted funds;
(c) personal sector - miscellaneous loans for education and other services;
(d) rest of the world sector - including net lending to rest of the world governments and to international lending bodies.

17.193 Also included is net lending by public corporations to the private sector, other than those secured on dwellings (F.422). This lending is assumed to be to the non-financial corporations sector.

17.194 The figures are taken from the detailed accounts of central government and public corporations and are of a good quality.

17.195 This also includes lending by local authorities other than that secured on dwellings.

F.429 Other long term loans by the rest of the world

17.196 This covers a variety of loans made by the rest of the world to residents of the United Kingdom.

17.197 For the central government sector it covers sterling liabilities to the rest of the world not included elsewhere. It includes sterling borrowing from governments or from central monetary authorities in the form of assistance with a sterling counterpart invested in Treasury bills (aid bills) and amounts swapped forward into later months against sterling with rest of the world monetary authorities in 1971 and 1972. Transfers from the Government's dollar portfolio into the reserves, gold subscriptions to the IMF, changes in IMF holdings of non-interest bearing notes other than those arising from drawings or subscriptions are also included. The Exchange Equalisation Account's loss on forward commitments entered into before devaluation in 1967 and part of official short-term capital transactions in the balance of payments, together with contributions to the European Coal and Steel Community reserves, are also recorded for the central government sector here.

17.198 It also includes loans to local government (both under the exchange cover scheme and non-guaranteed borrowing, excluding bonds), loans to UK corporations and loans by rest of the world banks (including the European Investment Bank) and other institutions. It does not include trade credit: this is included under instrument F.71: Other amounts receivable/payable.

17.199 Data come from the Bank of England for such loans made to: the private sector by banks in the rest of the world; public corporations; some monetary financial institutions and local government. Information on borrowing by securities dealers and insurance corporations comes from ONS surveys of them. Also included are borrowing by gold dealers and factoring companies. Estimates of loans received by public corporations and local government are of good quality whilst the figures for the private sector, MFIs and security dealers are less reliable.

F.5 Shares and other equity

F.51 Shares and other equity, excluding mutual funds' shares

F.514 Quoted UK shares

17.200 Quoted UK shares comprise shares listed on the London Stock Exchange, in the full market, the Unlisted Securities Market (until end-1996 when the USM closed) and the Alternative Investment Market (from the middle of 1995, when it opened).

17.201 Estimates are compiled for a variety of sources, including the London Stock Exchange, Bank of England and ONS statistical enquiries.

<u>Liabilities</u>

17.202 Estimates for new market issues, including issues listed on the Stock Exchange are compiled by the Bank of England (25.2.10). All figures are net of share buy-backs.

17.203 Issues by quoted UK corporations relate to new money raised through ordinary shares (public issues, offers for sale, issues by tender, placing and issues to shareholders and employees). Issues to shareholders are included only if the sole or principal share register is maintained in the UK. Estimates of issues are based on the prices at which shares are offered on the market. Subscriptions are included in the periods in which they are due to be paid.

17.204 Non-market issues include employee share schemes and the value of shares issued as part of the consideration during take-overs.

<u>Assets</u>

17.205 The share register survey (25.2.9) is an occasional sample survey of the share registers of listed UK corporations. It is an important source for assets. It was run annually between 1989 and 1994 and provides end-year estimates of the sectoral breakdown of holdings of quoted ordinary shares. It is used as a benchmark for estimates of insurance corporation and pension funds, industrial and commercial companies, households, NPISH and for the rest of the world. It is also used to derive a grossing factor for overseas portfolio transactions in ordinary shares. The figures used here in the financial account take account of the reliability of the survey estimates, as measured by sampling errors. The latest share register survey took place at the end of 1997.

17.206 Financial transactions recorded here include the take-up of new issues which are not separable from secondary transactions.

17.207 Transactions by private non-financial corporations include their estimated expenditure on acquiring independent corporations and subsidiaries from other sectors. It also includes their transactions in current assets (estimated from the ONS Financial Assets and Liabilities Survey (25.2.5)) and other identified transactions in the press. Figures for public corporations are taken from surveys of them and from their accounts.

17.208 Data for secondary transactions by banks, building societies and other financial corporations are net transactions (that is, acquisitions net of realisations (2.5.3)). Those for other financial corporations are, from 1968, adjusted for estimated costs incurred in share transactions and are based on quarterly returns of cash transactions.

17.209 Transactions by the public sector are taken from enquiries to local government and the accounts of central government. They include privatisations when these take the form of quoted shares.

17.210 Transactions in shares by the household and NPISH sectors are derived as a residual and are potentially subject to a wide margin of error. A survey of Charities in 1996 indicated that their holdings of UK and overseas equities amounted to £14.7 billion at the end of 1994.

17.211 Figures for the rest of the world sector transactions come from various sources, including the benchmark share register survey, short-term surveys on investment by non-residents using UK banks and securities dealers, and direct investment inquiries. The assets shown here of the rest of the world include profits generated by UK subsidiaries which are not remitted to the rest of the world parent corporation. They are treated as re-invested earnings.

F.515 Unquoted UK shares (including re-invested earnings on inward direct investment)

17.212 Unquoted UK shares are tradable or potentially tradable financial instruments issued by UK corporations. Figures are compiled from a variety of sources including the Bank of England and the ONS statistical inquiries (25.3).

Liabilities

17.213 Issues by UK companies are based on information from returns by financial corporations, the University of Nottingham's Centre for Management Buy Out Research and the Bank of England's monitoring of press and other sources. Investment in the share capital of UK subsidiaries by rest of the world parent corporations is also included here, based on information from the ONS direct investment inquiries (25.5.1). (However, investment by rest of the world parent corporations in loan capital is included in long term bonds F.3324). Holdings by the Rest of the World of UK real estate, deemed to be a financial investment, is included here.

17.214 The reliability of these data is influenced by the rough estimates for some unquoted issues.

Assets

17.215 Transactions in UK assets comprise all those which can be identified. They include the take-up of new issues, which are not separable from secondary transactions.

17.216 Figures for private non-financial corporations include their estimated expenditure on acquiring independent companies and subsidiaries from other sectors (including the public corporations sub-sector). It also includes their transactions in current assets (estimated from the ONS Financial Assets and Liabilities Survey (25.2.5)) and sales of subsidiaries to other sectors, including rest of the world acquisitions by rest of the world corporations (from the ONS statistics of domestic and cross border acquisitions and mergers). Data for transactions by public corporations are taken from their accounts.

17.217 Transactions by banks, building societies and other financial intermediaries are acquisitions net of disposals. Those for other financial intermediaries are, from 1968, adjusted for estimated costs incurred in share transactions and are based on quarterly returns of cash transactions.

17.218 Figures for the public sector are taken from enquiries to local government and the accounts of central government. They include privatisations when these take the form of selling unquoted shares, for instance through trade sales.

17.219 The estimates of transactions in unquoted shares by the household and NPISH sectors are derived as a residual and are subject to a wide margin of error.

17.220 Transactions by the rest of the world sector in unquoted shares come from various sources, including surveys run by the Bank of England and the ONS direct investment inquiries. They include inward direct investment in the share capital of unquoted subsidiaries (including re-invested earnings), acquisition of unquoted independent UK companies and subsidiaries from UK corporations and portfolio investment in venture capital funds and the unquoted shares of other financial institutions.

F.519 Rest of the world shares and other equity

17.221 This sub-category comprises shares and tradable or potentially tradable financial instruments issued by foreign corporations and quasi-corporations. It also includes subscriptions to international organisations which the ESA describes as government investments in the capital of international organisations which are legally constituted as corporations with share capital. An example is the European Investment Bank.

17.222 Transactions in non-UK real estate are also included here. The owners of land or buildings in a country are deemed to be notional residential units of the country in which the investment is located. The actual, foreign, owners of the real estate are regarded as having made a financial investment, akin to other equity, in the notional unit.

17.223 Figures for 1963 to 1979 relating to shares and tradable financial instruments were based partly on exchange control returns and partly on a Bank of England inquiry. Transactions by the central government comprise the sale of Suez Canal shares in 1976 and 1979.

17.224 Currently figures for transactions in shares and financial instruments for monetary financial institutions are obtained from a Bank of England inquiry (which is supplemented by an annual balance sheet inquiry). Estimates for other financial intermediaries, financial auxiliaries and insurance corporations and pension funds are based on quarterly ONS surveys of transactions. For data from 1997, ONS surveys to securities dealers, insurance corporations, pension funds, unit and investment trusts were extended to collect additional detail necessary under ESA1995 (whether the securities were issued by institutions in EU or non-EU countries) and that required under the new edition of the Balance of Payments Manual (BPM5). (A deduction from the reported figures which represents an estimate of the costs incurred in share transactions is made for all sectors.) The figures include estimates of direct investment in the form of shares.

17.225 Figures for private non-financial corporations are for direct investment in the form of shares as well as for other investment: the latter are derived from an ONS inquiry collecting balance sheet information.

17.226 The estimates for transactions by the household sector are based on Inland Revenue data but are subject to a wide margin of error. A survey of charities in 1996 indicated that their holdings of UK and overseas equities amounted to £14.7 billion at the end of 1994.

17.227 The rest of the world sector figures consist of total UK direct investment in the rest of the world in share capital plus portfolio investment in rest of the world shares and similar financial instruments.

17.228 Figures for foreign households' transactions in real estate in the UK are estimated by the Inland Revenue from stamp duty information. The transactions in real estate are believed to be carried out mainly by individuals resident abroad.

17.229 Estimates for UK investment in real estate in rest of the world are obtained from quarterly and annual inquiries carried out by the ONS. Most of the property investment is carried out by pension funds and insurance corporations. There is little information on UK households' purchases of real estate abroad.

F.52 Mutual funds' shares

17.230 This is sub-divided into UK mutual funds' shares (F.521) and rest of the world mutual funds' shares (F.529).

F.521 UK mutual funds' shares

17.231 These comprise unit trusts, investment trusts and property unit trusts and Open Ended Investment Companies (OEICs) and each is discussed separately in the following paragraphs.

Unit trust units

17.232 These statistics cover unit trusts authorised by the Securities and Investments Board under the terms of the Financial Services Act 1986. They do not cover other unitised collective investment schemes (e.g. unauthorised funds run on unit trust lines by, for example, securities firms and merchant banks, designed primarily for the use of institutional investors) or those based offshore (Channel Islands, Bermuda etc.) or in other EU Member States.

17.233 The majority of unit trusts are required to invest at least 90% of their assets in transferable securities traded on a recognised stock exchange. However, money market trusts are permitted to invest wholly in cash or near cash instruments (such as building society and bank deposits, Treasury bills, short dated British government securities etc.).

17.234 Unit trusts are set up under trust deeds, the trustee usually being a bank or insurance company. The trusts are administered not by the trustees but by independent management companies. Units representing a share in the trust's assets can be purchased by the public from the managers or resold to them at any time.

17.235 Figures for sales less purchases of authorised unit trust units are provided by the Association of Unit Trust and Investment Funds. Unitisations of investment trusts are excluded. Quarterly returns to ONS from insurance corporations, pension funds, unit and investment trusts (25.3.3) provide data for this sector's transactions: all other transactions are assumed to be with the personal sector. Sales are measured at market prices less an estimate of the initial charges that are included in the offer price of the units and cover the fund managers' expenses.

17.236 The ONS inquiries to insurance corporations and pension funds collect some data for these assets: for pension funds this includes quarterly transactions in unauthorised Unit Trusts; that collected from insurance corporations is just annual balance sheet data. Insurance corporations are also required to submit quarterly transactions data to ONS from 1997.

Investment trusts

17.237 Investment trust companies acquire financial assets with money subscribed by shareholders or borrowed in the form of loan capital. They are not trusts in the legal sense, but are limited companies with two special characteristics: their assets consist of securities (mainly ordinary shares), and they are debarred by their articles of association from distributing capital gains as dividends. Shares of investment trusts are traded on the Stock Exchange and increasingly can be bought direct from the company.

17.238 The figures cover investment trusts recognised as such by the Inland Revenue for tax purposes and some unrecognised trusts. Returns are received from a sample of companies accounting for about 60 per cent of the total assets of these trusts. Since 1992 quarterly and annual returns have been collected by the ONS (25.3.3). (Previously they were collected by the Bank of England.) As from end-1979 the data received are grossed up to include estimates for non-respondents.

Property unit trusts

17.239 The data are for all UK property unit trusts authorised under the terms of the Financial Services Act, 1986. The trusts are not allowed to promote themselves to the general public and participation is generally restricted to pension funds and charities. Property unit trusts invest predominantly in freehold or leasehold commercial property, yet may hold a small proportion of their investments in the securities of property companies. Their assets are held in the name of a trustee and are managed on a co-operative basis by a separate committee (elected by the unit holders) or company. Some investment is financed by long-term borrowing.

17.240 Data of the net sales of units by the trusts have been collected since 1966 and data of levels of assets and liabilities from end-1979. Returns are made by a sample of property unit trusts to the ONS (25.3.3) every quarter. The figures include estimates for non-response.

17.241 Data for transactions (acquisitions and realisations) by pension funds are collected each quarter in the ONS 'Transactions in financial assets' inquiry which is sent to a sample of UK pension funds (25.3.3). The transactions of charities, which are allocated to the NPISH sector, are estimated as the residual.

Open-ended investment companies

17.242 Open Ended Investment Companies (OEICs) gained legal status in the UK in January 1997 and the first was launched in May 1997. An OEIC is a cross between a unit trust and an investment trust, being similar to unit trusts in investment terms but governed by company law. It features single rather than dual pricing. A number of existing UK unit trusts are expected to convert to OEICs. During the initial months, of 1997, UK OEICs could only be set up as securities funds. However, later in 1997 the concept was extended to other types of fund, including cash funds, property funds, derivative funds and umbrella funds.

F.529 Rest of the world mutual funds' shares

17.243 No data are available for these at present. ONS are investigating whether information could be obtained from offshore institutions as a counterparty to the households sector. New questions are also being inserted on current ONS inquiries to insurance corporations, pension funds, unit and investment trusts to collect information on rest of the world mutual funds.

F.6 Insurance technical reserves

17.244 These reserves are the technical provisions of insurance corporations and pension funds against policy holders or beneficiaries. This category shows all transactions in them. It has two sub-categories: net equity of households in life insurance reserves and pension fund reserves (F.61) and prepayments of insurance premiums and reserves for outstanding claims (F.62).

F.61 Net equity of households in life insurance reserves and pension funds' reserves

17.245 For life assurance and pension funds this figure represents the net surplus of current income over current expenditure of these funds. In the national accounts the life assurance and pension funds are regarded as the property of the household sector and of households in the Rest of the World, but the investments of the funds (including purchases of property) are included in the transactions of the financial corporations (insurance corporations and pension funds sub-sector). In order to maintain the articulation of the account, the funds' surplus is shown as an increase in the financial claim of the household sector and the Rest of the World on insurance corporations and pension funds.

17.246 Estimates of the net increase in these funds are taken from their revenue account which is collected through ONS surveys each quarter to a sample of these institutions. Quarterly returns of investment by insurance companies and pension funds are similarly made by the sample of institutions to the ONS.

F.62 Prepayments of insurance premiums and reserves for outstanding claims

17.247 This represents the amounts already paid in insurance premiums against cover for periods that extend beyond the current year. This reflects the fact that, for example, most annual household payments of premiums do not fall due on 1 January each year. Thus a premium paid on 1 December will be received in the current year but most of the period during which risks are covered will be in the following calendar year. So the majority of the premium paid on 1 December will be part of the pre-paid premium reserves of insurance corporations at the end of the current year.

17.248 The prepayments of insurance premiums are the financial assets of policy holders. If these prepayments relate to life insurance the policy holders are resident or non-resident householders. If they relate to non-life insurance then the policy holders may belong to any sector of the economy or to the rest of the world. Similarly, insurance companies have other technical reserves including reserves relating to claims from policy holders which they have not yet paid out on. These reflect claims that are being processed and claims for that have not yet been made but which occurred during the current year. For example, they may relate to theft or damage that the policy holder has not yet discovered or events occurring at the very end of the year which have not yet been reported to the insurance company.

17.249 Data for this come from the ONS surveys to insurance corporations and pension funds (25.3.3) and the statutory returns made by insurance corporations to the Department of Trade and Industry: which is the supervisory authority for insurance corporations in the UK. Use is also made of the ONS Family Expenditure Survey (25.4.5) to identify the amounts that households are spending on the different types of insurance.

17.250 Detailed insurance product information from the Association of British Insurers is used to allocate this item to counterpart sectors.

F.7 Other accounts receivable/payable

17.251 This category consists of all transactions in other amounts receivable/payable, that is, financial claims which are created as a counterpart of a financial or a non-financial transaction in cases where there is a timing difference between this transaction and the corresponding payments. One component of this category is trade credits and advances (F.71); the other includes various items (F.79).

F.71 (part) Domestic trade credit

17.252 This relates to identified trade credit only. Omissions are likely to be substantial, thus reducing the quality of these data. The domestic trade credit shown is mainly that given and received by public corporations and certain financial institutions and central government. An estimate of trade credit given to local government is also included from 1989 and transactions in connection with advance payment of fuel bills are included from 1994.

17.253 Data for public corporations are derived from quarterly returns made by them to the ONS (25.2.3). Before 1972 the information was obtained from their annual accounts. The household sector allocation represents a proportion of unpaid bills to utilities prior to privatisation. The residual sector is private non-financial corporations. Transactions related to advance payment for British National Oil Corporation oil are included from the first quarter of 1980. Formalised credit given by retailers is included in F.424.

17.254 Identified trade credit given and received by insurance corporations and pension funds is included here too. Insurance corporations' outstanding premiums, claims admitted but not paid, agents balances, and reinsurance, co-insurance and treaty balances are included. For pension funds and long-term insurance funds the counterpart sector is the household sector, but for other insurance corporations the counterpart sectors are households and private non-financial corporations.

17.255 Local government trade credit figures are estimated (from the balancing item in the local government sector account); the counterparty sector is assumed to be private non-financial corporations.

17.256 The household sector allocation represented a proportion of unpaid bills to utilities prior to privatisation (up to the second quarter of 1991). There are no data for these after privatisation. It also includes, for 1994–1996, the trade credit arising from pre-payment of fuel bills in advance of the introduction of VAT on fuel, based on data from the ONS Financial Assets and Liabilities Survey (25.2.5). For this private non-financial corporations are the debtors and households the creditors.

17.257 No estimates are made for other types of trade credit, in particular that between the private non-financial corporations and household sectors, variations in which are probably sizeable. (This subject is discussed in *Economic Trends*, September 1992.)

F.71 (part) Import and export credit

17.258 These currently comprise suppliers' trade credit and associated advance and progress payments on exports and imports. These include credits financed by UK bank loans under ECGD buyer credit schemes, by the Crown Agents, by insurance companies and long-term credit linked to purchases of ships and commercial aircraft. It includes refinanced export credit. It does not include import credit of public corporations, but does include advance payments by such corporations.

17.259 However, trade credit between UK corporations and rest of the world parents, branches, subsidiaries and affiliates corporations are covered indistinguishably within outward and inward direct investment loans (F.4211 and F.4212).

F.79 Other amounts receivable/payable

17.260 These are financial assets which arise from timing differences between distributive transactions or financial transactions on the secondary market and the corresponding payment. It includes some financial claims stemming from income accruing over time, where such claims cannot be reallocated to the correct financial instrument from which they were derived.

Net acquisition of financial assets FA

17.261 This figures represent the net acquisitions less disposals of all the financial assets in the financial account.

Net incurrence of liabilities FL

17.262 This represents the total net incurrence of liabilities in the financial account.

B.9f Net lending/net borrowing

17.263 This represents the balance for the financial account: the difference between the net acquisition of financial assets and the net incurrence of liabilities.

Chapter 18

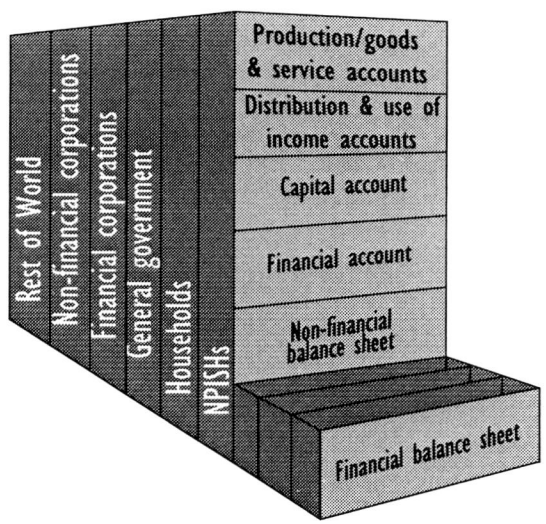

Financial balance sheets

Chapter 18 Financial balance sheets

General description

Introduction

18.1   This chapter describes

- what the financial balance sheets are
- how they fit into the national accounting framework
- uses of the financial balance sheets
- the different types of financial asset/liability

It discusses

- the data sources
- how a coherent data set is obtained
- the reliability and quality of the figures

For each instrument it

- describes the sources of the data
- gives details for particular sectors
- shows how data are calculated where no source exists for a sector

18.2   The financial balance sheets show the value of financial assets owned, and of financial liabilities outstanding, at a particular point in time. The financial balance sheets are drawn up for various sectors of the economy, the total UK economy and the rest of the world. For the rest of the world the financial balance sheet is also called the external assets and liabilities account: the rest of the world have no non-financial assets or liabilities in the UK. (The data sources and methods for non-financial balance sheets are to be covered in Chapter 16.)

18.3   The aim of the balance sheets is to give a picture of the assets, liabilities and net financial worth of institutional units at the start and the end of the accounting period and of changes between balance sheets. The changes in balance sheets are not at present shown explicitly in the UK accounts.

18.4   In general, the balance sheets are at current market value, so the changes they show from one time to another are not necessarily the same as transactions in the same interval. Price and other changes also impact on the outcome. Where there is no market valuation the nearest available approximation is used. The data on direct investment in particular are only at book value which can be substantially different from market value.

18.5   For each sector of the economy the financial balance sheets show the value of all financial assets and liabilities. The balancing item of the financial balance sheet – the difference between the sector's financial assets and liabilities – is referred to as the sector's 'net financial assets'.

18.6   The financial account on its own gives only a partial indication of the behaviour of the various sectors. In many cases the reason for flows of financial assets between sectors is to alter the composition of a balance sheet. For example, a transaction is often undertaken to change the owner's stock of an asset - perhaps because there has been a change in the value of that asset - and should be interpreted in that light. The financial balance sheets in a sense complete the picture given by the transactions accounts since they show the outstanding stocks of financial assets and liabilities for each sector.

18.7   Some of the sources of data on financial balance sheets are not ideal. For certain forms of asset the valuation is also a problem, because there may not be any market and thus no market price. The sources and methods used to compile the balance sheets are now more closely harmonised than in the past. Most of the financial transactions and balance sheets data are compiled using similar methods. Work in the ONS and Bank of England is continuing so that the consistency of the methodologies for compiling transactions and balance sheets can be improved and extended.

18.8   The accounts show for each sector the balance sheets at the end of each period. Also of analytical interest is the change in the balance sheet over the period. This change is explained by the financial account (described in Chapters 7 and 17) and the other changes in assets account (Chapter 8 - not yet implemented in the United Kingdom, although some elements are derivable as residuals).

18.9   The balance sheet records the value of assets and liabilities at the end of the accounting period. These items are categorised on the standard instrument classification used in the financial account and are valued at prices and exchange rates current at the end of the period. The difference between the total of financial assets and that of financial liabilities is the net financial worth at the end of the period.

Uses of the balance sheets

18.10   The balance sheets show the financial worth of each sector of the economy at a particular point in time. The changes from previous balance sheets illustrate both the change in the valuation of different instruments as stock markets move, currency exchange rates change etc. and the changing portfolios resulting from the financial transactions of the sectors. The balance sheets are thus closely linked with the financial accounts data.

Concepts

18.11   This section looks at issues relating to valuation, consolidation and accounting conventions.

Valuation

18.12   Throughout the balance sheets assets are in principle valued at their market value. That is, as if it were being acquired on the date to which the balance sheet relates. Where there are no observable prices – as may be the case if no purchases/sales of the instrument have been observed in the recent past – an attempt has to be made to estimate what the prices would be if the assets were acquired on the market on the date to which the balance sheet relates. However, direct investment is valued at book value.

18.13 For many of the financial assets market prices are usually available. They should be assigned the same value whether they appear as financial assets or liabilities. The values exclude service charges, fees, commissions and similar payments which are recorded as services provided in carrying out the transactions.

18.14 The value of assets and liabilities denominated in foreign currency should be translated into the national currency at the market exchange rate on the day to which the balance sheet relates. This rate should be the mid-point between the buying and the selling spot-rates for currency transactions.

Consolidation and accounting conventions

18.15 The ESA 1995 requires member states to submit their financial balance sheets on both a consolidated and a non-consolidated basis. Consolidation removes all assets with/liabilities to institutional units in the same sector or sub-sector to which the balance sheet relates.

18.16 In the UK the approach is to collect and calculate all balance sheet (and, for that matter, financial account) data on a non-consolidated basis. This enables the consolidated figures to be derived. In practice this means that loans between corporations within the same company group should be consolidated out when aggregate figures are supplied by the company group to the ONS but figures relating to loans, balances etc. with companies outside their group (though perhaps within the same sector) should be shown. This is different from what the UK did previously when there were some inconsistencies. For example, official holdings of gilts were netted off liabilities thus producing in effect a consolidated figure, but company securities figures were unconsolidated; bank deposits were shown gross whilst loans between building societies were netted out or consolidated.

18.17 The UK has developed a new system to handle the ESA 1995 based data. This enables it to meet the ESA requirements so that the financial balance sheets can be produced both unconsolidated and consolidated. A full breakdown of both assets and liabilities for each sector is calculated and produced.

18.18 Discontinuities occur over time in the balance sheets: for example, in recent years privatisations have led to switches between the public and private subsectors of non-financial corporations.

Classification of financial instruments

18.19 Financial assets are economic assets which comprise the following:

- means of payment
- financial claims
- economic assets which are close to financial claims in nature.

18.20 Means of payment consist of monetary gold, special drawing rights (SDRs), currency and transferable deposits.

18.21 Financial claims entitle their owners, the creditors, to receive a payment (or a series of payments) without any counter-performance from other institutional units, the debtors who have incurred the counterpart liabilities.

18.22 Economic assets which are close to financial claims in nature are shares and other equity and partly contingent assets. The institutional unit issuing such a financial asset is considered to have incurred a counterpart liability.

18.23   Contingent assets are those such as guarantees of payment by third parties, letters of credit, lines of credit, underwritten note issuance facilities and many derivative instruments. Thus contingent assets are contractual arrangements between institutional units, and between them and the rest of the world, which specify one or more conditions which must be fulfilled before a financial transaction takes place. A contingent asset is a financial asset when the contractual arrangement itself has a market value (i.e. because it can be traded or can be offset on the market) and then it is recorded as a financial asset. When a contractual arrangement does not have a value it is not recorded in the system.

18.24   The ESA 1995 system breaks down the types of asset into different instruments by their liquidity and legal characteristics. So the seven specified instruments of:

1   Monetary gold and special drawing rights
2   Currency and deposits
3   Securities other than shares
4   Loans
5   Shares and other equity
6   Insurance technical reserves
7   Other account receivable/payable

become further sub-divided. For example categories 3 and 4 (securities other than shares and loans) are divided into short-term or long-term; and currency and deposits (2) are split into currency (21) and transferable deposits (22).

18.25   In the UK the approach adopted is to measure short and long-term loans, for example, indirectly via a much more detailed classification of individual securities. This approach partly reflects the nature of London as one of the world's three main stock markets (with New York and Tokyo) in which many innovative financial instruments have been developed and are now regularly used, and new ones are continually being developed. Where such details are readily available the ONS will use them rather than putting on respondents the burden of aggregating their data into economic accounts/ ESA1995 based groups which may not be at all relevant to their particular business.

18.26   The UK thus has much greater details of certain instruments, at least for some sectors (notably in the financial corporations area), than are required by ESA. The UK has also taken into account the requirements of the European Central Bank in designing its system and in drawing up the requirements for data from institutions.

18.27   The instrument list for which data are published in the Blue Book is a condensed version of the full breakdown used for calculation purposes.

Data sources

18.28   The main data sources for the financial balance sheets are ONS inquiries, data collected by the Bank of England and the Building Societies Commission (25.3) and administrative data from the Treasury. Monetary financial institutions are important holders of financial assets and important suppliers of counterpart information and the ONS is assisted by the Bank of England in the production of these data.

18.29   The ONS quarterly inquiries to financial institutions (insurance corporations, pension funds, unit and investment trusts etc.) collect data for the balances held in certain short-term assets (e.g. cash, sterling commercial paper). For longer term instruments the data collected is that of flows which feed directly into the financial account. However, the ONS also send annual balance sheet inquiries to these institutions and these are used together with the quarterly financial transactions and information on changes in markets, to derive quarterly balance sheets for these institutions.

18.30 The ONS Financial Assets and Liabilities Survey (FALS) (25.2.5) is sent to non-financial corporations and collects quarterly balance sheet data from them. This is run each quarter (with an extended sample once a year).

18.31 The figures for other sectors and the rest of the world are collected in various ways. The Bank of England runs a quarterly balance sheet inquiry to all banks (and a monthly inquiry to the larger ones) (25.3.1). The Building Societies Commission similarly has a monthly balance sheet inquiry which covers all building societies (25.3.2). These give not only the quarterly balance sheet data for monetary financial institutions but also provide counterparty data relating to these sums. So, for example, the complete sectoral split of figures for bank deposits can be obtained from them. The Treasury also provides data from administrative sources which are used to compile the balance sheets.

18.32 The ONS Survey of Charities in 1996 yielded, for the first time, official data for this sector. This is to be used in the construction of the balance sheets for non-profit institutions serving households (the NPISH sector), currently amalgamated with households in the United Kingdom accounts.

18.33 For one or more sectors there may be no balance sheet data - for example, because data collection is not feasible (e.g. from households) and/or because periodic benchmark data shows that they hold and transact only very small amounts of the instrument concerned. Some surveys run by the ONS collect greater detail from corporations annually than they do each quarter. This is done to limit the form-filling burden that is placed upon the corporations sampled and reflects the fact that they may only collate detailed information themselves once a year (for their annual accounts). Other surveys such as FALS are made of only the largest corporations in a sector or industry group each quarter but once a year the forms may be sent to a larger number, including smaller corporations.

Assembling the financial balance sheets

18.34 The UK balance sheets (and financial transactions accounts) are a database of time series that build up from data sources to the matrix of the financial balance sheets by sector. For each time series the database records where it comes from, which cell in the sector accounts matrix it belongs to, and how it is aggregated with other series to produce the cell totals. The information held is actually at a greater detail than that required for the 1995 ESA. For example, there are hundreds of financial categories, or sub-categories, in the matrix, depending on the detailed nature of the data available in the source data. In some areas, however, the data obtained from the source each quarter may be less detailed than that required within the timetable for first publication or for the ESA. In some cases the ONS has no direct data sources and the derivation of the data in these circumstances is described below.

18.35 It is primarily the household and NPISH sectors where the ONS has almost no direct data sources. (The survey of charities in 1996 is an exception and has assisted production of figures for the NPISH sector.) However, depending on the instrument involved, this problem also affects the rest of the world sector and certain others. There are two techniques used to obtain data in these circumstances:

- counterparts
- residuals.

These both make use of the principle that each asset must have a matching liability.

18.36 Information is held on the database at a greater level of detail than that required by ESA because this gives a greater control over the allocation of counterparts and residuals. For example, in most tradable securities held in domestic currency the household and NPISH sectors are the residual ones but, for most tradable securities held in foreign currency the rest of the world sector is the residual. For each sector in the matrix there is a nominated statistician who is responsible for the methodology used to balance that sector account, who will monitor the plausibility of residuals.

Counterparts

18.37 Some balance sheet figures are estimated by using data from the other party - the counterparty. This method is particularly used for loans and deposits. It may be necessary because information from the sector concerned is only collected once a year, or because there is no regular source for information from them - as with the household and NPISH sectors. For example, all sectors' holdings of banks deposits are estimated using counterpart information from banks, who provide a full sector analysis of their deposit liabilities.

Residuals

18.38 The other way to estimate a cell for which there is no information is to treat that as the residual cell in the matrix for that particular financial instrument. This is based on the identity that the total of assets must equal the total of liabilities for any instrument. The observed data will not identify all holdings and so the residual sector is allocated the unidentified balance or the 'residual'. Under the old system the residual was usually allocated to a single sector - often the 'personal' sector - although it was sometimes apportioned between a number of sectors. Under the new ESA the old personal sector has been split, with most of it now shown in either the household or the NPISH sector. So it is common to have these two sectors as the residual ones under the current system. They are for example allocated the residual for UK government bonds. This technique is often used for shares and other equity.

The balancing process

18.39 Data suppliers who feed data directly into the financial balance sheet database are in the ONS and the Bank of England. They deliver source data on specified dates to the central co-ordinators at several points each quarter. The database of time series first balances each category of the matrix, according to agreed methodologies, and then aggregates to produce an account for each sector.

18.40 For each sector there is a nominated statistician who is responsible for the coherence of the accounts for that sector. These statisticians meet to agree adjustments to the allocation of the counterparts and residuals, or to the sample survey results, or to suggest that further investigations need to take place in respect of certain cells' data.

18.41 The balancing process also looks at how these data compare with the financial accounts data, the effect of any proposed adjustments on key economic indicators within the accounts, and whether it is consistent with other economic data : for example, the windfalls made when building societies demutualise, or the first TESSAs reaching their maturity.

18.42 After the meeting of statisticians to discuss the data set there is a redelivery of source data for the changes agreed at the meeting. Further changes may be made, in an iterative process at the end of which a set of accounts is produced which is fully consistent and coherent and is then published. The UK is one of the world-leaders in both the timeliness and the quality of its production of a coherent set of quarterly accounts - usually within 12 weeks of the end of a quarter.

Publication and key economic aggregates

18.43 A First Release is published about 12 weeks after the end of the reference period. At the same time full sector accounts are made available electronically on the databank. Full paper publication follows in the UK Economic Accounts, which includes a commentary linking together the economic stories for the different sectors.

Public Sector debt

18.44 Important economic aggregates, relating to the debt of the public sector, can be identified in the financial balance sheets. These, and their relationship to the public sector net cash requirement (PSNCR), are discussed in more detail in Chapter 21.

Money Supply (M4)

18.45 Money supply data are published primarily by the Bank of England, but its components and counterparts can be related to items in the balance sheets of the UK and its sectors. There several measures of money stock. The narrowest definition is M0 which represents notes and coins in circulation outside the Bank of England and also includes bankers' operational deposits at the Bank.

18.46 A broader definition of money is M4. M4 comprises holdings by the UK private non-monetary sectors (i.e. the M4 'private sector') of notes and coin, together with their sterling deposits at banks and building societies in the UK (including certificates of deposit and other paper issued by banks and building societies of not more than 5 years maturity) and, from December 1995, liabilities of UK banks and building societies arising from repos. M4 is recorded as F.2211 and the sterling part of F.21; Paper of 1–5 years is part of F.3315.

Quality and reliability of the data

18.47 Reference has already been made to the work that has occurred, and is still in progress, to harmonise the sources and methods used to compile the financial balance sheets. The quality of the data vary between both one instrument and another and between the different sectors holding or issuing an instrument. Where sectors are sent regular inquiries into financial assets and liabilities levels and/or flows their data are generally more reliable than those for the household sector and the non-profit institutions serving households (NPISH). Where a non-market instrument is held mainly by a single sector, such as the household sector's holdings of national savings, a reliable figure for the holding sector is usually available from the records of the issuing sector. In addition, issuing sectors can often provide sectoral details of outstanding loans; however, they cannot regularly supply data on sectoral holdings of marketable instruments.

18.48 Banks and (from 1987) building societies are the only sectors which supply complete balance sheets each quarter. Quarterly returns are completed by most other financial institutions, local government and public corporations, however these contain levels data only for short-term instruments (i.e. those with under one year from issue to maturity). For long-term instruments the data collected is of quarterly transactions (acquisitions and realisations) in the financial instruments. Quarterly levels have to be derived from these financial flows information. This is supplemented by end-year balance sheets returns to these institutions which cover all their assets. Thus the quality of the quarterly balance sheet data for long-term assets held by these institutions tend to be poorer. This is particularly the case for marketable instruments since price changes as well as flows have to be taken into account when the quarterly levels are calculated.

18.49   Thus the quality and reliability of the data depend upon factors such as:

- the frequency of collection of the data (quarterly or annual or more infrequent);
- whether they come from a full census (e.g. Bank of England figure for banks) or from a sample (most ONS surveys are sample ones);
- sampling error and bias;
- whether financial balances are collected directly each quarter (as from banks and building societies) or derived from financial transactions data (e.g. the ONS surveys of non-MFIs);
- the quality of the data: are respondents asked to supply data that they themselves use each quarter?
- whether there is a statutory obligation for the data to be supplied;
- the timeliness of data receipt;
- whether the sector is dominated by a few very large institutional units (e.g. the top eight or so insurance corporations write at least half of all insurance business in the UK) and whether those particular units have returned forms in time to the ONS/others;

and others.

The financial instruments

AF.1   Monetary gold and special drawing rights (SDRs)

18.50   These are recorded as assets of the UK. These financial instruments do not have counterpart liabilities; hence they do not appear in the liability half of the balance sheet. They consist of gold and special drawing rights which are held as assets by the Government in the Exchange Equalisation Account (EEA). The Treasury owns and manages the EEA but it is operated by the Bank of England.

18.51   For ESA95, the UK has a derogation under which gold held by monetary financial institutions (MFIs) as a financial asset will continue to be recorded as such (rather than with valuables in the non-financial accounts). However, such gold is regarded in the same way as currencies held for banking purposes and is recorded with Currency and deposits (F2) as foreign currency.

18.52   Special Drawing Rights (SDRs) are international reserve assets created by the International Monetary Fund (IMF) and allocated to its members to supplement existing reserves. This category records the UK official holdings of SDRs.

18.53   The ESA states that monetary gold is to be valued at the price established in organised gold markets. Gold is valued at the ruling official price of 35 SDRs per fine ounce until end-1977 and at end-year market rates from end-1978 onwards.

18.54   SDRs are valued at closing middle-market rates of exchange.

18.55   Holdings of this instrument are shown for central government only. The data comes from the Bank of England. Since the source of the data for both gold and SDRs are government administered sources they are of high quality.

AF.2 Currency and deposits

18.56 This comprises all sterling and foreign currency in circulation, gold owned by monetary financial institutions, transferable deposits and other deposits.

AF.21 Currency

18.57 Sterling currency notes issued by the Issue Department of the Bank of England are a liability of the central bank, while coin is issued by the Royal Mint and is regarded as a liability of central government. (The sector classification of the Bank of England Issue Department has changed with the introduction of the ESA95: it was formerly regarded as being within central government.) The Bank of England holds securities (mainly gilts) as backing for the note issue.

18.58 The data for central government holdings of notes and coin represent those by government for its own use. The assets of the monetary financial institutions sector consist of holdings of notes and coin by the sector including Bank of England notes held by Scottish and Northern Ireland banks as backing for their own notes. The liabilities of the sector comprise all notes in circulation (including Scottish and Northern Ireland bank notes). Data come from the banking statistics collected by the Bank of England.

18.59 The data for building societies come from the Building Societies Commission. The data for other financial intermediaries come from ONS inquiries.

18.60 Holdings by the local government sector are assumed to be negligible. The figures for public non-financial corporations' domestic (i.e. sterling) notes and coins are based on working balances held by the Post Office. The holdings of private non-financial corporations are derived as the residual.

18.61 A special adjustment allows for the fact that the amount of sterling notes and coins held by corporations is high on Thursdays (due to weekly wage payments in cash on Fridays) and the amount held by households is high on Fridays (reflecting withdrawals to cover weekend spending). Further adjustments are also made to allow for the build up of households' notes and coins over the Christmas and Easter holiday periods.

18.62 Data for holdings of foreign currency notes and coins come from banking statistics for monetary financial institutions; and for other sectors are crudely estimated from the transactions, derived from travel data collected by the ONS International Passenger Survey. They are currently attributed to the private non-financial corporations sector. The valuation is the nominal or face value of the currency translated to sterling at end-period exchange rates.

AF.22 Transferable deposits

AF.221 Deposits with UK Monetary Financial Institutions (MFIs)

18.63 For assets this is a single heading, for liabilities it may be split further, into:

- Sterling deposits with UK MFIs (AF.2211)
- Foreign currency deposits with UK MFIs (AF.2212)

AF.2211 Sterling deposits with UK MFIs

18.64 These comprise all types of deposit held with institutions authorised by the Bank of England or entitled to accept deposits in the UK. These include UK incorporated institutions and UK branches of European authorised institutions (briefly referred to as 'banks'). UK MFIs which can accept deposits are currently banks and building societies. Until the end of 1981 such deposits could also be made with UK financial corporations other than banks. However, the under Banking Act 1979, institutions not in the banking sector were no longer permitted to accept deposits. Their figures to 1981 are not separately identified in the accounts.

18.65 The figures include:

a. deposits with the Bank of England Banking Department, other than from the monetary sector (intra-sector deposits);
b. deposits from overseas offices of banks;
c. deposits and receipts under Save As You Earn (SAYE) and sharesave contracts, on which banks and building societies pay interest net of tax. These include interest credited to depositors' accounts;
d. from 1983, time deposits on which banks and building societies pay interest gross of tax;
e. liabilities under sale and repurchase agreements of British Government securities and other paper.

18.66 Both sight and time deposits are included in the above. Deposits evidenced by the issue of negotiable paper (certificates of deposit, commercial paper, promissory notes, bills and other paper) are not included in this sub-instrument but are within money market instruments (AF.331). The figures also exclude the institutions' liabilities in the form of shares and other securities, reserves and other internal accounts. Building society negotiable bonds are assumed to be entirely taken up by the monetary sector and are included in bank lending: sterling (AF.411).

18.67 The banks' figures are compiled by the Bank of England from returns provided by the institutions (25.3.1). The increase in banks' liabilities to other sectors is measured as the increase in other sectors' deposits as reported in the returns (allowing for transit and suspense items). The building society figures are based on monthly balance sheet returns made to the Building Societies Commission from the societies (25.3.2).

18.68 Figures for counterpart sectors are almost entirely obtained from the analysis provided in the banking sector returns. However, data for insurance corporations and pension funds come from the ONS surveys of a sample of these institutions (25.3.3).

18.69 Valuation of these deposits are at the amounts of principal that the debtors are contractually obliged to repay the creditors under the terms of the deposits if the deposits were to be liquidated on the date the balance sheet is set. The value will include accrued interest.

AF.2212 Foreign currency deposits with UK MFIs

18.70 The data for foreign currency deposits with UK MFIs are compiled by the Bank of England from returns provided by the institutions. The increase in banks' liabilities to other sectors is measured as the increase in other sectors' deposits as reported in the returns (allowing for transit and suspense items).

18.71 Figures for counterpart sectors are almost entirely obtained from the analysis provided in the banking sector returns. However, data from the ONS surveys of insurance corporations and pension funds is used here.

18.72 Valuation of these deposits is at the amounts of principal that the debtors are contractually obliged to repay the creditors under the terms of the deposits if the deposits were to be liquidated on the date the balance sheet is set. The value will include accrued interest.

AF.229 Deposits with rest of the world MFIs

18.73 This consists of transferable deposits, both sterling and foreign currency, held with MFIs in the rest of the world. From July 1979, official reserve holdings of European Currency Units (ECUs) are also included. ECUs were acquired from swaps with the European Monetary Co-operation Fund (EMCF) before 1 January 1994 and with the European Monetary Institute (EMI) afterwards. The UK's reserve position in the IMF appears in other deposits, AF.29.

18.74 Assets with banks abroad are derived from the international banking statistics of countries in the Bank of International Settlements (BIS) reporting area and cover the UK non-bank private sector's deposits with banks in the rest of the world. These data come from the BIS via the Bank of England. Some assets of general government and public corporations may also be included but are not separately distinguishable. There may be discontinuities in the levels series. Trends in the data have been informed by trends in rest of the world bank borrowing and deposits derived from the ONS Financial Assets and Liabilities Survey (25.2.5). Figures for UK MFIs come from inquiries carried out by the Bank of England. Data for securities dealers' holdings with banks abroad are their asset levels as reported in an ONS survey (25.3.3). Estimates for earlier years, back to 1986, are based on information from published annual accounts.

AF.29 Other deposits

18.75 National savings, tax instruments and the UK reserve position in the IMF and in the European Monetary Co-operation Fund (EMCF) before 1 January 1994 and with the European Monetary Institute (latterly the European Central Bank) afterwards are recorded here.

18.76 For national savings the following are included here:

a. deposits: outstanding deposits, including accrued interest, with the ordinary departments of the National Savings Bank (NSB), the NSB investment account from January 1981 and Trustee Savings Banks up to the third quarter of 1979 (these are subsequently recorded under MFIs).
b. certificates and bonds: income bonds, deposit bonds, premium savings bonds, British savings bonds (removed from sale on 31 December 1979), national savings stamps, gift tokens, national savings certificates, SAYE (Save-As-You-Earn) contracts and yearly plan agreements including interest accrued and increases due to index-linking.

18.77 Most national savings instruments are treated as assets of the household sector and liabilities of the general government sector. However, National Savings Bank investment accounts, income bonds and deposit bonds have in the past also been held by other sectors (notably other financial intermediaries and financial auxiliaries and non-financial corporations). Figures for public non-financial corporations are included from the second quarter of 1982: this represents the working balance held by the Post Office on account of its national savings business.

18.78 Tax instruments are also included in this category. They are instruments issued to tax payers in exchange for advance payments to meet future tax liabilities. They comprise certificates of tax deposit and, in earlier periods, tax reserve certificates and tax deposit accounts. (Holdings not identified by sector are entered under private non-financial corporations.)

18.79 The UK's reserve position in the IMF is also recorded here. It is a deposit with no maturity and equals the difference between the UK quota and the sterling holding by the IMF. The reserve position fluctuates as other countries draw or repay sterling to the IMF. The figures for this come from government administered sources and are of high quality.

18.80 Valuation of these deposits is at the amounts of principal that the debtors are contractually obliged to repay the creditors under the terms of the deposits if the deposits were to be liquidated on the date the balance sheet is set. The value will include accrued interest.

AF.3   Securities other than shares

18.81 These are broken down into

- short term securities: money market instruments (AF.331),
- long term securities: bonds (AF.332) and
- derivatives (AF.34).

Both short and long term securities are further sub-divided by sector of issuer. Accrued interest has been treated in the financial account as being reinvested in the corresponding securities and is thus included in the balance sheet with those securities. Securities other than shares (excluding derivatives) are valued at current market prices including the value of the accrued interest.

AF.331   Short term securities other than shares: money market instruments

18.82 For both assets and liabilities a detailed split is provided, by type of issuer, as follows:

- UK central government (AF.3311)
- UK local authority bills (AF.3312)
- UK Monetary Financial Institutions(MFIs) (AF.3315)
- Other UK residents (AF.3316)
- Rest of the world (AF.3319)

AF.3311 UK general government

18.83 Both sterling and foreign currency (ECU etc.) Treasury bills and Treasury non-interest bearing notes are recorded here. Bills held by the Bank of England as the sterling counterpart of foreign currency deposits arising from central bank assistance are excluded, but the foreign currency deposits are included in government foreign currency debt. Valuation is at face, or nominal, value.

18.84 For periods up to the end of 1986 the private non-financial corporations sector was the residual for this instrument. For the years 1987 to 1990 inclusive the transactions of this sector were assumed to be zero. From 1991 onwards the holdings recorded in the Financial Assets and Liability survey are used as the holdings of large corporations. Information for public corporations comes from the Bank of England.

18.85 Information is available from the Bank of England for the banks. For building societies the data come from the Building Societies Commission's quarterly balance sheet inquiry. Data for insurance corporations and pension funds are collected on the quarterly ONS 'Transactions in financial assets' inquiries: despite the title of the inquiries, the quarterly balances for these instruments are also collected. Holdings by other financial intermediaries and financial auxiliaries are calculated as fifty per cent of the instrument residual although some data are available from ONS and Bank of England inquiries.

18.86 The balance sheets of local government come from quarterly balance sheet inquiries on borrowing and lending carried out by the DoE, the Welsh Office, the Scottish Office and from Northern Ireland.

18.87 Holdings by the household and NPISH sectors are assumed to be small.

18.88 The rest of the world data relate to Central Monetary Institutions (CMIs) and non-CMIs and are provided by the Bank of England, based on information on UK banks' custody holdings on behalf of these non-residents. Total rest of the world sector holdings are now the sum of these, plus an amount equal to fifty per cent of the residual, to allow for the under reporting of non-CMIs.

AF.3312 UK local authority bills

18.89 For local government bills the total figure is taken from returns made quarterly by local authorities in the UK. The figures are sectorised using a combination of local authority and counterparty sector sources. This is described in more detail below.

18.90 Figures for local government short term money market instruments are taken from returns made by local authorities to the Department of the Environment, Scottish Office, Welsh Office and from central government accounts. Borrowing from public non-financial corporations is also that shown by local authorities. Similarly data for holdings of local government debt by other domestic sectors is derived from a combination of local government data and information from the appropriate sector (for example, the ONS quarterly inquiries of insurance corporations; returns made to the Bank of England by banks). For the rest of the world sector the figures come from data compiled by the Bank of England for the balance of payments. The non-monetary financial sector is the residual.

AF.3315 UK Monetary Financial Institutions(MFIs)

18.91 This covers UK bank and building society certificates of deposit (CDS - in both sterling and foreign currency) and UK corporate commercial paper issued by MFIs (sterling and foreign currency) along with promissory notes, bills and other short-term negotiable paper issued by banks in both sterling and foreign currency.

18.92 Reported holdings of certificates of deposit by private non-financial corporations are compiled by the Bank of England from the ONS quarterly survey of Financial Assets and Liabilities (25.2.5). In addition, part of the residual (ten per cent for sterling; two percent for foreign currency certificates of deposit and commercial paper) is attributed to private non-financial corporations.

18.93 The issues of money market instruments by banks are available from returns these institutions provide to the Bank of England (25.3.1). These returns began at the start of 1984. Holdings and issues are reported on the monthly balance sheet form to the Bank of England. Where the residual difference between issues by banks and reported holdings by other sectors has to be allocated across counterparty sectors, an agreement between the Bank of England and the ONS determines the sectoral allocations.

18.94 Similarly, building societies provide, via the Building Societies Commission, monthly data on their holdings of banks' short-term money market instruments (25.3.2). For non-monetary financial institutions, data come from the ONS inquiries (25.3.3) to insurance corporations, pension funds and Securities Dealers. Attributed data is allocated as various percentages of the residual sterling holdings and the total foreign currency holdings of each type of short-term money market instruments.

18.95 Local government figures for holdings of these banks' instruments are obtained from returns made to the DoE, Welsh Office and Scottish Office.

18.96 Household and NPISH sector holdings of banks' money market instruments are known to be very small. They are allocated two and eight per cent respectively of the residual level of holdings for sterling certificates of deposit and commercial paper and NPISH sector one per cent of the residual holdings of foreign currency versions of these two types.

18.97 Reported holdings of banks' sterling instruments by the rest of the world sector come from a monthly Bank of England balance sheet form. In addition, forty per cent of the residual assets of certificates of deposit and commercial paper are attributed to this sector. For foreign currency instruments, the rest of the world is allocated holdings of ninety-two percent of the residual for CDS and commercial paper.

18.98 Building society issues and holdings of building society short-term money market instruments are reported by the societies monthly to the Building Societies Commission (25.3.2). An agreement between the ONS and the Bank of England determines the sectoral allocation of the residual difference between issues by Building Societies and reported assets by the other sectors. The residual allocations are done separately for certificates of deposit and commercial paper.

18.99 The source of data on holdings of building society short-term money market instruments for private non-financial corporations is the ONS Financial Assets and Liabilities survey (for sterling CDS only) (25.2.5). In addition a portion of the residuals for certificates of deposit and commercial paper are allocated to this sector. Public corporations do not hold these instruments.

18.100 Banks' holdings of building society short-term money market instruments come from the balance sheet inquiry carried out by the Bank of England (25.3.1). Data for non-monetary financial institutions come from the ONS quarterly inquiries (25.3.3) to insurance corporations, pension funds and securities dealer: these collect end quarter holdings. A portion of the residual holdings are also allocated to this sub-sector, reflecting the data for types of institutions within this group that are not surveyed.

18.101 Central and local government do not hold short term money market instruments issued by building societies.

18.102 The households and NPISH sectors are allocated a small portion of the residual for certificates of deposit and commercial paper.

18.103 The rest of the world's holdings of building societies' short term money market instruments are collected by the Building Societies Commission and passed to the Bank of England. An allocation of the residual is made to this sector.

18.104 Banks' liabilities also include commercial bills which have been accepted by banks derived from information collected on Bank of England forms. Bills rediscounted with the Bank of England's Issue Department are recorded as an asset of the central bank. Information on other sectors' assets is obtained from inquiry sources. The lending associated with these acceptances is included under bank lending (AF.41).

18.105 The quality of the liabilities data for short-term money market instruments issued by UK MFIs is generally good as it is based on comprehensive surveys of MFIs. MFI asset figures will also be of good quality for the same reasons. The asset figures for other sectors will be less reliable since they include a portion of the residual holdings.

AF.3316 Other UK residents

18.106 This comprises the following: UK corporate commercial paper issued by non MFIs (both sterling and foreign currency) and UK local authority bills.

18.107 Data on issues of sterling commercial paper by non-monetary financial corporations are provided by the Bank of England based on returns to the Bank by commercial paper issuers. Asset holdings are based on inquiry sources with the residual being divided among private non-financial corporations (50%), other financial intermediaries (30%), households (2%), NPISH (8%) and rest of the world (10%).

18.108 Euroclear data are the source for issues of foreign currency commercial paper on the Euro-markets. This is divided between private non-financial corporations and non-monetary financial institutions in the ratio 95:5. Issues of foreign currency commercial paper from elsewhere are obtained from returns by the main issuers, supplemented by estimates; all are private non-financial corporations. The rest of the world is the residual sector.

18.109 Banks' holdings of these money market instruments (commercial paper, other short-term paper) issued by UK corporations are reported on one or more of the Bank of England forms. Building Societies' holdings are reported on returns made by the societies to the Building Societies Commission with the aggregate data being passed onto the Bank of England. Insurance corporations' and pension funds' holdings are collected on quarterly ONS inquiries to these institutions. (25.3.1, 25.3.2, 25.3.3).

18.110 Data quality for liabilities for sterling commercial paper is good. MFI asset figures are also of good quality because they are based on comprehensive surveys of these institutions. The asset figures for other sectors will be less reliable since they include a portion of the residual holdings. The foreign currency data, both liabilities and assets, are generally not as good, except for MFI assets.

AF.3319 Rest of the world

18.111 The data are derived from a variety of sources including the ONS security dealers inquiries and Financial Assets and Liabilities Survey, together with the Bank of England. New questions were added to the ONS inquiries to insurance corporations and pension funds to collect data from 1997 onwards on MMIs issued by the rest of the world. These should be available in 1998. Bank of England inquiries were amended, from 1997. This instrument includes sterling commercial paper where it has been issued by non-residents.

AF.332 Long term securities: bonds

18.112 These are divided by type of issue:

- Issues by UK central government AF.3321
- Issues by UK local authorities AF.3322
- Issues by UK MFIs AF.3325
- Issues by other UK residents AF.3326
- Issues by the rest of the world AF.3329

Long term bonds recorded on the assets side of the balance sheet will include bonds issued by the rest of the World, as well as by the UK government and other UK residents.

AF.3321 Issues by UK central government

18.113 Long term bonds issued by UK central government are commonly referred to as British Government securities (BGS) or gilt-edged securities (gilts). They are the main form of longer term government borrowing and may be of the conventional type or index-linked. They are marketable securities denominated in sterling or foreign currency. In earlier years they included sterling government-guaranteed marketable securities of nationalised industries. British Government foreign currency notes and bonds are included, together with government issues abroad.

18.114 UK government bonds may be conventional or index-linked. Conventional bonds pay out interest every six months until maturity. Index-linked bonds pay out interest depending upon the rate of increase in the retail prices index (RPI) – for example the interest per £100 gilt may be the rate of the RPI increase plus 2.5%. Some of the gilts that the government have issued are undated; that is, there is no fixed date for redemption.

18.115 The figures for central government liabilities are compiled by the Bank of England using their figures and those of the National Debt Commissioners.

Sterling bonds

18.116 For private non-financial corporations the figures are taken from holdings (at book value) reported by large corporations in the ONS Financial Assets and Liability Survey (25.2.5), and previously the Survey of Company Liquidity.

18.117 The figures for banks and for other financial corporations are compiled using the quarterly balance sheet returns made to the Bank of England, the Building Societies Commission and the ONS (25.3).

18.118 The figures for the rest of the world sector are based on custody holdings of sterling bonds held by non residents reported by banks, information from the Stock Register, occasional surveys and evidence from balancing the economic accounts.

18.119 Adjustments are made when necessary to convert the figures of external liabilities from nominal to market value. The residual difference between total general government sterling liabilities and identified holdings of other sectors is generally attributed to the household and NPISH sectors.

Foreign currency bonds

18.120 This instrument includes holdings of British government foreign currency bonds. These include an issue in 1991 of ECU 2,500 million HM Government 10 year bonds, issues in 1992 of DM 5,000 million and US $3,000 million HM Government 5 and 10 year bonds respectively; and issues commencing 1992 of 3 year HM Government ECU Treasury notes. Data for these are from official sources and thus of high quality. The counterparty holders of these bonds are banks, insurance corporations and pension funds, other financial intermediaries and the rest of the world sector. Counterparty data are obtained from Bank of England and the ONS inquiries; the overseas sector is the residual sector for part of the long term bonds issued by UK general government. The ONS is able to estimate the total value of foreign currency bonds issued by the UK government but is unable to measure exactly how much are purchased and held by non-residents. To balance the accounts it is assumed therefore that any holdings that cannot be directly attributed to UK domestic sectors must be holdings of the rest of the world.

AF.3322 Issues by UK local authorities

18.121 Long-term bonds issued by local government (Local Authority bonds) are generally similar to conventional gilts. Data from DETR, Welsh Office and Scottish Office inquiries to local government bodies form the estimates for private non-financial corporations' holdings of local government bonds.

AF.3325 Issues by UK monetary financial institutions

AF.3326 Issues by other UK residents

18.122 These categories include bonds issued by UK financial corporations and private non-financial corporations and others, together with UK corporate quoted and unquoted preference shares. They also contain Eurobonds, medium term notes, debentures and loan stock and preference shares issued by UK private sector corporations, as well as issues by non-profit making institutions such as universities and housing associations.

18.123 The sources for these data are the London Stock Exchange, the Bank of England and various ONS statistical surveys.

Liabilities

18.124 The London Stock Exchange provides figures for issues of Eurobonds, medium term notes and other debt securities listed on the London Stock Exchange. For building societies, statistical inquiries run by the Building Societies Commission provide good quality data. The Bank of England provide estimates for other market issues of bonds. These data are supplemented with information on non-market issues from some financial institutions and by estimates for unlisted preference shares and loan notes. Listed preference shares, eurobonds and domestic debentures and loan stock are generally valued at market prices. Most other securities are at nominal value.

18.125 Data for the loan capital of UK companies held by overseas corporations is based on information from the Bank of England and the ONS direct investment inquiries (25.5.1). The reliability of the bonds and preference share data as a whole is affected by the rough estimates made for some non-market issues. The residual is split equally between private non-financial corporations and non-monetary financial corporations.

Assets

18.126 Figures for non-financial corporations are obtained from the ONS Financial Assets and Liabilities Survey (25.2.5).

18.127 The sources for data for the financial sector are various inquiries run by the Bank of England, the Building Societies Commission and the ONS (25.3).

18.128 The accounts of central government and public corporations, together with inquiries run by the DETR and the Welsh Office and Scottish Office provide the data for the public sector holdings of these assets.

18.129 In the absence of any known source of information the holdings of the household and NPISH sectors of bonds and preference shares are estimated to increase by £10 million and £40 million a quarter respectively, in line with their estimated transactions.

18.130 The ONS surveys of cross border acquisitions and mergers and of overseas direct investment (25.5.1) yield the holdings of preference shares and loan capital by the rest of the world. This includes their holdings of UK subsidiaries' loan capital (including preference shares). Other non-resident holdings in this category are estimated by accumulating flows, adjusting for exchange rate movements and taking into account the currencies of new issues. The ESA95 defines direct investment as where a corporation owns over ten percent of another corporation's voting shares. This represents a definitional change from the previous ESA, of 1979, which defined it as a holding of over twenty per cent. However, early investigations by the ONS suggest that there will be little impact caused by this change to the threshold.

AF.3329 Issues by the rest of the world

18.131 These comprise bonds issued by foreign governments, municipal authorities and corporations as well as rest of the world corporate and unquoted preference shares insofar as they are held by UK residents. Eurobonds, medium term notes, debentures and loan stock issued by private and public sectors in the rest of the world are also part of this category.

18.132 From 1963 to 1979 figures were based partly on a Bank of England inquiry and partly on exchange control returns.

18.133 Figures from the ONS Financial Assets and Liabilities Survey (25.2.5) are now used to determine the holdings of private non-financial corporations. Public corporations' accounts yield their holdings data.

18.134 The figures for financial and monetary institutions come from annual balance sheet inquiries to them which are carried out by the Bank of England, the Building Societies Commission and the ONS (25.3). New questions were added to the inquiries to insurance corporations, pension funds, unit and investment trusts in order to collect the additional detail required for ESA1995 and the fifth edition of the balance of payments manual (BPM5). These will provide extra data relating to 1997 in 1998 with official foreign exchange holdings broken down into currency and deposits and securities and securities then split into equities, bonds and notes and money market instruments and derivatives.

18.135 Currently figures from the accounts of central government provide the holdings by central government. Data for local government holdings come from the returns made to the DETR, Welsh Office, Scottish Office and from Northern Ireland. Data for the UK government's holdings in the official reserves of these bonds come from the Bank of England. These data are of good quality.

18.136 The household and NPISH sectors holdings of overseas bonds are based on limited data from the Inland Revenue but are subject to a wide margin of error.

18.137 The rest of the world figures consist of total UK direct investment in share capital in the rest of the world plus portfolio investment. This liability figure is obtained as the total of the figures for the other sectors (which hold these bonds as assets).

AF.4   Loans

18.138 The detail provided here is of short term and long term loans, both further sub-divided by type of loan.

18.139 The values recorded in the balance sheets of both creditors and debtors are the amounts of principal that the debtors are contractually obliged to repay the creditors.

AF.41   Short term loans

18.140 The sub-division used here is as follows:

- Sterling loans by UK MFIs   AF.4111
- Foreign currency loans by UK MFIs AF.4112
- Loans by the rest of the world AF.419

Loans by other sectors are all assumed to be long term.

AF.4111 Sterling loans by UK MFIs

18.141 These are the UK banks and building societies' domestic currency (i.e. sterling) loans. The figures comprise lending by banks through advances, credit card lending, other loans, commercial bills and some other paper and also claims under sale and repurchase agreements of British Government securities and other paper, but excluding investment in securities. Building societies' loans recorded here represent advances, net of repayments, of loans not secured on dwellings. Loans in categories recognised as long-term are excluded here and shown under various sub-headings within long-term loans (AF.42). However, the lending included within this item will include some unspecified loans of more than one year original term.

18.142 The banks provide figures to the Bank of England (25.3.1). The counterpart sector analysis of loans and advances is obtained from the returns, subject to adjustment to allow for items in the course of clearance. Chapter 17 (17.81) describes the treatment of transit and suspense accounts and the inter-bank difference.

18.143 The greater part of the building society lending recorded here is to households, but loans to financial and non-financial corporations, including the societies' own subsidiaries, also feature. The figures, including all sector transactions, are drawn from returns submitted by the societies to the Building Societies Commission (25.3.2). The quality of these data is good.

AF.4112 Foreign currency loans by UK MFIs

18.144 These are the UK banks' and building societies' loans denominated in foreign currencies. As for sterling loans (AF.4111), the figures comprise lending by banks through advances, loans, commercial bills and some other paper and also claims under sale and repurchase agreements of British Government securities and other paper, but excluding investment in securities. Building societies' loans recorded here represent advances, net of repayments, of loans not secured on dwellings: foreign currency ones are far less frequent than sterling ones. Loans in categories recognised as long-term are excluded here and shown under various sub-headings within long-term loans (AF.42). However, the lending included within this item will include some unspecified loans of more than one year original term.

18.145 The banks provide figures to the Bank of England (25.3.1). The counterpart sector analysis of loans and advances is obtained from the returns, subject to adjustment to allow for items in the course of clearance.

18.146 Most of the building society lending recorded here is to households, however there are also some loans to financial and non-financial corporations. The figures, including all sector transactions, are drawn from returns submitted by the societies to the Building Societies Commission (25.3.2). The quality of these data is good.

AF.419 Short term loans (liabilities): Loans by the rest of the world

18.147 These comprise loans in both sterling and foreign currency by MFIs in the rest of the world.

18.148 For UK private non-financial corporations data are derived by difference between the total lending to the non-bank private sector (from the Bank for International Settlements (BIS) and identified lending to the non-monetary financial sector, derived from inquiries to the sector. The estimates are also influenced by the ONS Financial Assets and Liability Survey (25.2.5) which is carried out on a quarterly and annual basis to these corporations. The Bank of England supply information for borrowing by UK MFIs from the rest of the world. The other source of data for such borrowing by the financial sector is the quarterly ONS inquiry to securities dealers (25.3.3).

18.149 Estimates of UK public sector borrowing from commercial banks in the BIS reporting area are been derived from official records and deducted from the BIS data to give estimates for total UK private sector borrowing. None of this instrument is thought to relate to borrowing by households or the NPISH sector.

AF.42 Long term loans

18.150 These are split between the following categories:

- Outward direct investment (AF.4211)
- Inward direct investment (AF.4212)
- Loans secured on dwellings (AF.422)
- Finance Leasing loans (AF.423)
- Other loans by UK residents (AF.424)
- Other loans by the rest of the world (AF.429)

AF.421 Direct investment

AF.4211 Outward direct investment

18.151 These loans are an asset of the UK. They cover net investment by UK corporations in rest of the world branches, subsidiaries or associated corporations other than in the form of securities. A direct investment in a corporation means the investor has a significant and lasting influence on the operations of the corporation (normally represented by a holding of 10 per cent or more voting shares in the corporation). The 10 per cent was introduced to the ONS overseas direct investment inquiry forms with effect from the end of 1996 (for annual inquiries, first quarter of 1997 for the quarterly ones). The pre-ESA 95 threshold for ONS inquiries was 20 per cent. Early investigations by the ONS suggest that the inquiries based on the 20 per cent limit did not miss much direct investment, so the impact of this change is likely to be small.

18.152 Data here include fixed assets, stock building and stock appreciation and also other financial items such as working capital, other loans and trade credit. It excludes real estate (this is in AF.51, Shares and other equity).

18.153 Estimates for outward direct investment loans are derived from quarterly and annual inquiries carried out by the Bank of England (to monetary financial institutions) and the ONS (to other corporations and institutions). The outward direct investment inquiries run by the ONS are sent to all corporations in the oil, insurance and other financial institutions sectors. The remaining private corporations are sampled by value of assets: coverage is 99 per cent by value of assets, which represents 85 per cent of the population of corporations by number. Estimates for outward direct investment loans by MFIs are derived from quarterly and annual inquiries carried out by the Bank of England.

18.154 Public corporations, households and NPISH are not surveyed and are deemed to have no loans between UK parents and overseas subsidiaries.

AF.4212   Inward direct investment

18.155 These figures show the rest of the world's investment in the UK private sector (other than in UK corporation securities which are entered under AF.3 and AF.5). They are a liability of the UK. They include figures for direct investment which are obtained from the ONS direct investment inquiries and similar inquiries to banks run by the Bank of England. They also cover miscellaneous investment where this does not consist of identified transactions in real estate (the latter are included in F51).

18.156 The figures for banks are obtained from banking sector returns made to the Bank of England. Those for other financial institutions and private non-financial corporations are estimates obtained from inquiries run by the ONS to those institutions. The ONS inward direct investment inquiries are to private non-MFI corporations and they have a complete coverage of the oil, insurance and OFI sectors. The remaining corporations are sampled by value of assets: coverage is 92 per cent by assets (which represents 85 per cent of the population by number). Public corporations in the United Kingdom are wholly publicly owned and therefore not the subject of inward direct investment.

18.157 The figures for banks are obtained from banking sector returns made to the Bank of England (25.3.1). Building societies are not covered since, being owned by their members, they cannot be foreign owned.

18.158 Households and NPISH are assumed to have no loans between UK subsidiaries and foreign parents.

AF.422 *Loans secured on dwellings*

18.159 Loans secured on dwellings comprise loans made by UK banks (AF.4221), building societies (AF.4222) and other lenders, principally general government (AF.4223), mainly for house purchase and home improvement. It is assumed that all these loans are made to the household sector and to the NPISH sector, which includes housing associations.

18.160 Loans secured on dwellings are obtained by sector as follows.

- The figures for banks and building societies are obtained separately from returns made to the Bank of England and the Building Societies Commission.
- For insurance corporations and pension funds the figures come from the ONS quarterly inquiries to samples of these institutions.
- Data for other financial intermediaries' loans secured on dwellings relate mainly to lending by centralised mortgage lenders and the acquisition of mortgage portfolios by loan securitisation subsidiaries of banks and other lenders. Loans made by the Trustee savings banks (to end-1981) are also included here.
- Central government loans are those made by central government under the Housing Act, 1961 and the Housing (Scotland) Act, 1962 to private housing associations; loans made by the Housing Corporation to housing societies and loans by new town corporations. The data are obtained from official records.
- Mortgage advances for house purchase and improvements, including loans for the purchase of council houses by tenants and loans to housing associations form local authority loans secured on dwellings. Quarterly figures are obtained from returns made to the Department of the Environment (England) and the Welsh Office (Wales). Figures for Scotland are interpolated from financial year totals. There are no figures for Northern Ireland.

AF.423 Finance leasing loans

18.161 The figures show the amounts which have been lent under finance leases by banks and other financial intermediaries, primarily specialist finance leasing subsidiaries of banks. They are calculated as the value of assets leased out under new leases, net of repayments made through the estimated capital element of rental payments by lessees. The total lending of the lessor sectors is obtained from returns made by banks to the Bank of England, from surveys of other financial intermediaries carried out by the ONS and by reference to information collected by the Finance and Leasing Association from its members. The greater part of finance leasing is by non-financial corporations, but some is attributable to financial corporations and general government.

18.162 The coverage of the figures is limited to leases which are financial in substance. Other forms of lease are operating leases, which are not regarded as financial instruments.

AF.424 Other long-term loans by UK residents

18.163 This comprises loans by the remaining sectors of the UK economy: non-financial corporations (including credit extended by retailers); insurance corporations, pension funds and other non-monetary financial institutions; central government; local government; other loans by NPISH. Each sector is covered individually in the text below.

Loans by non-financial corporations(excluding credit extended by retailers)

18.164 These are picked up as counterparts of loans taken out by what are mainly non-monetary financial corporations. Data quality is poor because of the assumptions on which the estimates are based.

Credit extended by retailers

18.165 The figures relate to hire purchase agreements and to certain other forms of credit including sales on budget accounts, credit sale agreements and personal loans repayable by instalments. Credit advanced by retailers in the form of trading checks exchangeable only in their own shops is included, but credit advanced by check traders as such is not covered. The figures also exclude monthly accounts and sales on bank credit cards such as Mastercard and Visa (included in bank lending: AF.4111) as well as (probably large) amounts of unpaid bills (included in trade credit, AF.71, in so far as they are identified) and credit extended by other types of retailer).

18.166 The source of these data are the ONS Retail Sales Inquiry (25.1.10). It is assumed that all credit advanced by retailers is to the household sector.

18.167 The figures for the public non-financial corporations represent credit given by public corporations through gas and electricity showrooms (no longer applicable since privatisation).

Other loans by non-monetary financial institutions

18.168 This item includes instalment credit, other loans and advances (excluding loans secured on dwellings) by finance houses and other consumer credit companies, other than those included in the monetary financial institutions sector, and loans made to their parent organisation by public and private sector pension funds. Also included are loans made by insurance corporations and by special finance agencies. Counterpart figures for groups of institutions are allocated to households, general government or to non-financial corporations as appropriate.

18.169 Data comes from ONS surveys to such financial institutions (25.3.3), some of which collect an analysis by counterparty sector (for example, the ONS quarterly survey of the financial transactions of pension funds collects such loans split into: loans to parent organisation; loans to financial institutions; loans to financial corporations; loans to other than financial organisations; loans to other UK; loans to rest of the world.

Other loans by central government

18.170 These figures relate to net lending by central government to the following sectors:

a. building societies - under the House Purchase and Housing Act, 1959;
b. non-financial corporations - including, from September 1972, shipbuilding credits re-financed from voted funds;
c. personal sector - miscellaneous loans for education and other services;
d. overseas sector - including net lending to overseas governments and to international lending bodies.

18.171 The figures are taken from the detailed accounts of central government and are of a good quality.

Other lending by local government

18.172 These are loans to promote industrial development in the locality.

AF.429 Other long-term loans by the rest of the world

18.173 Loans made by the rest of the world of a number of different types are recorded here.

18.174 The UK central government sector's sterling liabilities to the rest of the world, which are not included elsewhere, are shown here. This includes sterling borrowing from governments or from central monetary authorities in the form of assistance with a sterling counterpart invested in Treasury bills (aid bills) and amounts swapped forward into later months against sterling with rest of the world monetary authorities in 1971 and 1972. Transfers from the Government's dollar portfolio into the reserves, gold subscriptions to the IMF, changes in IMF holdings of non-interest bearing notes other than those arising from drawings or subscriptions are also included. Part of official short-term capital transactions in the balance of payments, together with contributions to the European Coal and Steel Community reserves, are also recorded for the central government sector here.

18.175 This category also covers loans to local government (both under the exchange cover scheme and non-guaranteed borrowing, excluding bonds), loans to UK corporations and loans by rest of the world banks (including the European Investment Bank) and other institutions. It does not include trade credit; this is included under instrument AF.7: Other accounts receivable/payable.

18.176 The Bank of England provide data for loans made to: the non-bank private sector by rest of the world banks in the BIS reporting area; public corporations; and some monetary financial institutions (MFIs). Local government borrowing from abroad is taken from the returns of local authority borrowing. Information on borrowing by securities dealers and insurance corporations comes from ONS surveys of them (25.3.3). Also included are borrowing by gold dealers and factoring companies. Estimates of loans received by public corporations and local government are of good quality whilst the figures for the private sector, MFIs and security dealers are less reliable.

AF.5 Shares and other equity

18.177 This category is sub-divided into mutual funds' shares (AF.52) and other shares (AF.51). Quoted shares are valued at their current market prices for both the assets and liabilities side of the account, while unlisted shares and other equity are a combination of estimated market prices and accumulated flows (ie nominal value). Shares and other equity are not, legally, a liability of the issuer, but an ownership right over the liquidation value of the corporation, whose amount is not known in advance.

AF.51   Shares and other equity, excluding mutual funds shares

18.178 This is sub-divided into:

- Quoted UK shares, AF.514
- Unquoted UK shares and other equity, AF.515
- Other UK equity, AF.516
- Shares and other equity issued by the rest of the world, AF.519

AF.514 Quoted UK shares

18.179 Quoted UK shares comprise shares listed on the London Stock Exchange, in the full market, the Alternative Investment Market (from the middle of 1995) and the Unlisted Securities Market (until end-1996 when the USM closed).

18.180 Estimates are compiled for a variety of sources, including the London Stock Exchange, Bank of England and ONS statistical enquiries.

Liabilities

18.181 Market values of liabilities are provided by the London Stock Exchange.

Assets

18.182 The ONS's share register survey (25.2.9) is an important source for assets. It was run annually between 1989 and 1994 and at the end of 1997. It provides end-year estimates of the sectoral breakdown of holdings of quoted ordinary shares. It is used as a benchmark for estimates of insurance corporations and pension funds, industrial and commercial companies, households, NPISH and for the rest of the world. It is also used to derive a grossing factor for portfolio transactions in ordinary shares by the Rest of the World. The figures used take account of the reliability of the survey estimates, as measured by sampling errors.

18.183 Figures for private non-financial corporations come quarterly from the ONS Financial Assets and Liabilities Survey (25.2.5), and are benchmarked on the Share Register Survey. Data for public corporations comes from their annual accounts and surveys of them. Data for banks, building societies and other financial corporations are based on annual balance sheet returns which are made by these institutions to the Bank of England, the Building Societies Commission and the ONS (25.3.2, 25.3.3).

18.184 Data for the public sector are taken from inquiries to local government and the accounts of central government. They include holdings of quoted shares retained in privatised companies. These data are of good quality.

18.185 Holdings of the households and NPISH sectors are derived as a residual and in the light of the share register survey estimates. They are potentially subject to a wide margin of error. The 1996 ONS Survey of Charities indicated that their holdings of UK and overseas equities amounted to £14.7 billion at the end of 1994.

18.186 Figures for holdings by the rest of the world of quoted UK shares come from the benchmark share register survey, short-term surveys on investment by non-residents using UK banks and securities dealers and ONS direct investment inquiries (25.5.1).

AF.515 Unquoted UK shares

18.187 Unquoted UK shares are tradable or potentially tradable financial instruments issued by UK corporations. Figures are compiled from a variety of sources including the Bank of England and the ONS statistical inquiries (25.3).

Liabilities

18.188 The liabilities of banks and non-monetary financial institutions are derived from balance sheet returns while liabilities of independent unquoted industrial and commercial companies are estimated as a fixed proportion (23 %) of the value of quoted industrial and commercial companies. This proportion was derived from an analysis of the company accounts of a sample of independent UK companies over the period 1986–88 (later data were not available). This found a stable relationship between the total value of shareholders' funds in unquoted and quoted companies, implying a similar relationship between total market values. Investment in the share capital of UK subsidiaries by overseas parent corporations is also included here, based on information from the ONS direct investment inquiries (Investment by overseas parent corporations in loan capital is included in long term bonds, AF.332).

18.189 The reliability of these data is influenced by the rough estimates for some unquoted issues.

Assets

18.190 Figures for the public sector are taken from enquiries to local authorities and the accounts of public corporations.

18.191 Data for banks, building societies and other financial intermediaries come from annual balance sheet inquiries to these institutions.

18.192 The assets of private non-financial corporations are estimated as a fixed proportion (20%) of private non-financial corporations' liabilities and are intended to represent trade investments (25.3).

18.193 Figures for holdings by the rest of the world sector of UK unquoted shares come from various sources, including surveys run by the Bank of England and the ONS direct investment inquiries. They include inward direct investment in the share capital of unquoted subsidiaries (including re-invested earnings), acquisition of unquoted independent UK companies and subsidiaries from UK corporations and portfolio investment in venture capital funds and the unquoted shares of other financial intermediaries.

18.194 Holdings by the NPISH sector are assumed to be negligible. The estimates for the households sector's holdings of these shares are derived as a residual and are subject to a wide margin of error.

AF.516 Other UK equity

AF.519 Shares and other equity issued by the Rest of the World

18.195 This sub-category comprises shares and tradable or potentially tradable financial instruments issued by foreign corporations. It also includes subscriptions to international organisations which the ESA describes as government investments in the capital of international organisations which are legally constituted as corporations with share capital. An example is the European Investment Bank.

18.196 Holdings by UK residents of real estate in the rest of the world are also included here. Land or buildings in a country are always deemed to be owned by resident units in that same country. In this case the actual UK resident owner is regarded as having made a financial investment, akin to direct investment, in a notional resident of the foreign country and it is that investment which is shown in this category.

18.197 Currently figures for shares and financial instruments held by monetary financial institutions are obtained from a Bank of England inquiry (which is supplemented by an annual balance sheet inquiry). Estimates for other financial intermediaries, financial auxiliaries and insurance corporations and pension funds are based on quarterly ONS surveys of transactions. For data from 1997, ONS surveys to securities dealers, insurance corporations, pension funds, unit and investment trusts were extended to collect additional detail necessary under ESA95 (whether the securities were issued by institutions in EU or non-EU countries) and that required under the new edition of the Balance of Payments Manual (BPM5). The figures include estimates of direct investment in the form of shares.

18.198 Figures for private non-financial corporations are for direct investment in the form of shares as well as for other investment: the latter are derived from an ONS inquiry collecting balance sheet information.

18.199 The estimates for holdings by the household and NPISH sectors are based on Inland Revenue data on transactions and are subject to a wide margin of error. A survey of charities in 1996 indicated that their holdings of UK and overseas equities amounted to £14.7 billion at the end of 1994. This assisted the allocation of these assets between the Households and NPISH sectors, the ratio applied being 20:80.

18.200 The rest of the world sector liabilities figures consist of total UK direct investment in the rest of the world in share capital plus portfolio investment in rest of the world shares and similar financial instruments.

18.201 Estimates for UK investments in real estate in rest of the world are obtained from quarterly and annual inquiries carried out by the ONS. Most of the property investment is carried out by pension funds and insurance corporations. There is little information on UK households' ownership of real estate abroad.

AF.52   Mutual funds' shares

18.202 These comprise investments in unit trusts, investment trusts and property unit trusts. Open ended investment companies (OEICs) will be shown here in due course - see below. These are valued at their current stock exchange price, if they are quoted, or at their current redemption value, if they are redeemable by the fund itself.

AF.521 UK Mutual funds shares

<u>Unit trust units</u>

18.203 Holdings of unit trusts units, where the unit trust is authorised by the Securities and Investments Board under the terms of the Financial Services Act 1986, are shown here. Other unitised collective investment schemes are not covered here (e.g. unauthorised funds run on unit trust lines by, for example, securities firms and merchant banks, designed primarily for the use of institutional investors) neither are those based offshore (Channel Islands, Bermuda etc.) or in other EC Member States.

18.204 The majority of unit trusts are required to invest at least 90 per cent of their assets in transferable securities traded on a recognised stock exchange. However, money market trusts are permitted to invest wholly in cash or near cash instruments (building society and bank deposits, Treasury bills, short dated British government securities etc.)

18.205 They are set up under trust deeds, the trustee usually being a bank or insurance company. The trusts are administered not by the trustees but by independent management companies. Units representing a share in the trust's assets can be purchased by the public from the managers or resold to them at any time.

18.206 Figures are provided by the Association of Unit Trust and Investment Funds. Unitisations of investment trusts are excluded. Annual returns to ONS from insurance corporations and pension funds (25.3.3) provide data for this sector's holdings: all other holdings are assumed to be with the household and NPISH sectors.

18.207 Balance sheet data are collected by ONS annually from insurance corporations and pension funds for unauthorised unit trust units.

Investment trusts

18.208 Investment trust companies acquire financial assets with money subscribed by shareholders or borrowed in the form of loan capital. They are not trusts in the legal sense, but are limited companies with two special characteristics: their assets consist of securities (mainly ordinary shares), and they are debarred by their articles of association from distributing capital gains as dividends. Shares of investment trusts are traded on the Stock Exchange and increasingly can be bought direct from the company.

18.209 The figures cover investment trusts recognised as such by the Inland Revenue for tax purposes and some unrecognised trusts. Returns are received from a sample of companies accounting for about 60% of the total assets of these trusts. Since 1992 quarterly and annual returns have been collected by the ONS (25.3.3). (Previously they were collected by the Bank of England). As from end-1979 the data received are grossed up to include estimates for non-respondents.

Property unit trusts

18.210 The data are for all UK property unit trusts authorised under the terms of the Financial Services Act, 1986. The trusts are not allowed to promote themselves to the general public and participation is generally restricted to pension funds and charities. Property unit trusts invest predominantly in freehold or leasehold commercial property, yet may hold a small proportion of their investments in the securities of property companies. Their assets are held in the name of a trustee and are managed on a co-operative basis by a separate committee (elected by the unit holders) or company. Some investment is financed by long-term borrowing.

18.211 Returns are made by a sample of property unit trusts to the ONS every quarter. The figures include estimates for non-response.

Open Ended Investment Companies

18.212 Data for the market value of liabilities of these institutions, which began operations in the United Kingdom in 1997, are provided by the London Stock Exchange.

AF.529 *Rest of the world mutual funds' shares*

18.213 No data are available for these at present. ONS are investigating whether information could be obtained from offshore institutions as a counterparty to the personal sector. New questions have been added to ONS inquiries to insurance corporations, pension funds, unit and investment trusts to collect information on rest of the world mutual funds, with effect from end 1996. The new questions also aim to gather information on whether the mutual funds are based in EC or non-EC countries.

AF.6   Insurance technical reserves

18.214   These reserves are the technical provisions of insurance corporations and pension funds against policy holders or beneficiaries. There are two types of reserves: firstly the net equity of households in life insurance reserves and pension funds' reserves (AF.61) and secondly reserves comprising prepayments of insurance premiums and reserves for outstanding claims (AF.62).

AF.61   Net equity of households in life insurance reserves and pension funds' reserves

18.215   These are the provisions made by the corporations or the purpose of obtaining, once the established conditions are met, the claims and benefits foreseen. So for life insurance they are the reserves against outstanding risks and provisions for with-profits insurance that add to the value on maturity of with-profits endowment policies. For pension funds these are provisions set up to provide pensions for employees or the self-employed.

18.216   In the economic accounts the life assurance and pension funds are regarded as the property of the household sector (and, to a very small extent, of households in the rest of the world) but the investments of the funds (including purchases of property) are included in the transactions of the financial corporations (insurance corporations and pension funds sub-sector). In order to maintain the articulation of the account, the funds' provisions are shown as financial claims of the household sector and the rest of the world on insurance corporations and pension funds (25.3.3).

18.217   Quarterly estimates of the net increase in these funds are taken from their revenue account which is collected through ONS surveys to a sample of these institutions. Quarterly returns of investment by insurance companies and pension funds are similarly made by the sample of institutions to the ONS. Annual data for these funds are collected on the annual returns sent by the ONS to these institutions.

18.218   Liabilities of central government shown here relate to those public sector schemes for which a notional pension 'fund' is maintained, primarily the NHS and teachers' superannuation schemes. The figures are taken from the financial account of central government, and are regarded as part of the household sector's claims on central government.

AF.62   Prepayments of insurance premiums and reserves for outstanding claims

18.219   These are provisions established by insurance corporations to cover the following.

- The amount representing that part of gross premiums written which is to be allocated to subsequent accounting periods. For example, most annual household payments of premiums do not fall due on 1 January each year. Thus a premium paid on 1 December will be received in the current year but most of the period during which risks are covered will be in the following calendar year. So the majority of the premium paid on 1 December will be part of the pre-paid premium reserves of insurance corporations at the end of the current year.

- The total estimated ultimate cost of settling all claims arising from events which have occurred up to the end of the accounting period, whether they have been reported to the institution or not, less amounts already paid relating to such claims. For example, claims that are being processed; claims that have not yet been made but which relate to events occurring during the current year. Such claims may cover theft or damage that the policy holder has not yet discovered or events occurring at the very end of the year which have not yet been reported to the insurance company.

18.220 The prepayments of insurance premiums are the financial assets of policy holders. If these prepayments relate to life insurance the policy holders are resident or non-resident householders. If they relate to non-life insurance then the policy holders may belong to any section of the economy or to the rest of the world. Similarly, insurance companies have other technical reserves including reserves relating to claims from policy holders on which they have not yet paid out.

18.221 Data for these series come from the ONS surveys to insurance corporations and pension funds (25.3.3) and the statutory returns made by insurance corporations to the Department of Trade and Industry. DTI is the supervisory authority for insurance corporations in the UK. Use is also made of the ONS Family Expenditure Survey (25.4.5) to identify the amounts that households are spending on the different types of insurance.

AF.7 Other accounts receivable/payable

18.222 This category comprises trade credits and advances and other items receivable/payable. They are valued for both creditors and debtors at the amount the debtors are contractually obliged to pay the creditors when the obligation is extinguished.

Domestic trade credit (AF.71)

18.223 This relates to identified trade credit only. Omissions are likely to be substantial, thus reducing the quality of these data. The domestic trade credit shown is mainly that given and received by public corporations and certain financial institutions and central government. An estimate of trade credit given to local government is also included from 1989 and transactions in connection with advance payment of fuel bills are included from 1994.

18.224 Data for public corporations are derived from quarterly returns made by them to the ONS. Before 1972 the information was obtained from their annual accounts. The households sector allocation represents a proportion of unpaid bills to utilities before privatisation. The residual sector is private non-financial corporations.

18.225 Identified trade credit given and received by insurance corporations and pension funds is included here too. Insurance corporations' outstanding premiums, claims admitted but not paid, agents balances, and reinsurance, co-insurance and treaty balances are included. For pension funds and long-term insurance funds the counterpart sector is the household one, but for other insurance corporations the counterpart sectors are the household and the private non-financial corporations ones.

18.226 Central government data here represents credit given by central government trading bodies. The figures are obtained from annual accounts. The counterparty sector (the debtors) are assumed to be private non-financial corporations. Local government trade credit figures are estimated; the creditors are private non-financial corporations.

18.227 The household sector allocation represents a proportion of unpaid bills to utilities prior to privatisation (up to the second quarter of 1991), there are no data for these after privatisation. It also includes, for 1994–1996, the trade credit arising from pre-payment of fuel bills in advance of the introduction of VAT on fuel, based on data from the ONS's Financial Assets and Liabilities Survey. For this private non-financial corporations are the debtors and households the creditors.

18.228 No estimates are made for other types of trade credit, in particular that between the private non-financial corporations and household sectors, variations in which are probably sizeable. This subject was discussed in *Economic Trends*, September 1992.

Import and export credit (AF.71)

18.229 Import and export credit currently comprise suppliers' trade credit and associated advance and progress payments on exports and imports. These include those financed by UK bank loans under ECGD buyer credit schemes, by the Crown Agents, by insurance corporations and long-term credit linked to purchases of ships and commercial aircraft. It includes refinanced export credit. It excludes import credit of public non-financial corporations, but it does include advance payments by such corporations.

18.230 However, trade credit between UK corporations and the rest of the world parents, branches, subsidiaries and affiliates corporations are covered indistinguishably within outward and inward direct investment loans (AF.421).

Other accounts receivable/payable (AF.79)

18.231 Other items recorded here relate to financial claims which arise from timing differences between distributive transactions or financial transactions on the secondary market and the corresponding payment. They relate to:

- taxes on production
- PAYE taxes on income
- Council tax / community charge / rates
- social security contributions and benefits
- subsidies
- central government final consumption expenditure.

Note that the difference between the accrual and payment of interest is not included here, but is deemed to be invested in the original financial instrument.

Chapter 19

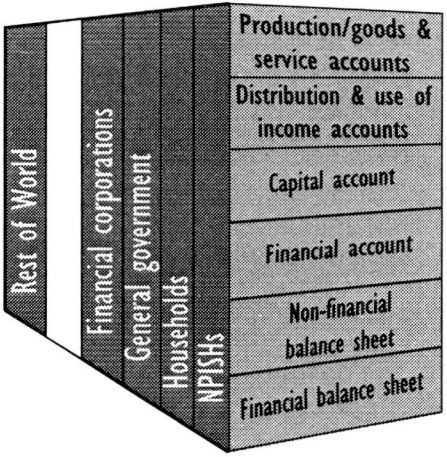

Non-financial corporations

Chapter 19 Non-financial corporations

Introduction

19.1   This chapter deals with the non-financial corporations sector in terms of the institutions it covers in the United Kingdom, including public and private corporations and quasi-corporations. A complete ('top-to-bottom') account is set out and the derivation of the component series from their sources is described.

Definition of the sector

19.2   Non-financial corporations are institutional units whose transactions are distinct from those of their owners, are market producers and have the production of goods and non-financial services as their main activity. They have a quite different economic function from financial corporations whose income comes mainly from interest and dividends on, or transactions in, financial assets.

19.3   The institutional units in the sector are of the following types:

- private corporations (mainly companies);
- public corporations;
- co-operatives and partnerships which are recognised as independent legal entities;
- government market producers which are recognised as independent legal entities;
- non-profit institutions or associations serving non-financial corporations, which are them selves recognised independent legal entities (e.g. trade associations);
- holding corporations where the preponderant activities of the group they control fulfil the above criteria;
- quasi-corporations (*see* paragraphs 19.15–19.16 below).

The economic activities of the self-employed (unincorporated businesses) are excluded and are classified to the household sector.

19.4   The coverage of this sector and its various sub-sectors has changed substantially over time because of transfer of activities between the public and private sectors, liquidations, establishment of new corporations and acquisitions of unincorporated businesses. However, a detailed description of these changes is beyond the scope of this chapter.

19.5   The sector only includes units resident in the United Kingdom. Thus it excludes overseas branches and subsidiaries of UK corporations but includes UK branches and subsidiaries of foreign corporations. Transactions between UK parent corporations and their branches and subsidiaries abroad are not netted out on consolidation but are included as transactions with the Rest of the World sector, as are transactions of UK branches and subsidiaries of foreign parent corporations with those parents.

19.6   The sector is conventionally divided into two subsectors:

- public non-financial corporations (S.11001);
- private and foreign controlled non-financial corporations (S.11002/3).

Public corporations (S.11001)

19.7   These are corporate enterprises which are publicly owned and controlled but which have a substantial degree of independence in the conduct of their day-to-day business. They may have been created by either a central or local government authority, which appoints most or all of the board of management, and the enterprise's borrowing is subject to limits laid down by Parliament.

19.8   Public corporations comprise not only the bodies managing publicly-owned industries (e.g. the Post Office) but also a number of other institutions which are active in other branches of economic activity (e.g. the British Broadcasting Corporation).

19.9   A major expansion of the subsector occurred with the nationalisation of industries after the Second World War, involving coal, gas, electricity, much of the transport industry and most of iron and steel. Between 1953 and 1963 iron and steel enterprises were progressively sold back into private hands but in 1967 they were taken into public ownership again under the British Steel Corporation. Some road haulage enterprises were also denationalised after 1953 though large parts of the road freight industry continued to be publicly owned until 1982.

19.10   The sub-sector has also expanded because of changes in the status and organisation of certain general government trading bodies, e.g. the Civil Aviation Authority, Her Majesty's Stationery Office and British Nuclear Fuels.

19.11   The bulk of the national accounts data needed for public corporations is taken from quarterly ONS inquiries to the largest corporations, supplemented by information from published accounts, other government departments and counterpart sources or by extrapolation.

19.12   Since the early 1980s there has been a large contraction in the public corporations subsector with denationalisation or privatisation of many bodies which used to be publicly owned but now form part of the private corporations subsector. The major privatisations have been (in chronological sequence):

- British Telecommunications (privatised November 1984);
- British Gas (December 1986);
- British Airways (February 1987);
- British Airports Authority (July 1987);
- British Steel (December 1988);
- the regional and Welsh water authorities (December 1989);
- the regional electricity companies (December 1990);
- National Power and PowerGen (March 1991);
- British Coal mines (December 1994);
- British Rail (1994–1997 in stages).

Private national and foreign owned corporations (S.11002 and S.11003)

19.13   These are corporations in the private sector whose income is mainly derived from the production of goods, the provision of non-financial services and other trading activities. Under a derogation from the ESA the United Kingdom (in common with various other EU member states) is recording financial auxiliaries within non-financial corporations until 2002 because of difficulties of satisfactorily identifying this sub-sector.

19.14   The main sources of data here are surveys of businesses carried out by the ONS, based on samples of corporations drawn from the Inter-Departmental Business Register (IDBR). Other sources are the dividends and interest matrix (DIM), central government data on tax payments, counterpart data comes from banks and other financial corporations, the balance of payments and information on acquisitions and mergers.

Quasi-corporations

19.15   These are bodies which have no independent legal status but keep a complete set of accounts and exhibit economic and financial behaviour which is distinct from that of their owners and more similar to that of corporations. In the United Kingdom partnerships used to be classified to the household sector, in common with all other unincorporated businesses, but are now classified as quasi-corporations under the above criteria. On the other hand sole proprietors' accounts are not genuinely separable from those of the households of which they form a part and in many cases their businesses do not have autonomy of decision-making. The United Kingdom has therefore decided to classify them to the household sector rather than to quasi-corporations.

19.16   Quasi-corporations are sectorised as either financial or non-financial. To be in the non-financial corporations sector they must be market producers mainly concerned with the production of goods and non-financial services (e.g. professional partnerships and independent schools). This sector also includes all notional resident units.

Output (P.1)

19.17   This is broken down into market output (P.11) - the largest element - and output for own use (P.12).

Market output (P.11)

19.18   This is defined as the output that is sold or disposed of on the market, normally at prices that are economically significant. It can be derived from commercial accounts as the sum of:

- sales of own production;
- changes in inventories of finished goods and work-in-progress;
- output not sold on the market.

Market output is valued at basic prices (i.e. excluding taxes on products).

Sales of own production

19.19   Data on these come from annual ONS surveys of businesses, mainly in production industries, construction and the distributive and service trades. (*See* Chapter 25, paragraph 25.1 for details.) Inquiries conducted by the Ministry of Agriculture, Fisheries and Food and DETR are also drawn on for agriculture and construction respectively. Commercial accounting practice and national accounting standards largely coincide in this area so the information is readily available and easily usable. The Inter-Departmental Business Register (IDBR) contains codes which provide a sector breakdown.

19.20   Quarterly data on market sales come from further ONS inquiries, mainly those carried out for the monthly index of production, the monthly index of retail sales and monthly and quarterly surveys of the service trades (25.1). The coverage of quarterly surveys is less comprehensive than that of annual ones, so some quarterly figures are estimated by extrapolation. When annual data are produced the quarterly estimates are all aligned to them.

Changes in inventories

19.21   Inventories for this purpose are produced goods and services that are being held for sale, for use in production or for some other purpose at a later date. They consist of materials and supplies, work-in-progress, finished goods and goods for resale. For the national accounts they must be valued at replacement cost as at the time of production. However, commercial accounting uses historic cost or 'book values' (i.e. as at the time of purchase) which, given changes in prices, may be very different.

19.22   For national accounts purposes survey data on a historic basis therefore need to be revalued using price information from the ONS producer and retail price indices. Estimation of inventories is more fully described in Chapter 15.

Output not sold on the market

19.23   This is the final component of market output. Estimates of sales to units within the same enterprise group are collected through various business surveys from which the data on sales are derived. They are combined with payments to employees in the form of income in kind (e.g. company cars, tied accommodation and meal vouchers), information on which is obtained from the Inland Revenue, Family Expenditure Survey and surveys of labour costs. Income in kind also appears in output, in mixed income and in household consumption.

Output for own final use. (P.12)

19.24   This consists of goods or services that are retained by the enterprises in which they are produced. Corporations have no final consumption so output for their own final use relates only to own-account fixed capital formation. Data for this are collected separately in the ONS inquiries into sales of own production (25.1).

Intermediate consumption   (P.2)

19.25   This is the value of the goods and services consumed by processes of production, excluding fixed assets. It is estimated from annual purchases inquiries (*see* 25.1.3. and 25.1.7–25.1.9).

Value added, gross (B.1g)

19.26   This is the difference between the value of output and that of intermediate onsumption. When summed over all sectors of the economy, and adjusted by the addition of taxes (*less* subsidies) on products, it equates to gross domestic product (GDP).

Consumption of fixed capital (K.1)

19.27   Estimates of this are derived from the perpetual inventory model, as described in Chapter 16.

Value added, net (B.1n)

19.28   This is value added *after* taking account of capital consumption. It is therefore calculated as the value of output *less* both intermediate consumption and consumption of fixed capital.

Compensation of employees (D.1)

19.29   This is the total remuneration (in cash or in kind) payable by a corporation to its employees in return for work. It is in two parts: wages and salaries (D.11) and employers' social contributions (D.12).

Wages and salaries (D.11)

19.30   This comprises basic pay, overtime, bonuses, holiday pay and commission, together with income tax, national insurance contributions, etc, payable by employees, including those that are in practice withheld by the employing corporation and paid directly to the Inland Revenue, pension scheme, etc, on behalf of the employee. Wages and salaries may be paid in many ways, including goods or services provided as remuneration in kind instead of, or in addition to, cash. The item appears as a *use* within the generation of income account of corporations.

19.31   The ultimate source of data on wages and salaries is tax records augmented by quarterly data as described in Chapter 14.

19.32   Separate estimates are made for benefits in kind. When a corporation calculates its trading profits it deducts from its total income both the wages and salaries paid in cash and the cost of providing income in kind. The individual employee is subject to tax on most such income and, though it is not generally collected through the PAYE system, employers are required to give details to the Inland Revenue. A breakdown by industry is not provided by this source but is available from occasional surveys of labour costs. Further information on income in kind is obtained from the Family Expenditure Survey. Some types of income in kind can be allocated wholly to this sector; others are allocated approximately.

19.33   The largest component of income in kind paid by non-financial corporations is company cars, which are subject to tax to the extent that they are for private use.

Employers' social contributions (D.12)

19.34   This is the value of the social contributions incurred by employers to obtain social benefits for their employees, and is viewed as a supplement to wages and salaries. National insurance and other social security contributions are included, using data prepared by the Government Actuary's Department. For employers' contributions to superannuation schemes there are annual figures, quarterly estimates being obtained by interpolation and projection. Funded pension schemes are also covered by an ONS survey. The aggregate contributions are allocated to sectors by reference to their wage and salary bills.

Taxes on production (other than those on products) (D.29)

19.35   As output is recorded at basic prices (excluding taxes on products) only 'other taxes on production' (D.29) are included here. They include business rates and motor vehicle duties paid by corporations. Also included are franchise payments by television companies to the Independent Television Commission; the Independent Broadcasting Levy, part of company registration fees and fees charged by industry regulators. All such payments are deemed to be compulsory payments, the size of which exceeds the cost of any service supplied by government to the individul taxpayer.

Other subsidies on production received (D.39)

19.36   These are current unrequited payments that government makes to corporations on the basis of the levels of their production. They do not include subsidies on products, or grants made by government to finance the capital formation of corporations or compensate them for damage to capital assets. They do include subsidies for employment and training. The main source of subsidy data is the central government quarterly GEMS return.

Operating surplus, gross and net (B.2n and B.2g)

19.37   This, the balancing item for the generation of income account, is made up of:

> value added
> *less* compensation of employees payable
> *less* taxes on production payable
> *plus* subsidies received.

This is carried down into the allocation of primary income account as a resource.

19.38   This calculation illustrates the fact that national accounting and commercial accounting practices differ. The main difference is that commercial accounting practice accounts net interest received as profit whereas national accounts treat interest flows as property income. National accounts also distinguish between profit and holding gains so that profit is calculated on a replacement cost basis, though few corporations do so. The net surplus is the gross figure *less* consumption of fixed capital.

Property income   (D.4)

19.39   Income is receivable by the owner of a financial asset or tangible non-produced asset in return for providing funds to, or putting the asset at the disposal of, another institutional unit. It is composed of five elements:

- interest;
- distributed income of corporations;
- reinvested earnings on direct foreign investment;
- attributed property income of insurance policy holders;
- rent.

Property income is shown as a resource and as a use of non-financial corporations in their allocation of primary income account, since it represents both the property income they receive (resource) and that which they pay out (use).

Interest   (D.41)

19.40   Data for the interest payable and receivable by non-financial corporations come from the dividends and interest matrix (DIM), which is described in detail in Chapter 14. The DIM uses data on the financial assets and liabilities of each sector together with some known flows to estimate the dividends and interest income receivable and payable by each sector. Data for non-financial corporations' financial assets and liabilities come from a special ONS survey (*see* 25.2.5).

Distributed income of corporations   (D.42)

19.41   Dividends and shares issued to shareholders are shown here, along with any dividends on shares owned in other corporations, data coming from the dividends and interest matrix. Non-financial corporations' financial assets and liabilities are obtained from the special ONS survey (25.2.5, 25.2.3) while the special case of the distributed income of quasi corporations owned by government is covered by information from the government accounts.

Reinvested earnings on direct foreign investment   (D.43)

19.42   These equate to:

operating surplus of direct foreign investment corporations
plus property incomes or current transfers receivable
less any property incomes or current transfers payable (including actual remittances to foreign direct investors and any current taxes payable on the income, wealth, etc, of direct foreign investment enterprises).

These figures may be either positive or negative.

19.43   The figures come from the ONS survey of direct investment (25.5.1). Earnings from direct investment abroad by non-financial corporations are shown as resources here, and their re-investment is included as the acquisition of an asset in the financial account (F.519). The contra flow, amounts earned by UK subsidiaries etc and deemed to be remitted abroad and re-invested in the United Kingdom, are included as uses here and their re-investment recorded as the acquisition of a liability in the financial account (F.515).

Property income attributed to insurance policy holders (D.44)

19.44 Property income is received by the corporate holders of insurance policies from the investment by the insurers of insurance technical reserves held in financial assets, land and buildings. (*See* Annex 2 to Chapter 20.)

19.45 Data come from ONS surveys of insurance enterprises and pension funds and from the ONS dividends and interest matrix, which all provide quarterly figures. The matrix provides a total property income figure for insurance corporations and pension funds and data from insurance corporations themselves are used to deduct property income on their own shareholders' reserves.

Rent (D.45)

19.46 In principle this relates only to rent on land and sub-soil assets. However, in practice it is not possible in most cases to separate out the rent on land itself from that of buildings erected there. In practice individual series are allocated to whichever category (rent or receipt / payment for building services) is thought to predominate. Rent is both a resource (rent receivable) and a use (rent payable) in the allocation of primary income account. Since 1985 rent figures have been based on Inland Revenue tax records. For recent years, for which complete data are not available from this source, estimates are derived from information on trends in rental income from portfolios of commercial property. Quarterly estimates of rent are obtained by interpolation between and projection of annual data.

Balance of primary incomes, gross (B.5g)

19.47 This is simply carried down to appear as a resource in the secondary distribution of income account.

Social benefits other than social transfers in kind (D.62)

19.48 These relate to unfunded social insurance schemes, none of which have been identified in this sector.

Other current transfers (D.7)

19.49 These consist of:

- net non-life insurance premiums (D.71);
- claims (D.72);
- miscellaneous current transfers (D.75).

Net non-life insurance premiums and miscellaneous current transfers appear in the *uses* section of the secondary distribution of income account while non-life insurance claims are *resources* within that account.

Net non-life insurance premiums (D.71)

19.50   These are the total of the non-life premiums payable after deduction of service charges paid for the arrangement of the insurance *plus* the premium supplements treated as payable out of the property income attributed to insurance policy holders, as described in paragraph 19.44 above. (*See also* Annex 2 to Chapter 20.)

19.51   The source for these data is ONS surveys of insurance corporations (25.3.3) which collect premiums receivable by type of insurance product, supplemented by data from the Association of British Insurers for purposes of sectorisation.

Non-life insurance claims (D.72)

19.52   These are the amounts receivable in settlement of claims that become due during the current accounting period. The settlement of a non-life insurance claim is shown as a transfer to the claiming non-financial corporation and thus appears in the *resources* section of their secondary distribution of income account.

19.53   Again the ONS surveys of insurance corporations (25.3.3) provide the necessary data on claims payable by broad class of insurance while sectoral information from the Association of British Insurers enables non-financial corporations to be distinguished.

Miscellaneous current transfers (D.75)

19.54   These are current transfer taking place between non-financial corporations and other institutional units. They are shown in the secondary distribution of income account as one of the *uses* of the non-financial corporations sector. The identified uses comprise fines and penalties paid to general government.

Current taxes on income and wealth (D.5)

19.55   These are taxes on corporate profits and wealth that are payable regularly every tax period. A breakdown of this into taxes on income (D.51) and other current taxes (D.59) is given in the secondary distribution of income account, as *uses* of non-financial corporations.

Taxes on income (D.51)

19.56   These are taxes on the profits (including capital gains) of non-financial corporations, and are based on Inland Revenue records. The use of corporation tax data is described in Annex 1 to Chapter 14. Taxes comprise corporation tax, capital gains tax and petroleum revenue tax.

Other current taxes (D.59)

19.57   No taxes in this category are paid by non-financial corporations.

Disposable income, gross (B.6g)

19.58   This is the balancing item for the secondary distribution of income account. It also appears as a resource in the use of disposable income account.

Saving, gross (B.8g)

19.59   This is the balancing item on the *uses* side of the use of disposable income account, and equals the net disposable income for the non-financial corporations sector. It is also carried down to appear in the 'changes in liabilities and net worth' section of the capital account.

Capital transfers (D.9)

19.60   These are characterised by the disposal or acquisition of assets by one or both parties, tending to be large and infrequent. They are broken down in the tables into capital taxes (D.91), investment grants (D.92) and other capital transfers (D.99). Investment grants and other capital transfers appear in the 'changes in net worth' section of the capital account whereas capital taxes and other capital transfers appear as 'changes in assets' in that account. General government is the payer or recipient of the vast majority of capital transfers. Consequently government accounts are the principal source of data.

Capital taxes (D.91)

19.61   These are taxes which are levied on the values of the assets or net worth of corporations, or on the value of assets transferred between them, and are treated as changes in assets. They tend to be irregular and infrequent with just a few large entries, though there is a small continuous element in development land tax. The tax on windfall profits of utilities is included here for 1997 and 1998.

Investment grants (D.92)

19.62   Non-financial corporations receive grants to finance all or part of the cost of acquiring fixed assets. Such capital transfers are usually in cash in the United Kingdom, though in principle they could be in kind. They are generally shown as changes in liabilities and net worth but transfers to a public corporation from its sponsoring department are treated as equity injections. Data are obtained from the public expenditure monitoring system, in particular the quarterly GEMS returns.

Other capital transfers (D.99)

19.63   All capital transfers other than capital taxes and investment grants are included here, including the cancellation of debt by mutual agreement between creditor and debtor, except that the cancellation by central government of public corporations' debts in the context of privatisation is regarded as the exchange of one financial instrument (a loan) for another (equity). The main data source is the public expenditure monitoring system, in particular the quarterly GEMS returns.

Change in net worth due to saving and capital transfers (B10.1)

19.64   This is the balancing item on the capital account. After deduction of consumption of fixed capital it represents the positive or negative amount available to the non-financial corporations sector for the acquisition of non-financial and financial assets.

Gross capital formation (P.5)

19.65 The main categories of gross capital formation are :

- gross fixed capital formation (P.51);
- changes in inventories (P.52);
- acquisitions *less* disposals of valuables (P.53).

Each of these is described in the following paragraphs.

Gross fixed capital formation (P.51)

19.66 This is the total value of a producer's investment in fixed assets *plus* improvements to land and the costs associated with the transfer of assets. The investment is in assets which are used in the production process for more than a year. The main sources of data are ONS surveys of corporations in the production, construction, distribution and service industries (25.2.1, 25.2.2), returns made by public corporations (25.2.3) and a DETR survey of construction output. These data are supplemented by estimates from the commodity flow approach which results in an improved quarterly profile. (*See also* Chapter 15.)

Changes from former system of accounts

Under the new SNA this category includes intangible fixed assets such as computer software and mineral exploration, which were formerly excluded. *See* Chapter 15 for more details.

Changes in inventories (P.52)

19.67 Inventories are goods and services that are held for sale or other use at a later date, and consist of materials and supplies, work-in-progress, finished goods and goods for resale. Sources are ONS inquiries (25.1.3, 25.1.5).

19.68 Difficulties in measuring changes in inventories for the national accounts arise because the valuation procedures required are different from commercial accounting practice. The methodology is described in Chapter 15.

Changes from former system of accounts

In the previous system of national accounts inventories were known as stocks and changes in them as stock building. Under the new SNA this category now includes work-in-progress on cultivated assets and crops and in certain service industries such as computer software and films, all of which were formerly excluded.

Acquisitions *less* disposals of valuables (P.53)

19.69 This records the net purchases (or sales) of valuables. Valuables are goods that are not used primarily for production or consumption but are acquired and held mainly as stores of value. Estimates of corporations' purchases and disposals of precious metals, precious stones, antiques and other art objects are included here.

19.70 The 'whole economy' estimate is calculated from imports *less* exports of valuables, obtained from trade data, and estimates of dealers' margins, with a somewhat arbitrary allocation between non-financial corporations and the other sectors involved (*viz.* households and government).

Acquisitions *less* disposals of non-produced non-financial assets (K.2)

19.71   This shows corporations' purchases and sales of land and other non-financial assets that come into existence other than through processes of production, including the costs of ownership transfer and any major improvements to the assets. As discussed in Chapter 5 the separation of the cost of land from the cost of buildings is extremely difficult and therefore in practice the borderline between this category and gross fixed capital formation is approximate.

Net lending (+) and net borrowing (-) (B.9)

19.72   This, the balancing item of the non-financial assets account, should equal the net acquisition of financial assets and liabilities in the financial account, but with sign reversed. In practice it does not, the statistical discrepancy between the two being an indication of net errors and omissions. The statistical discrepancy is shown at the end of the financial account.

Financial account

19.73   The main sources of data for private non-financial corporations are:

- Financial Assets and Liabilities Survey (FALS), using quarterly returns to the ONS from the largest corporation groups (25.2.5);
- overseas direct investment inquiries, with quarterly and annual returns to the ONS;
- counterpart data from banks and other financial corporations (25.3);
- counterpart data from the balance of payments;
- other information from counterpart sectors (e.g. loans from central government);
- data collections within UK on cross-border acquisitions and mergers;
- quarterly returns of the financial transactions of the largest public corporations (25.2.3).

Securities other than shares (F.3)

19.74   This heading includes a variety of elements: money market instruments, corporation bonds and preference shares, government bonds, bonds issued by other sectors (including the Rest of the World) and derivatives. Figures are estimated for these instruments individually for each sector, the sources and methods involved being covered in greater detail in Chapter 17.

19.75   For private non-financial corporations the changes in holdings recorded by the ONS Financial Assets and Liabilities Survey (25.2.5) are the main source used to estimate total transactions in the different instruments.   Some information is also available from counterparty sectors, notably banks and building societies, often on a quarterly basis. These counterparty data are usually balance sheet figures from which estimates of transactions are derived.

19.76   Data provided to the Bank of England by commercial paper issuers (including non-financial corporations) and Euroclear data from the Bank for International Settlements (BIS) are also used. Issues of foreign currency commercial paper elsewhere are estimated from returns made by the main issuers, all of whom are private non-financial corporations. Again the figures for transactions in the financial account are derived from balance sheet data.

19.77   The data on *issues* of sterling commercial paper are of good quality but asset figures are less reliable as a part of the residual is allocated to them. Foreign currency data are generally not as good as those denominated in sterling.

19.78 Figures for local authority debt instruments come from returns by individual authorities to the DETR.

19.79 Figures for private non-financial corporations' transactions in gilts are grossed-up estimates based on changes in holdings (at book value) reported by large corporations in the Financial Assets and Liabilities Survey (25.2.5), and previously the Survey of Company Liquidity.

19.80 Public corporations' transactions in securities are derived from the balance sheet data collected from the largest of them through the Financial Assets and Liability Survey, together with information from counterparty sector sources and from the corporations' own accounts. Generally the figures are of better quality than those for private non-financial corporations.

Loans (F.4)

19.81 Most of these figures are compiled by the Bank of England from returns provided by banks, by the Building Societies Commission from societies' returns and by the ONS from surveys addressed to other financial institutions (25.3). Certain data for United Kingdom private non-financial corporations are also received from the European Investment Bank, the Bank for International Settlements and the Finance and Leasing Association. Information on loans by central government and public corporations comes from their detailed accounts. Counterparty sector data are used but are sometimes insufficiently detailed, in which case a convention for sectorisation is agreed between the Bank of England and the ONS.

19.82 Loans relating to direct investment (inward and outward) are obtained from ONS surveys (25.5.1). These cover all corporations in the oil sector *plus* a high proportion of other private corporations. Public corporations are deemed not to have such loans.

19.83 The ONS retail sales survey (25.1.10) provides data on new credit extended by retailers and on repayments of such credit. It is all assumed to be advanced to the household sector.

Shares and other equity (F.5)

19.84 Estimates are compiled from a variety of sources, including the London Stock Exchange, Bank of England and ONS statistical enquiries. Transactions in real estate and equities by the Rest of the World are included here.

Liabilities

19.85 Estimates for market issues, including issues listed on the Stock Exchange, are compiled by the Bank of England. Issues by quoted UK corporations relate to new money raised through ordinary shares, i.e. public issues, offers for sale, issues by tender, placing and issues to employees and shareholders (though the last of these are included only if the sole or principal share register is maintained in the United Kingdom). Estimates are based on the prices at which shares are offered on the market. Issues of unquoted equity for United Kingdom companies are based on information from the University of Nottingham's Centre for Management Buy Out Research and the Bank of England's monitoring of press and other sources. Data from the ONS direct investment inquiries (25.5.1) is used for reinvested earnings on direct investment. (*See* paragraphs 19.42 to 19.43 above.)

Assets

19.86　Transactions by private non-financial corporations include their estimated expenditure on acquiring independent corporations and subsidiaries from other sectors. They also include their transactions in current assets (estimated from the Financial Assets and Liabilities Survey, 25.2.5) and other transactions identified from the press. Sales of subsidiaries to corporations abroad are taken from ONS statistics of cross-border acquisitions and mergers. Figures for public corporations are taken from surveys of them and from their annual accounts.

Prepayments of insurance premiums and reserves for outstanding claims (F.62)

19.87　This represents the amounts already paid in insurance premiums against cover for periods that extend beyond the current year, together with provisions for outstanding claims. These are both financial assets of policy holders. Data come from ONS surveys of insurance corporations and the statutory returns made by insurance corporations to the Department of Trade and Industry. Detailed insurance product data from the Association of British Insurers are also used to allocate the figures to counterpart sectors.

Other accounts receivable (F.7)

19.88　This category consists of financial claims which are created as a counterpart of a financial or non-financial transaction when there is a timing difference between this and the corresponding payment. Trade credits and advances come into this category.

Domestic trade credit (F.71, part)

19.89　The credit shown is that given and received by public corporations, based on quarterly returns made by them. No estimates are made for other types of trade credit, in particular that between the private non-financial corporations and the household sector, variations in which are probably sizeable. (See *Economic Trends*, September 1992.)

Import and export credit (F.71, part)

19.90　Trade credit between United Kingdom corporations and Rest of the World parents, branches, subsidiaries and affiliated corporations is covered indistinguishably within outward and inward direct investment loans (part of F.4).

Other amounts receivable/payable (F.79)

19.91　These are financial assets which arise from timing differences between distributive or financial transactions on the secondary market and the corresponding payment. However, interest accruing is recorded as being re-invested in the instrument which gives rise to it.

Net lending (+) and net borrowing (-) (B.9)

19.92　This represents the balance on the financial account: the difference between net acquisition of financial assets and net incurring of liabilities. The discrepancy between this and the net lending or borrowing derived from the capital account is shown at the end of the financial account.

Balance sheets

19.93 For private non-financial corporations information for the balance sheet come mainly from the Financial Assets and Liabilities Survey (25.2.5), counterpart data (from banks, other financial corporations (25.3) and the balance of payments) and figures for acquisitions and mergers.

19.94 The largest public corporations render returns to the ONS each quarter while material on smaller ones is taken from their annual accounts. Information from counterpart sources such as banks and central government is also used.

19.95 More detailed descriptions of the balance sheet accounts, the instruments they show and the data sources are given in Chapter 18.

Currency and deposits (AF.2)

19.96 Figures relating to deposits with United Kingdom institutions are obtained from the counterpart analysis provided by banks and building societies to the Bank of England and the Building Societies Commission respectively. Estimates of the part of this which relates to deposits with foreign monetary financial institutions are derived from Bank for International Settlement data. For the public corporations sector figures for working balances held by the Post Office are also used. The methodology corresponds to that described in paragraphs 19.75–19.76 for transactions.

Securities other than shares (AF.3)

19.97 These include money market instruments (e.g. commercial paper) and bonds (including government bonds), both of which are discussed in detail in Chapter 17. For private corporations the main source is the Financial Assets and Liabilities Survey (25.2.5) which gathers balance sheet data for different instruments at the end of each quarter, and some information is available from counterpart monetary financial institutions. Balance sheet data are also provided to the Bank of England and the Bank for International Settlement. Figures for holdings of local authority debt instruments come from individual authorities' returns to the DETR. In addition, private non-financial corporations are usually allocated part of the residual for different components of this instrument. The London Stock Exchange provides data for issues of Eurobonds, medium term notes and other debt securities listed on the Stock Exchange for all corporations.

19.98 Public corporations' balance sheet data are obtained from the Financial Assets and Liabilities Survey (25.2.5) which is sent to the largest of them. Other information comes from counterpart sector sources, as described in the previous paragraph. The corporations' own accounts are also used.

Loans (AF.4)

19.99 These data come from balance sheet returns submitted by banks and building societies (*via* the Building Societies Commission) to the Bank of England and from ONS surveys of other financial institutions (25.3). The returns collect some counterpart information and are supplemented by information from the European Investment Bank, the Bank for International Settlements and the Finance and Leasing Association. Data on loans to corporations by central government and by public corporations come from these institutions' detailed accounts. The ONS direct investment surveys (of inward and outward flows) (25.2.1) cover all corporations in the oil sector and a high proportion of other private corporations involved in such investment. Public corporations, on the other hand, do not have, and are not the the subject of, direct investment. Finally the ONS retail sales survey (25.1.10) supplies data on the amount of credit extended by retailers, which constitutes a balance sheet asset for corporations.

Shares and other equity (AF.5)

19.100 The data for this come from various sources including the Bank of England (25.3), the London Stock Exchange, the Financial Assets and Liabilities Survey (25.2.5) and ONS inquiries into overseas direct investment (25.2.1). For unquoted equities information comes from returns by financial corporations, the University of Nottingham's Centre for Management Buy Out Research and the Bank of England's monitoring of press and other sources. For public corporations data from their annual accounts are also used.

Insurance technical reserves (AF.6)

Net equity of households in life insurance reserves and in pension fund reserves (AF.61)

19.101 This liability, which relates to corporations operating non-autonomous pension funds, does not apply in the United Kingdom.

Prepayments of insurance premiums and reserves for outstanding claims (AF.62)

19.102 This represents the insurance premiums already paid against cover that extends beyond the date of the current balance sheet and the proportion of the risks involved in relation to the time remaining on the contract, along with the insurance corporations' assessment of the amounts they expect to pay out to settle claims. Data come from the ONS insurance corporations surveys (25.3.3) (of quarterly and annual income and expenditure and annual balance sheets) and the annual statutory returns made by insurance corporations to the Department of Trade and Industry. Detailed insurance product data from the Association of British Insurers are also used, to allocate the figures to counterpart sectors.

Net financial assets (BF.90)

19.103 This is the final entry in the accounts for non-financial corporations and represents their net worth.

Chapter 20

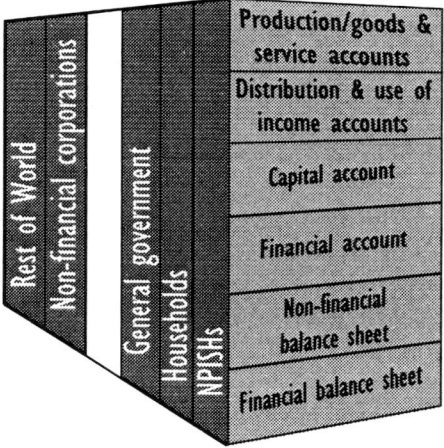

Financial corporations (S12)

Chapter 20 Financial corporations (S12)

Introduction

20.1 This chapter covers financial corporations. It starts by recapitulating the conceptual definition (given in detail in Chapter 10) and describes the financial corporations sector in terms of the institutions it covers in the UK. It then describes the derivation of series from their sources. Non-financial corporations are dealt with in Chapter 19.

This chapter describes:

- the financial corporations sector
- the institutions it covers in the UK
- sub-sectors available in the data

It looks at

- each set of accounts in detail
- the derivation of gross value added for the sector
- the need to impute service charges for financial intermediaries (FISIM)

For each series within the accounts it gives

- a description of the series
- its position within the accounts
- . a description of data sources.

Coverage of this sector

20.2 Financial corporations are all corporations and quasi-corporations which are principally engaged in financial intermediation (financial intermediaries) and/or in auxiliary financial activities (financial auxiliaries).

- **Financial intermediation** is the activity by which an institutional unit acquires financial assets and at the same time incurs liabilities on its own account by engaging in financial transactions on the market. The assets and liabilities of financial intermediaries have different characteristics, so that the funds are transformed or repackaged with respect to maturity, scale, risk etc. in the financial intermediation process.

- **Auxiliary financial activities** are ones closely related to financial intermediation but which are not financial intermediation themselves. For example, activities such as the management of investment funds on behalf of others or insurance broking. Financial auxiliaries do not set themselves at risk by acquiring financial assets or incurring liabilities; they only facilitate financial intermediation.

Thus financial corporations have a quite different economic function from that of non-financial corporations whose income comes from the production of goods and non-financial services.

20.3 Through the financial intermediation process, funds are channelled between third parties with a surplus on one side and those with a lack of funds on the others. A financial intermediary does not merely act as an agent for these other units but instead puts itself at risk by acquiring financial assets and incurring liabilities on its own account.

20.4 The main units classified to this sector include:

- banks (including the central bank of the UK, the Bank of England)
- building societies
- corporations engaged in finance leasing and other forms of personal and commercial finance
- securities and derivatives dealers (on own account)
- mutual funds including money market funds, unit trusts, investment and other collective investment schemes
- insurance corporations and pension funds
- financial auxiliaries including securities brokers, fund managers, investment advisers, insurance brokers etc
- holding corporations, if the group of subsidiaries within the UK economic territory as a whole is mainly engaged in financial intermediation or auxiliary activities
- partnerships which are principally engaged in financial activities
- financial quasi-corporations.

A full list was given in Chapter 10.

20.5 Certain financial bodies are classified to other sectors. For example, the National Savings Bank and the Exchange Equalisation Account are classified to central government because they are an integral part of its financial operations. The insurance activities of Lloyd's UK personal underwriters are classified to the household sector because of the difficulties in distinguishing them from the rest of the household sector. Lloyd's corporate members (allowed to underwrite business for the first time in 1994) are classified as financial corporations. Apart from the Bank of England, no public corporations have been classified to this sector in the UK since Girobank was privatised in 1990.

Quasi-corporations

20.6 Financial quasi-corporations have been moved into this sector as a result of the change to ESA 95. The ESA distinguishes quasi-corporations from their owners who are primarily households in the private sector. They are bodies which keep a complete set of accounts but which are without independent legal status. However they have an economic and financial behaviour which is different from their owners and similar to that of corporations. In the United Kingdom, they comprise:

- unincorporated enterprises, including partnerships, principally engaged in financial intermediation or auxiliary activities and owned by UK residents
- branches principally engaged in financial intermediation or auxiliary activities operating in the UK and which belong to non-resident institutional units

However, sole proprietorships are treated as part of the household sector because their accounts are not generally separable from those of the household of which they are part and in many cases they may not be deemed to have autonomy of decision. In view of this the United Kingdom decided to classify them to the household sector rather than, as the ESA specifies, to quasi-corporations. For example, Lloyd's UK personal underwriters are regarded as part of the household sector.

20.7 There are limited data sources on financial quasi-corporations. Estimates of their operating surpluses are derived by ONS from Inland Revenue's survey of personal incomes. Bank of England returns identify their bank deposits and borrowing from which interest flows can be derived. ONS overseas direct investment inquiries collect information on profits of branches and transactions with their overseas parents.

Sub-sectors available in the UK data

20.8 For the production account this sector is not split. However, for the other accounts and the financial balance sheets it is sub-divided as follows:

- monetary financial institutions (MFIs): (the central bank (S.121) plus other monetary institutions (banks and building societies) (S.122);
- other financial intermediaries (except insurance corporations and pension funds) (S.123) and financial auxiliaries (S.124);
- insurance corporations and pension funds (S.125).

Monetary financial institutions (S.121 and 122)

20.9 This sector consists of the central bank (the Bank of England, sub-sector S.121) and other monetary financial institutions (sub-sector S.122). The central bank comprises the Banking and Issue Departments of the Bank of England, but not the Exchange Equalisation Account (the government account at the Bank of England which contains the official reserves). (Other EU countries treat their official reserves as assets of the central bank.) The other monetary financial institutions sub-sector (S.122) currently comprises banks and building societies and is equivalent to the other depository corporations' sub-sector as defined in the System of National Accounts 1993 (SNA93). However, the latter includes credit unions and money market mutual funds which are very small in the UK - to the extent that data are collected they are included in sub-sector S.123 rather than this sector. Sub-sectors S.121 and S.122 are not published separately. Combined they are equivalent to monetary financial institutions for statistical purposes as defined by the European Central Bank.

Central bank and banks

20.10 Data on the UK central bank are supplied by the Bank of England itself. Data relating to the production, distribution and use of income, capital accounts and financial assets and liabilities for other banks are also compiled by the Bank of England from returns made by banks (25.3.1). The Bank collects data on financial assets and liabilities monthly and quarterly. These include sector analyses of deposits, advances and loans which are also used by the ONS to calculate interest flows with other sectors. Estimates of sterling denominated financial flows (eg net lending or borrowing in the period) are derived from levels data, with adjustments made for population changes, revaluations and write-offs from 1986. The figures for financial flows in foreign currency additionally involve adjustments to take account of exchange rate movements and are less certain. Monthly figures are collected from the larger banks (generally those with eligible liabilities of £30 million or more or total liabilities of £300 million or more). At end-December 1997 monthly returns were being submitted by 369 of the 468 banks; accounting for the great majority of this sector's business.

20.11 Reviews of banking sector statistics are carried out periodically. The most recent which was largely implemented at the end of September 1997 reflects the data requirements of the 1995 ESA.

Building societies

20.12   Since January 1987, data for building societies have been compiled from a system of returns made by the societies to the Building Societies Commission (25.3.2) Aggregated statistics are passed to the Bank of England for inclusion in the monetary statistics and onward transmission to the ONS for inclusion in the financial accounts. Other data for the non-financial accounts are compiled by the Commission from annual profit and loss data and are supplied to the ONS direct.

Other financial intermediaries, except insurance corporations and pension funds, and financial auxiliaries

Other financial intermediaries, except insurance corporations and pension funds (S.123)

20.13   ONS collects quarterly and annual data from the main groups of other financial intermediaries mainly by way of statutory sample surveys. A full list of inquiries is given in Chapter 25 (25.3.3 and 25.3.4). They cover finance leasing companies, non-monetary sector credit grantors, unit trusts, investment trusts, property unit trusts, and securities dealers. The inquiries mainly cover transactions in financial assets and liabilities but income and expenditure data are collected for securities dealers. These are used in the production and primary distribution of income accounts. However, coverage of sub-sector S.123 is incomplete. Data on factoring companies and venture capital and development companies are estimated from external information based on published reports and accounts and no data are collected from the following (although counterpart data from, for example, banks, would be included): bank holding corporations, credit unions, financial holding companies, financial trusts, insurance holding corporations and unauthorised unit trusts. There is a continuing programme of work to extend coverage of this sub-sector and to improve data quality.

Financial auxiliaries (S.124)

20.14   The UK has a derogation for this sub-sector, as do a number of other EU member states. Until 2002, or earlier if sufficient data become available, the UK will combine data and estimates on financial auxiliaries with those of non-financial corporations. Previously, auxiliaries were treated as part of the non-financial corporate sector but few data were available. Estimates of their income from employment are derived from Inland Revenue, employment and average earnings data. Insurance brokers' and fund managers' overseas earnings are derived from ONS inquiries.

Insurance corporations and pension funds (S.125)

20.15   Data for insurance corporations and pension funds come from a suite of quarterly and annual inquiries conducted by the ONS (25.3.3). The inquiries cover samples of corporations writing life insurance business and non-life business. In 1996 coverage was in excess of 80% in terms of premium income written in the UK. A similar suite of inquiries is sent to private and public sector self-administered pension funds. The figures collected form a complete "top-to-bottom" account for these corporations and pension funds. The published figures are grossed up to UK sector totals from the sample data collected and other external data.

Production Account

Output (P.1)

20.16   This is broken down into two parts: market output (P.11), which is the largest, and output for own use (P.12).

Market output (P.11)

20.17   The estimates of sales and inventories which are made by corporations using commercial accounting practices enable this to be derived as the sum of: sales of own production plus changes in inventories of finished goods and work-in-progress plus output not sold on the market.

Sales of own production (and the problems of defining output for financial corporations)

20.18   There are problems defining comprehensively the output of the financial services sector. These are discussed in detail in Annex 1 to this chapter, and in Chapter 2. The problems can be illustrated by considering the example of banks and building societies. Although these financial institutions levy some charges, fees or commissions for borrowing or lending services, they derive most of their income from lending money at a higher rate of interest than they pay on money deposited. However, interest receipts and payments are classified as transfer payments in the economic accounts rather than as receipts and payments for a financial service or output. As such, they do not contribute to the measured output of the economy. For such institutions the economic accounts impute a sale of services (P.119 in the resources side of the production account) equal to the difference between interest receivable and interest payable. This output is deemed to be wholly intermediate consumption of a notional industry/sector (P.119 in the uses side of the production account) with no final sales, so that total domestic product or value added is left unchanged. Counterpart adjustments are also included in property income in the distribution of income account.

20.19   The information on output of services explicitly charged is gathered through the annual and quarterly returns made to the Bank of England, the Building Societies Commission and the ONS surveys of other financial institutions referred to above (25.3).

Changes in inventories

20.20   Inventories (formerly called stocks) are of little importance for financial corporations because physical inputs and outputs are minimal. Financial corporations can have work-in-progress relating to services carried out over a period of time such as executorships and work on company flotations etc but again the amounts involved are thought to be small. There are no regular data sources for this item and the figures provided are no more than token estimates loosely based on an ONS survey of the legal services industry.

Output not sold on the market

20.21   Output not sold on the market is the final component of market output. This covers sales to units within the same enterprise group and products used for payments in kind, including compensation of employees in kind. Separate estimates for this item are not available for financial corporations.

Output for own use (P.12)

20.22 This consists of services that are retained for their own final use by the owners of the enterprises in which they are produced. For corporations generally, this comprises own account capital formation. For financial corporations it is assumed to be negligible.

Intermediate consumption (P.2)

20.23 Intermediate consumption is the value of the goods and services consumed as inputs by a process of production, excluding fixed assets (whose consumption is recorded as consumption of fixed capital - see K.1 below). The goods and services may be either transformed or used up in the process of production.

20.24 For financial corporations, estimates of intermediate consumption are calculated as the difference between total production and gross value added for the sector. The income and expenditure inquiries to financial companies provide aggregate data but no product information is available because the biennial inquiries into the purchases of services industries do not cover financial corporations.

Value added, gross (B.1g)

20.25 This is the balancing item for the production account: equalling the value of output less intermediate consumption. For financial corporations it is derived directly, by summing the various components of value added that are available from survey sources and Inland Revenue estimates.

Consumption of fixed capital (K.1)

20.26 Buildings, telecommunications, and computer hardware and software are among the main fixed assets for financial corporations. Capital consumption also appears in the acquisition of non-financial assets account. The method of calculation is described in Chapters 15 and 16.

Value added, net (B.1n)

20.27 This forms the last line in the production account and is carried forward, as a resource, into the generation of income account. It is value added after taking into account the consumption of fixed capital.

Distribution and use of income accounts

Compensation of employees (D.1)

20.28 This is the total remuneration in cash or in kind, payable by a corporation to employees in return for work done by the latter during the accounting period. It is a use within the generation of income account for corporations and a resource within the allocation of primary income account for households. This is shown in two parts: wages and salaries (D.11) and employers' social contributions (D.12).

Wages and salaries (D.11)

20.29 The main sources for data on wages and salaries are tax records and the ONS monthly and quarterly surveys of numbers in employment and the quarterly survey of average earnings (25.4). The allocation of wages and salaries to this sector derives from the industry classification used in those sources. (Further details of the derivation of wages and salaries are given in Chapter 14.)

20.30 Estimates for income in kind, provision of company cars, beneficial loans etc, are obtained by applying assumptions about the industrial distribution of such benefits to totals derived by the Inland Revenue.

Employers' social contributions (D.12)

20.31 This is the value of the social contributions incurred by employers in order to obtain social benefits for their employees. It is viewed as a supplement to wages and salaries. It is described in more detail in Chapters 5 and 14 (14.39–14.41). The element attributable to the financial corporations sector is obtained by applying the whole economy ratio of employers' contributions to wages and salaries, to the wage and salary bill of financial corporations. These are uses within the generation of income account.

Taxes and subsidies on production and imports, paid (D.2 and D.3)

Taxes on products

20.32 Taxes on products are payable on goods and services when they are produced, delivered, sold, transferred or otherwise disposed of by their producers. They include taxes and duties on imports. Output is measured at basic prices which exclude taxes and subsidies on products. Consequently these are not charged to the generation of income account of the sector.

Production taxes other than on products (D.29)

20.33 These are all taxes incurred by corporations as a result of engaging in production other than taxes on products. They are recorded as uses within the generation of income account. They exclude taxes on profits, and include local business rates (NNDR) and vehicle excise duty.

Subsidies received, other than on products (D.39)

20.34 No subsidies on production other than on products are received by financial corporations.

Gross operating surplus (B.2g)

20.35 This is the balancing item for the generation of income account.

20.36 Net operating surplus is a measure of the 'profit' on the sector's trading accounts after taking account of the consumption of fixed capital. Note however that national accounting and commercial accounting practices differ, particularly on the measurement of output and capital consumption (depreciation on a replacement cost basis).

Property income (D.4)

20.37 This is the income receivable by the owner of a financial asset or a tangible non-produced asset in return for providing funds to, or putting the asset at the disposal of, another institutional unit. It is paid by the other institutional unit. It is composed of five elements: interest, the distributed income of corporations (dividends), reinvested earnings on direct foreign investment, attributed property income of insurance policy holders and rent. Property income is shown as both a resource (the property income they receive) and a use (that which they pay out) in the allocation of primary income accounts for financial corporations.

Interest (D.41)

20.38 Data for the interest payable and receivable by financial corporations come from the ONS dividends and interest matrix (DIM). This has been described in detail in paragraphs 14.75–14.88. The DIM uses data on representative interest rates and stocks of financial assets and liabilities of each sector and certain flows to calculate the dividends and interest income receivable by each sector (and hence payable by the counterpart sectors). Data for financial corporations come from the quarterly and annual returns they make to the Bank of England, the Building Societies Commission and the suite of ONS inquiries to financial institutions (25.3).

Financial Intermediation Services Indirectly Measured (FISIM)

20.39 This is an indirect measure of the value of services for which financial intermediaries do not explicitly charge but obtain their reward from the difference between interest receivable and interest payable. (It replaces the financial services adjustment previously used in the UK accounts.) It is the counterpart to the amount included in the production account of financial corporations and needs to be excluded from property income to avoid double counting. In the UK FISIM is currently calculated as the total property income receivable by all financial intermediaries (except insurance and pension funds) *less* their total property income payable. The adjustment, with negative sign, is made to financial corporations' resources. FISIM is not, however, allocated to user sectors. So a notional sector is used, with negative value added, and a positive FISIM adjustment in the distribution of income account. By this means the entries for interest are those for actual interest receivable and payable for all sectors. This method of calculating FISIM, while recognised by ESA 95, is not regarded as a satisfactory long-term solution. It is described in more detail in Annex 1 to this chapter. The method will be reviewed in the light of the on-going international discussions on the methodology for allocating FISIM to user sectors.

Distributed income of corporations (D.42)

20.40 This item records dividends paid and received on shares and other equity during the accounting period plus estimates of the value of shares issued to shareholders in lieu of payments of actual dividend payments (scrip dividends). Financial corporations make dividend payments on their own issued share capital and also receive dividends on their considerable holdings of shares of other corporations. The data come from the ONS dividends and interest matrix (DIM). The DIM is described in detail in paragraphs 14.75–14.88. The DIM uses data on the financial assets and liabilities of, each sector, and figures for total dividends and interest paid/received, to calculate the dividends and interest income receivable by each sector. Data for financial corporations comes from the quarterly and annual returns they make to the Bank of England, the Building Societies Commission and the ONS (25.3).

20.41 Since financial corporations control a large proportion of shares quoted on the London Stock Exchange, as well as having substantial investments in equities issued by the rest of the world, the sums shown here are large.

Reinvested earnings on direct foreign investment (D.43)

20.42 These are equal to:

the operating surplus of the direct foreign investment corporation
plus any property incomes or current transfers receivable
less any property incomes or current transfers payable, including actual remittances to foreign direct investors and any current taxes payable on the income, wealth etc. of the direct foreign investment enterprise.
These figures may be either positive or negative.

20.43 Figures are derived from the ONS overseas direct investment surveys (25.5.1). Reinvested earnings of direct foreign investment received are a resource and such earnings paid are a use within corporations' allocation of primary income account.

Attributed property income of insurance policy holders (D.44)

20.44 This is the net property income received from the investment of insurance technical reserves held in financial assets, land and buildings. Technical reserves are the assets of policy holders. Under ESA 95 receipts of property income from them are shown as being paid by insurance corporations and pension funds to the policy holders on this line in the accounts. It is subsequently treated as being paid back to the insurance enterprises and pension funds in the form of premium and contribution supplements in addition to the actual premiums and contributions payable. These flows need to be imputed. They do not occur in practice because such property income is not distributed by insurance corporations. This is a change from earlier national accounting systems, which did not impute such flows of property income.

20.45 Data for this come from the ONS surveys of insurance enterprises and pension funds (D25.3.3) and the ONS dividends and interest matrix (DIM), which all provide quarterly figures. The DIM provides property income figures for insurance corporations and pension funds. Data from insurance corporations are used to deduct property income attributable to their shareholders from earnings on shareholders' reserves and that part of the earnings on policyholders' reserves that can be legally allocated to shareholders. The attributed property income is apportioned between policy holder sectors as described in Chapter 14 (14.84). See also Annex 2 to this chapter.

Rent (D.45)

20.46 This relates to rent on land and sub-soil assets. Rentals for buildings are in principle treated as the purchase and sale of services. However, in practice, it is not always possible to make the distinction in data collections and the split has to be estimated. Rent is both a resource (rent receivable) and a use (rent payable) in the allocation of primary income account. Data are derived from Inland Revenue tax records but supplementary information is also available from some of the inquiries to financial corporations.

Balance of primary incomes (B.5)

20.47 This balance of primary incomes, operating surplus plus net property income, is carried down to appear as a resource in the secondary distribution of income account. It may be shown gross or net of fixed capital consumption.

Social contributions (D.61)

20.48   For financial corporations these are resources in the secondary distribution of income account, where they mainly represent the sums received by insurance corporations and pension funds. No imputed contributions occur in this sector. Payments by financial corporations as employers are included in compensation of employees (D.12).

20.49   The main sources for this information are the Inland Revenue and ONS inquiries to life assurance and pension funds (25.3.3).

Other current transfers (D.7)

20.50   These are: net non-life insurance premiums (D.71) and claims (D.72) and miscellaneous current transfers (D.75).

Net non-life insurance premiums (D.71)

20.51   These relate to insurance policies. The gross non-life insurance premiums comprise:

- the actual premiums receivable from policy holders to obtain insurance cover over the accounting period (premiums earned); and
- the imputed premium supplements attributed to insurance policy holders D.44 (see paragraph 20.46).

Deducting the insurance service charge (the amount taken by the insurance corporation for arranging the insurance, which is shown as output of the insurance corporation and intermediate consumption by policy holding corporations) gives net non-life insurance premiums. These are shown as uses in the policy-holder sectors and as resources for insurance corporations.

20.52   The source for these data are the ONS surveys of insurance corporations which collect premiums receivable by type of insurance product (25.3.3). The ONS surveys are supplemented by data from insurance corporations' statutory returns to the DTI and certain data from the Association of British Insurers. These are used to help split the ONS total premium figures between UK counterparty sectors.

Non-life insurance claims (D.72)

20.53   These are the amounts payable in settlement of claims that become due as a result of events during the current accounting period. The settlement of a non-life insurance claim is recorded as a use in respect of insurance corporations. It is shown as a transfer to the claiming sector and thus appears in the resources section of that sector's secondary distribution of income account.

20.54   The sources for these data are the ONS surveys (25.3.3) of insurance corporations which collect claims payable by broad classes of insurance. Data from insurance corporations' statutory returns to their supervisory authority and the Association of British Insurers are used to help obtain the split into UK counterpart sectors.

Miscellaneous current transfers (D.75)

20.55   These are various different kinds of current transfers which take place between financial corporations and resident and non-resident units. They are shown in the secondary distribution of income account as one of the uses of this sector. The amounts involved are small.

Current taxes on income and wealth (D.5)

Taxes on income (D.51)

20.56   These are taxes on the profits (including capital gains) of financial corporations. The source is the Inland Revenue.

Other current taxes (D.59)

20.57   No taxes in this category are paid by this sector.

Social benefits other than social transfers in kind (D.62)

20.58   These consist of all social insurance benefits provided by financial corporations under private funded social insurance schemes. They are intended to provide for the needs that arise from certain events or circumstances: for example, sickness, unemployment, retirement, housing, education. The figures mainly relate to pensions paid. They are shown as a use in the secondary distribution of income account for financial corporations and as a resource for the household sector and for households in the Rest of the World.

20.59   Data are derived from ONS inquiries to life assurance and pension funds (25.3.3).

Disposable income (B.6)

20.60   This is the balancing item for the secondary distribution of income account. It also appears as the resource in the use of disposable income account. It may be shown before or after deduction of fixed capital consumption.

Adjustment for change in net equity of households in pension funds (D.8)

20.61   The increase in pension funds reserves represented by contributions *less* pensions paid is regarded as an asset of the households sector. However, the treatment of these items in the distribution of income accounts leads to it being recorded in the pension providing sector (usually financial corporations). This correction is necessary to put the savings of the households and pension providing sectors on to the correct basis.

Gross saving (B.8g)

20.62   This is the balancing item on the uses side of the use of disposable income account. It is carried down to appear as the opening balance in the capital account as an entry in the changes in liabilities and net worth section.

Accumulation accounts

Capital accounts

Capital transfers (D.9)

20.63   These are characterised by the disposal or acquisition of assets by one or both parties. They are broken down in the tables into capital taxes (D.91), investment grants (D.92) and other capital transfers (D.99).

Capital taxes (D.91)

20.64   The most recent tax of this kind to affect the financial corporations sector was the one-off tax on bank deposits in 1981–82. The data source is the Inland Revenue.

Investment grants (D.92)

20.65   Investment grants consist of capital transfers by governments or the rest of world, including institutions of the European Union, to other institutional units to finance the costs of acquiring fixed assets.

20.66   Data sources are the government accounts underlying the public expenditure monitoring system. No amounts have been recorded for financial companies in recent years.

Other capital transfers (D.99)

20.67   Other capital transfers comprise transfers, except capital taxes and investment grants, that redistribute savings, rather than income, between sectors. Included here, amongst other items, is the cancellation of debt by mutual agreement between creditors and debtors but not the write-off of debt. Under ESA the latter should be recorded in the "other changes in the volume of assets accounts".

20.68   These comprise payments by central government to the Post Office Superannuation Fund, and the cancellation by monetary financial institutions of debt of the rest of the world. The sources are government accounts and the banking statistics.

Change in net worth due to saving and capital transfers (B.10.1)

20.69   This is the balancing item for this account. It is shown in the changes in liabilities and net worth side of the acquisitions of non-financial assets' account and in the changes in assets portion of the change in net worth due to saving and capital transfers' part of the capital account. Gross of capital consumption it represents the positive or negative amount available to the sector for the acquisition of non-financial and financial assets.

Consumption of fixed capital (K.1)

20.70   Consumption of fixed capital represents the capital resource 'used up' in the course of production. It is also recorded in the production account. The derivation of this series is described in Chapters 15 and 16.

Gross capital formation (P.5)

20.71 The components of gross capital formation are :

- gross fixed capital formation (P.51)
- changes in inventories (P.52)
- acquisitions less disposals of valuables (P.53).

More information on each of these is given in Chapter 15.

Gross fixed capital formation (P.51)

20.72 This is the total value of investment by the financial corporations sector in fixed assets, plus improvements to land and the costs associated with the transfer of assets. The main sources of data are: ONS surveys to financial corporations (25.3) and the Bank of England. However the data collected are not comprehensive and are supplemented by estimates from the commodity supply approach.

Changes in inventories (P.52)

20.73 Inventories (formerly called stocks) are of little importance for financial corporations because physical inputs and outputs are minimal. Financial corporations can have work-in-progress relating to services carried out over a period of time such as executorships and work on company flotations etc but again the amounts involved are thought to be small. There are no regular data sources for this item and the figures provided are no more than token estimates loosely based on an ONS survey of the legal services industry.

Acquisitions less disposals of valuables (P.53)

20.74 This should record the net purchases (or sales) of valuables. There are no direct sources for financial corporations and transactions are currently estimated from overseas trade and dealers' margins.

Acquisitions less disposals of non-produced non-financial assets (K.2)

20.75 Information on net investment in land, property and ground rents by life assurance and pension funds is available from ONS inquiries.

Net lending(+)/net borrowing(-) (B.9)

20.76 This is the balancing item on the acquisitions of non-financial assets account. By definition this item should equal the financial account bottom line figure for net acquisition of financial assets and liabilities, but with sign reversed. In practice the raw data do not balance. The statistical discrepancy between the two, which reflects the net sum of all the errors and omissions in the sector account, is shown at the bottom of the financial account.

Financial account

20.77 The financial account records transactions in financial assets and liabilities between institutional units, and between them and the rest of the world. For financial corporations the main data sources are the monthly and quarterly returns to Bank of England, the Building Societies Commission and the ONS. A full list is given in Chapter 25 (25.3). The inquiries cover the main sub-sectors within financial corporations but there are some gaps and no information on financial transactions is collected from financial auxiliaries (see para 20.15). The inquiries collect detailed figures on financial transactions and balance sheet positions (quarterly for short-term assets and liabilities and annually for longer term-investments). In some cases the inquiries also provide detail on the counterpart sector to various transactions eg bank lending to non-financial corporations, bank lending to households etc. The individual lines in the account are described below.

Monetary Gold and Special Drawing Rights (F.1)

20.78 The Exchange Equalisation account which contains the United Kingdom official reserves is a central government account, and consequently transactions in this instrument are recorded in the central government sector.

Currency and deposits (F.2)

20.79 The figures for sterling currency transactions by financial corporations include issues of notes by the Issue Department of the Bank of England and changes in holdings of notes and coin (including Scottish and Northern Ireland bank notes)held by the sector. Data come from returns from financial corporations and in addition data from the Bank for International Settlement are used to derive estimates for holdings of foreign currency and deposits with foreign monetary financial institutions (MFIs). By definition deposits are held as liabilities only by monetary financial institutions (central bank, other banks and building societies). Most of the figures, including the split of deposit assets by sector, are obtained from the monthly and quarterly returns by these institutions. Gold held by UK MFIs is included here. (A UK derogation allows it to be treated as a financial asset rather than a valuable.)

Securities other than shares (F.3)

20.80 This heading includes a variety of sub-instruments: money market instruments, public corporation bonds, preference shares, government bonds, bonds issued by other sectors and the rest of the world, and derivatives. Figures are estimated for these instruments individually for each sector. The methods and data sources are covered in detail in Chapter 17. The inclusion of transactions in derivatives is a change under the new system of accounts. Previously they were excluded because they were regarded as contingent assets and liabilities. The Bank of England have developed a new inquiry form to collect information on derivatives from banks, who conduct the majority of transactions in derivatives. ONS inquiries are being changed to collect similar information from non-bank securities dealers and the main investing institutions such as life assurance and pension funds. No data will be available prior to 1998.

20.81 Most of the information on transactions for this item is collected from institutions holding such assets. But coverage is not complete. Therefore some elements of the residual transactions in this instrument are allocated to this sector to improve the coherence of the sector accounts.

Loans (F.4)

20.82   The figures include lending secured on dwellings and other long-term secured lending together with short-term overdrafts and loan accounts. Most of these figures are compiled by the Bank of England from returns provided by banks, from returns submitted by the building societies to the Building Societies Commission and from ONS surveys to other financial institutions including non-bank credit grantors and life assurance companies. These institutions are the major providers of loans in the UK economy.

Shares and other equity (F.5)

20.83   Estimates are compiled from a variety of sources, including the London Stock Exchange, Bank of England and statistical enquiries by ONS. Transactions with the rest of the world in real estate are also included here.

Liabilities

20.84   Estimates for market issues, are made using data from the London Stock Exchange and the Bank of England (25.2.10). All issues by quoted UK corporations raising new money via ordinary share issues (public issues, offers for sale, issues by tender, placing and issues to shareholders and employees) are included. Estimates of issues are based on the prices at which shares are offered on the market. Issues of unquoted equity by UK companies are based on information from the University of Nottingham's Centre for Management Buy Out Research and the Bank of England's monitoring of press and other sources. Data from the ONS direct investment inquiries (25.5.1) is used for re-invested earnings on direct investment. These sources all distinguish financial corporations.

Assets

20.85   Transactions by corporations include their estimated expenditure on acquiring independent corporations and subsidiaries from other sectors. Sales of subsidiaries to corporations abroad are taken from the ONS statistics of domestic and cross border acquisitions and mergers. Banks, building societies and other financial institutions provide figures for quarterly transactions in both quoted and unquoted shares. Those for other financial institutions are, from 1968, adjusted for estimated costs incurred in share transactions and are based on quarterly returns of cash transactions.

Insurance technical reserves (F.6)

Net equity of households in life insurance reserves and in pension fund reserves (F.61)

20.86   This item records changes in the technical provisions set aside by life insurance corporations and pension funds to meet future claims and benefits of policy holders. It represents the net surplus of their income over expenditure during the accounting period which is then regarded as an increase in the household sector's financial claim on them. Estimates of the net increase in these reserves are made using information from quarterly and annual income and expenditure returns by life assurance and pension funds to the ONS (25.3.3).

Prepayments of insurance premiums and reserves for outstanding claims (F.62)

Liabilities

20.87   For insurance corporations these are liabilities, representing the net increase in the amounts they have in their reserves for cover relating to risks in the future and sums set aside for the payment of outstanding claims. The data come from the ONS inquiries to insurance corporations (25.3.3).

Assets

20.88 This represents for policyholders the amounts payable in the current period in insurance premiums against cover for subsequent periods. It also covers provisions for sums which have not yet been paid to the policy holder, relating to claims for events which have happened but which the insurance corporations have not yet settled. These are financial assets of policy holders. Data come from ONS surveys to insurance corporations and the statutory returns made by insurance corporations to the Department of Trade and Industry, who supervise insurance corporations in the United Kingdom. Detailed insurance product data from the insurance trade association, the Association of British Insurers, are also utilised in order to obtain the allocation of these insurance corporations' technical reserves to counterpart sectors.

Other accounts receivable/payable (F.7)

20.89 This category consists of changes in financial claims arising from the timing difference between transactions in goods and services, distributive transactions or secondary trade in financial assets and the corresponding payments which can be either early or late. For financial corporations these are mainly related to accrued interest flows and amounts due to and from stockbrokers/ securities dealers.

Domestic trade credit

20.90 No estimates are made for trade credit involving financial corporations. This subject is discussed in Economic Trends, September 1992.

Import and export credit

20.91 Trade credit between UK corporations and rest of the world parents, branches, subsidiaries and affiliates corporations are covered indistinguishably within outward and inward direct investment loans (part of F.4).

Other amounts receivable/payable

20.92 For financial corporations these amounts mainly relate to the difference between interest earned in the period, including index linking, and that actually received, and amounts due to and from UK stockbrokers/securities dealers and overseas financial institutions.

Net lending (+)/net borrowing (-)

20.93 This represents the balance for the financial account: the difference between the net acquisition of financial assets and the net incurrence of liabilities. It should be the same as the balance on the acquisitions of non-financial assets account but with sign reversed.

Financial balance sheets

20.94 For financial corporations the data for the balance sheets come mainly from monthly and quarterly returns to Bank of England, quarterly returns to the Building Societies Commission and quarterly and annual returns to ONS. A full list of inquiries is given in Chapter 25 (25.3). In addition, figures are sometimes obtained from the counterparties (e.g. the public sector in respect of its deposits with and borrowing from financial institutions).

Currency and deposits (AF.2)

20.95 The figures mainly relate to financial corporations' deposit liabilities and assets but also include issues and holding of notes and coin. Data on issues and holdings of notes and coins by financial corporations relate to banks and building societies. Holdings of currency by other financial corporations are negligible. They are not separately identified in data collections. Estimates of insurance corporations' holdings are made based on a 1982 benchmark using DTI statutory returns. Data on deposits, including the split by sector, also come from the Bank of England and the Building Societies Commission. Information on deposits is also collected on ONS inquiries to financial corporations (25.3) and this is used, together with data from the Bank for International Settlement, to derive estimates of holdings of foreign currency deposits with foreign monetary financial institutions.

Securities other than shares (AF.3)

20.96 This heading includes a variety of sub-instruments as described in paragraph 20.84. Individual estimates are made for holdings of each sub-instrument. The main sources are the monthly and quarterly inquiries to financial corporations carried out by the Bank of England, the Building Societies Commission and the ONS (25.3). For money market instruments and other short-term assets the balances at the end of each month or quarter are collected. Similar information is also collected for issues of money market securities and other short-term liabilities. The London Stock Exchange also provides data on all corporate issues of Eurobonds, medium-term notes and other debt securities listed on the exchange.

Loans (AF.4)

20.97 Data come from the monthly and quarterly balance sheet returns submitted by banks and building societies and the ONS inquiries to financial institutions (25.3).

20.98 The ONS direct investment surveys (of inward and outward direct investment) (25.5.1) cover all corporations in the insurance sector and a high proportion of other corporations involved in such investment.

Shares and other equity (AF.5)

20.99 The data for this come from sources including the Bank of England, the London Stock Exchange (25.2.10), the ONS Financial Assets and Liabilities Survey (25.2.5) to non-financial corporations and ONS overseas direct investment inquiries. The counterpart of UK holdings of real estate in the rest of the world is included here as an asset of domestic sectors, while the corresponding investment by residents of other countries in UK real estate is a liability of domestic sectors in this line. Many series within this category are estimated from quarterly transactions data using a perpetual inventory method. Quarterly data are subsequently revised in light of annual balance sheet data. Some data on financial corporations' holdings of unquoted shares are available from ONS inquiries. Use is also made of information from the University of Nottingham's Centre for Management Buy Out Research and the Bank of England's monitoring of press and other sources.

Insurance technical reserves

Net equity of households in life insurance reserves and in pension fund reserves (AF.61)

20.100 These reserves are the technical provisions of insurance corporations and pension funds held to pay future benefits to policy holders or beneficiaries. For insurance corporations they are reserves against outstanding risks on life insurance policies they have written and provisions for with-profits insurance that add to the value on maturity of with-profits endowment policies. For pension funds these are provisions set up to provide pensions for employees or the self-employed. Thus this sum represents the total claim by the household sector on the assets held by insurance and pension funds in respect of their long-term business. The estimates are based on annual balance sheet data supplemented by quarterly data on the income, expenditure, and financial transactions of these institutions. The data are collected by the ONS (25.3.3).

Prepayments of insurance premiums and reserves for outstanding claims (AF.62)

<u>Liabilities</u>

20.101 These are liabilities of insurance corporations and comprise the provisions representing:

- that part of gross premiums written which is to be allocated to the following accounting period;
- the estimated total cost of settling all claims arising from events that have occurred up to the end of the accounting period, less amounts for those claims already settled.

20.102 Data come from the quarterly and annual inquiries to insurance corporations conducted by the ONS (25.3.3).

<u>Assets</u>

20.103 These assets represent the share of insurance corporations' liabilities described in the previous paragraph that are attributable to policy holders in the financial corporations sector. The share is derived using a variety of sources, DTI statutory returns, ONS Family Expenditure Survey, Association of British Insurers, to allocate these reserves across all policy-holding sectors.

Other accounts receivable (AF.7)

20.104 This covers financial claims that are created as a counterpart of a transaction where there is a timing difference between the transaction and the corresponding payment.

Domestic trade credit

20.105 No estimates are made for trade credit relating to financial corporations. This subject is discussed in Economic Trends, September 1992.

Import and export credit

20.106 Trade credit between UK corporations and parents, branches, subsidiaries and affiliated corporations in the rest of the world are covered indistinguishably within outward and inward direct investment loans (part of F.4).

Other amounts receivable/payable

20.107 For financial corporations these amounts mainly relate to the difference between interest earned in the period, including index linking, and that actually received and amounts due to and from UK stockbrokers/securities dealers and overseas financial institutions.

Net financial assets (BF.90)

20.108 This is the final entry in the accounts/tables for financial corporations. It represents their net financial worth.

Annex 1

Financial intermediation services indirectly measured (FISIM)

20a.1 FISIM receipts are earned by institutions whose main activity is financial intermediation, except insurance companies, i.e. S.121–123, and relate to those services for which the institution is compensated by interest rate differentials.

20a.2 For most industries the economic accounts measure output by the value of sales of goods and services adjusted for changes in inventories. For the financial intermediation sector this measure is incomplete because many of the institutions obtain the majority of their net income not from direct charges for the service provided, but by the difference between the interest they receive and the interest they pay. The value of sales for which an explicit charge is made is relatively small, and so is value added on the general definition. When, in the generation of income account, compensation of employees is charged, this leaves a negative operating surplus.

20a.3 To resolve this the economic accounts imputes sales of financial intermediation services. To put a value on the output is not straightforward. When a bank accepts deposits from A and lends them on to B, it is providing services to both. A measure of the total value of the service is the difference between the interest received and that paid. If we take that value and add it to the sales of financial intermediaries we obtain a reasonable estimate of the output and value added of the financial intermediary industry or sector. The total output of financial intermediation for which no explicit charges are made is measured, by convention, as the total property income received by the institutions (excluding income received from investment of their own funds) minus their total interest payments.

20a.4 However, this leaves us with two problems: it is inconsistent with total gross domestic product and, having imputed a flow into the financial intermediary, we have to impute a flow out to balance the account.

20a.5 A solution is to impute a notional sector and industry which "purchases" the imputed output of the financial intermediaries, but has no output of its own. Therefore its value added is negative, but it receives a resource of the same amount from financial intermediaries in the allocation of primary income account. Thus we have the four entries required for the quadruple entry system of the economic accounts:

	Financial intermediaries	Notional sector
Production account	+X	-X
Primary income account	-X	+X

20a.6 This is the solution, based on earlier international guidelines adopted in the past, and in the 1998 Blue Book. (The notional sector is included in the 'not sectorised' column in the 1.8 series of tables.) However, under ESA95 it is expected that a different treatment will be implemented eventually. This would carry the imputation of output to its logical conclusion by imputing the purchase of services to the customer sectors and industry. Since some of those customers are final consumers (households, NPISH, government and the Rest of the World) this would increase measured gross domestic product.

20a.7 It is proposed that for each type of intermediary, the difference between the interest rate charged to borrowers and a "reference" rate carrying no element of implicit service charge would reflect the price of the service provided to the borrower, and the difference between the same reference rate and the rate paid to depositors or lenders would reflect the price of the service paid to them. These margins would need to be allocated among the sectors and industries as (a) purchases by them of financial intermediation services and (b) offsetting interest flows in the dividends and interest matrix.

20a.8 This proposal, attractive in principle, gives rise to practical difficulties, particularly in terms of the calculation of the appropriate reference rate and the allocation of FISIM consumed by households between final use as consumers and intermediate use as traders. Implementation of this treatment has therefore been delayed pending international discussions on appropriate methodology.

Annex 2

Insurance

20b.1 The activity of insurance is intended to provide individual institutional units with financial protection against various risks or contingencies. It is also a form of financial intermediation in which funds are collected from policy holders to be invested in financial (or other) assets which are held as *technical reserves* against the eventualities specified in the insurance policies. The monies collected from policy holders collectively are eventually channelled back to them, although significant amounts of property income (dividends and interest etc.) may be earned from the investment of the funds (and also holding gains or losses accrue) between when the time funds are received as premiums and when they are paid out in claims or on the maturity of the policies.

20b.2 Insurance is thus regarded as a form of financial intermediation since insurance corporations invest funds held for policy holders and this activity has considerable influence on the level of premiums charged: keeping premiums lower than they would otherwise be. The management of its investment portfolio (of financial and non-financial assets) is an integral part of the insurance activity.

20b.3 This annex summarises the less obvious aspects of the treatment of insurance in the accounts:

Output

20b.4 As a form of financial intermediation, there is no explicit service charge for insurance. However, for the national accounts the output of insurance is imputed as:

> Output = total actual premiums earned
> *plus* total premium supplements (= D.44, the income from the investment of the insurance technical reserves - see below)
> *less* total claims due
> *less* the change in technical reserves against outstanding risks

Holding gains and losses are excluded - they are appropriate to the 'other changes in assets' account. In the case of private funded social insurance schemes, for 'premiums' read 'social contributions'.

20b.5 The counterpart of output is in the intermediate or final consumption of policy holding sectors.

Accruals

20b.6 Receipts of premiums, social contributions and payments of claims are recorded on an accruals basis. Premiums accruing are those earned; that is, the amounts that cover the risks incurred during the current period. That part of premiums receivable which covers risks occurring in subsequent periods is recorded as a financial transaction "pre-paid premium reserves" (F.62) at the end of the current period. Similarly, claims are recorded in the period in which the event occurs that gives rise to the claim, irrespective of when the insurance company accepts it. Social insurance schemes are those in which social contributions are paid by employees or other individuals, or by employers on behalf of their employees, in order to secure entitlement to social insurance benefits for the employees or other contributors, their dependants or survivors. The claims accrue in the period in which the employment takes place that gives rise to them. The difference between accruals and the cash flow of social contributions is included in Accounts receivable/payable (F.79) in the financial account.

Premium and contribution supplements

20b.7   The technical reserves of insurance companies are regarded as assets of the policy holders. In practice the property income from investment of these reserves serves to reduce the premiums or social contributions required. In the national accounts two counterbalancing flows are imputed: of property income (D.44) from insurance companies to households; and premium supplements (part of D.71) and contribution supplements (part of D.611) in the opposite direction. These supplements are also use in the calculation of output of insurance. Technical reserves do not include the insurance companies' own (shareholders') reserves.

Adjustment for the change in net equity of households in pension fund reserves D.8

20b.8   Above this point in the accounts, pension contributions are recorded as a use of households and a resource of insurance corporations. Conversely, pension payments are a resource of households and a use of insurance companies. Consequently an excess of contributions over pensions would result in saving being recorded as positive for financial corporations and negative for households.

20b.9   This adjustment ensures that any excess of pension contributions over pension receipts (i.e. of 'transfers' payable over 'transfers' receivable) does not affect household saving. It is a resource of households. This reflects the underlying economic reality. Consistently with this, households are treated in the financial and balance sheet accounts as owning the reserves of private funded schemes.

20b.10   The adjustment (D.8) equals:

> total value of actual social contributions in respect of pensions payable into private funded pension schemes
> *plus*   total value of contribution supplements payable out of the property income attributed to insurance policy holders (D.44)
> *less*   the value of the associated service charges (P.3)
> *less*   the total value of the pensions paid out as social insurance benefits by private funded pension schemes.

Extraordinary payments (part of D.99 - 'Other capital transfers')

20b.11   Extraordinary payments made by employers into social insurance funds are included here as other capital transfers when their reason is to increase the actuarial reserves of these funds. This happened on a significant scale in the 1970s in the UK when many such funds were found to be under-funded in relation to their actuarial liabilities (to meet future pension payments).

Net equity of households in life insurance reserves F.611

20b.12   This item appears in the financial account and balance sheet. It represents provisions against outstanding risks and provisions for with-profits insurance that add to the value on maturity of with-profit endowments or similar policies. Households are treated in the financial and balance sheet accounts as owning these reserves; F.611 is a liability of the insurance corporations and pension funds sub-sector (S.125).

Net equity of households in pension fund reserves F.612

20b.13   This item is the counterpart of 'Adjustment for change in net equity of households in pension fund reserves' (D.8), in the use of disposable income account.

Chapter 21

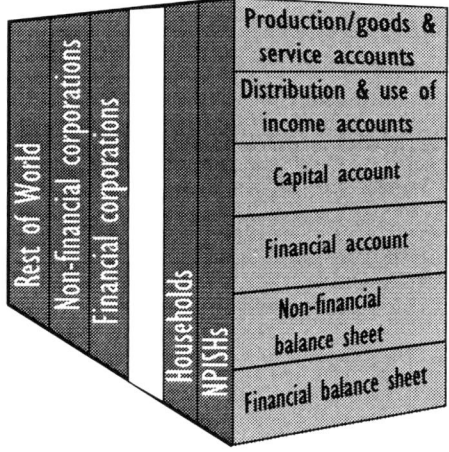

General Government (S13)

Chapter 21 General government (S.13)

Coverage of sector

21.1   In the European System of Accounts, the general government sector is defined as follows:

'The sector general government (S.13) includes all institutional units which are other non-market producers whose output is intended for individual and collective consumption, and mainly financed by compulsory payments made by units belonging to other sectors, and/or all institutional units principally engaged in the redistribution of national income or wealth.'

21.2   In the United Kingdom the sector exists at two levels: central government (S.1311) and local government (S.1313).

Central government (S.1311)

21.3   This sector covers:

- The Consolidated Fund and the National Loans Fund, the two main accounts of central government at the Bank of England

- Departments and agencies which are answerable to a minister of the Crown or other responsible person, who is in turn answerable to Parliament.

- Bodies not administered as government departments but subject to ministerial or departmental control. Regional health authorities are included but not National Health Service trusts, which are public corporations. Non-departmental public bodies that are not public corporations are included.

- Extra-budgetary non-trading funds and accounts controlled by departments, such as the National Insurance Fund, the Exchange Equalisation Account and the Contingencies Fund.

21.4   It should be noted that, unlike in some other countries, the National Insurance Funds are classified to this sector and not to the Social Security sector (S.1314) as provided for in the ESA. To be classified as a separate institutional unit, and thus form a separate social security sector, the fund would have to have autonomy of decisions in its activities. Decisions about rates of contributions and benefits are, however, taken in the context of the Budget and general public expenditure planning. For example, in 1995 only 54 per cent of social security benefits were paid from the fund, the remainder being paid mainly out of Department of Social Security voted money.

21.5   In contrast, government trading funds do have autonomy in their businesses and are engaged in the production of market output, and so classified to the public non-financial corporations sector. Before 1997 some trading funds were classified to central government but this was changed from the 1997 Blue Book.

Changes from previous system

The general government sector covers the same units as under the former system, except that:

- The Issue Department of the Bank of England has been reclassified as part of the Central Bank, which means that bank notes are no longer shown as a liability of central government.

- Payments to and receipts from the European Union which had been shown as channelled through central government are now recorded as transactions of the EU with the appropriate sector directly.

- Debt cancellations are now shown in the flow accounts; previously most were not recorded. They are shown as the repayment of borrowing with an equivalent counterpart transaction - usually an injection of equity if government is writing off the debt of a public corporation soon to be privatised; otherwise a capital transfer from the creditor to the debtor.

- Some payments to government previously recorded as distributive transactions are now shown as payments for services, such as passport fees and driving licence fees.

- The presentation of taxes changes; for example household rates change from a tax on expenditure to a tax on income and wealth.

- Some tax reliefs, hitherto netted off tax receipts, are now recorded as transfer payments, reflecting the weaker link between the original tax liability and the benefit.

- The treatment of public sector pension schemes changes. For unfunded schemes the compensation of employees includes the accruing cost of future pension rights earned from current employment, rather than pensions paid. Pensions paid are shown as a social benefit.

Local government (S.1313)

21.6   Local authorities (LAs) are public authorities of limited geographical scope having power to raise funds through local taxation. They are required to make annual returns of their income and expenditure to central government. They comprise county, district, regional, unitary and borough councils; services run by joint authorities such as police, fire, civil defence and waste regulation; residuary bodies, magistrates courts, probation committees, drainage boards, market and national park authorities.

Some local public bodies that trade to make profit are treated as public corporations and are not included in the local authority income and expenditure statistics (even if they are controlled and partly financed by local authorities). Local authority trading bodies that are not sufficiently independent to be counted as public corporations are included in the local authority sector as market bodies. This includes LA Housing Revenue Accounts.

21.7   Companies controlled by local authorities are included in the public non-financial corporations sector.

21.8   Lists of bodies included in central and local government are given in ONS publication *Sector Classification for the National Accounts* (1998 edition).

General government, and the public expenditure planning process

21.9   The measure Expenditure of General Government (EGG) differs from General Government Expenditure (GGE) shown in previous Blue Books. It comprises the following items from the *uses* side of the accounts:

- Final consumption expenditure
- Subsidies (shown in accounts as a negative resource)
- Property income *
- Social benefits other than social transfers in kind
- Other current transfers *
- Gross capital formation
- Acquisition *less* disposals of non-produced non-financial assets
- Capital transfers *

These differ from GGE in previous Blue Books in the following ways:

- Consumption of fixed capital in final consumption expenditure has increased by virtue of the wider definition of capital assets
- Property income is now measured on an accruals basis
- Passport fees and some other fees for services have been netted off final consumption expenditure
- The accruing cost of civil service pensions, and actual pensions paid, are both included
- Subsidies and investment grants funded by EU institutions are recorded as Rest of the World expenditure, rather than government expenditure
- It includes some imputed flows for insurance
- It includes pensions paid in respect of notionally funded schemes such as those for teachers and NHS staff and most significantly
- Net lending (acquisition of financial assets) and net transactions in company securities are excluded

A new version of GGE based on ESA95 components has been defined. This is shown in Blue Book Supplementary Table 6 and in *Financial Statistics*. It differs from the expenditure of general government specified by Eurostat in that it consolidates out certain internal payments and eliminates some duplication inherent in the accounting structure.

21.10   In July 1998 a Comprehensive Spending Review set out the Government's plans for public expenditure in the three-year period to 2001–2002, further information being published in the annual *Financial Statement and Budget Report*, in *Public Expenditure Statistical Analysis* and in individual Departmental reports.

21.11   The plans from 1999-2000 are in terms of Total Managed Expenditure (TME) which is the sum of public sector current expenditure and public sector net investment. For control purposes TME is divided into:

- Departmental Expenditure Limits (DEL) covering about half of expenditure. Firm multi-year deals have been set up, with capital expenditure managed and controlled separately from current. Most of DEL is allocated to programmes of individual Departments but *Welfare to Work* expenditure financed by the Windfall Tax is controlled separately.

* payments within and between central and local government under these headings have been consolidated out.

Table 21.1: General government expenditure, 1995

		1995 £ million
	CURRENT EXPENDITURE	
	Current uses	
P.3	Final consumption	140,406
D.3	Subsidies (negative resources)	5,278
D.41	Debt interest	26,305
D.621	Social security benefits in cash	42,142
D.623	Unfunded employee social benefits	12,429
D.624	Social assistance benefits in cash	
	Central government	41,592
	Local authorities	13,719
D.71	Net non-life insurance premiums	377
D.74	Current international co-operation	2,018
D.75	Miscellaneous current transfers	11,369
	Less current resources netted off	
D.44	Insurance holders' property income	-32
D.61n2	Actual voluntary social contributions	
D.61112	Employers	-1,957
D.61122	Employees	-2,445
D.612	Imputed social contributions	-5,279
D.72	Net non-life insurance claims	-377
D.74	Current international co-operation	-1,283
	Total current expenditure	284,312
	CAPITAL EXPENDITURE	
	Capital uses	
D.92	Investment grants	7,269
D.99	Other capital grants	1,624
P.51	Gross capital expenditure net of sales	14,005
P.52	Net increase in inventories	-154
P.53	Acquisition less sales of valuables	-
K.2	Net acquisition of non-produced non-financial assets	
	less capital resources netted off	
D.92	Investment grants	-219
D.99	Other capital transfers	-138
	Total capital expenditure	22,224
	FINANCIAL EXPENDITURE PLUS TRANSACTIONS IN FINANCIAL ASSETS	
F.3	Acquisition of long-term securities less sales of long-term securities	
F.4	New lending less repayment of loans	-408
F.5	Acquisition less sales of shares and other equity	-2,158
F.6	Net increase in insurance technical reserves	73
	Total financial expenditure	-2,493
	Grand total expenditure	305,403

- Annually Managed Expenditure (AME). This is expenditure which cannot reasonably be made subject to firm multi-year deals, including social security benefits, expenditure under the Common Agricultural Policy and locally-financed expenditure.

More information on TME, DEL and AME was set out in the *Economic and Fiscal Strategy Report* (EFSR) (Cm. 3978, June 1998).

Measures of government deficit

21.12 The ways in which government raises revenue and the pattern of public expenditure have an impact on the whole economy. The surplus or deficit on public sector transactions also has a significance of its own. It reflects the net impact of the public sector on the economy and influences other economic variables such as growth and inflation.

21.13 Four measures of surplus or deficit were identified in the EFSR:

- Public sector surplus on current budget (net saving *plus* receipts of capital taxes, B.8n and D.91r).

- General government net lending or borrowing (B.9) (formerly the financial surplus or deficit), which is one of the convergence criteria set out in the Maastricht Treaty.

- Public sector net lending or borrowing (B.9).

- The public sector net cash requirement (PSNCR), which used to be called the PSBR.

- Equivalent measures can be calculated separately for central and local government and public corporations but the above formulations are those most frequently used.

21.14 The definition and measurement of the PSNCR are the joint responsibility of the Office for National Statistics and the Treasury. It is consistent with national accounts data. The distinctive elements are as follows:

- It is available monthly.

- Policy lending (e.g. to students) adds to the PSNCR, repayments (e.g. privatisation proceeds) reduce it.

- It is a cash figure. This enables it to be calculated simply and quickly (the first estimate being published on the 12th working day after the end of the month to which it relates). In contrast the national accounts series are accruals based.

- It relates to the whole public sector. That is to say it covers the borrowing requirement of public corporations as well as of central and local government. It is consolidated; that is, borrowing by central government to fund lending to local authorities or public corporations is counted only once. Similarly, purchases of gilts by local authorities are netted off the local authority borrowing requirement.

- It records borrowing plus the net rundown in financial assets held by government to manage its liquidity. The public sector may finance its activities either by borrowing or by running down short-term assets. The two are regarded as equivalent for measuring the PSNCR.

Figure 21.1

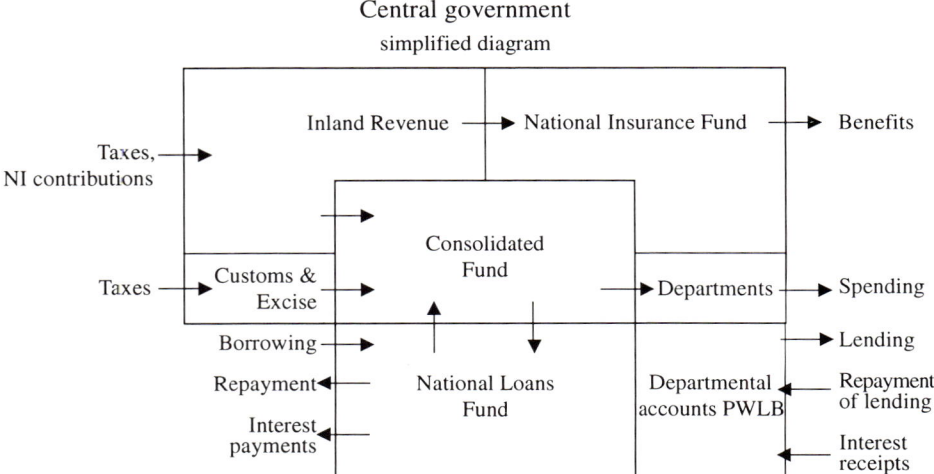

Central government
simplified diagram

- Net lending to the private sector and abroad and net cash expenditure on company securities are among the determinants of PSNCR. That is to say they affect the size of the PSNCR rather than the way in which it is financed. This treatment follows international practice. The large negative transactions in company securities in recent years reflect the proceeds of privatisation. There has been some debate as to whether privatisation proceeds should be regarded as reducing the PSNCR in this way or whether they should be treated as financing it. These large lumpy privatisation transactions can obscure the underlying balance so an alternative measure of PSNCR excluding privatisation proceeds is also published.

General government net lending/borrowing (B.9)

21.15 This is the balance on the capital account. As an internationally accepted national accounts concept it can be produced for most OECD countries; it was selected with gross debt at nominal value as an indicator of excessive government deficit under the Maastricht Treaty. Its distinctive characteristics are:

- It is available quarterly.

- A surplus (i.e. net repayment of debt) is shown positive, a deficit (net borrowing) negative.

- It is on an accrued basis.

- It relates to general government (that is, central and local government). It is by its nature consolidated; central government grants to local authorities will not affect it, though they will affect the financial surplus of the separate sub-sectors.

- All financial transactions are 'below the line'. That is, they finance the surplus or deficit rather than determine its size. Capital expenditure is, however, 'above the line' and therefore adds to the deficit.

General government saving (B.8)

21.16 This is the surplus or deficit on the use of disposable income account, and is the amount available for acquisition of physical or financial assets, or the reduction of liabilities. It represents the surplus or deficit of current resources over current uses. The public sector surplus on current budget equals public sector net saving (the surplus on the *use of disposable income account* after allowing for depreciation) *plus* receipts of capital taxes.

Outstanding debt

21.17 The measures of surplus or deficit described above relate to government performance in the current year. Also important are measures of the level of outstanding liabilities built up as a result of deficits in previous periods. These liabilities are mainly formal debt obligations, but there are other liabilities such as accounts payable. The financial balance sheets of the general government sector accounts show gross outstanding liabilities, mainly long-term central government debt. There are other concepts of debt in use. This section is intended to help reconcile the different measures. Key elements in defining outstanding debt are:

Market price or nominal value? The national accounts balance sheets are at market prices on the day on which the balance is struck (i.e. end-quarter). This reflects the need for a valuation that is meaningful both for the government as issuer and the creditor, who in many cases will buy and sell government debt in the market. For the government's own accounting purposes, and the Maastricht reporting requirements, nominal (or redemption) values are appropriate. However, since it is not possible to predict the exchange rate that will apply to the redemption of debt denominated in foreign currency the convention is that it is revalued at the current exchange rate.

Gross or net? In certain circumstances it can be prudent for governments to use a temporary surplus to buy short term assets rather than to redeem debt. Similarly a deficit might be financed by selling such assets. Just as PSNCR offsets transactions in short-term assets against borrowing, the concept of net debt offsets holdings by the government of short-term assets against its gross debt. Changes in net debt therefore correspond closely to the PSNCR flow. (*See* Table 21.2 below).

Consolidated or unconsolidated? Consolidated debt does not include debt held within the sector; again this is consistent with the PSNCR.

Date. For many years the Bank of England has produced analyses of public sector debt at the end of financial years. For consistency with other EU member states end-calendar year data have been produced from end-1991. The national accounts data at market values are estimated quarterly.

Sector coverage. Data for the public sector, general government or the three subsectors are available.

Other features to be considered are whether accounts payable or contingent liabilities are included.

The longest running series of outstanding debt figures relates to the national debt.

The national debt

21.18 The national debt comprises the total liabilities of the National Loans Fund. It is gross debt, and since it includes debt held by other government funds it is unconsolidated. Published data relate to the end of financial years. The total excludes accrued interest (such as index-linked increases) on National Savings, Consolidated Fund liabilities (including contingent liabilities, e.g. coin), liabilities of central government funds (notably the Bank of England Issue Department's note liabilities, Northern Ireland government debt and stocks issued by certain government funds) and sundry other contingent liabilities and guaranteed debt.

21.19 The national debt includes the whole nominal value of all issued stocks, even where there are outstanding instalments due from market holders; in such circumstances a counter entry is included in public sector liquid assets. The nominal value of index-linked gilt-edged stocks has been raised by the amount of index-related capital uplift accrued to 31 March each year where applicable. Definitive figures for the national debt are published annually in the Consolidated Fund and National Loans Fund Accounts Supplementary Statements.

Net public sector debt

21.20 Broader in concept is the net debt of the public sector, which is the balance sheet equivalent of PSNCR.

Both Bank of England publishes annually in its November *Quarterly Bulletin* an article on the net debt of the public sector. Data are also published as a supplementary table in *Financial Statistics*. Although the key totals are consolidated, figures of intra-sector holdings are included in the article.

21.21 Net public sector debt differs from net financial liabilities (BF.90) in the ESA because:

- It does not net off financial assets. ('Net' means that it is consolidated within the public sector.)
- It excludes F.7 trade credits and accounts receivable and payable.
- It excludes finance lease liabilities.
- It is at nominal values rather than market prices.

Relationship between PSBR and change in outstanding gross debt of general government

21.22 The reconciliation between the public sector borrowing requirement (as it then was) and the change in outstanding gross debt is shown in the following table. The difference can be seen to be due to two coverage differences (borrowing requirement of public corporations and general government transactions in liquid assets) and several valuation adjustments affecting the levels figures. Blue Book balance sheet figures are also affected by the change in the market prices of government securities.

Table 21.2

	1995 (£ million)
General government gross debt outstanding (nominal value)	
At beginning of year	336,395
At end of year	378,292
Change in outstanding debt	41,897
Public sector borrowing requirement (calendar year total)	35,119
less	
Market and overseas borrowing by public corporations	-2,238
equals	
General government borrowing requirement	37,357
plus	
Transactions in assets	1,001
Revaluation of foreign currency debt from changes in exchange rate	673
Gilts (index linked uplift)	1,383
Other (including valuation changes)	1,483
equals	
Change in outstanding debt	41,897

Main sources of central government data

Central government

21.23 Most central government expenditure is supply expenditure. That is, it is voted annually by Parliament and spent by Departments and agencies. Figures are collected quarterly from all major departments through the Treasury Government Expenditure Monitoring System (GEMS). Prior to 1996 a system called APEX was used, which produced detailed analyses as a by-product of the Paymaster General's computerised accounting system. However, random lags and leads in the recording of transactions led to the system producing series that were unrealistically volatile.

21.24 Expenditure on social security is monitored in great detail by the Department of Social Security, and the national accounts figures derive from that source.

21.25 The Government Actuary's Department provides quarterly estimates of social security contributions.

21.26 The Consolidated Fund and the National Loans Fund are the core accounts of central government at the Bank of England. Most receipts of government end up in one these accounts, and most payments originate there, but very few transactions of these accounts are with non-government sectors.

21.27 The quarterly accounts of the Consolidated Fund are the source for expenditure on the Consolidated Fund Standing Services. These are services which the Consolidated Fund is authorised by Act of Parliament to pay for without an annual vote. They include the Civil List and the pay of judges, etc.

21.28 The National Loans Fund Account is the source of central government borrowing data.

21.29 Quarterly returns from the Ministry of Finance, Northern Ireland supply income and expenditure in the province.

21.30 The Inland Revenue and Customs and Excise provide details of taxes collected. These are published monthly in *Financial Statistics*.

21.31 The Department of Culture, Media and Sport provides quarterly figures on the operation of the National Lottery.

Local government

21.32 Most local authority data is annual, relating to financial years. Detailed annual returns of expenditure and income are compiled by local authorities and collected by the Department of the Environment, Transport and the Regions (DETR), Scottish Office, Welsh Office and the Northern Ireland departments. The aggregates are published in *Local Government Financial Statistics* for each country. Adjusted to national accounting definitions, they are used here for items of expenditure and income not attributed elsewhere. Final figures are based on authorities' audited accounts.

21.33 The housing subsidy claim forms are the basis for the housing element of the production account.

21.34 A quarterly return collected by DETR gives wages and salaries, interest and dividend receipts and council tax receipts. Interest and dividends are modified by the operation of the Dividends and Interest Matrix.

21.35 DETR and Scottish and Welsh Offices collect quarterly returns of capital expenditure.

21.36 DETR collects monthly for the whole United Kingdom returns of borrowing and lending by local government.

General government accounts

21.37   The following table shows the main features of the accounts of general government:

Table 21.3

Resources	Uses
Production account P.11 Market output P.12 Output for own final use P.13 Non-market output P.1 Total output	P.2 Intermediate consumption B.1g Gross value added Consumption of fixed capital B.1n Net value added

Generation of income account
 B.1g Gross value added
	D.1 Compensation of employees
	      D.11 Wages and salaries
	      D.12 Employers' social contributions
	 D.29 Production taxes not on products
	 D.39 *less* Subsidies not on products
	       Total uses
	 B.2g Operating surplus, gross

Allocation of primary income account
 B.2g Operating surplus, gross
 D.21 Taxes on products
 D.29 Other taxes on production
 D.2   Total taxes on production
 less
 D.31 Subsidies on products
 D.39 Other subsidies on production
 D.3   Total subsidies on production
 D.41 Interest
 D.42 Dividends
 D.44 Income attributable to policy holders
 D.45 Rent
      Total resources
	D.41 Interest
	 D.45 Rent
	      Total uses
	
	 B.5g Balance of primary incomes, gross

Secondary distribution of income account
 B.5g Balance of primary incomes, gross
 D.51 Taxes on income
 D.59 Other current taxes
 D.5   Total taxes on income and wealth
      Actual social contributions
      Imputed social contributions
 D.61 Total social contributions
 D.72 Net non-life insurance claims
 D.73 Current transfers within government
 D.74 Current international co-operation
 D.75 Miscellaneous current transfers
 D.7   Total current transfers
      Total resources
	D.5 Current taxes on income and wealth
	 Social benefits other than in kind
	 D.621 Social security benefits in cash
	 D.623 Unfunded employee benefits
	 D.624 Social assistance in cash
	 D.62 Total benefits
	 Other current transfers
	 D.71 Net non-life insurance premiums
	 D.73 Current transfers within government
	 D.74 Current international co-operation
	 D.75 Miscellaneous current transfers
	 D.7 Total current transfers
	      Total uses
	
	 B.6g Disposable income, gross

Table 21.3 - *cont...*

Resources	Uses

Use of disposable income account
B.6g Disposable income, gross
	P.31 Individual consumption expenditure
	P.32 Collective consumption expenditure
	P.3 Total consumption expenditure
	D.8 Adjustment for net equity of house-
	             holds in pension funds
	             Total uses
	B.8g Saving, gross

Capital account - change in net wealth due to saving and capital transfers
B.8g   Saving, gross
            Capital transfers receivable
D.91   Capital taxes
D.92   Investment grants
D.99   Other capital transfers
D.9   Capital transfers payable
D.92   Investment grants
D.99   Other capital transfers payable
D.9   Total capital transfers payable
B.10g Change in net worth due to gross
            saving and capital transfers	

Acquisition of non-financial assets account
B.10g Change in net worth due to gross
            saving and capital transfers
            Change in liabilities and net worth
P.51   GFCF
P.52   Inventories
P.53   Valuables
P.5   Capital formation
K.2   Non-produced non-financial assets
            Change in assets
B.9 Net lending/borrowing	

Liabilities	Assets
Financial transactions account *Transactions in financial liabilities* F.2 Currency and deposits F.3 Securities other than shares F.4 Loans F.7 Other accounts payable	*Transactions in financial assets* F.1 Monetary gold and special drawing rights F.2 Currency and deposits F.3 Securities other than shares F.4 Loans F.5 Shares and other equity F.6 Insurance technical reserves F.7 Other accounts receivable

Production account: Resources

Market output (P.11)

21.38   This item consists of goods and services disposed of in the market or intended to be. It therefore comprises:

- the total trading receipts (for example sales, fees, charges and rent) of market units (local government housing and other trading services, ECGD)

- all trading receipts of non-market units.

Note that non-market units can have market output; a unit only becomes a market unit when its trading receipts exceed 50 per cent of its production costs.

Output for own final use (P.12)

21.39   This consists only of the production by central and local government of fixed assets for their own use (own account gross fixed capital formation). Recently this has been mainly capitalised computer software.

Other non-market output (P.13)

21.40   Other non-market output of government is that provided free. It equals government final consumption expenditure.

It is calculated as the sum of resources consumed by non-market units (intermediate consumption, compensation of employees, capital consumption and taxes on production *less* subsidies on production) *less* the market output of non-market units and output for own final use.

Production account: Uses

Intermediate consumption (P.2)

21.41   This is the goods and services consumed in producing general government output. It is equivalent to the procurement element of final consumption expenditure (gross of receipts) plus the corresponding (i.e. non-pay) element of output for own final use.

Consumption of fixed capital (K.1)

21.42   Represents the capital resources used up in the course of production. It derives from the perpetual inventory model (*see* Chapter 16). It comprises capital consumption by market units (ECGD, local authority housing and other trading services) and by non-market units. The capital consumption by non-market units is a component in the valuation of their output.

Gross value added (B.1g)

21.43   The balance on the production account, carried forward into the generation of income account as the only resource.

Generation of income account: uses

Compensation of employees (D.1)

Wages and salaries (D.11)

21.44   This item comprises the whole wages and salaries bill for the sector, including those in GFCF and locally engaged staff overseas. It includes pay in both cash and kind.

Employers' contributions (D.12)

D.121   Actual contributions

21.45   This covers contributions by general government as employer to National Insurance and the National Health Service, to funded schemes for government employees (typically local authority employees) and to unfunded schemes for central government employees. Contributions to National Insurance and the NHS are estimated from the total receipts of such contributions by allocating an amount to each sector in proportion to its wages and salaries.

This series includes contributions to the NHS and teachers' notionally funded pension scheme in respect of NHS employees not in NHS trusts and teachers in central government institutions. (The compensation of employees in NHS trusts is a use of the public corporations sector.) These amounts are the actuarially-assessed contributions charged to employers.

Payments under the Pensions Increase Act in respect of teachers are included in current transfers to sectors employing teachers. Those sectors include the payments in D.121.

D.122   Imputed contributions

Contributions to unfunded schemes for the Civil Service and Armed Forces are the Assessed Superannuation Liability Contributions (ASLCs) paid by departments in respect of current employees (a change from the previous system, which scored pensions paid *less* employees' contributions).

Compensation of employees (D.1)

21.46   These series, which relate to the total sector, are derived from the sources above.

Production taxes other than on products (D.29)

21.47   These are taxes paid by government and comprise motor vehicle duty (MVD) and national non-domestic rates (NNDR, the 'business rate'). The NNDR figures are from government administrative records. MVD is estimated.

Subsidies other than on products (D.39)

21.48   These will include employment subsidies paid under the *welfare to work* programme. Some payments under the programme are paid to government employers and will appear as resources.

Net operating surplus (B.2n)

21.49   The operating surplus, net of consumption of fixed capital, relates to the surplus of output over costs for market units. By definition there is no net operating surplus on non-market units.

Allocation of primary income account: Resources

21.50 Apart from the operating surplus, the resources on this account consist of taxes on production (D.2) *less* subsidies (D.3) and property income (D.4).

Taxes on production (D.2)

21.51 These are compulsory and unrequited payments levied on the production and importation of goods and services, the employment of labour, the ownership and use of land, buildings or other assets used in production. They are payable whether or not profits are made.

21.52 Taxes on production are divided into two groups: taxes on products (D.21) which are payable per unit of output and other taxes on production (D.29) which are incurred as a result of entering into production but are not dependent on the quantity of goods and services produced or sold.

21.53 Taxes on production are recorded on an accruals basis. Cash receipts are lagged by average delays in receiving the duty. Further details of tax receipts are shown monthly in *Financial Statistics*. The difference between accrued taxes and cash receipts represents a financial asset of general government and is recorded under accounts receivable (D.79). Table 21.4 shows the particular taxes included in taxes on production and the relationship to the former category 'Taxes on expenditure'.

Table 21.4: Taxes on production

£million

	1995	Central government	European Union	Total
	Taxes on expenditure (old system)			103 444
less	Domestic rates[1]			-139
less	Vehicle excise duty paid by households			-2 641
plus	ITC franchise payments			398
plus	Various fees paid by businesses			190
D.2	Taxes on production	93 894	7 358	101 252
of which:				
D.21	Value added tax	43 622	4 845	48 467
D.21	Customs duties		2 308	2 308
D.21	Agricultural levies		150	150
D.21	Other customs and excise duties	30 529		30 529
D.21	Other EU levies[2]		55	55
D.21	Fossil fuel levy	1 306		1 306
D.21	Gas levy	161		161
D.21	Stamp duties	1 924		1 924
D.21	Payments to National Lottery Distribution Fund	1 360		1 360
D.29	National non-domestic rates	12 994		12 994
D.29	Local authority rates paid by sectors other than households[1]	65		65
D.29	Vehicle excise duty paid by businesses	1 313		1 313
D.29	Consumer Credit Act fees			
D.29	ITC franchise payments	398		398
D.29	Other licence fees paid by businesses	228		228

[1] Northern Ireland in 1995

[2] Sugar levy in 1995 and ECSC levies in earlier years

Subsidies (D3)

21.54 Subsidies are defined as current unrequited payments made to resident producers with the objective of influencing their level of production, their prices or the remuneration of the factors of production. They are subdivided into subsidies on products (D.31), which are subsidies payable per unit of a good or service produced or imported, and other subsidies on production (D.39). The ESA provides for subsidies on products to be further divided into subsidies on imports and 'other', but the United Kingdom has not subsidised imports over the period covered by the Blue Book.

Changes from former system of accounts

In the former system subsidies from the European Union were recorded as expenditure of central government, and the receipts from the EU recorded as negative current grants abroad.

21.55 Subsidies are recorded on an accruals basis. The difference between this and cash receipts is included in amounts payable in the financial account. The sources of the data are central and local government records and public corporations' annual reports, including the returns described above.

Property income (D.4)

21.56 The forms of property income that are resources of general government are:

- Interest (D.41)
- Distributed income of corporations (D.42)
- Property income attributed to insurance policy holders (D.44)
- Rent on land (D.45)

Changes from former system of accounts

Central government payments of interest on gilts, which are made every six months, are shown as accruing in each quarter.

Interest (D.41)

21.57 Cash receipts of interest are derived from central government accounts and quarterly returns from local authorities. They are converted to an estimated accrued basis by taking into account, via the Dividends and Interest Matrix, changes in holdings of particular debt instruments.

21.58 The main components are interest on central government lending to local government and public non-financial corporations, interest on the reserves and interest on internally held government securities and Treasury bills. Details are shown in the Dividends and Interest Matrix tables in the Blue Book. Interest arising from the shipbuilding credit scheme and the export credit scheme is included.

21.59 In addition to interest on loans to local authorities this item also includes premia receivable by central government from local authorities in respect of the Exchange Cover Scheme, and income received by central government funds managed by the National Debt Commissioners from investment in local authorities' debt. Income from public dividend capital of public corporations is included in D.42.

Exchange cover scheme

The scheme is run by central government and covers local authorities' and public corporations' borrowing abroad, which can be undertaken, direct from overseas or through United Kingdom banks. In either case, the local authority or public corporation may make use of the Exchange Cover Scheme, under which the central government guarantees to cover any addition to the sterling cost of interest payments and repayment of the loan arising from a fall in the exchange rate between the time the loan is negotiated and the time it is repaid. Conversely the central government would benefit if sterling appreciated. Losses and gains arising on interest payments are included as central government payments of interest (F.424). Only the cost of guaranteeing repayments of principal is recorded as a financial transaction. It is important to note that, in practice, no payments as such occur, i.e. the exchange losses or gains are purely notional. Under the Exchange Cover Scheme, the beneficiary - a local authority or public corporation - is entitled to purchase the foreign currency needed to repay the loan (interest) at the original exchange rate from the Exchange Equalisation Account. However in the local authorities' and public corporations' accounts repayments of borrowing are shown at the current market exchange rate in line with the balance of payments statistics. This item is needed as an adjustment to bring the figures back to an original exchange rate basis. A charge is made by central government for the guarantee, which is included under central government receipts of interest (D.41). Certain other public bodies such as the Northern Ireland central government and British Nuclear Fuels Limited are also included as well as corporations which have been subsequently privatised.

Distributed income of corporations (D42)

21.60 Distributed income of corporations comprises dividends (D.421). The source is the quarterly government expenditure monitoring system (GEMS) and National Loans Fund accounts. Government holdings of company securities have diminished in recent years as residual holdings in privatised companies have been disposed of. Dividend receipts have consequently declined.

Property income attributed to insurance policy holders (D44)

21.61 Since the technical reserves of insurance companies are assets of the policy holders, receipts from investing them are shown as property income of the policy holding sectors. Since in reality this income is retained by the insurance companies a payment back is imputed in the form of premium supplements (recorded in P.2, intermediate consumption). The information is derived from insurance company sources.

Rent (D45)

21.62 Most central government rent is oil royalties but this series also includes rent on land received by the Department of Transport and Ministry of Defence.

Allocation of primary income account: uses

21.63 Uses of general government on this account comprise interest (D.41) and rent (D.45).

Interest (D.41)

21.64 Interest is measured on an accruals basis. The consequences of this for the accruals are:

- Interest on gilts, paid out six monthly, is spread over the quarters in which it accrues.

- For index-linked bonds and National Savings Certificates, the increase in principal outstanding is recorded as interest, as well as any index-linked increase in the coupon.

- For bonds the difference between the redemption price and the issue price is recorded as interest spread over the life of the bond in equal amounts.

- The discount on Treasury bills at issue is distributed over its life as interest. Since these are short-term instruments in practice this means the interest is recorded at issue.

- The excess of interest accrued over that paid out is treated as being re-invested in the same instrument.

21.65 Interest on the *uses* side includes interest accruing on gilts, Treasury bills, National Savings instruments and other central and local government borrowing. It also includes the interest element of charges for finance leasing and payments under the Exchange Cover Scheme (*see* above). Cash payments come from the quarterly National Loans Fund accounts and the GEMS returns for central government, and a quarterly sample return for local government.

Balance of primary incomes (B.5g)

21.66 This is the balancing item on the allocation of primary income account, and summarises the amount due to government from its own economic activities and investments. The secondary distribution account shows how income is redistributed by way of taxes, social security contributions and benefits and other current transfers.

Secondary distribution of income account: Resources

21.67 The three main components of general government resources on this account are:

- Current taxes on income and wealth (D.5)
- Social contributions (D.6)
- Other current transfers (D.7)

Current taxes on income and wealth (D5)

21.68 The ESA definition of these is 'all compulsory unrequited payments in cash or in kind, levied periodically on the income and wealth of institutional units, and some periodic taxes which are assessed neither on income nor wealth'. More detail is published monthly in *Financial Statistics*. They are divided into:

- taxes on income (D.51)
- other current taxes (D.59)

Table 21.5 shows the main taxes included in each category and a reconciliation with the category taxes on income in the former system.

Table 21.5: Taxes on income and wealth (D.5)

£million

1995	Central government	Local government	General government
Taxes on income (old system)	**90 668**		**90 668**
plus Refunds now classified as expenditure	2 898		
plus Taxes on capital gains	1 628		1 628
D.51 Taxes on income	**92 147**		**92 147**
of which:			
Income tax	67 252		67 252
Petroleum Revenue Tax	834		834
Corporation tax	22 433		22 433
Tax on capital gains	1 628		1 628
D.59 Other current taxes	**2 737**	**9 204**	**11 941**
of which:			
Domestic rates/community charge/ council tax		9 209	9 209
Vehicle excise duty paid by households	2 641		2 641
Other licence fees paid by households	77		77

The distinction between negative taxes and expenditure

21.69   Households may receive assistance from government by way of cash benefits, or by allowances against tax. In an increasing number of instances, of which mortgage interest relief is the most long standing, the same benefit is given to taxpayers by way of tax allowances and to non-taxpayers by way of cash benefit. It is important therefore to set out the principles by which the national accounts record the benefit as expenditure or as negative taxation.

21.70   A tax allowance is treated as negative taxation if it meets all of the following criteria:

- the benefit to individual taxpayers does not exceed the amount of tax paid by them;
- it is made as a matter of economic policy; and
- the allowance is an integral part of the tax system.

This last criterion means that the allowance should fall naturally out of the normal rules for calculating tax payable. If a refund or allowance can be calculated independently of the procedure for calculating the tax payable, and uses different criteria, and does not depend on the actual amount of tax to be paid, or the marginal rate of tax, then it is not integral to the tax assessment.

21.71   In the case where the same benefit is given to taxpayers by way of tax allowances and to non-taxpayers by way of cash benefit, it may be possible to split the benefit such that government expenditure includes:

- the whole of the payments to non-taxpayers
- the rebate to taxpayers in excess of the tax paid by the individual in respect of the period to which the rebate relates

and the rebate to taxpayers up to the limit of the tax paid by the individual in respect of the period to which the rebate relates is classed as negative taxation. However, the availability of the rebate to non-taxpayers is normally a strong indicator that it is not negative taxation.

Taxes on income (D.51)

21.72 These consist of current taxes on income, wealth, etc. (D.5) and cover all compulsory, unrequited payments, in cash or in kind, levied periodically by general government and by the Rest of the World on the income and wealth of institutional units, and some periodic taxes which are assessed neither on the income nor the wealth. This includes income tax, corporation tax, capital gains tax, stamp tax, and petroleum revenue tax. The source is Inland Revenue records.

21.73 Income tax is deducted under PAYE from company distributions by virtue of Schedule 9 of the Income and Corporation Tax Act (1970) (ICTA), and deducted from company payments other than distributions under Schedule 20 of the Finance Act (1972). It is recorded in a way that is consistent with the method of recording in the personal sector (i.e. at the time that the tax deductions from personal income are actually made). Income tax deducted by building societies under section 343 of the ICTA, and on local authorities' interest under section 53 of the ICTA, is recorded on a similar accruals basis. The difference between these accruals series and the actual cash received by the Inland Revenue is shown in 'accounts receivable' in the financial transactions table. All other taxes on income are recorded on a cash basis.

Other current taxes (D.59)

21.74 Other current taxes (D.59) include:

- Current taxes on capital which consist of taxes that are payable periodically on the ownership or use of land or buildings by owners, and current taxes on net wealth and on other assets (jewellery, other external signs of wealth), except for taxes mentioned in D.29 (which are paid by enterprises as a result of engaging in production) and those mentioned in D.51 (taxes on income).

- It includes local authority taxes levied on households (rates, community charge, council tax), motor vehicle duty paid by households and payments by households for certain licences. The latter include licences that have fees that are out of all proportion to the cost of providing the licence, and licences that do not involve any checking of the applicant's suitability or any other service to the applicant. The data come from government departments.

Changes from former system of accounts

Taxes on income under the new definition includes capital gains tax (a tax on capital in the former system), local authority rates (a tax on expenditure) and the council tax (a tax sui generis in the former system). The Community Charge is classified to D.59. Dog and gun licences and motor vehicle licence duty paid by households, classified to D.59, were taxes on expenditure in the former system. Mortgage interest relief has been reclassified as expenditure from 1991–92 when relief first became restricted to the standard rate, as has private medical insurance relief (from 1994–95) and life assurance premium relief at source.

Social contributions (D.61)

21.75 These are contributions to social insurance schemes. Social insurance is a broader concept than social security alone. The resources of general government include the receipts of:

- Social security schemes of government

- Unfunded schemes run by general government such as the Civil Service and Armed Forces pension schemes. Departments' contributions are assessed as the accruing superannuation liability.

- Unfunded schemes run by local authorities for the fire and police services.

- Notionally funded schemes such as those for teachers and NHS staff. There is no fund but contributions are made by employers and employees to central government.

21.76 The receipts of social contributions comprise actual contributions (D.611) and imputed contributions (D.612). Imputed contributions are the counterpart of unfunded social benefits paid directly by employers (D.122). These form part of compensation of employees, and D.612 represents the employee's imputed payment shown on the *uses* side of the household account.

Actual contributions (D.611)

21.77 These are all uses of the household sector and comprise social contributions paid by employees and employers (because the latter are included within the compensation of employees), both compulsory and voluntary. They comprise:

- National insurance contributions

- Contributions by employers and employees to the teachers' and NHS pension schemes plus some small amounts relating to atomic energy

- Employee contributions to unfunded pension schemes

21.78 Accruals of National Insurance and Health contributions are estimated by the Government Actuary's Department. Other items are derived from GEMS and the government accounts. The difference between accrued contributions and cash receipts represents government assets in the form of accounts receivable.

Imputed contributions (D.612)

21.79 These are accruals of assessed superannuation liability contributions from government departments in respect of the Civil Service and HM Forces pension schemes, and unfunded schemes for the police and fire services.

Current transfers (D.7)

Net non-life insurance claims (D.72)

21.80 This item represents claims due in respect of the current year's insurance policies taken out by local authorities. It is derived from insurance company data.

Current transfers within general government (D.73)

21.81 This comprises central government current grants to local government.

It includes Revenue Support Grant, redistribution on national non-domestic rates, part of the cost of teachers' pension increase payments, specific grants such as those for mandatory student awards and rent allowances, and an imputed payment in respect of the cost of pension increase payments. The latter is part of the compensation of local authority employees (D.122) that households pay to government through D.612.

Under the new ESA these are not consolidated. They appear in the general government account as both resources and uses.

Data are from GEMS.

Current international co-operation (D.74)

21.82   This comprises receipts from the European Union (EU). The main receipt is the Fontainebleau abatement. This is a payment to the UK from the EU to reduce the UK's net contribution to the EU that arises because of the UK's small agricultural industry relative the rest of the economy. Other EU receipts include negotiated refunds and research council receipts. From 1990 to 1992 this category includes contributions from other nations to the cost of the Gulf War. Receipts from the European Regional Development Fund (ERDF) and European Social Fund are routed directly from the EU to the ultimate recipients.

Data are from government accounts.

Miscellaneous current transfers (D.75)

21.83   For general government, resources comprise fines and penalties (except those in relation to taxation, which are classified with the tax concerned).

The source is central government accounts.

Secondary distribution of income account: Uses

Social benefits in cash (D.62)

21.84   This item is made up of three components for general government:

- social security benefits in cash   **D.621**
- unfunded employee social benefits   **D.623**
- social assistance benefits in cash   **D.624**

Change from former system of accounts

In the former system social security benefits paid abroad were shown separately from those paid to UK residents. They are now grouped together, but can be separated by reference to the accounts for the counterpart sectors. Social security benefits in cash now include mortgage interest relief, private medical insurance relief and life assurance premium relief, formerly treated as tax refunds.

D.621 Social security benefits in cash

21.85   Social security benefits include benefits paid from the National Insurance Fund and other social security funds. The benefits can depend on an individual's National Insurance contributions. They include:

state retirement pensions
widow's and guardian's allowances
unemployment benefit
job seekers' allowance

sickness benefit
invalidity benefit
incapacity benefit
maternity benefit
death grant
injury benefit
disablement benefit
industrial death benefit
statutory sick pay
Redundancy Fund benefit
Maternity Fund benefit
Social Fund benefit

D.623 Unfunded employee social benefits

21.86 Unfunded employee social benefits are the counterpart to the imputed contributions from government as employer in the resources of this account. They include:

- pensions paid in respect of the civil service and armed forces pensions schemes
- fire and police service pensions paid by local authorities
- pensions paid by central government in respect of other unfunded employee schemes such as those for teachers and NHS staff (even though central government does not itself employ most of these staff directly), including pension increase payments funded by central government

The data sources are monitoring information of the Department of Social Security and GEMS.

D.624 Social assistance benefits in cash

21.87 Social assistance covers the same range of needs as social security but does not depend on contributions for eligibility. It does not include payments in response to events or circumstances that are not normally covered by social insurance schemes - for example natural disasters, which are recorded under D.75 - other current transfers and other capital transfers. In the United Kingdom social assistance benefits in cash include:

- child benefit
- one parent benefit
- widow's benefit
- guardian's allowance
- mandatory student awards
- scholarships
- allowances to patients in hospital
- assistance to the disabled, for example with the cost of running invalid chairs and cars
- non-contributory benefits
- mobility allowances
- family income supplement
- rent rebates and allowances
- grants in connection with measures to combat unemployment; for example training allowances, fees for training courses, etc
- mortgage interest relief at source (MIRAS)
- life assurance premium relief at source (LAPRAS)
- private medical insurance relief (PMI)

21.88   Under the previous system the last three benefits listed above were treated as negative taxation when given to tax payers. The distinction in the current system between those 'tax reliefs' to be offset against taxation and those to be classed as expenditure is described in paragraph 21.69 above. In addition to the benefits listed the category is expected to include the cost of working families tax credit.

The sources are the GEMS returns and returns from the Northern Ireland government.

Other current transfers (D.7)

Net non-life insurance premiums (D71)

21.89   This is the component of insurance premiums paid by local authorities that covers the cost of paying claims. (The remainder of insurance premiums paid covers the service provided - in P.2 - and the acquisition of financial assets - F.62.)

It excludes ECGD which is treated as a market unit such that the difference between its premium receipts and claims paid is treated as market output.

Current transfers within central government (D.73)

21.90   These comprise grants from central government to local authorities to finance their current expenditure. The distribution of the proceeds of national non-domestic rates is included here, as are payments under the Pensions Increase Act to local government as employers of teachers. The data source is GEMS.

Current international co-operation (D.74)

21.91   This comprises mainly foreign aid, military aid and social security benefits paid abroad. The source is the quarterly GEMS returns and the Consolidated Fund accounts (*see* above).

Miscellaneous current transfers (D.75)

21.92   In the United Kingdom these transfers include among the uses of general government:

- GNP fourth resource contribution to the EU
- grants to higher and further education institutions and to City Technology Colleges, all of which are classified as non-profit institutions serving persons
- research grants
- grants to other bodies including Training and Enterprise Councils
- grants to employers (other than local government) of teachers and NHS staff in respect of pensions increase payments

Fourth resource payments are taken from government accounts. Other sources are the GEMS returns, Department of Environment, Scottish and Welsh Office returns mentioned above and the Department for Education and Employment.

Disposable income (B.5)

21.93   This is the balance of the secondary distribution of income account and is carried forward into the capital accounts.

Use of disposable income account: Uses

Final consumption expenditure (P.3)

21.94   This comprises expenditure incurred by general government on goods and services that are directly used for the satisfaction of individual needs or wants, or the collective needs of members of the community. Final consumption may take place on the domestic territory or abroad.

21.95   Final consumption expenditure is divided into two categories:

- individual consumption
- collective consumption

Individual consumption expenditure is on goods and services which are supplied to individual households to satisfy the needs of that household. By convention all consumption expenditures under the following COFOG headings (*see* Annex 1 to this chapter) are treated as individual.

- education
- health
- social security and welfare
- recreational and cultural activities

Individual consumption expenditure of government may be added to that of households and the NPISH sector to obtain the actual final consumption of households (P.4).

21.96   Collective consumption services are provided simultaneously to all members of the community, or all members of a particular section of the community. The distinctions between individual and collective consumption are summed up in the following table.

Table 21.6

Individual consumption (= social transfers in kind)	Collective consumption
It must be possible to observe the acquisition of the good or service by an individual household or member thereof, and also the time at which it took place.	These services can be delivered simultaneously to every member of the community, or of a particular section of the community.
The household must have agreed to the provision of the good or service, and take whatever action is necessary to make it possible.	The use of such services is usually passive, and does not require the explicit agreement or active participation of the individuals concerned.
Acquisition of the good or service by one person, household or small group precludes its acquisition by others.	The provision of the service to one person or household does not reduce the amount available to others in the same community.

21.97 The source of general government final consumption is the quarterly GEMS system for central government (*see* 21.26 above) and the annual returns of local government expenditure (*see* 21.35 above). Quarterly returns of local government wages and salaries are combined with interpolated figures for intermediate consumption (procurement). Annual figures for consumption of fixed capital are obtained from the Perpetual Inventory Model (*see* Chapter 16). Quarterly figures are interpolated.

21.98 Final consumption equals the other non-market output of government.

Changes from former system of accounts

The coverage of final consumption has changed from the old system. It now excludes the following, reclassified as fixed capital formation:

- computer software and large database development designed to be used in production for over a year
- defence expenditure on military buildings, vehicles and equipment, except weapons and their supporting systems

However, offsetting these is the increase in capital consumption arising from these reclassifications.

The presentation of final consumption has changed also. In the former system, general government final consumption was directly analysed into its four components: the services of its employees (pay); goods and services purchased from other bodies (procurement); non-trading capital consumption; *minus* receipts from sales and reimbursements. In the new system these transactions take place in the production account, and general government (in the use of disposable income account) purchases the output of its own production.

21.99 In principle final consumption should be recorded on an accruals basis but in practice the total accrued final consumption expenditure for a financial year is equal to cash. Within financial years expenditure is an estimate of the accruing expenditure constrained to the financial year total.

Saving (B8)

21.100 This is the balancing item on the use of disposable income account and represents the amount available for the net acquisition of capital and financial assets, and capital transfers. Subject to changes in coverage described above, it is broadly equivalent to the current surplus in the pre-1998 system. Net saving is net of depreciation. Net saving plus receipts of capital taxes is the 'surplus on current budget' used to monitor the Treasury's "golden rule" which requires a positive surplus. More information is given in the *Economic and Fiscal Strategy Report* (June 1998).

Change in net worth due to savings and capital transfers account

Capital transfers receivable (D.9)

Capital taxes (D.91)

21.101 These comprise receipts by government of death duties and taxes on other capital transfers. It excludes stamp duties and capital gains tax. The figures come from Inland Revenue records. Monthly receipts are shown in *Financial Statistics*. The components are shown in the following table.

Table 21.7: Taxes on capital (D.91)

1995 £ million

	Central government
Death duties	1 409
Taxes on other capital transfers	31
Development land tax	1
Total	1 441

Investment grants (D.92)

21.102 There are no central government receipts of investment grants.

Local authority receipts consist mostly of grants from central government to local government to finance capital expenditure and Regional Development Fund grants from the EU. Central government grants to local government include grants from development corporations, the corporation being assumed to be acting as an agent of central government.

The data come from the GEMS system and local authority capital expenditure surveys undertaken by DoE, Welsh Office and Scottish Office.

Other capital transfers (D.99)

21.103 These are contributions from the private sector to local government capital formation. They are derived from the local government returns of expenditure and income.

Capital transfers payable

Investment grants (D.92)

21.104 These comprise payments by general government to finance fixed capital formation. As the general government account is not consolidated, investment grants from central government to local government are included as both receivable and payable. Investment grants also include grants to public corporations, housing associations and universities. They derive from GEMS and public expenditure monitoring data. Grants to public corporations are only included when they are given for specific capital expenditure.

Other capital transfers payable (D.99)

21.105 This comprises grants to public corporations reflecting the cancellation of debt. However, such grants in the context of a privatisation are excluded, and recorded as the acquisition of equity in the financial account. The figures here reflect the following cancellations of debt to the government in 1980–1995.

1980	£160 million debt of British Airways Board under the Civil Aviation Act (1980)
	£60 million debt of British Aerospace under the Civil Aviation Act (1980)
1981	£3,509 million debt of the British Steel Corporation
1982	£1,000 million debt of the British Steel Corporation
	£33 million debt of the Development Board for Rural Wales
1990	£1,368 million debt of Scottish Nuclear plc under the Electricity Act (1989)
	£366 million debt of the Northern Ireland Housing Executive, agreed by the Department of Finance, Northern Ireland
1995	£1,598 million debt of British Coal
1996	£1,800 million debt of the New Towns Development Corporations

Also included are debt commutation grants given by central government to cancel local authority debts.

Change in net worth due to savings and capital transfers (B.10)

21.106 This is the balance carried forward into the acquisition of non-financial assets account.

Acquisition of non-financial assets account

21.107 The concepts of this account are described more fully in Chapters 6 and 15.

Gross fixed capital formation (P.51) (GFCF)

21.108 GFCF comprises the acquisition less disposal by general government of produced fixed assets, both tangible and intangible. Tangible assets include buildings and other structures, machinery and equipment (including vehicles) and cultivated assets. They exclude most military weapons and their supporting systems but take in other defence equipment, including light weapons and armoured vehicles purchased by non-military parts of government such as the police. Also included are assets acquired under financial leases.

21.109 Intangible assets include mineral exploration, computer software and entertainment, literary or artistic originals.

Changes from former system of accounts

In the pre-1998 system fixed capital formation excluded all military expenditure except for forces married quarters. It also excluded intangibles.

21.110 The data come from public expenditure monitoring, largely the GEMS system and quarterly returns of capital expenditure from local authorities.

Inventories (P.52)

21.111 Data available on inventories of general government are mainly those of the Intervention Board for Agricultural Produce. Data are derived through the GEMS system, annual accounts for central government and the annual returns of expenditure and income of local authorities.

Acquisitions less disposals of valuables (P.53)

21.112 Valuables comprise the following types of good:

- precious stones and metals, excluding monetary gold
- antiques and other works of art
- other valuables, such as jewellery, made from precious stones and metals, and collectors items

It is assumed that net acquisition is zero for years up to 1997. Subsequent estimates will be made for general government on the basis of information that will be collected from Departments in the future.

Acquisitions *less* disposals of non-produced non-financial assets (K.2)

21.113 It has not yet been possible to estimate expenditure by general government on assets other than land.

Net lending (+) and borrowing (-) (B.9)

21.114 This is the balance on the acquisition of non financial assets account. It represents the amount becoming available for investment in financial assets or reduction of financial liabilities.

It is the measure used for assessing government deficits under the Excessive Deficit Procedure of the Maastricht Treaty.

Financial account

21.115 Except for accounts receivable and payable and where stated otherwise, the sources of data for this account are:

1. National Loans Funds accounts and associated Departmental accounts.
2. GEMS returns for lending by Departments
3. local government quarterly returns of borrowing and investment
4. DETR records of local government lending for house purchase
5. surveys of banks by the Bank of England (25.3.1)
6. the Building Societies Commission (25.3.2)
7. ONS surveys of financial corporations. (25.3.3)
8. Bank of England data on government transactions.
9. Paymaster

More general definitions of the instruments are given in Chapters 7 and 17.

Monetary gold and special drawing rights (SDRs) (F.1)

21.116 This comprises the gold and SDRs held in the Exchange Equalisation Account (EEA) as part of the United Kingdom official reserves. These are the only assets in the system for which there are no counterpart liabilities (unlike the pre-1998 system where all the reserves had a counterpart in the overseas sector). The source of the data is the EEA. SDRs are international reserve assets created by the IMF, but members do not have an actual unconditional liability to repay their allocations.

Currency and deposits (F.2)

21.117 This category comprises currency in circulation and all types of deposits, in both sterling and foreign currency. It is divided into three categories:

Currency (F.21)

21.118 This comprises changes in notes and coin held by government for its own purposes (assets), and issues of coin by the Royal Mint (a central government liability). The issue of bank notes is a function of the Issue Department of the Bank of England, and is recorded as transactions of the central bank sector.

Coin is a liability of central government.

In practice the government's needs for currency are negligible, and this item reflects the net issue of coin by the Royal Mint.

Changes from pre-1998 system

In the previous system the Issue Department of the Bank of England was regarded as the agent of central government, and new issues of notes were recorded as increases in government liabilities. However, since the Issue Department is required to hold securities equivalent to the note issue, the effect of this change on the government financial account is mainly to replace the increase in notes as a direct liability by an increase in government securities.

Transferable deposits(F.22)

21.119 Assets comprise all deposits by government in sterling or foreign currencies which are immediately convertible into cash or which are transferable by cheque, banker's order, direct debit or the like without any significant restriction or penalty. The series includes deposits with the Bank of England and with commercial banks and building societies. The data for both central and local government come from the banking statistics (25.3.1), and building society returns (25.3.2).

Other deposits (F.29)

21.120 Liabilities comprise deposits with central government and include National Savings (provided by DNS), tax instruments and short-term deposits by local authorities, short-term deposits by public corporations, the EC No. 1 Account, the Insolvency Services Account, and court funds. Data on National Savings are published monthly in *Financial Statistics*.

21.121 Assets comprise all deposits by general government other than those in F.22. The data come from the banking statistics (25.3.1).

Repurchase agreements (repos), whereby government deposits funds with a bank in return for a tradeable financial asset as collateral, would be included here as government assets and bank liabilities.

Securities other than shares (F.3)

21.122 These are defined as bearer instruments that are usually negotiable and traded on a secondary market, which do not grant the holder any ownership rights. Sale and repurchase transaction agreements are classified to F.4, or to F.2 if the counterparty is a bank.

Assets

21.123 Assets of general government comprise:

<u>Securities other than financial derivatives (F.33)</u>

Short-term (F.331): transactions in commercial bills; local authority transactions in Treasury bills; Treasury non-interest bearing notes. The source of data is HM Treasury.

Long Term (F.332): transactions in debentures (in recent years relating to privatised companies); local authority transactions in gilts (British government securities). Data are from the Treasury and from DETR returns of local authority borrowing for the PSBR. The category also includes some foreign securities previously included in official reserves.

<u>Financial derivatives (F.34)</u>

None are currently reported for general government.

Liabilities

21.124 Liabilities of general government under this heading comprise:

<u>Securities other than financial derivatives (F.33)</u>

Short-term (F331): net issues of Treasury bills, local authority bills
Longer-term (F332): British government securities, local authority securities

These securities can be issued in sterling or foreign currency.

Data are from the Treasury and from DETR returns of local authority borrowing for the PSNCR.

21.125 Borrowing includes the increase in accrued interest outstanding on National Savings and index-linked government securities. The increase in principal of index linked securities is also included.

<u>Financial derivatives (F.34)</u>

None are currently reported for general government.

Loans (F.4)

Short-term loans (F.41)

21.126 Central government short-term borrowing from banks is derived from the banking statistics and the accounts of central government.

Long-term loans (F.42)

21.127 Assets of general government in category F.42 include:

- lending by central government to local government
- lending by central and local government to public corporations
- lending by central and local government for housing (or, in recent years, net repayments), the central government lending being through housing associations
- lending by central government to other governments, mainly relating to overseas aid
- lending by central government to universities
- lending by central government to students

Data come from the central government GEMS returns and local authority returns to the Department of the Environment. (*See also* Chapter 17.)

21.128 Liabilities of general government include borrowing through finance leasing. Finance leasing consists of transactions by which a lessee leases an asset for all (or virtually all) of its productive life and takes on all ownership responsibilities, such as the maintenance of the asset. The involvement of the lessor is purely a financial one. A loan is imputed equivalent to the value of the asset and the rental is divided into repayments of principal and payments of interest. A few Private Finance Initiative schemes count as finance leases but most are recorded as operating leases (for provision of a service.)

21.129 Other central government liabilities include "ways and means" advances from the Issue Department of the Bank of England, short-term borrowing from UK and overseas banks, foreign currency debt and transactions in ECS liabilities.

Shares and other equity (F.5)

21.130 This covers transactions in ordinary and preference shares but not debentures. Privatisation sales and related debt restructuring have dominated this series in recent years. Data are derived from the accounts of central government and financial returns from local authorities.

21.131 F.516 relates to the cancellation by central government of public corporations' debt in connection with their privatisation. Major entries include:

1986	£1,591 million	New Towns Development Corporations
1988	£3,890 million	British Steel
1989	£5,028 million	Regional Water Companies
1991	£418 million	Scottish Power
1996	£414 million	Railtrack
	£2,465 million	British Energy
	£849 million	European Passenger Services

21.132 F.519, transactions in Rest of the World securities, comprises subscriptions to international lending bodies such as the European Bank of Reconstruction and Development.

Insurance technical reserves (F.6)

Prepayments of insurance premiums and reserves for outstanding claims (F.62)

21.133 For general government the item comprises solely local authority assets in F.62 relating to local authorities as policy holders. The data are taken from a sector analysis of insurance companies data. The amount is netted off local authority procurement expenditure reported to ONS since this includes the full cost of insurance premiums net of claims.

Accounts receivable and payable (F.7)

21.134 Accounts receivable and payable mainly represent the difference between cash receipts or payments by the general government sector and the accrued figures shown in the accounts. Transactions in assets (accounts receivable) represent the excess of income due over cash receipts. Cash receipts in general are obtained from central government accounts. Accounts receivable relate to taxes and social contributions.

21.135 Accounts payable, the change in liabilities of general government, represent the excess of amounts due for payment over amounts actually paid.

Financial balance sheets

21.136 Balance sheet data are obtained from

accumulated net transactions data
Bank of England administrative records of government transactions
Bank of England and ONS surveys of financial institutions
National Loans Fund administrative records
DETR surveys of local authorities

as described in the corresponding financial accounts paragraphs above.

Annex 1

ESTIMATES AT CONSTANT PRICES

For the estimation of gross domestic product at constant prices the estimates of general government expenditure on goods and services are revalued at constant prices. Gross fixed capital formation is revalued as part of the series for the whole economy, as described in Chapter 15.

Blue Book Table 1.4 shows general government final consumption expenditure at constant prices in two parts: individual consumption and collective consumption. These series are built up from components as indicated in the table at the end of this annex.

There is a particular difficulty in valuing the non-market output of general government, since by definition the selling price to the user (normally zero) does not cover the majority of the production costs. Furthermore, the services provided often are not easily divided into homogeneous units for which a price can be established. In Blue Books up to 1997 the method was to revalue the inputs to the services, either by the deflation of expenditure figures by indices of pay and prices or by the use of volume indicators. In general the methods of revaluation implied that there was no increase in the productivity of those employed in public services. This assumption has become increasingly unsatisfactory.

With the introduction by the government of output and performance assessment (OPA) many more indicators are becoming available, and with effect from the 1998 Blue Book ONS has begun to extend the use of direct measures of government output and consumption expenditure. The output measures have been taken back to 1986 in the published figures.

The methods of revaluation now fall into three groups:

- indicators of the volume of output
- indicators of the volume of inputs
- the deflation of expenditure by indices of pay and prices

The estimates of taxes on production at constant rates of tax present little difficulty. About one-sixth of the total - the gas levy and the taxes on alcoholic drink and hydrocarbon oils - is estimated by multiplying the physical quantities of the taxable goods by the rates of tax which applied in the base year. For ad valorem duties the ratio of tax to market price in the base year is applied to the expenditure at constant prices in other years on the items concerned.

Value added tax (VAT) is calculated by deflating the current price figures to constant prices. A special deflator is used for this purpose which is calculated as follows. Estimates of the amount of VAT payable at current prices are produced by applying appropriate ratios of tax to market price expenditure in each year to the appropriate categories of expenditure. Similarly estimates are made at constant prices by applying the ratio of tax to market price expenditure in the base year to the constant price estimates of the same categories of expenditure. The required deflator is obtained as the ratio of the current and constant price totals produced from these calculations.

Estimates for subsidies are more of a problem and the methods used are approximate. The largest item, agricultural subsidies, consists mainly of payments to farmers of the difference between guaranteed prices for various agricultural products and actual prices in the market. The estimate at constant prices is obtained by applying the subsidy per unit for each product in the base year to the quantities marketed in other years. Central government and local authority subsidies for housing are revalued by reference to the subsidy per house in the base year and subsequent changes in the stock of such houses. The revaluation is carried out in rather less detail quarterly than annually.

Method of revaluation	
Individual services	
National Health Service:	Output volume indicators:
Education	Output volume indicators:
Social security	Output volume indicators:
Cultural and recreational	Expenditure deflated by an appropriate price index.
Collective services	
Military defence and forces' pay, including all employer's contributions and pensions paid	Volume indicators: separate strengths for the army, navy and airforce, subdivided into male and female for officers and for other ranks.
Civilian wages and salaries, including all employer's contributions and pensions paid	Volume indicator: civilian staff employed by the Ministry of Defence.
Intermediate consumption	The expenditure is deflated in 15 commodity groups *plus* administration expenses. Each commodity group is deflated by a specially constructed base-weighted price index derived mainly from producer price indices and adjusted for changes in the rate of VAT. Administration expenses are deflated by a specially constructed base-weighted price index for central government administrative costs.
Other central government: pay, including all employer's contributions and pensions paid	Input volume indicator: numbers employed in relevant Departments.
Other central government: intermediate consumption	Expenditure deflated by specially constructed base-weighted price index for central government administration costs.
Police pay	Volume indicator: index of strength of police forces in Great Britain.
Other local government pay and intermediate consumption	Expenditure deflated by specially constructed base-weighted price indices.

Annex 2

Functional classification: COFOG - The international classification of functions of Government

General government expenditure is classified according to the international Classification of the Functions of Government (COFOG). In this annex a brief summary of the international definition is given for each heading, followed by a list of the main United Kingdom government agencies whose expenditure is classified to the heading.

1. General public services

Executive and legislative organs, financial and fiscal affairs, external affairs including foreign aid, fundamental research of all kinds and general services not connected with a specific function. Applied research or administration connected with a specific function (for example, defence) is classified to the appropriate major group. In the United Kingdom this group includes the following bodies:

Foreign and Commonwealth Office *Cabinet Office*
Department for International Development *Privy Council Office*
Public Record Office *Northern Ireland Office*
Scottish Record Office *Crown Estate Office*
General Register Office for Scotland *Government Actuary's Department*
Welsh Office (part) *Royal Mint*
HM Treasury *Paymaster General's Office*
Customs and Excise *Central Office of Information*
Inland Revenue *House of Commons*
Department for National Savings *House of Lords*
National Investment and Loans Office *National Audit Office*
Office for National Statistics *Office of Parliamentary Commissioner for Administration*
Office of Public Service *Local government agencies and departments with similar functions*

2. Defence affairs and services

Administration, supervision, and operation of military and civil defence affairs and forces: land, sea, air and space defence forces; engineering, transport, communications, intelligence, materiel, personnel and other non-combat forces and commands; reserve and auxiliary forces of the defence establishment; provision of equipment, structures, supplies and so forth; offices of military attaches stationed abroad and military aid; defence-related applied research and experimental development. Administration and operation of military schools and colleges where curricula resemble those of civilian institutions are classified to Education even though attendance may be limited to military personnel and their families. Base hospital administration and operation are classified to Health but field hospitals are included here. Payment of pensions for military personnel is classified under Social Security. In the United Kingdom this comprises expenditure by the *Ministry of Defence*

3. Public order and safety affairs

This comprises police and fire protection; law courts; administration and operation of prisons; other public order and safety affairs. Police include regular and auxiliary police forces supported by public authorities and port, border, coastguard and other special police forces. Police duties include traffic regulation, alien registration, the operation of police laboratories, the maintenance of arrest records and statistics related to police work, and the provision of equipment and supplies for police work (including vehicles, aircraft, and vessels). Police training is included but police colleges offering general education in addition to police training are classified in Education. In the United Kingdom this group includes the following bodies:

Home Office
Lord Chancellor's Department
Northern Ireland Court Service
Crown Prosecution Service
Serious Fraud Office
Treasury Solicitor's Department
Scottish Office (part)

Scottish Courts Administration
Crown Office, Scotland
Welsh Office (part)
Royal Ulster Constabulary
Police forces
Local courts
Fire Brigades

4. Education

The International Standard Classification of Education (ISCED) of the United Nations Educational, Scientific and Cultural Organisation (UNESCO) defines education as consisting of organised and sustained communication to bring about learning. It may be conducted within or outside an official school system or institutional arrangement. It covers education for all types of students and for all age groups, including adult education. Education includes activities that in some countries and in some languages may be described as 'training' or 'cultural development'. It excludes types of communication that are not designed to bring about learning or are not planned in a pattern or sequence with established aims; for example, leisure-time activities. Military schools and colleges where curricula resemble those of civilian institutions are included here even though attendance may be limited to military personnel and their families. Operation, support, etc, of sporting and recreational facilities associated with educational institutions are classified here. In the United Kingdom this group includes the following bodies:

Department for Education and
  Employment (part)
Department of HM Chief Inspector of
  Schools in England
Office for Standards in Education

Scottish Office (part)

Office of HM Chief Inspector of Schools in Wales

Local government schools and education departments

5. Health

Includes hospitals (both specialised and general), medical centres, nursing and convalescent homes, medical, paramedical, and dental practitioners, public health services, and supply of prescribed medicines, prostheses, etc. Military base hospitals are included but field hospitals are classified to defence. Public health services are included. Medical research is also included. In the United Kingdom this group includes the following bodies:

Department of Health
Health and Safety Commission
Scottish Office (part)

Health Service Commissioners
Local government public health departments
Welsh Office (part)

6. Social protection

Social security services are defined as chiefly transfer payments (including payments in kind) to compensate for reduction or loss of income or inadequate earning capacity. Welfare services are defined as assistance delivered to clients or groups of clients with special needs, such as the young, the old, or the handicapped. In the United Kingdom this group includes the following bodies:

Department of Social Security   *Redundancy Funds*
National Insurance Funds   *Local welfare services*

7. Recreation, culture and religion

Administration of sporting and recreational affairs; management, operation, support, provision, etc, of facilities for active or passive sporting pursuits or events and for recreational activities. Facilities for spectator accommodation are included. Support includes outlays for national, regional, or local team representation in sporting events and for equipment, coaching, training, and other things needed to field a team or player. Also included are subsidies to professional teams or individual competitors. Operating, support, etc, of sporting and recreational facilities associated with educational institutions are in Education.

Administration of cultural affairs; management, operation, support, provision, etc, of facilities for cultural pursuits, such as libraries, museums, art galleries, theatres, exhibition halls, monuments, historic houses and sites, zoological and botanical gardens, aquariums, arboreta and so forth; production, operation, support, etc, of cultural events including concerts, stage and film productions, radio and TV productions, art shows, etc; transfers or other types of support to individual artists, writers, designers, composers and others working in the arts and to organizations engaged in promoting cultural activities. Also included are outlays in support of national, regional or local celebrations provided they are not intended chiefly to attract tourists, in which case the outlays are classified to group 11. Outlays in support of cultural events intended for presentation beyond national boundaries are classified to group 1.

In the United Kingdom this group includes the following bodies:

Department of Culture, Media and Sport   *Welsh Office (part)*
Department of the Environment, Transport   *Local government agencies and departments with similar*
and the Regions   *functions*
Scottish Office (part)

8. Housing and community amenities

Housing comprises administration of housing affairs and services; development, monitoring, and evaluation of housing activities whether or not the activities are under the auspices of public authorities; development and regulation of housing standards (other than construction standards which are classified to group 11); administration of rent controls and eligibility standards for publicly supported dwelling units; provision of housing for the general public or for people with special needs, i.e. construction, purchase, remodelling and repairing of dwelling units; acquisition of land needed for construction of dwellings; slum clearance related to provision of housing; subsidies, grants or loans for increasing, improving, or maintaining the housing stock (other than rent subsidies paid to households which are considered a form of income assistance and are classified in Social security and welfare); offices, bureaux or units producing and disseminating public information about housing; applied research into, and experimental development of, housing standards and design (but not construction methods or materials, which are classified in group 11).

The category of overall community development affairs covers chiefly the planning of new communities or of rehabilitated communities in which, typically, there would be a broad spectrum of physical facilities that would improve the quality of life of the people involved. In general such plans involve not only housing and industries but also facilities for the health, education, culture and recreation of the community. The plans may also include schemes for financing actual construction and, in the case of urban renewal projects, arrangements for removal of existing populations. The category of overall community development affairs also covers the offices, bureaux or units engaged in administering zoning laws including regulations on land use and building standards other than standards covering housing.

This group also covers government non-trading expenditure on water supply affairs and services and the licensing of private firms engaged in this work.

In the United Kingdom this group includes the following bodies:

Department of the Environment, Transport and the Regions (part)
Scottish Office (part)
Welsh Office (part)

Land Registry
Office of Water Services
Local authority housing departments

9. Environment protection

Comprises refuse collection, treatment and disposal; sewage systems and waste water treatment; pollution abatement; nature conservation; related research and development; other environmental protection services.

The group also covers licensing of private firms engaged in this work. Operation by public authorities of refuse collection and disposal systems, including refuse of all types (benign or noxious materials, including radioactive wastes, solids, liquids or gases, including rainwater) from whatever source (households, factories, mines), collected by any method (trucks, piping, storm sewers), treated (incinerated, chemically transformed) or untreated and disposal by dumping at sea, on land or by other means. The administration of regulations on the generation and release of pollutants to the environment is also included. In the United Kingdom this group includes the following bodies:

Department of the Environment, Transport and the Regions (part)
Scottish Office (part)
Ministry of Agriculture, Fisheries and Food
Welsh Office (part)
Office of Water Services (part)
Local authority refuse collection, street cleaning departments.

10. General economic affairs

This group includes the formulation of general economic policies and the regulation or support of general economic activities such as export and import trade as a whole, international financial affairs, commodity and equity markets, overall income controls, supervision of the banking industry, general trade promotion activities, general regulation of monopolies and other restraints on trade and market entry, and economic and commercial matters that cannot be assigned to a specific industry. It also includes offices, bureaux and units operating such institutions as the patent, trademark and copyright offices; weather forecasting service, standards institution and various survey institutions (such as hydrological survey and geodesic survey). In general, economic and commercial affairs of a particular industry are classified under group 11. In the United Kingdom this group includes the following bodies:

Department of Trade and Industry (part)
Office of Fair Trading
Department of the Environment, Transport and the Regions (part)
Department for Education and Employment (part)
Advisory, Conciliation and Arbitration Service (ACAS)
Charity Commission
Registry of Friendly Societies
Ordnance Survey
Local government agencies and departments with similar functions

11. Sectoral economic affairs

Administration of affairs and services relating to particular industries: offices, bureaux and units engaged in activities designed to develop, expand and generally improve the position of establishments in those industries by means of loans or transfers, by subsidizing outputs or inputs, by tax rebates or by assistance in other forms; units that develop and administer regulations, that inspect premises for conformity with safety regulations and that exercise other kinds of control (for example to protect consumers against dangerous products). However, environmental protection is included in group 9. Also included are outlays on, or support for, applied research for specific industries; dissemination of information; collection and compilation of statistics of special interest to particular industries; maintaining liaison with trade associations and other interested organizations. In the United Kingdom this group includes the following bodies:

Department of the Environment, Transport and the Regions (part)
Department of Trade and Industry (part)
Forestry Commission
Health and Safety Commission
Intervention Board for Agricultural Produce
Ministry of Agriculture, Fisheries and Food
Office of Gas Supply
Office of National Lottery
Office of Passenger Rail Franchising
Office of the Rail Regulator
Office of Electricity Regulation
Office of Telecommunications
Passenger Transport Authorities
Scottish Office (part)
Welsh Office (part)
Local government agencies and departments with similar functions

12. Expenditure not classified by major group

Interest payments and outlays for underwriting and floating government loans. (Administrative costs of public debt management are classified to group 1.) Transfers of a general character between different levels of government; other expenditures not classified by major group. In the United Kingdom this group includes the following items:

National Loans Fund (interest)
Department of the Environment, Transport and the Regions (part)
Local authority interest payments

Chapter 22

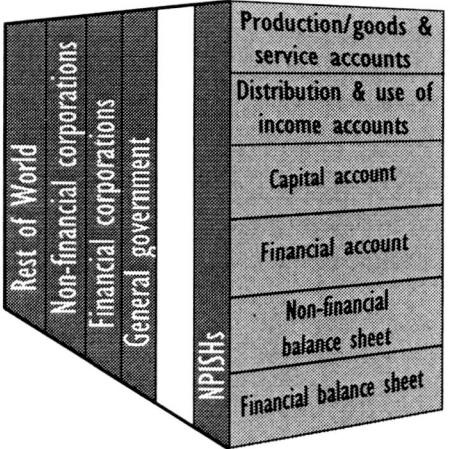
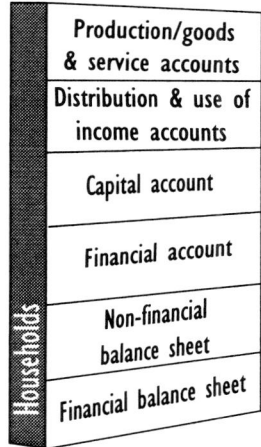

The household sector

Chapter 22 The household sector

Coverage of sector

22.1   This chapter describes the estimation of the range of variables relevant to the accounts for the household sector. It also includes a description of the methodology used for compiling the commodity estimates of household consumption.

22.2   The chapter covers the production account and the distribution and use of income account (these, together, were known as the current account in the previous system), and the capital account and the financial account. It also covers the financial balance sheet.

22.3   The definition of the household sector is given in Chapter 10. The sector covers people living in traditional households, as well as those living in institutions. The latter, who number only about $1\frac{1}{2}$ per cent of the UK population, include those living in retirement homes, hostels, boarding houses, hotels or prisons. The household sector also embraces sole traders, any non-profit institutions serving households (NPISH) which do not have separate legal status and private trusts. These are explained further below.

Changes in coverage

In the previous accounts, the household sector was included within what was termed the personal sector. The personal sector also included NPISH, which now constitute a separate sector (*see* Chapter 23). Much of the analysis in the old accounts related to the personal sector, although certain information was also available solely for the household sector. In the new system, the definition of the household sector shows two main changes from what constituted households within the personal sector of the previous system. First, while the personal sector embraced all unincorporated businesses, in the new system, only sole traders are included in the household sector (partnerships are now part of the private corporations sector). Secondly, the property income received by life assurance companies and pension funds on the investment of pension funds, previously scored directly within the personal sector which was deemed to own the funds, is now attributed in the first instance to the receiving institutions and then redistributed to policy-holders. There are thus a number of changes in the way transactions are recorded for the households and financial corporations sectors (*see* 22.6).

Sole traders

22.4   The activity of sole traders is regarded as an integral part of the households to which they belong. The income generated by such activity is partly a remuneration for work done and partly a return on capital - essentially profit or, in the new system, operating surplus. It is clearly difficult to separate what are otherwise two components of the income account. The surplus arising from the activity of sole traders therefore covers returns to labour and capital indistinguishably. In the new accounts this is called 'mixed income'. Its UK equivalent in previous terminology was self-employment income (specifically, here, that part relating to sole traders).

Private trusts

22.5   These are trusts created for individuals, usually with named beneficiaries. They include, for example, trusts created to provide for dependent relatives. Trusts created with charitable status are separate institutional units, included in the NPISH sector.

Life assurance and pension schemes

22.6   In the previous accounts, the life funds of life assurance and pension schemes were regarded as the collective property of households, and receipts of dividends and interest were included as part of the income of the personal sector. In the new accounts, receipts of property income (dividends, interest etc) accrue to the investing life insurance or pension fund but are then deemed to be paid to the policy holders, ie households, in the primary income account. The households then 'pay them back' to the funds in the form of contribution supplements appearing in the secondary distribution of income account.

Sub-sectors of the household sector

22.7   In the ESA95, the household sector may be divided into sub-sectors according to various criteria such as main source of income. This is not being done in the UK accounts.

Summary of the accounts for the household sector

22.8   An overview of the whole sequence of accounts for the household sector helps to show some of the important variables and principal features. The main components of the accounts are given below, with figures for 1995. Most variables are further disaggregated in the full accounts. Further details on the structure of the accounts is given at the beginning of each section, and the full detail of the accounts is also explained, along with the methodology for the estimation of the variables.

Main components of the household sector accounts, 1995

£million

Production account

Output (P.1)	174 183
Intermediate consumption (P.2)	59 166
Value added (gross) (B.1g = P.1 - P.2)	115 017
Consumption of fixed capital (K.1)	19 102
Value added (net) (B.1n = P.1 - P.2 - K.1)	95 915

Distribution and use of income accounts

Value added (gross) (B.1g)	115 017
of which	
Operating surplus (B.2)	39 321
Mixed income (B.3)	40 239
Compensation of employees (D.1) (receipts *less* payments)	385 101
Property income (D.4) (receipts *less* payments)	62 909
Balance of primary income (B.5g = B.2 + B.3 + D.1 + D.4)	527 570
Current taxes on income etc (D.5)	86 225
Social contributions *less* social benefits	
and other current transfers (D.61 + D.62 + D.7)	53 399

	£million
Disposable income (B.6g = B.5 - D.5 + [D.61 + D.62 + D.7])	494 744
Adjustment for change in equity reserves (D.8)	11 690
Final consumption expenditure (P.3)	454 171
Saving (B.8g = B.6 + D.8 - P.3)	52 263

Accumulation accounts

Capital account

Saving (B.8g = B.6 + D.8 - P.3)	52 263
Capital transfers (D.9) (receipts *less* payments)	2 749
Change in net worth due to saving and capital transfers (B.10.1g = B.8 + D.9)	55 012
Gross fixed capital formation (P.51)	27 507
Change in inventories (P.52)	262
Acquisitions *less* disposals of valuables (P.53)	44
Acquisition *less* disposals of non-produced non-financial assets (K.2)	-81
Net lending/borrowing (B.9 = B.10.1g - P.51 - P.52 - P.53 - K.2)	27 280

Financial account

Currency and deposits (F.2) (receipts *less* payments)	31 395
Securities other than shares (F.3) (receipts *less* payments)	3 150
Loans (F.4) (receipts *less* payments)	-23 711
Shares and other equity (F.5)	-9 365
Insurance technical reserves (F.6)	32 478
Other accounts receivable *less* other accounts payable (F.7)	-1 448
Net lending/borrowing (B.9 = F.2 + F.3 + F.4 + F.5 + F.6 + F.7)	32 499

22.9   There is also an alternative presentation which uses a different definition of income (adjusted disposable income) and consumption (actual final consumption), as explained in paragraphs 22.91–22.92. The balance for saving is, however, the same. The various components of the accounts are explained in more detail below.

The sequence of the accounts for the household sector

22.10   This section shows the structure and detail of the accounts for the household sector, and describes the methodology used for estimating the variables. Each sub-section will include a summary presentation of the particular account being considered, with figures for the year 1995. As mentioned earlier, three broad types of account are considered:

> Production account
> Distribution and use of income accounts
> Accumulation account

22.11   The estimation methodology for the figures which go into the household sector accounts falls into three broad groups. First, for some variables, such as compensation of employees and final consumption, estimates relate uniquely to the household sector. In the second group, specific estimates are made for the household sector as part of the process of deriving the aggregate figure. Finally, for other variables, the sector figures are derived by breaking down the aggregate as well as possible. Where this is done, the estimate for the household sector will often be derived as a residual. This applies particularly to a number of series in the financial account. (*See* 17.44 *et seq.*)

22.12 Generally speaking, where this third approach is used, the description of the methodology will have been included in another part of the publication. This also applies to the variables in the first group, other than for household consumption, which is described in this chapter. In both cases, only brief mention or a cross-reference will be given in this section. However, for estimates which are made directly, a fuller description of the methodology will normally be included here.

22.13 One important source of the sector figures is the dividends and interest matrix (DIM), which provides estimates of payments and receipts, for all the main institutional sectors, of the whole range of dividends and interest components. A description of the derivation of the matrix is provided in Chapter 14.

Production account

22.14 The production account shows the output and intermediate consumption of the household sector. The difference between these two variables gives the sector's contribution to total GDP, value added. The account, in summary form, with estimates for 1995, is shown below.

Production account, 1995

£million

Uses		Resources	
Intermediate consumption	59 166	Output	174 183
Value added, gross	45 017		
Consumption of fixed capital	19 102		
Value added (net)	95 915		

22.15 The variables making up the production account are estimated as described below.

Resources

Output (P.1)

22.16 Production in the household sector arises in two forms, as described below. The first reflects the activity of sole traders, for example, farmers, and professional people. The second relates to the production of goods and services for own consumption. It should be noted that some goods and services produced by unincorporated businesses may be consumed by the household, for example, consumption by farmers of their own produce. In practice, it may not always be easy to separate out these two components in the estimates.

Market output (P.11)

22.17 Output under this head includes, for example, production of services by sole traders (eg solicitors) and production by farms which are not incorporated. Estimates are based on the sector allocation of business units on the IDBR, ensuring that the allocation across all sectors is consistent with the largely industry-based aggregate. The estimates for the household sector relate predominantly to small firms, which are less well covered on the register. Further details on the estimates are included in the output section in Chapter 13.

Output for own final use (P.12)

22.18　This covers, mainly, agricultural production (within final consumption expenditure), and household construction of new dwellings and the renovation of existing dwellings (within GFCF). Also included is imputed rent of owner-occupiers - who are deemed to produce housing services for which they pay themselves - and payments for domestic services (both within final consumption). All these four components also appear in the income measure of GDP. The estimates are made from the expenditure side of the accounts, as described later in this chapter.

Uses

Intermediate consumption (P.2)

22.19　Intermediate consumption relates to the cost of goods and services purchased by sole traders for use in producing output. It also includes the expenditure made by owner-occupiers on decoration, maintenance and repair of dwellings. In respect of the activities of unincorporated businesses, estimates are based on the sector allocation on the IDBR, again ensuring that the allocation across all sectors is consistent with the largely industry-based aggregate. Here again, the estimates for the household sector relate predominantly to small firms, which are less well covered on the register. Further details of the estimates are given in the section on intermediate consumption in Chapter 13.

Value added (gross) (B.1g)

22.20　This is the difference between output and intermediate consumption, and represents the contribution of the household sector to GDP, before allowing for consumption of fixed capital. In practice, it should be noted that the estimate of value added, in particular, is made consistent with the estimate of GDP for the whole economy, through the balancing process (*see* Chapter 11).

Consumption of fixed capital (K.1)

22.21　Estimates for the household sector, which relate mainly to dwellings but also the fixed assets of farmers, are derived from the model-based approach used for estimating capital stock, as described in Chapter 16.

Value added (net) (B.1n)

22.22　This is the difference between gross value added and consumption of fixed capital, and represents the contribution of the household sector to net domestic product.

Distribution and use of income accounts

22.23　These accounts show, through a number of separate accounts, the distribution and use of household income. The accounts, some of which are further sub-divided, are discussed separately below. The main accounts are as follows.

　　　II 1.1　*Generation of income account*

　　　II 1.2　*Allocation of primary income account*

Generation of income account (II.1.1)

22.24   The generation of income account for households shows how the value added of the production account is allocated to the factors of production. It shows, first, the compensation of employees paid by unincorporated enterprises owned by households (that is, essentially, sole traders), for example, farmers and retailers. It is important to note that the former reflects what is *paid* by the household sector, rather than the major component of household income, namely what is *received* by employees for the services of their labour, which is covered in the allocation of primary income account. Given this, the main description of the compensation of employees variable appearing in this chapter is given in this latter account (*see* paragraph 22.38). The balance of value added *less* compensation of employees less net taxes is defined as operating surplus (*see* paragraph 22.32) and, for households only, mixed income. The form of the account is shown below.

Generation of income account, 1995

	£million
Resources	
Value added (net)	95 915
Consumption of fixed capital	19 102
Uses	
Compensation of employees	35 402
Taxes less subsidies on production	55
Operating surplus	39 321
Mixed income	40 239

22.25   The variables making up the account are estimated as follows.

Resources

Value added (gross) (B.1g)

22.26   This is the balance from the production account

Uses

Compensation of employees (D.1)

22.27   Compensation of employees is the remuneration, in cash or in kind, payable by households as employers for the services of their employees in a particular period. In the household account, this covers payments by sole traders to their staff and payments to private domestic staff. The aggregate has two main components - wages and salaries, and employers' social contributions.

Wages and salaries (D.11)

22.28   Wages and salaries is the payment to employees, in cash or kind, for work done. Annual estimates are based on Inland Revenue sources (*see* Chapter 14); quarterly figures are derived by interpolation and extrapolation on the basis of ONS inquiry data on employment and average earnings (25.4.1).

Employers' social contributions (D.12)

22.29   These are contributions payable by employers to social security and privately- funded social insurance schemes. A distinction is made between actual and imputed payments, as described below.

Employers' actual social contributions (D.121)

22.30   These contributions reflect the payments by employers to social security funds or private funds for the benefit of employees. Estimates are derived from central government and from ONS income and expenditure inquiries to life assurance corporations and pension funds. (25.3.3). The estimates for this sector are obtained by applying the whole economy ratio of employers' contributions to wages and salaries to the wage and salary bill of the sector.

Operating surplus (B.2)

22.31   The operating surplus appearing in the household sector account relates, by definition, to the household sector's rental income from buildings, including the imputed rental of owner occupied dwellings. The derivation of this estimate is described in paragraph 22a.82.

Mixed income (B.3)

22.32   This represents rewards to labour (compensation of employees) and to capital (operating surplus) which cannot realistically be distinguished. Mixed income is, essentially, that part of income from self-employment relating to sole traders, as liable for tax under Schedule D. Although it appears in the account as a balancing item, mixed income is, in practice, estimated independently within the UK economic accounts, as part of the income measure of GDP. The estimate included here in the generation of income account is derived, as for all estimates, from an assessment and balancing of all relevant data available.

22.33   The methodology for estimation of mixed income is described fully in Chapter 14. In brief, annual estimates are compiled separately for farmers, based on MAFF information, and for other self-employed, using the Inland Revenue Schedule D information, together with certain adjustments for coverage and the hidden economy. Estimates on a quarterly basis are made as interpolations and projections of the annual figures, using various indicators of movements in economic activity and prices.

Allocation of primary income account (II.1.2)

22.34   The allocation of primary income account relates to households as recipients rather than producers. Primary incomes are those arising as part of the production process or through ownership of assets required for production. The account shows the income **received** by households for their role in the production process, and also property income (rent, dividends and interest) received and paid. It is shown, in summary in the following table. Some of the variables are disaggregated for the Blue Book table, as detailed below. The balance of operating surplus plus mixed income plus compensation of employees plus property income is defined as the balance of primary income.

Allocation of primary income account, 1995

	£million
Resources	
Operating surplus	39 321
Mixed income	40 239
Compensation of employees	385 101
Property income	102 997
Uses	
Property income	40 088
Balance of primary incomes	527 570

22.35   The variables making up the account are estimated as follows.

Resources

Operating surplus (B.2)

Mixed income (B.3)

22.36   These are as defined in the generation of income account, above.

Compensation of employees (D.1)

22.37   Compensation of employees under this head is the remuneration, in cash or in kind, **received** by employees in the household sector from employers as payment for the services of labour in a particular period. The aggregate has two main components - wages and salaries, and employers' social contributions. The variables have been discussed in detail in Chapters 5 and 14, which also cover the issue of the boundary between employees and self-employed.

Wages and salaries (D.11)

22.38   The estimates include remuneration for regular activity, together with payments for overtime and bonuses, and also various allowances for housing, cost of living and transport to and from work, but excluding 'expenses'. The figures are before deduction of income tax and employees' social contributions. Payments in kind include, for example, free or reduced price meals or accommodation, and private use of a company car.

22.39   Annual estimates of wages and salaries are based on IR PAYE information (25.4.2), together with certain coverage adjustments, for example, for pay below the tax threshold, income in kind, locally engaged staff working abroad and an evasion adjustment. The quarterly figures are interpolations and extrapolation of the annual data, based on ONS information on numbers in employment and average earnings, by industry, (25.4.1). A full description is included in Chapter 14.

Employers' social contributions (D.12)

22.40   These relate to contributions payable by employers to social security and privately- funded social insurance schemes. A distinction is made between actual and imputed payments, as described below. The contributions provide employees with entitlement to social benefits, for example for sickness, accident, redundancy and retirement. Although the contributions are paid, actually or notionally, directly to a fund, they are included in compensation of employees. Subsequently, in the secondary of distribution account, these amounts are treated as payments by the household sector into the funds, which are in the financial corporations and general government sectors.

Employers' actual social contributions (D.121)

22.41   These contributions reflect the payments by employers to social security funds and private schemes for the benefit of employees. Estimates are derived from information on pension contributions collected in ONS annual inquiries to insurance companies and pension funds (25.3.3) and from the Government Actuary's Department in respect of social security contributions.

Employers' imputed social contributions (D.122)

22.42   These correspond to benefits paid to employees from employers' own current resources, rather than from a fund to which they would have made contributions. The main examples of such schemes are those operating within central government for civil servants, NHS employees, the armed forces and teachers. The amounts to be imputed should be equal to what would be needed to meet the entitlement were a funded scheme to be used. They are calculated in this way for the main government schemes. However, in some cases (e.g. fire and police) the value of pensions currently paid out *less* employees' contributions is taken as a proxy. Data are obtained from central government accounts.

Property income (D.4)

22.43   Property income, under the resources head, relates to receipts from other sectors in respect of ownership of financial and tangible non-produced assets. It is split into five main components, as shown below.

Interest (D.41)

22.44   This covers interest receipts by households related to, for example, holdings of British government securities and national savings, and also bank and building society deposits. Estimates, generally, are based on the levels of interest-bearing assets held by the household sector, and relevant rates of interest. The estimates are derived as part of the establishment of the 'DIM' matrix of dividends and interest flows across all sectors, as described in Chapter 14.

Distributed income of corporations (D.42)

22.45   This has two sub-components.

Dividends (D.421)

22.46   This embraces receipts of dividends on shares and similar securities. There is, essentially, no direct information on such receipts by households. The estimates are derived largely as a residual, within the DIM analysis.

Withdrawals from income of quasi-corporations (D.422)

22.47   This item reflects income withdrawn from the profits of partnerships by the owners within the household sector for their own use, and is based on counterpart data. (*See* paragraph 5.42.)

Property income attributed to insurance policyholders (D.44)

22.48   This item reflects the treatment of insurance technical reserves as belonging to the policy holders in the various institutional sectors (*see* paragraph 22.6). The figure for households represents the sector's share of the total income arising, allocated as described in Chapter 14.

Rent (D.45)

22.49   Rent under this head relates to land and sub-soil assets. In principle it does not include rental receipts from buildings, which are payments for a service and part of output. For households, receipts of rent will relate mainly to farming land.

Uses

Property income (D.4)

22.50 Property income, under the *uses* head, relates to payments to other sectors in respect of their ownership of financial and tangible non-produced assets. It is split into two components - interest and rents.

Interest (D.41)

22.51 This covers interest payments by households, principally in respect of loans for house purchase, and other lending by banks and other financial intermediaries. As with interest received, estimates of payments are based on the levels of interest-bearing liabilities held by the household sector, and the relevant rates of interest. The estimates are derived as part of the establishment of the DIM matrix of dividends and interest flows across all sectors, as described in Chapter 14.

Rent (D.45)

22.52 Rent relates to land and sub-soil assets. In principle it does not cover rental payments on buildings, which are part of household final consumption. For households, receipts and payments of rent relate mainly to agricultural land.

Balance of primary incomes (B.5)

22.53 This is the balance of resources less uses, and represents income available to households as a result of participation in the production process.

Secondary distribution of income account (II.2)

22.54 The secondary distribution of income account shows how the balance of primary income of households is modified by redistribution of payments of current taxes, payments of social contributions and receipts of benefits (other than in kind) and net other current transfers. It is given, in summary form, below. Most variables are disaggregated for the published table, as detailed below. The balance of primary income less taxes paid, plus net receipts of social contributions, benefits and current transfers, is defined as disposable income.

Secondary distribution of income account, 1995

£million

Resources		Uses	
Balance of primary income	527 570	Current taxes on income etc	86 225
Social contributions	448	Social contributions	104 800
Other social benefits	149 234	Other social benefits	918
Other current transfers	28 507	Other current transfers	19 072
		Disposable income	494 744

22.55 The variables making up the account are estimated as follows.

Resources

Balance of primary income (B.5)

22.56 This is the balance from the allocation of primary incomes account.

Social benefits other than social transfers in kind (D.62)

22.57 These benefits, referred to as 'other social benefits' in the above table, may be considered under four headings. They exclude reimbursements from social security funds for households to purchase goods and services; these are included as part of social transfers in kind.

Social security benefits in cash (D.621)

22.58 These embrace cash benefits received by households from social security funds. Estimates are derived from Department of Social Security records.

Private funded social benefits (D.622)

22.59 These are receipts by the household sector from privately funded and related social insurance schemes. The payments received will relate to benefits similar to those mentioned above. Estimates are based on information collected in ONS's quarterly income and expenditure inquiry to insurance companies and pension funds (25.3.3).

Unfunded employee social benefits (D.623)

22.60 This head covers receipts by households from unfunded schemes. Again they will cover benefits similar to those set out above. The estimates are based on government accounts.

Social assistance benefits in cash (D.624)

22.61 These benefits, provided by government and the NPISH sector to households, are not made under specific social security schemes. Estimates are derived from government expenditure and Friendly Societies' returns.

Other current transfers (D.7)

22.62 These include a variety of different transfers (ie payments which are unrequited, with nothing received in exchange) serving different purposes. Two components of household receipts are identified.

Non-life insurance claims (D.72)

22.63 This item is the amount due in respect of claims made under non-life insurance policies. The recording relates to the time at which the insured event took place, rather than, for example, the time of payment. Third party claims are treated as being paid directly by the insurer to the beneficiary and not via the policy holder. Estimates, which tie in with the figures of net non-life premiums on the uses side of the account, are based on the sector allocation of the total.

Miscellaneous current transfers (D.75)

22.64 This head consists of a variety of current transfers. The main household items on the receipts side of the account are transfers from the Rest of the World (e.g. gifts), NPISH (e.g. grants) and central government. All current transfers between households should also be included, but it is not generally practicable to record such payments. Estimates are made from a variety of sources.

Uses

Current taxes on income, wealth etc (D.5)

22.65 These are compulsory, unrequited payments made by the household sector to the government sector. Two broad headings are included (see below). The concepts and other issues have been discussed in Chapters 5 and 21.

Taxes on income (D.51)

22.66 These embrace taxes on incomes, profits and capital gains. In respect of the household sector, the main components are

(i) personal income taxes (schedule E),
(ii) taxes on unincorporated enterprises (schedule D),
(iii) taxes on capital gains, arising from gains made on disposable of financial and non-financial assets, and
(iv) taxes on winnings from gambling and lotteries. (These are not yet included in the UK accounts.)

In the previous accounts, taxes on capital gains were regarded as taxes on capital not on income. Estimates are based on Inland Revenue sources (25.4).

Other current taxes (D.59)

22.67 This group covers a variety of taxes including, in particular, motor vehicle excise duty, council tax and payments by households to obtain certain licences. The estimates are obtained from central and local government accounts.

Social contributions (D.61)

22.68 On the uses side of the accounts, these are payments to social insurance schemes (eg pension schemes) to make provision for social benefits such as pensions. The payments, which may be actual or imputed, may be made by employers on behalf of their employees, or by employees, self-employed and non-employed on their own behalf. They may also be compulsory or voluntary. The payments are analysed by these various categories, as shown below. The concepts and other issues related to social contributions have been discussed in Chapters 5 and 14.

Actual social contributions (D.611)

22.69 Actual social contributions are made up of three broad components which are described below. The estimates, including the breakdown into three groups, are derived from information collected in the ONS quarterly inquiry to pension funds (25.3.3) and the Government Actuary's Department.

<u>Employers' actual social contributions (D.6111)</u>

22.70 These are payments made by employers on behalf of employees to social security or various other pension funds. They are regarded as part of compensation of employees (wages, salaries etc) in the allocation of primary income account, where they are a resource of the household sector. The figures under this heading, which are the same as appearing there, record payment by the household sector to the insurance funds in the financial corporations sector and to the government social security scheme (or back to the employer in the case of unfunded schemes).

Employees' social contributions (D.6112)

22.71 These are social contributions paid by employees to social security funds and private funded social insurance schemes. They include the contribution supplements payable out of the property income of funds, attributed to policy holders/scheme members (D.44). The contributions are net of service charges for running the funds.

Social contributions by self-employed and non-employed persons (D.6113)

22.72 This covers only National Insurance costs paid by self-employed and non-employed persons. As for employees (above) these contributions should also include contribution supplements relating to property income accruing to the policy holders but data are not yet available.

Imputed social contributions (D.612)

22.73 This item represents imputed contributions made by employers in respect of unfunded social benefits. The estimates are imputed since payments of benefits are made out of current resources rather than a funded scheme. As with employers' actual social contributions, this variable is identical with that appearing as part of compensation of employees in the allocation of primary income account, where they are a resource. The item is here regarded as a use, essentially a payment to the employer's sector. Currently, only central government schemes are identified.

Social benefits other than social transfers in kind (D.62)

22.74 These benefits, referred to as 'other social benefits' in the above table, are zero on the uses side of the accounts.

Other current transfers (D.7)

22.75 The general head includes a number of different transfers serving different purposes. Two components of household payments are identified.

Net non-life insurance premiums (D.71)

22.76 This relates to policies taken out by households on their own initiative, for their own benefit, and are separate from any employer or government scheme. The figures going into the accounts are the total non-life premium less the service charge of administering the scheme. The premium element includes a supplement arising out of the reserves of the fund which is attributable to policy holders. Estimates, which tie in with the figures of net non-life claims on the resources side of the account, are based on the sector allocation of the total, as described in Chapter 20 Annex 2.

Miscellaneous current transfers (D.75)

22.77 This consists of a variety of current transfers. When it becomes possible to separate out the NPISH sector, the main household items on the uses side of the account will be contributions, in cash or kind, to various bodies such as charities, subscriptions to trades unions, clubs and similar organisations but currently only payments of court fines, certain government fees and transfers to and from the Rest of the World have been identified. In theory, all current transfers between households should be included. However, as mentioned for the resources side of the account, this is not practicable. Estimates are made from various sources.

Disposable income (B.6)

22.78 This item - the balance between resources and uses - represents the amount available to the household sector for spending on consumption goods or saving.

Redistribution of income in kind account (II 3)

22.79 The purpose of this account (described in principle in Chapter 5) is to provide a broader definition of household consumption, embracing also the individual consumption expenditure of government and NPISH. It records the receipt of social transfers in kind (D.63) from general government and the NPISH sector (equal to their individual final consumption expenditure) to produce adjusted disposable income (B.7g). Further description of the government element is given in Chapter 21.

Social transfers in kind (D.63)

Social benefits in kind (D.631)

22.80 These correspond to the individual consumption expenditure of general government, described in detail in Chapter 21. They are taken from the public expenditure monitoring system.

Adjusted disposable income (B.7)

22.81 This is the only entry on the uses side of the account, equal to the sum of disposable income and social transfers in kind.

Use of income account (II.4)

22.82 There are two versions of the use of income account, corresponding to the two concepts of disposable income of the previous accounts. These are the 'Use of disposable income account' and the 'Use of adjusted disposable income account'. Common to both accounts is an item representing the adjustment for the change in net equity of households in pension funds' reserves. Household sector saving in the two accounts is therefore the same.

Use of disposable income account (II 4.1)

22.83 This account relates to the concept of disposable income and the corresponding expenditure, which is household final consumption expenditure.

Use of disposable income account, 1995

	£million
Resources	
Disposable income	494 744
Adjustment for change in net equity of households in pension fund reserves	11 690
Uses	
Final consumption expenditure	454 171
Saving	52 263

The items making up the account are estimated as follows:

Resources

Disposable income (B.6)

22.84   This is the balance from the secondary distribution of income account.

Adjustment for the change in net equity of households in pension funds reserves (D.8)

22.85   The inclusion of this item reflects two particular measurement concepts of the accounts. The first is that the reserves of private funded pension schemes are treated as being owned by the household sector which has claim on the funds. Secondly, the pension contributions to and pensions paid by the private funded schemes are included, as transfers, in the secondary distribution of income account. Reflecting this latter treatment, the 'balance' of contributions less receipts would be included as part of disposable income of the pension schemes rather than of households. Thus, in order to ensure that the saving is correctly attributed to households, it is necessary to include this balance, as a resource, in the use of income account. The financial account shows the same amount being re-invested in the pension funds in F.61.

Uses

Final consumption expenditure (P.3)

Individual consumption expenditure (P.31)

22.86   These items, which are one and the same, are the total current expenditure of the household sector on goods and services, including spending abroad. The main issues of definition and coverage are discussed later in this chapter. All household consumption is individual, as opposed to collective. This major variable in the accounts is estimated from the sum of estimates of spending on individual goods and services. The methodology for the separate commodities is set out in the annex to this chapter.

Saving (B.8)

22.87   This is the balance of resources (household disposable income plus the adjustment for the change in net equity reserves of households in pension funds reserves) and uses (household final consumption) in the account.

22.88   There are a number of important points which need to be made about household saving.

- The estimate is presented both gross and net of consumption of fixed capital. In the previous accounts, the estimate for the personal sector was presented on a gross basis.

- The accuracy of the figure for saving clearly depends on the accuracy of all items of household income and spending. However, the figure is also the comparatively small difference between two very large variables. As such, errors in the income and spending data will have a relatively much larger impact on the estimate for saving. This and problems of recording, such as timing differences, will mean that the quarterly series, in particular, will often move erratically.

22.89 It should also be noted that estimates of saving are also provided by the financial and capital transactions of the household sector. Put another way, saving plus the balance of the capital account, which leads to net lending/borrowing (*see* paragraph 22.121), should be equal to net lending/borrowing, as emerging from the financial accounts (paragraph 22.136). In practice, given that the accuracy of the financial account data for the household sector is less than that for the production, distribution and use of income, and capital accounts, the former is used more for broad validation of the estimate of saving derived from the latter accounts.

22.90 The level of saving is often expressed as a ratio - the saving ratio - defined to be saving as a percentage of the sum of disposable income and the adjustment for change in net equity of households in pension fund reserves.

Use of adjusted disposable income account (II.4.2)

22.91 This account relates to the wider concept of income and consumption of households, as referred to in paragraph 22.79.

Actual individual consumption (P.41)

22.92 This records the value of goods and services households actually consume, irrespective of which sector (households, NPISH or general government) makes the expenditure.

Accumulation accounts

22.93 There are three accumulation accounts:

- the capital account,

- the financial account and

- the other changes in assets account.

The first two are considered here. The other changes in assets account, which records changes in assets and liabilities, other than through saving and changes in wealth, and is explained in Chapter 8, is not published in the UK economic accounts.

The capital account (III.1)

22.94 The capital account shows how saving from the previous accounts and capital transfers are used to finance the acquisition of non-financial assets. The account is made up of two sub-accounts, as described below.

Change in net worth due to saving and capital transfers account (III.1.1)

22.95   This account shows how saving from the current account is modified by capital transfers.

Change in net worth due to saving and capital transfer account, 1995

	£million
Changes in liabilities and net worth	
Capital transfers (receivable *less* payable)	2 749
Saving	52 263
Changes in assets	
Change in net worth due to saving and capital transfers	55 012

22.96   The components of the account are estimated as follows

Changes in liabilities and net worth

Saving (B.8)

22.97   This represents the balance brought forward from the use of disposable income account.

Capital transfers (D.9)

22.98   Capital transfers are transactions in which one institutional unit provides a good or service or asset, in cash or kind, to another unit without receiving any counterpart. Further information on such transfers, including the distinction between them and current transfers, is given in Chapter 21. The main capital transfers for the household account are identified below.

Capital transfers, receivable (D.9)

22.99 Two components of receipts of capital transfers by households are identified.

Investment grants (D.92)

22.100 These include payments by local authorities for the construction of and improvement to dwellings. Estimates are based on information from the public expenditure monitoring system.

Other capital transfers (D.99)

22.101 This covers migrants' transfers from overseas and payments by government to top up the British Telecommunications pension fund at the time of privatisation.

Capital transfers, payable (D.9)

22.102 Two components of payments of capital transfers by households are identified.

Capital taxes (D.91)

22.103 These are taxes levied on net worth or on the value of the sale of an asset. For households, these currently comprise mainly inheritance tax, plus a small amount of tax on other capital transfers. It should be noted that taxes on capital gains, previously treated as a capital tax, are now included in taxes on income and wealth. Figures are derived from Inland Revenue sources.

Other capital transfers (D.99)

22.104 These payments include, for households, legacies and migrants' transfers.

Changes in assets

Changes in net worth due to saving and capital transfers (B10.1)

22.105 This is the balance between saving and receipts less payments of capital transfers. It represents amounts available for the purchase of financial and non-financial assets.

Acquisition of non-financial assets account (III.1.2)

22.106 This sub-account shows how saving and net capital transfers are used to acquire non-financial assets. If the available financing is greater than the value of the assets acquired, then a balance - net lending - is struck. If the converse obtains, then the balance is net borrowing. The account is shown here in summary form. Some of the variables are disaggregated for the published table, as indicated in the text.

Acquisition of non-financial assets account, 1995

	£million
Changes in liabilities and net worth	
Change in net worth due to net saving and capital transfers	35 910
Consumption of fixed capital	19 102
Changes in assets	
Gross fixed capital formation	27 507
Change in inventories	262
Acquisitions less disposals of valuables	44
Acquisitions less disposals of non-produced non-financial assets	-81
Net lending/borrowing	*27 280*

Changes in liabilities and net worth

Changes in net worth due to saving and capital transfers (B10.1)

22.107 This is the balance from the previous account, net of capital consumption.

Consumption of fixed capital (K.1)

22.108 Consumption of fixed capital (or capital consumption) is essentially the amount of resources used up in the productive process. The figure is the same as that included for the production account (*see* paragraph 22.21).

Changes in assets

Gross fixed capital formation (GFCF) (P.51)

22.109 GFCF represents investment expenditure in tangible and intangible fixed assets. A full description of the concepts and related issues has been included in Chapter 6. The methodology for the estimates of GFCF is contained in Chapter 15. This included a description of how estimates are made for the various institutional sectors.

22.110 The estimates of GFCF for the household sector are based largely on the classification information included on the IDBR.

Acquisitions less disposals of tangible fixed assets (P.511)

22.111 The main household sector component of GFCF included under this head is investment in dwellings (this also includes capital improvements to existing dwellings). As indicated in Chapter 15, estimates are based on the Department of the Environment survey of construction output. Also included under this head will be smaller amounts of investment by farmers, professional people and sole traders generally, in other assets, such as machinery and equipment, other buildings and structures, and also cultivated assets.

Acquisitions less disposals of intangible fixed assets (P.512)

22.112 Investment by households in intangible fixed assets relates mainly to entertainment, literary and artistic originals. In practice, it has not yet been possible to produce estimates as there are no obvious data sources.

Additions to the value of non-produced non-financial assets (P.513)

22.113 This heading is sub-divided into two groups

Major improvements to non-produced non-financial assets (P.5131)

22.114 This relates to improvements to land and sub-soil assets, and is minimal in the United Kingdom.

Costs of ownership transfer on non-produced non-financial assets (P.5132)

22.115 Reflecting problems in separating land and buildings, transfer costs relate to both land and buildings, rather than just land (*see* Chapter 15). Household investment under this head relates principally to transfer costs related to the sale and purchase of dwellings. The figures are derived from the sector allocation of the total, as described in Chapter 15.

Changes in inventories (P.52)

22.116 Inventories are goods which are held by producers prior to further processing or sale, or which will enter intermediate consumption or be sold without further processing. A full description of the concepts and related issues has been included in Chapters 6 and 15. In respect of households, the contribution of inventories to the capital account relates to stocks held for business purposes by farmers, professional people and other sole traders. The household estimates are based on information on the sector classification contained on the IDBR and on Inland Revenue data.

Acquisitions less disposals of valuables (P.53)

22.117 Valuables are goods which are not used in the productive process, and which are generally held as a store of value. They include precious stones and metals, antiques and works of art. A description of the concepts of this new component of GFCF has been included in Chapters 6 and 15. It is likely that some valuables would have appeared in the previous accounts as part of household final consumption. The methodology for producing sector estimates is still being developed, and the quality of the published estimates is quite poor, being based on overseas trade in jewellery etc., and antique dealers' margins.

Acquisitions less disposals of non-produced non-financial assets (K.2)

22.118 These comprise non-produced, non-financial assets, mainly land, used in the production process, and intangible assets, as defined for two components below. The methodology for the estimates for the institutional sectors has been set out in Chapter 15, and is mentioned only briefly here.

Acquisitions less disposals of land and other tangible non-produced assets (K.21)

22.119 This component currently covers inter-sector transactions in land only. These are derived from the Inland Revenue Survey of Property Incomes and other Inland Revenue survey data.

Acquisitions less disposals of intangible non-produced assets (K.22)

22.120 This heading covers transactions in patented entities, leases on land and buildings or other transferable contracts, and purchased goodwill.

Net lending/net borrowing (B.9)

22.121 This is the balancing item of the capital account for households. Net lending/borrowing shown here should be equal to net lending/borrowing as emerging from the financial accounts (paragraph 22.139). The statistical discrepancy between the two measures of net lending/borrowing is an indicator of net errors and omissions in the sector account.

Financial account and balance sheets

22.122 The final two accounts covered in the Blue Book sector presentation relate to financial variables. They are, first, the financial account, covering financial transactions which complement those associated with the production, distribution and use of income, and capital accounts, described above, and secondly, the financial balance sheet, which shows the 'stock' of financial assets and liabilities The two are linked through a third account - the 'other changes in the volume of assets account', as described in Chapter 8. While much direct information is available on financial transactions, for example from banks, building societies and other financial institutions, and government, estimates of transactions are also based on the 'stock' data from balance sheets.

22.123 Initially, the UK is not producing separate accounts for households and NPISH. Many entries for the combined sector account are derived by residual from other sectors. Consequently the reader should refer to Chapters 17 and 18 for further information about individual instruments.

Financial account (III.2)

22.124 The financial account shows the changes in financial assets, such as holdings of bonds and shares, deposits with banks and loans made, and in financial liabilities, for example the issue of securities and borrowing. The structure of the financial account for households is shown below.

Financial account, 1995

	£million
Net acquisition of financial assets	
Currency and deposits	31 395
Securities other than shares	3 514
Loans	10
Shares and other equity	-9 365
Insurance technical reserves	32 478
Other accounts receivable	2 989
Net acquisition of financial liabilities	
Securities other than shares	364
Loans	23 721
Other accounts payable	4 437
Net lending/borrowing	32 499

22.125 The variables within the financial account are estimated as described below. A summary description only is given, covering aspects of the account of particular relevance to households. Full details are contained in Chapter 17. There is virtually no direct financial information collected from households. The estimates are made from counterpart information, allocation of a residual, or 'guesstimate'.

Net acquisition of financial assets

Currency and deposits (F.2)

22.126 Estimates for households for currency are derived as the allocation of a residual, and for deposits, both bank and buildings societies, mainly from counterpart data from these sources.

Securities other than shares (F.3)

22.127 This heading covers money market instruments, bonds, preference shares and financial derivatives. Holdings of this sector are small, and are estimated as residuals.

Shares and other equity (F.5)

22.128 Information on household transactions in UK shares and other equity is mostly derived as the allocation of a residual. However, for some components, for example unit trust and investment trusts, the bulk of the investment is attributed to households and NPISH. For overseas shares and other equity, estimates are based on Inland Revenue information.

Insurance technical reserves (F.6)

22.129 In respect of the household sector, two components are identified The data are derived from inquiries to the industry (25.3.3).

Net equity of households in life insurance reserves and pension funds (F.61)

22.130 This transaction reflects the treatment of transactions by life insurance and pension finds, and the fact that such funds are regarded as the property of the household sector. Derivation of the estimates is described in Chapter 17.

Prepayments of insurance premiums and reserves for outstanding claims (F.62)

22.131 This relates to prepayments of non-life insurance premiums and reserves for outstanding claims. The estimates are described in Chapter 17.

Other accounts receivable (F.7)

22.132 The main assets under this head for households are amounts due to sole traders.

Net acquisition of financial liabilities

Loans (F.4)

22.133 Estimates for household sector loans, from general government, banks and buildings societies are obtained mainly from counterpart data from these sources. Included in particular are loans for house purchase. Other loans to households relate to advances made by finance houses and other consumer credit companies. Estimates for these other loans are based on the relevant ONS inquiries (25.3).

Other accounts payable (F.7)

22.134 The main liabilities under this head for households are short-term credit from retailers. (25.1.10).

Net lending/net borrowing (B.9)

22.135 This represents the difference between net acquisition of assets and net acquisition of liabilities. It is identical in principle with net lending/net borrowing as estimated from the current and capital accounts (*see* paragraph 22.121)

Financial balance sheet (IV.3)

22.136 The structure of the financial balance sheet for households and the data sources are fundamentally the same as for the financial account. A significant asset unique to households is their net equity in life assurance and pension funds' reserves.

Insurance technical reserves (AF.6)

22.137 The estimates for balance sheets for (i) the net equity of households in life insurance and pension funds' reserves is the counterpart to the prime source, and for (ii) the prepayments of insurance premiums and reserves for outstanding claims is based on the sector allocation of the total, as described in Chapter 18.

Net financial assets/ liabilities (BF.90)

22.138 This is the difference between total financial assets and total financial liabilities. It represents the net worth of the sector at the end of the period.

Annex 1

Final consumption expenditure

22a.1 This annex describes the methodology for the estimation of household final consumption. An outline of the approach for the derivation of the aggregate estimate, together with some key features of the compilation, was given in Chapter 12. This section expands on the earlier summary presentation and covers, in particular, the compilation methodology for the estimates of spending on the individual goods and services which make up the total.

22a.2 Before looking at the estimation methodology there are a number of issues of coverage, definition and presentation which should be raised.

Coverage changes

22a.3 The coverage of household consumption involves a number of changes compared with the previous account. The main one is the separate treatment of spending by NPISH, which ESA95 treats as a separate sector. Also excluded are motor vehicle excise duty and domestic rates, which are now classified in *taxes on income, wealth*, etc. A proportion of sales of antiques and jewellery is now deducted from final consumption expenditure and included in valuables within gross capital formation. (*See* Chapter 12.) There are some new inclusions in household consumption, including miscellaneous government fees and licences, which are treated as payments for services rather than taxes as previously. There is also a revised treatment of insurance, described in Chapter 20 Annex 2. Further, a tighter definition of the United Kingdom residence criteria now means that students are now always regarded as residents of their own country while studying overseas. In total, the balance of inclusions and exclusions (other than the removal of NPISH) reduces household consumption. In addition, a new analysis of spending by purpose (COICOP) has been introduced.

Own account production

22a.4 Household production of goods and certain services for own consumption is within the production boundary. The main items included in household consumption are owner-occupied housing services, domestic service and agricultural produce. The first two activities are explained below. The estimation of own-account agricultural production is covered under *food* in the section on the commodity figures. It should be noted that, where own-account activity takes place, it is recorded not only in household final consumption as part of expenditure GDP but also, for consistency, in the output and income measures.

Domestic service

22a.5 The employment of paid domestic staff, such as servants and gardeners, is regarded as part of household consumption. However, where such work is undertaken by members of the household for the household the activity is not part of consumption.

Owner-occupied dwellings

22a.6 One particular feature of the household sector is that owner-occupiers are deemed to be unincorporated businesses producing housing services which they then consume. This requires the imputing of a rental payment to the owner-occupied sector, analogous to the activity of payment of market or public sector rentals. The main reason for this approach is to minimise distortions to the consumption figures which would otherwise arise when there were changes in housing tenure, for example from rented to owner-occupation. The methodology for imputing owner-occupied rentals is described later in this chapter.

Investment in dwellings

22a.7 Expenditure on the purchase of new dwellings (together with any ownership transfer costs) and on major improvements to existing dwellings is *not* part of household consumption, but is included in gross fixed capital formation.

Mortgage payments

22a.8 Mortgage payments are not included in household consumption. Their interest component is a payment of property income (D.41) while the repayment of principal is a financial transaction.

Interest payments

22a.9 Interest payments in general, though part of a household's spending, are treated as distributive transactions of property income in the economic accounts, not as part of household consumption. This applies to all interest payments on loans or credit purchase arrangements except where the credit is extended by the retailer and is included in the amount charged for the goods.

Business expenses

22a.10 Household final consumption excludes any business expenditure incurred by the household as part of the economic activity of unincorporated businesses. Such expenditure is included in household intermediate consumption. Also excluded from household final consumption is any business expenditure which might otherwise be claimed for tax purposes on, for example, cars.

Income in kind

22a.11 Household consumption includes a wide range of goods and services received by households as income in kind, in lieu of cash. The main items are the use of company cars for private purposes and free or subsidised accommodation or meals. The estimates are also included, for consistency, as part of output for the production measure and within compensation of employees in the income estimates. The methodology for the estimation of income in kind is described in Chapter 14.

Second-hand goods

22a.12 The estimates of household consumption are net of sales of second-hand goods. Where sale of such goods is between households neither the sale nor the purchase is recorded in household consumption (since they cancel each other out). However, where purchases are made from dealers, the figures include the value of the dealers' margins. Where transactions are with other sectors, such as the purchase of a second-hand company car, the full value of the purchase is recorded in household consumption (while the sale value would have been included as negative GFCF in the corporations sector).

Fees paid to government

22a.13 Households make certain payments to government to obtain various kinds of licence, permits, certificates and passports. In the accounts some of these payments are treated as taxes and the rest as household consumption. In broad terms, if licences are granted automatically on receipt of the due fee then the payment is regarded as a tax. However, where government uses the issue of a licence to undertake some regulatory function the payments are treated as household expenditure on a service. Thus payments for owning and using vehicles, boats and aircraft, and also for licences to hunt, shoot and fish, are all regarded as taxes. All other payments are included within household consumption.

Durable goods

22a.14   With the exception of gross fixed capital formation (GFCF) on dwellings, household consumption embraces spending on all other durable goods, such as motor cars. In certain circumstances such expenditure might be thought of, not as final consumption but as GFCF. However, to treat the expenditure in this way would require, say, the imputation of an annual amount representing the value of the benefit derived by the household from the goods, essentially as is done for owner-occupied dwellings. Such imputation would be far from easy. Expenditure is treated as household consumption spending.

Betting and gaming

22a.15   Household consumption on betting and gaming, including the national lottery, is defined as the amount staked less the amount returned in winnings, representing, essentially, the cost of the service. Estimates are explained in paragraphs 22a.115–22a.118.

Life assurance and pension funds

22a.16   The way in which these funds are treated in the new accounts has been explained in paragraph 22.6.

Insurance

22a.17   The important changes to the treatment of insurance in the new accounts have been covered in Chapter 20 Annex 2. The estimates for the various components of insurance are described later in this chapter.

Valuables

22a.18   Valuables are now included as a separate category, as part of gross capital formation. In the previous accounts some valuables were included in household consumption.

FISIM

22a.19   The treatment of financial intermediation services indirectly measured is described in Annex 1 to Chapter 20.

Hidden economy

22a.20   Methodology for measuring illegal activity is under consideration, some such transactions being already covered indistinguishably in the accounts.

Household consumption overseas and spending in the UK by non-residents

22a.21   The aggregate estimates of household consumption in the accounts include spending overseas by United Kingdom residents and exclude spending in the UK by non-residents - the so-called 'domestic concept'. However, the data collected to estimate the commodity components of spending mostly include expenditure in the UK by non-residents, but not spending abroad. The estimates for the aggregates, on the proper UK residence basis - the so-called 'national concept' - are arrived at by adding to the component-based figures an estimate of total UK household spending overseas and deducting the total of non-residents' spending in the UK. UK household spending overseas also appears as part of travel expenditure in the figures for imports of services: similarly, spending by non-residents is part of the travel component of exports of services. Other issues related to the estimation of this type of spending are discussed later in this Annex (paragraph 22a.130 *et seq*).

Actual final consumption

22a.22   Household final consumption expenditure includes only the direct spending by resident households. Some of this spending is on goods and services, such as health and education, which is provided by both the market and the state. Household consumption includes the market spending. However, spending by government (and NPISH) on such goods and services is part of the provision of individual (as opposed to collective) goods and services by these sectors. This spending is included in the final consumption expenditure of the providing sector. For these and other 'individual' components of spending changes in the pattern of provision will clearly affect the level of the figures for household final consumption. Thus an alternative, wider, concept of consumption has been established. This is called actual final consumption, and is defined as household and NPISH final consumption expenditure *plus* government and NPISH individual consumption expenditure. Figures for actual final consumption are not affected by arrangements as to who provides the good or service and are of use for certain comparative analyses of spending, over time and between countries.

The classification of consumption

22a.23 The detailed components of household consumption may be analysed in a number of ways. Two approaches are used in practice. The first is a commodity breakdown into types of good and service. It has the advantages that it makes sense when comparing final demand transactions with industrial activity and it tends to match the actual data flows from their various sources. Although useful in practice the commodity approach has no basis in international standards. The consumption classifications are not based on industrial product standards (such as the Central Product Classification, CPC), since these tend to identify products which consumers cannot recognise.

22a.24 The other approach is a *purpose* classification (sometimes called *function*) and for this an internationally agreed standard, the Classification of Individual Consumption by Purpose (COICOP) is used. COICOP groups together consumption according to the purpose of its use. Thus the heading *clothing*, for example, indicates amounts spent on keeping oneself clothed, grouping together garments, clothing materials, laundry and repair. It can be seen that the *purpose* classification groups together items measured using different statistical sources and produced by disparate industries, which may not be desirable for some purposes.

22a.25 For these reasons the full detail of household final consumption expenditure is presently classified according to commodity. This is published regularly in *Consumer Trends*. The detailed source notes which follow later in this chapter are ordered according to commodity headings.

22a.26 Some problems arise in applying these classifications in practice. To begin with, the quality and purity of the estimates will depend very much on the data sources used. These are discussed below. One of the main problems in making estimates for individual goods or services relates to the measurement of household spending abroad and spending by non-residents in the UK. As mentioned in paragraph 22.163 above, estimates for separate commodities do not include spending abroad but they do embrace some spending by non-residents in the UK. The correct concept for total spending by the household sector ('national concept') is achieved by adding in total spending by UK residents abroad and subtracting total spending by non-residents in the UK. This approach for the commodities largely reflects the practical problems of deriving component figures on the same basis as the total, in particular of obtaining detailed commodity information on household spending abroad and deducting non-residents' expenditure from the estimates based on retail sales data. One important consequence of this treatment of commodity spending is that, where household budget surveys are used, figures for spending need to be augmented by estimates of non-resident expenditure in the UK. Such estimates are based on periodic information from the International Passenger Survey (25.5.4). Thus, in practice, the detailed figures shown in the table do not exactly represent expenditure by UK residents on the particular goods and services forming the analysis.

22a.27 A second problem arises with composite goods and services. The prime example concerns expenditure in hotels, cafes and restaurants. In principle this category embraces both food and drink. However, in the UK, statistics of all expenditure on alcoholic drink are derived from a single highly reliable source and appear under the category *alcoholic beverages*. Thus spending in hotels, cafes and restaurants reflects only the broad group of food and accommodation, excluding drink. Problems also arise with expenditure on package holidays. Here it is far from easy to obtain the information to enable the separate elements relating to air fares, expenditure abroad, and the travel agents' commission to be combined to provide the required figure. Thus these elements are estimated from totals and allocated to the separate relevant categories.

22a.28   The impact of such problems is to reduce the purity of the estimates for the individual items of household consumption.

The estimation of household consumption

22a.29   This section looks, in broad terms, at the various sources used for making the estimates of household consumption. An indication is given of the kind of commodities estimated by the various sources. Further details are included later, in the section which describes the approach for individual goods and services.

22a.30   There are five main sources which are used to estimate household spending, which have certain advantages and disadvantages. The main issues are discussed below. The sources are:

    (i)   sample surveys of spending by households and individuals;
    (ii)   statistics of retail and other traders' turnover;
    (iii) other statistics of supply or sales of particular goods and services;
    (iv)   administrative sources;
    (v)   commodity flow analysis.

Sample surveys of spending by households and individuals

22a.31   The main form of survey considered under this head is the household budget survey in which households record their spending on goods and services, normally for a one-week or two-week period. One advantage of this kind of approach is that the information collected will usually cover spending across a comprehensive range of items. Further, the information will generally exclude business spending, and be on the valuation basis required for the accounts.

22a.32   However, there are certain disadvantages to this approach.

- Response may be differential rather than generally uniform across the sample of households. Where this occurs there is usually an understatement at the top end of the income range, which may be only partly offset by under-representation in the lower income ranges. Again this may lead to bias in the estimates, both in total and for particular categories of spending.

- Bias is also likely to arise in the estimates for certain items of expenditure, such as alcoholic drink and tobacco, which for obvious reasons are likely to be under-reported in the surveys. However, the important point should be made that, generally, the effect of bias in the survey is likely to be considerably less in relation to the *movement* in spending than to its level.

- There are also problems with the estimates of spending on infrequently-purchased items such as cars, furniture and other major durables, where the incidence of recording in the survey reporting period will tend to be low. Thus the estimates for these goods will be subject to a comparatively large statistical error as well as possible bias. In some surveys improved estimates for these items are derived by collecting information on spending retrospectively, over a longer time-scale than the conventional recording period used in the inquiry.

- For certain items (for example financial services and insurance) it is impossible to measure transactions in a way that is consistent with economic accounts concepts.

- Household surveys do not cover spending by those people living in institutions. Although this spending is comparatively small in aggregate it will be of greater importance for particular commodities.

- Spending may be influenced by participation in the survey, and thus the level and pattern of spending recorded by the household may be distorted.

22a.33   For the estimates of household consumption made in the UK three continuous surveys of households or individuals are used. These are the ONS Family Expenditure Survey (FES) and International Passenger Survey (IPS) and the MAFF National Food Survey (NFS). Details of these surveys are given in Chapter 25 but some main points can be noted here. The FES and NFS are both surveys of households. The former collects information on the value of spending over the whole range of consumer goods and services. The NFS covers essentially food items, obtaining information on both the value and 'volume' of purchases. The IPS is a survey of individuals, both resident and non-resident, conducted at airports and seaports and seeking information on holiday and travel expenditure. All three surveys are affected to varying degrees by the problems described above. A key part of the work of the statistician is to identify the importance of such problems and allow for them, as well as possible, in the estimates.

22a.34   There are a number of aspects about the use of survey data for estimates of household consumption which might be raised. They are presented here largely in the context of the FES but will also apply, to varying degrees, to the NFS and IPS. To begin with, grossing factors are calculated to convert the expenditure per household recorded in the survey to national totals. The grossing factors are based on estimates of the household population from ONS and the average household size from the FES. Secondly, in order to mitigate the effects of sampling variation in the survey, estimates for most categories of spending are subject to an element of smoothing. For annual estimates the smoothing process uses a three-year moving average of the constant price estimates. For the quarterly series the estimates are based on the average quarterly pattern evident in the series in recent years. The procedure is applied judiciously. If a 'raw' series exhibited erratic movement which was thought or known to be genuine then smoothing would not be undertaken. Thirdly, where differential response is considered to be a problem, upward adjustments - often comparatively large - are made to a number of commodity groups. Finally, various additions are made to the commodity estimates to cover the expenditure of people living in institutions and, where applicable, foreign tourists. (Expenditure by foreign tourists is subtracted in aggregate.) Expenditure by juveniles is now covered in the FES; previously adjustments were included for this form of spending.

22a.35   The above three surveys are used to estimate household consumption on the following broad commodity groups. Further details are given in the section on the estimation methodology for the individual goods and services.

Family Expenditure Survey:

<u>Goods</u> - Books, newspapers and magazines; horticultural goods; photographic goods and film processing; household cleaning materials; components of recreational goods, pharmaceutical products, medical equipment and spectacles (excluding NHS payments)

<u>Services</u> - Housing maintenance (other than DIY); catering; private medical expenses; hairdressing and beauty care; insurance not elsewhere included; components of running costs of motor vehicles, travel, household services and domestic services

National Food Survey:

Food consumed within the home

International Passenger Survey:

Household spending abroad; spending by non-residents in the UK.

22a.36 Estimates at constant prices are mostly made by deflating the value estimates for each category by appropriate price indices, generally from the retail prices index. Further details are given below and for the individual commodity groups.

Statistics of retail and other traders' turnover

22a.37 The second source for statistics of household consumption is information collected in surveys directed at businesses which sell goods and services to consumers. Within this the main source is the retail trades, though wholesalers and producers also sell direct to the public. Information from service industries mainly serving consumers, such as hairdressing and cinemas, may also be incorporated in the estimates. The main benefit from the business survey approach is that coverage of spending can be large compared with what might be achieved for a household budget survey of practicable size, thus yielding more reliable results for the same cost. The surveys also cover spending by foreign tourists and by the institutional population.

22a.38 In contrast, there are certain disadvantages. In particular, the surveys do not cover anything like the full range of goods and (particularly) services purchased by consumers. Further, despite improvements in technology, the commodity information which can realistically be collected from businesses is far less detailed than that available from budget surveys. Where commodity information is collected - essentially almost solely from retailing and only rarely from other industries - detailed information is available only annually, though figures for some broad commodity groups are provided in the short-period retailing inquiries. Two further problems arise with this source. Firstly, the data collected will include spending by businesses, which should be part of intermediate consumption. Secondly, within the general problems which affect the use of statistical registers for these kinds of inquiry, particular difficulties arise with covering *small* businesses, which represent an important segment of the consumer market.

22a.39 The main business surveys used in the estimation of household consumption are the ONS annual inquiries into the distribution and service trades, particularly retailing, and also the monthly retail sales inquiry. All these surveys collect information on the total value of sales. In addition, commodity information is collected regularly in the annual and monthly retailing inquiries, and occasionally in other distribution and service sector surveys such as wholesaling and motor trades. *Ad hoc* information has also been collected from time to time on the value of direct sales to the public by manufacturing industry.

22a.40 In order to use information from the retailing and other inquiries in the estimation of household consumption a number of adjustments need to be made to the survey results.

- First, as the inquiries cover only Great Britain, an across-the-board adjustment is made for retail sales in Northern Ireland.

- Secondly, prior to 1995 and the introduction of the Inter-Departmental Business Register (IDBR), the grossing-up of the survey results makes some allowance for sales of small traders who fell below the VAT threshold.

- Adjustments are made to allow for sales by non-retailers (such as hairdressers or garage forecourts) where the sale of goods is a secondary but important activity.

- Finally, deductions are made for the estimated level of *business* expenditure (part of intermediate consumption) recorded in the figures. The deductions are based on information collected from a telephone survey of both small and large businesses.

22a.41 The methodology described above relates essentially to the derivation of annual estimates, where a detailed commodity analysis of retail sales is available. For quarterly information, and before the annual results are available, commodity estimates are based on information collected in the ONS's monthly retail sales inquiry. This embraces the direct collection of limited commodity information, recorded monthly and quarterly, and projection of the latest annual data for commodity groups. The approach adopted for the individual categories of spending is to break down expenditure according to the different kinds of business - for example butchers, jewellers and mail order - in which spending occurs, and to project the estimate by the movement in the total turnover for the kinds of business selling this particular commodity. The success of this method depends much on the extent to which the sales are concentrated in specialised shops, and the stability of spending patterns over the different kinds of retail outlet.

22a.42 The retail sales and related survey data are used to estimate household consumption on the following broad categories of spending:

Clothing and footwear; furniture and floor coverings; household textiles and soft furnishings; durable goods; DIY goods; hardware; sports goods, toys, games and camping equipment; records and tapes; perfumery; jewellery and silverware; watches and clocks; components of recreational goods, pharmaceutical products and other goods generally.

Further details are contained in the section on the methodology for individual commodities.

22a.43 Estimates at constant prices are made by deflating the value estimates for each category by appropriate price indices, generally from the retail prices index. Again, further details are given below and for the individual commodity groups.

Other statistics of supplies or sales of particular goods and services

22a.44 This source relates to goods and services which are supplied by a single provider or a few providers, or the information comes from other representative bodies such as a trade association. The advantages of this kind of source are the comprehensiveness of coverage and the fact that any unusual movements in the figures can normally be readily discussed and resolved. On the other hand the information provided may not always be on the basis required for the accounts. For example, business expenditure may be included in the estimate or valuation may be different from what is required in the accounts. How these problems are dealt with, and other issues, are discussed further in the context of the estimation for particular components of spending.

22a.45 The main items using these sources are gas, electricity, communications, rail travel and bus travel. Estimates at constant prices are derived mostly by deflating value figures by suitable price indices or by valuing volume indicators at base year prices (particularly passenger kilometres travelled).

Administrative sources

22a.46   The use of information collected for administrative purposes, for example certain data on licences and taxes, has similar advantages and disadvantages to the source described above. This source is used in some form or other for estimates of spending on alcoholic drink, tobacco, betting and gaming. The estimates at constant prices are a mix of direct volume data and deflated values.

Commodity flow approach

22a.47   The commodity flow approach is a means of making estimates using a statistical model rather than directly from collection of data. In general terms the methodology aims to derive estimates for the various components of demand within the economic accounts (*namely* intermediate consumption, household consumption, government consumption, GFCF, inventories and exports) from figures for supply. The estimates are made for detailed product groups. A key aspect of this approach is to determine the allocation of supply over the various components of demand. This is normally based on the supply/demand structure of some previous period, which will generally have been assembled as part of the compilation of input-output tables. A slightly fuller description of this approach is included in Chapter 15 in respect of estimates of GFCF, where the methodology is of more relevance.

22a.48   Within the estimation of household consumption the commodity flow approach is used more to validate estimates made from the various sources described above than to provide direct figures.

Sources - the overall picture

22a.49   It is clear, therefore, that there is a range of sources for estimating household consumption. In practice the way in which the figures are compiled is to determine the best estimate for each good or service from the various sources - described above - which might be available. Regular comparisons are made between, for example, estimates from FES and retail inquiry-based sources for a number of commodity groups and also, for both these sources, with NFS data for food. Use is also made of VAT receipts as a check on the level of overall spending, though the impact of the hidden economy needs to be borne in mind here. Once the best individual estimates are established the aggregate is then just the sum of the components. However, the balancing process is likely to lead to adjustments being made to the initial estimates for the individual items and hence to the total.

Constant prices

22a.50   As with the current price figures, the aggregate estimate of household consumption at constant prices for both the annual and quarterly accounts is essentially derived by summing estimates for individual commodities or services. Some indication of the methodology for compiling the constant price figures for broad commodity groups has been given above and further detail is included later in respect of individual goods and services. The methodology is a mixture of deflated values (used for about two-thirds of total spending) and direct or indirect volume estimates, essentially the revaluation of present quantity data at base year prices, used for the other one-third. The price information used for deflation is provided mostly by the ONS retail prices index (25.4.7) but certain other price information is used, particularly for services. Deflation is undertaken in as much detail as possible, consistent with maintaining an adequate level of accuracy in both the value and price data. The main areas where volume series are used relate to spending on housing services, drink, tobacco, motor vehicles and certain transport services.

22a.51   An important feature of the estimation at constant prices which affects all parts of the economic accounts is how to allow for quality changes in goods and services. Some general discussion of the problem has been included in Chapter 13. Within household consumption on *goods* the main problems are likely to arise with durable goods, in particular computers. For *services* quality is difficult to define, let alone measure. Where it can be identified it appears in various guises, for example convenience of shopping facilities or comfort in means of transport. In general price series for goods do attempt to make some allowance for quality changes, though this is not really possible at present for services. The broad principle followed is that where the increase in the price of a good includes some improvement in quality this part of the change is regarded as a volume increase, not a price increase.

Sources and methods in detail

22a.52   The following pages describe in detail the sources and methods of estimation currently used for individual components of household consumption. References are made to major changes in methodology which have occurred in recent years. Further details may be found in the 1985 and 1968 editions of *Sources and Methods*. The sections are arranged in line with the commodity classification (*see* paragraph 22a.25) as follows.

> ***Durable Goods***
> *Motor and other vehicles*
> *Other durables*
> ***Non-durable goods***
> *Food*
> *Alcohol and tobacco*
> *Clothing and footwear*
> *Energy*
> *Other goods*
> ***Services***
> *Rent and water charges*
> *Catering*
> *Motor vehicle services*
> *Travel and communications*
> *Monetary services*
> *Household and domestic services*
> *Recreational and cultural services*
> *Miscellaneous services*
> *Medical services*
> *Tourism adjustments.*

Durable goods

Vehicles

22a.53   Motor Vehicles. Estimates of household spending on the purchase of new cars are based on information on the V55 Registration Document. This provides a direct measure of the number of vehicles sold to private individuals. Information is available for a range of different makes and models. The current-price estimates are derived by multiplying these data by trade information on prices, together with estimates for 'on the road' costs. At constant prices the estimates are obtained by revaluing the numbers of different makes and models by appropriate base year prices.

22a.54   This approach was introduced in 1996 and applied back to 1993. Previously estimates were derived by subtracting the estimate of fixed capital formation on new cars from the estimate of total spending based on valuing new car registrations. An annual benchmark of the total value of sales was also available from the ONS annual motor trades inquiry (25.1.7).

22a.55   For second-hand cars previously owned by householders the estimates required under this head relate to dealers' margins, that is essentially the difference between sales and purchases of used vehicles. The information is obtained from the ONS annual motor trades inquiry (25.1.7). An adjustment is made for cars previously owned by other sectors. In the absence of any appropriate data quarterly estimates are derived by extrapolation and interpolation. Constant-price estimates are obtained by deflating the figures for margins using the retail prices index (25.4.7) component for second-hand cars.

22a.56   Estimates for motor cycles are also based on numbers of registrations, with some breakdown by engine capacity. These data are revalued to current prices using the retail prices index component for motor cycles. The constant-price figures are derived by multiplying the numbers of different types of motorcycle by the appropriate base year prices.

22a.57   <u>Other Vehicles</u>. Expenditure on pedal cycles and cycle accessories is available from the retailing inquiries (25.1.8, 25.1.10) (*see* paragraphs 22a.37–22a.43). The estimates are interpolated and projected quarterly. Estimates at constant prices are derived using a component of the retail prices index.

22a.58   For boats, estimates are based on information on UK sales from the British Marine Industries Federation, including allowance for imports and exports. For aircraft the figures are estimates. The quarterly estimates are interpolated and projected. Estimates at constant prices are derived as for caravans.

22a.59   Expenditure on new caravans is based on information from the ONS quarterly sales inquiry, adjusted for margins, VAT and delivery and with allowance for exports and imports. Some deduction is made for business expenditure. Expenditure at constant prices is obtained by multiplying the number of caravans attributed to consumers by the estimated average value in the base year. The dealers margin on household trade in second-hand caravans is based on a certain proportion of new caravan sales being so traded and the estimated average dealer's margin. Estimates at constant prices are based on these numbers multiplied by the estimated average dealer's margin in the base year.

Other durable goods

22a.60   The other durable goods categories (furniture, floor covering, audio-visual equipment, photographic and optical goods and major appliances) are all derived from the retail sales inquiries which are described in paragraphs 22a.37–22a.43 above. The following table shows the items which go to make up each of these categories.

Furniture	All furniture (except office and garden furniture), nursery furniture, perambulators, mattresses, dealers' margins on second-hand furniture and picture frames
Floor coverings	Carpets, rugs, mats, carpet tiles, vinyl and linoleum floor coverings and floor tiles
Audio-visual equipment	All audio-visual goods (e.g. radios, TVs, video recorders, CD players), personal camcorders, computers, calculators, musical instruments and other audio-visual goods. (The purchase and processing of photographic film are included under *Photos* and camcorder, audio and video tapes are included under *Records*.)
Photographic and optical goods	Cameras, and other photographic goods, (excluding the purchase and processing of photographic film which are included under 'Photos'), optical goods (excluding spectacles and contact lenses)

	Major appliances	Domestic electrical and gas appliances, e.g. cookers, washing machines, refrigerators, dishwashers, heaters, vacuum cleaners, sewing machines, power tools, lawnmowers, etc, but excluding central heating which is regarded as capital expenditure.

Constant-price estimates are derived by deflating the current price figures by appropriate price components of the retail prices index.

Non-durable goods

Food

22a.61 Food (household expenditure) comprises all food and non-alcoholic beverages bought by households for consumption in the home or elsewhere. However, food bought in catering establishments is excluded, as is that bought by those living in institutions where catering is provided, such as Armed Forces messes. These items are included in catering. Food withdrawn by farmers and other commercial food producers for their own consumption is included.

22a.62 The estimates make extensive use of MAFF's National Food Survey (25.4.6), which provides quarterly information on the quantity and value of food brought into the home. This is supplemented by certain information from ONS quarterly production inquiries (25.1) - for soft drinks - and from manufacturers' associations - for chocolate and sugar confectionery. In making the household expenditure estimates for the economic accounts the survey data are adjusted upwards in respect of some food items (for example cakes, biscuits, fresh fruit and nuts) where retail purchases by other members of the household for consumption outside the home occur. Also, a downwards adjustment is made to reflect the reduction in household expenditure on food by those on holiday in the UK or overseas. The information from the NFS is essentially on a *per household* basis, and is grossed up to the estimate of the household population. Estimates of production for own consumption are based on the value of such food brought in to the household in the reporting period.

22a.63 Estimates at constant prices, with two exceptions, are derived by deflating the current-price figures, estimated as above, by appropriate components of the retail prices index. This improved methodology was introduced in 1991 and carried back to the estimates for 1986. It replaced estimates largely based on the quantities consumed, which did not adequately reflect changes in the *quality* of the products. The deflation is carried out in as much detail as possible. It also differentiates between similar products - for example within fruit and vegetables - available at different times of the year. For such items the price information used relates essentially to the relevant period of the base year, not to the base year as a whole. Estimates for soft drinks and confectionery are based on quantity information.

Alcoholic beverages and tobacco

22a.64 The group *alcoholic drink* covers that bought by consumers from retailers and in public houses, hotels, restaurants, other commercial catering establishments and non-profit-making clubs. The inclusion of catering establishments reflects problems in obtaining separate estimates of spending (*see* paragraph 22.169). The drink is valued at the prices paid by the consumer including any charge for serving it. Separate series exist for beer, spirits, wine, cider and perry.

22a.65 Estimates, at current and constant prices, are based on the volume of sales and average prices of individual types of alcoholic drink for both 'off' and 'on' trades. The information, which includes prices and volumes for a detailed list of products for the 'off' and 'on' trades separately, is obtained bi-monthly from a survey of retail outlets carried out by a private research company. The volume information from the survey is grossed-up to be consistent with annual clearance figures from HM Customs and Excise.

22a.66   Prior to 1994 estimates were largely based on quantity data for different types of drink - for beer on the volume of production, and for wines and spirits on volumes on which duty was paid. The volume data were inflated to current prices using information from the retail prices index and other sources.

22a.67   *Tobacco.* Estimates for cigarettes and other tobacco products are based on quarterly information from Customs and Excise on the volume of tobacco clearances released from bond for home consumption. The current-price figures are derived by inflating the volume data by appropriate price components of the retail prices index. At constant prices the estimates are simply the volume information expressed at the appropriate prices of the base year.

Clothing and footwear

22a.68   Estimates at current prices for both these groups are based on information collected in the retailing and related inquiries, as described in paragraphs 22a.37–22a.43. Constant-price estimates are derived by deflating the current-price figures by appropriate price components of the retail prices index. Separate series exist for men's and boys' wear, women's, girls' and infants' wear and footwear.

Energy

22a.69   *Petrol and oil.* Estimates of consumers' expenditure on petrol are based on information provided by the trade to the DTI (Energy). The data relate to deliveries to petrol stations and other customers of various grades of petrol. Estimates are made for the proportion of these supplies used for private motoring and these are converted to current values by estimates of the average prices of the various grades. Expenditure on oil is derived from the FES (25.4.5). At constant prices estimates for petrol are obtained by applying average base-year prices to the corresponding volumes for various grades, and for oil by deflating the value figure by information from the RPI.

22a.70   *Electricity.* Estimates at current prices are based on the quarterly value of supply to consumers as provided by the electricity companies to the DTI (Energy). Constant price estimates are derived by deflating the value figures by the electricity price component of the retail prices index. The deflation makes allowance for different tariffs.

22a.71   *Gas.* Estimates at current prices are based on the quarterly value of supply to consumers as provided by the gas supply companies to the DTI (Energy). At constant prices estimates are obtained by deflating the value figure by the gas price component of the retail prices index. Again some allowance is made for different tariffs.

22a.72   *Coal and coke* Estimates at current prices are based on quarterly returns provided by the supply companies to the DTI (Energy). Expenditure on each of the categories of coal and coke is revalued at constant prices by applying base year average prices to the quantities sold.

22a.73   *Other fuels.* Estimates of the quantity and value of paraffin, fuel oil and liquid gases bought by consumers are provided by suppliers. Expenditure on fuel oil and liquid gases at constant prices is obtained by applying the average price in the base year to the quantities bought. There are no statistics of expenditure on wood and the figures are broad estimates; the amount involved is thought to be relatively small.

Other goods

22a.74   Within household goods, estimates for textiles and soft furnishings, hardware and DIY goods are based on information collected in the retailing and related inquiries (25.1.8, 25.1.10), as described in paragraphs 22a.37–22a.43. Constant-price estimates are derived by deflating the current-price figures by appropriate components of the retail prices index. Estimates for household soap and cleaning materials are based on the Family Expenditure Survey (25.4.5) and deflated by components of the retail prices index.

22a.75 Estimates of spending on newspapers, magazines and books are based on the FES (25.4.5). The figures are augmented by estimates of spending by the non-household population using, in particular, information from the Publishers' Association in respect of books. Estimates for stationery are derived from retail sales. For constant prices deflators are based on appropriate price series, including retail price indices and, for books, information from the Publishers' Association.

22a.76 *Chemists and opticians' goods and spectacles (non-NHS).* Estimates are based on information collected in the FES. Constant-price figures are based on deflation of value data by suitable price series.

22a.77 *Medication.* Current-price figures are derived from information collected in the ONS retailing and related inquiries. The constant-price estimates are obtained by deflating the current price figures by appropriate price components of the retail prices index.

22a.78 *Toiletries.* Estimates are based on information from the retailing and related inquiries. Constant-price figures are derived by deflating value estimate by appropriate retail price indices.

22a.79 *Miscellaneous goods.* The sources and methods for compiling the estimates included in this section can be divided into two parts: those based on retail sales and those based on the Family Expenditure Survey. Estimates for horticultural goods, photos and pets are based on the FES while the remainder are based on the various retailing inquiries.

Services

Rents and water charges

22a.80 Estimates of quarterly rentals paid are derived in two broad groups. First, estimates of public sector rents are provided, annually, by local authorities and housing associations. Secondly, in respect of rentals paid to private sector landlords, estimates are based on information from the FES, covering both furnished and unfurnished properties. This second component also includes an estimate for 'other' rents, such as on caravans, and ground rent. The survey-based elements are grossed up to totals from DETR housing stock data.

22a.81 At constant prices estimates for all kinds of paid rentals are derived by expressing the volume data (housing stock) at base-period prices.

22a.82 For imputed rentals separate estimates, using similar methodology, are made for owner-occupiers and those living rent free. The broad principle involved is to impute to a given owner-occupied property a rental value which is the same as the rental which would be paid for a similar property in the private rented sector. The existing methodology was introduced in the 1996 *Blue Book* as part of a European Union Council decision under the GNP Directive. The approach involves establishing a benchmark for imputed rentals which is then projected forward. The benchmark figures are derived from an econometric regression model which uses certain characteristics of dwellings (such as location, type, size and age) to provide estimates of imputed rental based on rentals actually paid. The benchmark was established for the first quarter of 1991, the period when data were available from English House Condition Survey, and the decennial Census of Population. Estimates for subsequent periods are based on movements in the number of owner-occupied dwellings and average rentals paid for comparable properties.

22a.83 At constant prices estimates are based on the number of dwellings, with a small 'quality' adjustment representing improvements to existing properties.

22a.84 *Sewerage and water charges.* Estimates based on information provided by the Office of Water Services (in respect of England and Wales) and by the Scottish Office (for Scotland). Constant-price estimates are obtained by deflating the value data by appropriate price indices.

Catering

22a.85   *Catering (meals and accommodation).* Estimates under this head include various categories of spending on meals and accommodation in catering and related and certain other establishments. Included is spending on meals and accommodation in hotels, restaurants, clubs and canteens, on board and lodging for those not living in households, by HM Forces and by staff living in hospitals. The figures do not include expenditure on meals and accommodation by general government (for example in hospitals, prisons, homes for the aged) or on school meals, which are all part of general government final consumption. Expenditure on alcoholic drink is excluded but expenditure on other drinks is included. Expenditure in residential and nursing homes is included.

22a.86   Various sources are used for the estimates. The main one is the FES (25.4.5), to which certain adjustments are made in order to achieve the proper coverage of the household sector (*see* paragraph 22a.21). These include additions to cover the expenditure of persons not living in households and foreign tourists (the latter based on the International Passenger Survey, 25.5.4), payments by HM forces for accommodation and canteen and NAAFI meals (based on information collected by the MoD) and payments by medical staff to National Health Service hospitals for board and lodging.

22a.87   Also included are payments by students for board and lodging at universities and private schools, and on school meals by students (both based on information collected by the DfEE), the expenditure of those living in boarding houses and hostels (derived as estimated numbers multiplied by an average figure for spending, based in part on the level of income support) and the expenditure of those living in residential and nursing homes, both private and local authority. The latter are based on the results of a special survey carried out for ONS in 1993 (*see Economic Trends*, November 1993). The survey provides a benchmark estimate for this form of spending, which has been projected by estimates of numbers and average fees. Estimates at constant prices are derived from numbers and base-year prices.

22a.88   *Food in kind* represents the value of subsidised food provided to employees, residents of residential and nursing homes, hostels and other communal establishments.

22a.89   Government accounts data are used to make estimates for the payments to local authorities social services departments for residential fees and services rendered to individuals in local authority homes.

22a.90   The estimates are revalued at constant prices by means of components of the retail prices index and a specially-constructed index of hotel accommodation prices. School meals are separately revalued in the same way as general government expenditure on school meals.

Motor vehicle services

22a.91   *Motor vehicle repairs.* Estimates are made up of two components: repairs financed directly by households and those financed directly by insurance claims. The former are based on the FES (25.4.5) as described in paragraph 22.175. In the case of repairs financed by insurance claims payments are almost invariably paid by insurance companies direct to garages; such payments are not recorded in the FES. For these repairs estimates are based on information from the Association of British Insurers, taking account of the split of claims as between personal injury and damage to vehicles and, for the latter, a further split as between repairs and replacement. An addition to the gross premium is made to allow for the 'premium supplement' which is calculated from insurance companies' accounts. Constant-price estimates are obtained by deflating value data by prices from the RPI. For motor vehicle insurance an estimate of the administrative cost, that is the expenditure on premiums *minus* payment of claims, is included.

22a.92 *Motor vehicle spares, other motor vehicle costs and self drive car hire.* The main source of information is the FES (*see* paragraph 22a.35). The latter figure is augmented by an estimate of foreign tourist expenditure on car hire, obtained from the IPS. The constant-price estimates are values deflated by the appropriate index from the RPI.

22a.93 *Company cars.* Estimates of the value of the use of company cars for private purposes are part of household consumption. They relate to the taxable value of company cars, as provided by the Inland Revenue. These are also included, as income in kind, in the income measure of GDP. The value figures are revalued to constant prices using the price index for self-drive hire.

22a.94 *Driving lessons.* Estimates at current prices are based on the projection of a benchmark figure (for 1990) using movements in the number of driving tests and information from the retail prices index. The constant-price figures use the above volume projection and the average cost in the base year. Quarterly estimates are interpolated.

22a.95 *Driving tests.* Annual estimates are obtained from information from the Department of Transport, while quarterly estimates are interpolations and extrapolations. Constant-price data are derived by deflating the value figures by a price index of fees.

Travel and communications

22a.96 *Rail travel.* Estimates of household consumption on rail travel are based on quarterly information on receipts provided to the Department of Transport by the rail operating companies. The information is provided separately for a number of fare categories (such as ordinary and season tickets). Additions are made for travel in Northern Ireland. Deductions, based on the National Travel Survey, are made for business expenditure and for general government expenditure. Constant-price figures are derived by deflating the separate current price estimates by indices of estimated average receipts per passenger mile.

22a.97 *Buses and coaches.* Current-price estimates of household spending are based on quarterly information on receipts provided to the DETR by the bus and coach operating companies. Additions are made for travel in Northern Ireland. Deductions, based on DETR information, are made for concessionary fares, for business expenditure and for general government expenditure. Constant-price figures are derived by deflating the current-price estimates by price indices constructed by DETR.

22a.98 *Air travel.* Separate estimates are made for international air travel (excluding travel to the Irish Republic) and for other 'domestic' travel. For the former, estimates are based on quarterly information on numbers of tourists and average fares (including those on inclusive tours) from the ONS International Passenger Survey (25.5.4). For 'domestic' travel estimates for travel to the Irish Republic are based on numbers of travellers (converted to passenger-kilometres) and estimates for travel within the UK on direct data about passenger-kilometres provided by the Civil Aviation Authority. The combined total of these two components is multiplied by average British Airways fares per passenger-kilometre, calculated annually and interpolated quarterly. Finally, an estimate is deducted for the large part of this total which represents business expenditure. For revaluing air expenditures at constant prices estimates of average fares in the base year have been calculated per passenger mile for three destination categories: Europe; America and Oceania; and Africa, Asia and the Middle East. These prices are applied to quarterly data on passenger-miles travelled in each category. For emigrants constant-price estimates are based on numbers emigrating to Europe and to the rest of the world multiplied by average fares in the base year.

22a.99 *Taxis*. Estimates at current prices are based on data from the FES (25.4.5) and, for foreign tourist expenditure, the IPS (25.5.4). The constant-prices estimates are derived from the deflation of the current price data using component price information from the RPI.

2a2.100 *Sea travel*. Estimates are made for a variety of types of travel. For international travel by sea, value figures are based on the IPS. However, because the Survey does not cover travel to the Irish Republic or pleasure cruises, separate estimates are made for these two forms of sea travel, based on Department of Transport information about the number of passengers. The current-price figures are obtained by inflating the numbers data by estimates of average fares. (For the Irish Republic a typical single fare is taken to be the average.) For cruises a benchmark estimate of an average fare is extrapolated using an index of hotel accommodation prices. For internal and other minor items of sea travel little information is available and estimates are based on the projection of a benchmark. The figures for sea travel also include estimates for expenditure on board ship as well as fares paid. Estimates at constant prices are derived mostly by revaluing figures for the numbers travelling by the various forms of sea transport at the estimated average fares for the base year.

22a.101 *Miscellaneous travel*. Travel agents' commissions represents one part of total household expenditure on international package tours. This is in three components - air/sea fares, expenditure abroad and travel agents' commission. Estimates of travel agents' commission are based on a special analysis of the IPS relating to people travelling on package tours. Ten per cent of expenditure on package tours (excluding fares) is assumed to be commission. Constant-price estimates are obtained by multiplying the numbers of travellers by the estimated average charge in the base year. Removals and storage expenditure starts with constant-price estimates derived from an indicator based on particulars delivered to the Inland Revenue for stamp duty purposes following transfer of ownership of land and dwellings, adjusted for local authority housing sales and sale for other tenure types. This information is revalued by the average price of a removal in the base year. Current-price data are obtained by multiplying the constant-price series by the component of the retail prices index for miscellaneous household services.

22a.102 *Postal services and telecommunications*. For the former, quarterly estimates are based on the FES (*see* paragraph 22a.35). At constant prices the value data are deflated by RPI components. For telecommunications, estimates are based on quarterly information provided by BT, other operators and the industry regulator. Estimates at constant prices are the current-price figures deflated by appropriate indices from the RPI.

Financial services

22a.103 *Life insurance and pension funds*. The estimates are based on information on service costs and, for self-administered schemes, from the ONS quarterly and annual surveys of the income and expenditure of pension schemes (25.3.3). Quarterly figures are based on a corresponding but smaller quarterly inquiry. The constant-price figures are derived by deflating the value estimates by a deflator of costs in the industry.

22a.104 *Stamp duties*. This item consists of stamp duties incurred by persons on the transfer of bonds and shares. Stamp duties incurred in the transfer of ownership of land and buildings are excluded; they are treated as fixed capital formation. Quarterly information is available from the Inland Revenue, to which some adjustments are made to obtain that part relevant to the household sector. Estimates at constant prices are obtained by deflating using appropriate share indices.

22a.105 *Monetary services n.e.c.* This includes a variety of charges for financial services. Bank charges are derived from estimates, made by the Bank of England, of fees, charges and commissions paid to banks by the personal sector. Foreign currency exchange charges are derived by taking a fixed proportion of household expenditure abroad and of overseas visitors spending in the UK. Where the exchange is made in bulk, for instance by the travel industry when arranging package tours, the exchange costs may be somewhat less than 1%. However an individual's exchange costs can be considerably higher and 1% is assumed to represent the overall rate. Constant-price figures are the value data deflated by appropriate RPI components. Unit trust management charges relate to the initial investment (typically around 5%) and an annual 'maintenance' charge (around 1%). ONS estimates of charges are based on around a half of all unit investments being held by the household sector. The constant-price estimates are derived by deflating the current-price figures using the FT all-share index. Charge card membership covers the costs to personal holders of charge cards such as American Express and Diners Club. Estimates are based on the number of personal holders multiplied by an estimate of the average annual charge. Constant-price estimates are based on deflation by RPI components. A number of other miscellaneous charges are also covered.

22a.106 *Stockbrokers' charges.* Estimates are based on information provided by the Stock Exchange, covering both explicit fees charged and earnings made by dealers on the exchange rate 'spread'. The estimates at constant prices are obtained from numbers of transactions and average base period prices.

Household and domestic services

22a.107 *Contractors' charges and structural insurance.* Estimates of contractors' charges for routine maintenance are derived for two components. First, figures of direct spending by the household are based on information from the FES (*see* paragraph 22a.35) and from the periodic English House Condition Survey. Secondly, for spending on work financed directly from insurance claims, estimates are based on information from the Association of British Insurers. Estimates at constant prices are obtained by deflating the value figures by appropriate RPI components. An estimate for the service charge on structural insurance is then added. This is obtained by applying an estimated ratio of service costs to premiums, provided by the Association of British Insurers, to the level of premiums derived from FES data. An addition to the gross premium is made to allow for the 'premium supplement' which is calculated from insurance companies' accounts. Constant-price figures are obtained by deflating the value estimates by appropriate price indices.

22a.108 *Laundry, dry cleaning, shoe repairs, other repairs and service in kind.* These estimates are derived from the Family Expenditure Survey (25.4.5) and deflated by appropriate components of the retail prices index.

22a.109 *Contents insurance.* This represents the administrative costs of house contents insurance and is derived in an identical fashion to the service charge on structural insurance.

Recreation and cultural services

22a.110 *Television and video hire.* Estimates are based on information from the retailing and related inquiries. Constant-price figures are derived by deflating value estimates using appropriate retail price indices.

22a.111 *Television licences.* Quarterly figures on the value and number of licences, separately for colour and monochrome, are provided by the National Television Licence Records Office. The constant-price data are the projections.

22a.112 *Television, VCR and radio repairs.* Estimates of household consumption on these are based on information from the FES. Constant-price figures are obtained by deflating values by appropriate price information.

22a.113 *Cinema.* Estimates are based on quarterly information on the value and number of cinema admissions from a voluntary panel of contributors within the cinema industry in Great Britain. An estimate is added in for Northern Ireland. The constant-price figures are derived from the numbers of admissions and the average price in the base year.

544

22a.114 *Other admissions and social subscriptions.* These cover respectively admissions to theatres, dances, participant and spectator sports events, theme parks, zoos and museums, *plus* subscriptions to clubs, trade unions and friendly societies. Estimates of household consumption on these categories are based on information from the FES (25.4.5). Constant-price figures are obtained by deflating values by appropriate price information.

22a.115 *Betting and gaming.* Household consumption on betting is measured by the amount staked *less* the part returned in the form of winnings, that is the net loss incurred. This net loss is equivalent to the net takings of persons engaged in the industry *plus* the amount taken by the government in the form of betting duties.

22a.116 For the National Lottery estimates are based on data from OFLOT. The constant-price estimates have been obtained using the retail prices index excluding housing costs, RPI(X), as the deflator. For the various forms of betting a broadly similar approach is adopted for the estimates for football pools, fixed-odds football betting, horse and greyhound race betting with bookmakers on and off-course (including betting shops) and horse and greyhound racing totalisators. For these, figures of the amounts and rates of betting duty, provided by Customs and Excise, can be used to estimate total takings for each type of betting. Estimates of the amounts retained are obtained by applying 'retention' factors based on Gaming Board and other sources, to total takings. Quarterly figures are interpolated.

22a.117 For the remaining areas, various sources are used. For casinos and gaming machines, estimates again make use of information from the Gaming Board, in the former case using information on amounts staked and operators' take, and in the latter the numbers of different types of machine and the average retention figure. In both cases duties are included in the estimates. Figures for spending on bingo, including admission and participation fees, are based on Customs and Excise estimates of duty and numbers participating. Quarterly figures are interpolated.

22a.118 For most of the above, expenditure is revalued at constant prices first by adjusting for changes, since the base year, in rates of tax and in retention rates, and secondly by deflating by the general consumer price index.

22a.119 *Education.* Expenditure under this head relates to tuition fees paid to universities, schools and other educational establishments. Expenditure on food and accommodation is included under in catering. Annual information for universities, covering both British and overseas students, is provided by the University Funding Council; separate figures are obtained for the Open University. Quarterly estimates are interpolated. Constant price figures are derived by deflating the value estimates by a UFC price index.

22a.120 For independent schools, figures are based on data on annual average fees provided by the Independent Schools Information Service, together with estimates of numbers of pupils from DfEE and ISIS sources. Quarterly estimates are interpolated. The constant price estimates are derived from the pupil numbers and figures of base year fees.

22a.121 For the third component, estimates are compiled from information provided by local authorities, together with estimates of fees from private sector educational establishments. Quarterly figures are interpolated. Constant price figure are calculated by deflating the value data by a price index based on the cost of wages and salaries and of goods and services procured for education.

Medical services

22a.122 *NHS payments*. NHS payments for goods and services embrace payments for hospital, specialist and ancillary services (including private beds in national health service hospitals); pharmaceutical services (including prescription charges); general dental services and supplementary ophthalmic services. Estimates at current prices are based on quarterly information supplied by the Department of Health. Estimates at constant prices are obtained by deflating the value data by a specially constructed price index, as determined for use in deflating government final consumption on NHS goods and services.

22a.123 *Private medicine*. Estimates, relating to payments by patients for the cost of private treatment, are based on information collected in the FES, as mentioned in paragraphs 22a.35. Constant price figures are based on deflation of value data by suitable price series.

22a.124 *Medical insurance*. Estimates of household spending on medical insurance are based on information provided to the Department of Health by the service providers. Quarterly estimates are interpolations and extrapolations of the annual data. Estimates at constant price figures are obtained by deflating the value data by the price index for goods and services used to revalue NHS final consumption.

Miscellaneous services

22a.125 *Hairdressing and beauty care*. The estimates are based on information collected in the FES and are deflated to constant prices by the relevant component of the retail prices index.

22a.126 *Accident insurance*. This represents the administrative costs of accident insurance and is derived in the same way as building insurance described above.

22a.127 *Undertaking*. Figures are derived by multiplying the number of deaths and the National Association of Funeral Directors estimated average cost of a funeral. Expenditure at constant prices is obtained from the same numbers and the average base year cost.

22a.128 *Survey fees*. Estimates for survey fees on mortgage loans are based on Department of the Environment data. Current price estimates are obtained by multiplying together the total value of mortgages advanced, the number of surveys carried out per household and the fee for a survey as a proportion of the mortgage value. In order to construct constant price estimates, the number of new mortgages is used as a volume indicator.

22a.129 *Other miscellaneous services*. This item includes a range of services not elsewhere covered, including legal fees (other than those for the transfer of ownership of land and buildings, which are treated as part of fixed capital formation), advertising and pawnbrokers' charges, services of photographers, auctioneers, etc., cloakroom fees, subscriptions to libraries, etc. There is very little information about these items and the amounts included are guesses.

Tourism adjustment

22a.130 Spending abroad by UK residents and spending in the UK by foreign tourists is treated in a particular way in the economic accounts. This is explained in the following paragraphs.

22a.131 Expenditure by UK residents abroad is part of household sector consumption, but spending by non-residents in the UK is not. However, the sources used for the estimates of spending and the way the estimates of total consumption are made - by adding together expenditures on individual goods and services - requires that spending by UK residents abroad needs to be added as an aggregate and spending by non-residents in the UK must be deducted as an aggregate to obtain the proper aggregate estimate. The aggregates for each of UK spending abroad and foreign spending in the UK also appear in travel services in the balance of payments account - the former as an import, the latter as an export. Although the total is correct, what this means for the individual commodity figures is that they essentially reflect spending by both UK residents and non-residents in the UK, but do not cover spending by UK residents abroad.

22a.132 The estimates of total *expenditure* by UK residents abroad cover spending by overseas travellers and also by members of the Forces and government employees stationed abroad. The estimates of expenditure abroad by travellers are based on quarterly information collected in the ONS International Passenger Survey (25.5.4). The survey provides an aggregate figure of spending overseas, from which is deducted an estimate of business spending. This spending estimate excludes fare payments for travel to and from the United Kingdom, which are included in the travel category. The estimates of spending by members of the Forces in foreign countries and in NAAFI's and other United Kingdom institutions abroad are based on quarterly information on the pay of military personnel serving abroad. Information on the pay and allowances of government staff based in the United Kingdom and serving abroad is available annually from departments. Assumptions are made about the proportion of pay and allowances which is spent abroad, and quarterly figures are interpolated and projected.

22a.133 Estimates at constant prices are derived by deflating the value figures by a specially-constructed price indices. The indices are obtained by weighting together information on retail prices in overseas countries, adjusted for exchange rates, by the estimated pattern of purchases by UK travellers. The weighting also takes account of the geographical pattern of spending. Separate indices are used for the expenditure of United Kingdom military forces and United Kingdom Government staff serving abroad. Again these are based upon local consumer prices adjusted for changes in exchange rates, with the weights reflecting the distribution of country of service.

22a.134 The estimates of total *expenditure in the UK by non-residents* covers spending by travellers to the UK and also by United States Forces stationed in the United Kingdom, by employees of overseas governments, and by foreign students and journalists (who for economic accounts purposes are deemed to be non-residents). The estimates of expenditure in the UK by travellers are based on quarterly information collected in the ONS International Passenger Survey (25.5.4). Estimates of spending by United States Forces stationed in this country is supplied by their military authorities. Estimates of the expenditure in the UK of employees of overseas governments based in this country, of overseas students studying here and of foreign journalists, are derived by multiplying their numbers by estimates of average expenditure per head. These figures are based on various sources, including the Foreign and Commonwealth Office and the British Council.

22a.135 Estimates at constant prices are derived by deflating value figures by specially-constructed price indices. These are based on appropriate RPI components, with expenditure by foreign students using the all-items index.

Chapter 23

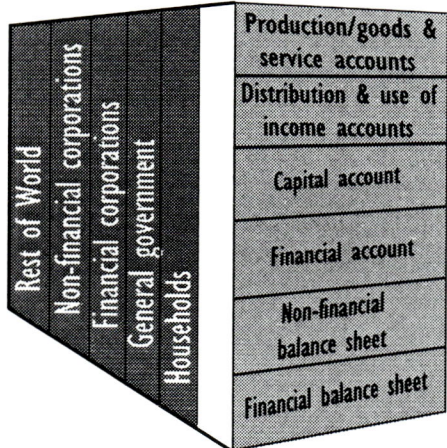

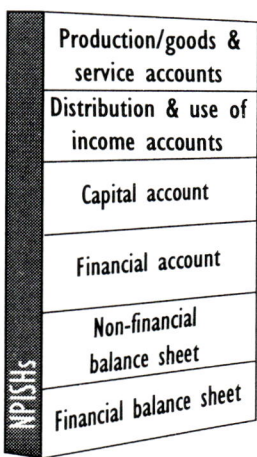

NPISHs-
Non profit-making
institutions
serving households

Chapter 23 Non-profit institutions serving households

23.1   The non-profit institutions serving households (NPISH) sector covers all such units which have independent legal status apart from those controlled and financed mainly by government). Three broad types of NPISH can be identified:

- Academic establishments, principally universities (including their colleges), the Open University, higher education and further education colleges;
- Associations which provide benefit primarily for their members and are financed mainly by subscriptions, including professional and learned societies, trade unions, churches and religious societies, housing associations, non-collecting friendly societies, social or recreational organisations and sports clubs;
- Bodies which serve the interests of people other than their members, including charities and similar relief and aid organisations financed by donations from the public, government and business.

The second group excludes those bodies where membership gives a right to a predetermined range of goods or services; for example book clubs and the Automobile Association.

23.2   These institutions form a distinct sector in the ESA, having previously been included by the United Kingdom within what was then the Personal sector. However, at the time when the new system of accounts was introduced generally for the United Kingdom the data and methodology for the NPISH component were not sufficiently well-developed for it to be separately identified. For the present, therefore, it is being consolidated with the Households sector, but there is every intention of establishing it as a sector in its own right as soon as possible.

23.3   The estimates which go into the NPISH sector, whether or not it is separated out, fall into three broad categories:

- Those which relate uniquely to such institutions, such as their final consumption and receipts from sales of goods and services;
- Specific estimates made for the NPISH sector alongside others as part of a common process for deriving sectoral figures;
- Estimates derived by breaking down a broader aggregate on the basis of whatever *ad hoc* information is available.

23.4   By way of example, the first category includes information on academic establishments from the Higher Education Statistics Agency and information on charities from benchmark surveys conducted by the ONS covering their sources and types of income, expenditure and certain balance sheet data, these being projected forward using sources which are less detailed but available more regularly. There is also good information for trade unions and friendly societies but for some other types of institution the estimates are less firmly based. Within the second of the above categories an important source of figures for NPISH, as for other sectors, is the Dividends and Interest Matrix (DIM) described in Chapter 14. Another example of methodology applying across sectors is the information on wages and salaries, which form a major part of NPISH turnover. Finally, where the third approach has to be used the essence of the methodology will be found described in other parts of this publication. However, in some cases the NPISH estimates represent little more than the allocation of the residuals found once all other sectors have been accounted for.

23.5   Given that separate figures cannot yet be published for the NPISH sector it would be inappropriate and misleading to present here a systematic description of the various ESA accounts such as appears in Chapters 19 to 22 and 24 for other sectors, but this form of presentation will be made available once the remaining technical difficulties have been overcome.

Chapter 24

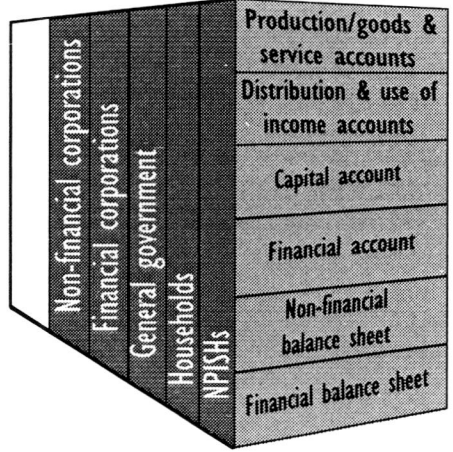

Rest of the World Sector

Chapter 24 Rest of the World sector

Introduction

24.1   This chapter describes the sector covering transactions between the UK and the rest of the world. It defines the rest of the world sector and how it relates to other sectors of the economy. It explains the relationship between the rest of the world sector within the economic accounts and the balance of payments account. It gives details about the tables which make up both sets of accounts and the sources of data used.

24.2   It is convenient to treat the rest of the world as if it were a sector of the economic accounts even though it is not necessary to produce a complete set of accounts for the rest of the world. The other sectors in the economic accounts are obtained by disaggregating the total economy into homogeneous groups of resident institutional units which are similar in their economic behaviour, their objectives and their functions. The rest of the world sector records transactions, flows and economic relationships between residents of all sectors of the economy combined (i.e. non-financial and financial corporations, non-profit institutions serving households, households and general government) and non-resident institutional units.

Definition of the Sector

24.3   The total economy is defined in terms of resident units. A unit is said to be a resident unit of a country when it has a centre of economic interest on the economic territory of that country, i.e. when it engages for an extended period (usually one year or more) in economic activities on the territory. Resident units engage in transactions with non-resident units i.e. units which are resident in other economies. These transactions are the external transactions of the economy and are grouped in the rest of the world accounts. Consequently, as far as classification is concerned, a specific item for the rest of the world sector is included at the end of the classification of sectors. As the rest of the world plays a role in the accounting structure similar to that of an institutional sector, its account is viewed from the point of view of the rest of the world. A resource for the rest of the world is a use for the total economy and vice versa. If the balance is positive, it means a surplus for the rest of the world and a deficit for the whole economy and the reverse if the balance is negative.

24.4   Residents of the United Kingdom comprise:

- private individuals living permanently in the UK;
- the UK central government and local authorities;
- all business enterprises and non-profit making organisations located in the UK, including branches and subsidiaries of non-resident companies located in the UK, but excluding branches and subsidiaries of UK companies located overseas;
- agencies of the UK government operating abroad such as embassies or military units which are regarded as residents of the UK (and, conversely, the agencies of foreign governments and international organisations located in the UK are regarded as non-resident).

24.5   Although the definition of residence for an individual is relatively straightforward, some confusion can arise between migrants who change residence and visitors who do not. A person leaving the UK for one year or more is regarded as an emigrant and thus a non-resident, and a person arriving in the UK intending to stay one year or more is regarded as an immigrant and thus a UK resident. There has been a change in the new system regarding students and those travelling for medical treatment. These individuals will now continue to be regarded as residents of their normal country even if they stay away for more than one year.

24.6   Some non-resident units are treated as resident units. These include:

- Non-resident units in their capacity as owners of land or buildings on the economic territory of the United Kingdom (but only in respect of transactions affecting such land or buildings);

- Non-residents who have a centre of economic interest on the economic territory of the United Kingdom (i.e. in general, those who engage in economic transactions for a year or more). Exceptions to this rule are that:

    Non-residents involved in the installation of equipment are treated as non-residents even if the work lasts for more than a year, and

    Non-residents carrying out a construction activity for a period of less than a year are treated as residents if the output constitutes gross fixed capital formation.

    These exceptions are changes from the general rule applied before the introduction of the European System of Accounts (ESA 95).

24.7   There are certain transactions included in the rest of the world sector which do not involve transactions between residents and non-residents. These are:

i) For imported goods, the cost of any transport services (up to the border of the exporting country) provided by UK-resident units is included in import values declared free on board (f.o.b.). These services are also included in the export of services.

ii) Transactions in foreign assets between residents belonging to different sectors are shown in the detailed financial accounts for the rest of the world. Although they do not affect the country's financial position vis-a-vis the rest of the world they affect the financial relationships of individual sectors with the rest of the world. Examples of these assets are shares issued by corporations in the rest of the world.

iii) Transactions in the country's liabilities between non-residents belonging to different geographical zones are shown in the geographical breakdown of the rest of the world accounts. Although these transactions do not affect the country's overall liability to the rest of the world, they affect its liabilities to different parts of the world. Data on such transactions are not currently available.

24.8   A convenient breakdown for analysis of the rest of the world sector is:

a) the European Union (EU) (S.21);

- the member countries of the EU (S.211)
- the Institutions of the EU (S.212)

b) third countries and international organisations (S.22).

However, for many items, data are only available at a total level.

24.9   The economic territory of the United Kingdom is defined as:

a) the geographic territory administered by the UK government within which persons, goods, services and capital move freely;

b) any free zones, including bonded warehouses and factories under customs' control;

c) the national air-space, territorial waters and the continental shelf lying in international waters, over which the United Kingdom enjoys exclusive rights;

d) territorial enclaves (e.g. embassies, consulates, military and scientific bases etc.) situated in the rest of the world and used, under international treaties and agreements between states, by UK general government agencies;

e) deposits of oil, natural gas, etc. in international waters outside the continental shelf of the country, worked by units resident in the territory as defined in the previous categories.

24.10   Under ESA95, UK offshore islands - the Channel Islands and the Isle of Man - are excluded from the economic territory of the UK because, although there is free movement of goods and people between the UK mainland and these islands, fiscal policy is administered independently by the islands.

Changes from former system of accounts

In the past, for balance of payments purposes at least, the economic territory of the United Kingdom has been defined to include the Channel Islands and estimates were made to adjust the data for use in the national accounts. Action has now been taken to classify these offshore islands as part of the rest of the world in all inquiries undertaken by the Office for National Statistics (ONS) and the Bank of England. Work by ONS and the Bank is in progress to try to estimate the past flows between the UK and the islands and adjust other sources to the new basis.

Relationship with balance of payments

24.11   There is a close relationship between the rest of the world sector of the economic accounts and the balance of payments account as both measure transactions between residents and non-residents of the economy. Residence is defined in the same way for both purposes.

24.12   Although the two accounts use different formats, the components used to make up both accounts can be reconciled. A broad reconciliation table is given below.

24.13   The two presentations provide alternative but complementary views of the economy. The economic accounts (including the rest of the world sector) based on ESA 95 provide a comprehensive and systematic framework for the collection and presentation of the economic statistics of an economy. They form a closed system incorporating both ends of every transaction recorded, with the rest of the world account being constructed from the viewpoint of the rest of the world rather than the compiling economy.

24.14   The balance of payments accounts are intended to analyse the economic relationship between the UK economy and the rest of the world; the accounts are therefore viewed from the reverse direction to the national accounts. In principle, each transaction is recorded twice – both as a credit and a debit entry – with the accounts balancing across in the current, capital and financial accounts. However, as estimates are derived from various sources (some of which are more reliable than others), the two entries may not match exactly; also, timing differences both in recording transactions and in converting from foreign currency can lead to imbalances. In order to balance the accounts, a balancing item (called net errors and omissions) is included in the tables.

Rest of world sector account	Balance of payments account
External account of goods and services	Imports/exports of goods and services
External account of primary incomes and current transfers	Income and current transfers
Allocation of primary income account	Investment income
Secondary distribution of income account	Current transfers
Use of disposable income account	
Change in net worth due to saving and capital transfers account	Capital transfers
Acquisition of non-financial assets account	Acquisition/disposal of non-produced non-financial assets
Financial account	Financial account
Financial balance sheets	International investment position

Note: *Detailed reconciliation tables can be found in the fifth edition of the IMF Balance of Payments manual (BPM5)*

24.15   The UK presentation of the balance of payments account is consistent with the fifth edition of the International Monetary Fund's (IMF) Balance of Payments manual which, in turn, is consistent with the System of National Accounts 1993 and hence ESA 95. In general, the balance of payments accounts are analysed in greater detail than is the rest of the world sector in the national accounts. Descriptions of the components in both accounts are given below.

The sequence of accounts for the rest of the world account

24.16   This follows the same general pattern as that used in the institutional sector accounts:

A. Current accounts
B. Accumulation accounts
C. Balance sheets

© National Accounts Concepts, Sources and Methods

A. Current accounts

External account of goods and services (V.I)

External account of goods and services	
Uses	**Resources**
P6 Exports of goods/services B11 External balance of goods/ services	P7 Imports of goods/services

Resources: Imports of goods and services (P.7)

Uses: Exports of goods and services (P.6)

24.17   These are transactions in goods and services (purchases, sales, barters etc.) between residents and non-residents. Transactions in goods occur generally when there is a change of ownership whether or not the goods move across frontiers. Exceptions to this general rule include:

a) goods for significant processing or repair which do not change ownership are recorded in both imports and exports;

b) goods delivered between affiliated enterprises are assumed to change ownership and are recorded;

c) no imports or exports are recorded for merchanting whereby resident merchants or commodity dealers buy from non-residents and sell again to non-residents within the same accounting period. Earnings from merchanting are included in services;

d) goods in transit through a country, goods shipped to or from a country's own embassies, military bases, etc., and transportation and other equipment or goods which leave a country temporarily are excluded although they move across the frontier.

A full description of sources and coverage is given below paras 24.63 *et seq* in the section describing the balance of payments for which a more detailed breakdown is required.

External account of primary incomes and current transfers (V.II)

24.18   With the exception of the balances B.11 and B.12, all major categories in this account may appear in both resources and uses.

External balance of goods and services (B.11)

24.19   This is the difference between imports (resources) and exports (uses) from the previous table.

Compensation of employees (D.1)

24.20 This is the total remuneration, in cash or in kind, payable by employers to their employees for work done during the accounting period, including both wages and salaries and employers' social contributions. It covers:

a) cross-border working - this is likely to be insignificant for the UK with the exception of workers moving between Northern Ireland and the Republic of Ireland. Limited information is available from the International Passenger Survey (IPS) [25.5.4] and from the Irish Central Statistical Office (CSO);

b) UK workers working in foreign embassies and military bases situated in the UK; and local workers working in UK embassies and military bases abroad. Information is obtained from government departments (Ministry of Defence (MOD), Foreign and Commonwealth Office (FCO)), US military bases and ONS inquiries to foreign embassies.

Compensation of employees is recorded as resources (in respect of foreign residents employed in the United Kingdom or UK embassies etc.); and as uses (UK residents employed abroad or by foreign embassies etc.)

Taxes on production (D.2)

24.21 Resources include taxes on production (particularly in the agricultural and energy sectors) and imports paid to the institutions of the European Union. Information is obtained from government sources.

less subsidies (D.3)

24.22 Uses are current unrequited payments which the Institutions of the European Union make to resident producers (e.g. Set-aside payments to UK farmers), with the objective of influencing their levels of production or their prices. Information is obtained from government sources (HM Treasury). No other subsidies affect this sector account.

Property income (D.4)

24.23 This is the income receivable by the owner of a financial or a tangible non-produced asset in return for providing funds to, or putting the asset at the disposal of, another unit. Data are available from ONS inquiries (overseas direct investment (25.5.1) and inquiries to banks and other financial institutions (25.3). The total is broken down by using the matrix of dividends and interest flows (DIM) across sectors (using actual and estimated relationships) into the following categories:

Interest (D.41)

24.24 This is an amount that a debtor becomes liable to pay to a creditor over a given period of time without reducing the principal outstanding under the terms of a financial instrument agreed between them. Uses include interest on bank deposits held in the rest of the world and interest on loans by UK banks to non-residents. Resources include interest paid by UK residents on loans from banks in the rest of the world or interest paid by UK banks to non-residents.

Distributed income of corporations (D.42)

24.25 These are dividends received by owners of shares and withdrawals from income of quasi-corporations, i.e. the amounts which entrepreneurs withdraw for their own use from the profits earned. Income transferred to resident parent companies from non-resident units (branch-offices, agencies, etc.) (and vice versa) is included. The net operating surplus received by residents as owners of land and buildings in the rest of the world (and vice versa) is also included.

Re-invested earnings on direct foreign investment (D.43)

24.26   This is equivalent to: the operating surplus of a foreign direct investment enterprise **plus** any property income or current transfers receivable **minus** any property income or current transfers payable, including actual remittances to foreign direct investors and any current taxes payable on the income, wealth etc. of the foreign direct investment enterprise. Figures come from the ONS and Bank of England surveys of overseas direct investment (25.5.1 and 25.3.1).

Property income attributed to insurance policy-holders (D.44)

24.27   This corresponds to the property income received from the investment of insurance technical reserves. These reserves are invested by insurance enterprises and pension funds in financial assets or land (from which net property income after deducting any interest paid is received) or in buildings (which generate net operating surpluses). These reserves are deemed to be owned by the policyholders (*see* Chapter 20 Annex 1). Resources of the rest of the world comprise payments by UK insurance companies to non-resident policyholders. Uses of the rest of the world comprise payments by rest of the world insurance companies to UK resident policyholders. Chapter 14 describes the allocation of these flows to sectors.

Current taxes on income, wealth, etc. (D.5)

24.28   These cover all compulsory unrequited payments, in cash or in kind, levied periodically by general government and by the rest of the world on the income and wealth of institutional units and some periodic taxes which are assessed neither on their income nor wealth. Information is obtained from government sources (HM Treasury). They include:

Taxes on income (D.51)

24.29   These comprise taxes on incomes, profits and capital gains. Company taxes are assumed to be nil. Investigations are underway to identify sources for taxes paid by non-residents working in the UK and UK residents working abroad.

Other current taxes (D.59)

24.30   These are considered to be nil or unmeasureable.

Social contributions (D.61)

24.31   Uses relate to actual or imputed contributions made by non-resident employees to UK social insurance schemes. Resources relate to actual or imputed contributions made by UK resident employees to foreign social insurance schemes. No resources have yet been identified.

Social benefits other than social transfers in kind (D.62)

24.32   These cover payments of benefits under various social insurance schemes (including social security benefits and private funded and unfunded social insurance schemes). Resources include payments of social benefits (e.g. pensions) paid to former residents now living abroad. Information is obtained from government sources (Department of Social Security). Uses include payments by foreign social insurance schemes to beneficiaries resident in the United Kingdom.

Other current transfers (D.7)

24.33   These cover a variety of transactions including:

Net non-life insurance premiums (D.71)

24.34   The premiums comprise both the actual premiums payable by policy holders to obtain cover (premiums earned) and premium supplements payable out of the property income attributed to insurance policy holders after deducting the service charges of enterprises arranging insurance. They include premiums to cover against damage to goods or property or harm to persons or against financial losses. Information is obtained from ONS inquiries on overseas trade in services (25.5.2).

Non-life insurance claims (D.72)

24.35   These represent amounts insurance enterprises are obliged to pay in settlement of injuries or damage suffered by persons or goods (including fixed capital goods) during the current accounting period. They are obtained from ONS inquiries on overseas trade in services (25.5.2).

Miscellaneous current transfers (D.75)

24.36   These cover a wide range of transfers including among others:

- the transfer of all voluntary contributions (other than legacies), membership subscriptions and financial assistance from households to non-resident institutions, and from non-residents to NPISH;
- voluntary contributions including gifts of food, clothing, medicines, etc. by charities for distribution to non-resident households;
- remittances by emigrants or workers settled permanently abroad to members of their families or by parents to children in another country;
- payments of compensation for injury to persons or damage to property etc.
- UK contributions to the EU budget based on the '4th resource', currently using GNP as measured under ESA 79.

Some administrative information is available from government departments.

Adjustment for the change in net equity of households in pension funds

24.37   This item relates to households in the rest of the world. The inclusion of this item reflects two particular measurement concepts of the accounts. The first is that the reserves of private funded pension schemes are treated as being owned by the household sector (and households in the rest of the world) which have a claim on the funds. Secondly, the pension contributions to and pensions paid by the private funded schemes are included, as transfers, in the secondary distribution of income account. Reflecting this latter treatment, the 'balance' of contributions *less* receipts is included as part of the disposable income of the pension scheme. Thus, in order to ensure that the saving is correctly attributed to the households sector, and to (households in) the rest of the world, it is necessary to include this balance, as a resource, in the use of income account for these sectors. The financial account shows the same amount being reinvested in the pension fund in F.61. Further details, including how the balance is defined, are given in chapter 14 and Annex 2 to Chapter 20.

B. Accumulation account

Capital account (III.1.)

Capital transfers (D.9)

24.38 These involve the acquisition or disposal of an asset or assets by at least one of the parties to the transaction. Whether made in cash or in kind, they should result in a commensurate change in the financial or non-financial assets shown in the balance sheets of one or both parties to the transaction. They include:

Investment grants (D.92)

24.39 These consist of transfers in cash or kind made by general government to non-residents (resources) or by the rest of the world (including those paid directly by the institutions of the EU) to residents (uses) to finance all or part of the costs of their acquiring fixed assets. Investment grants to the rest of the world are restricted to transfers with the specific objective of financing capital formation by non-resident units. They include unrequited transfers for the construction of bridges, roads, factories, hospitals or schools in developing countries, or for constructing buildings for international organisations. They may comprise instalment payments over a period of time or single payments. The heading also covers the supply of fixed capital goods free of charge. Data are provided by HM Treasury, the Department for International Development (DID) and the FCO.

Other capital transfers (D.99)

24.40 These cover transfers other than investment grants and capital taxes which do not themselves redistribute income but redistribute saving or wealth between the different sectors of the UK economy and the rest of the world. They include: payments by general government or by the rest of the world to the owners of capital goods destroyed by fire, flood, war, etc.; the counterpart transaction of cancellation of debts by agreement between units belonging to different sectors (e.g. the cancellation by government of a debt owed by a foreign country); and major payments in compensation for extensive damage or serious injury not covered by insurance policies e.g. major explosions, oil spills etc. Some information is available from HM Treasury and the DID.

Changes in net worth due to saving and capital transfers (B.10.1)

24.41 This is the balance between saving and receipts less payments of capital transfers. It represents the amount available for the purchase of financial or non-financial assets.

Changes in assets

Acquisitions less disposals of non-produced non-financial assets (K.2)

24.42 This records the purchases and sales of intangible non-produced assets such as patents, copyrights, franchises, leases and other transferable contracts, goodwill etc. Information on copyrights and patents is collected in the Overseas Trade in Services inquiry (OTIS) (25.5.2), but other information is likely to be minimal.

Net lending(+)/net borrowing (-) (B.9)

24.43   This is the balancing item of the account.

Financial account (V.III.2)

Financial account

Net acquisition of:

Financial assets	Financial liabilities
F1 Monetary gold and special drawing rights	
F2 Currency and deposits	F2 Currency and deposits
F3 Securities other than shares	F3 Securities other than shares
F4 Loans	F4 Loans
F5 Shares and other equity	F5 Shares and other equity
F6 Insurance technical reserves	F6 Insurance technical reserves
F7 Other accounts receivable	F7 Other accounts payable

Net acquisition of financial assets and liabilities (*see also* **Chapter 17**)

Monetary gold and special drawing rights (SDRs) (F.1)

24.44   These are the only assets for which there are no counterpart liabilities in the system and therefore always involve changes in ownership of financial assets. Monetary gold is gold held as a component of foreign reserves by monetary authorities or by others who are subject to the effective control of the authorities. SDRs are international reserve assets created by the IMF and allocated to its members to supplement existing reserve assets. Details are provided by the Bank of England.

Currency and deposits (F.2)

24.45   All sectors and the rest of the world can hold currency and deposits. This item measures all transactions in currency i.e. notes and coins in circulation used to make payments and transactions in all types of deposits in national or foreign currency. Also included here are the UK reserve position with the IMF, with the European Monetary Co-operation Fund before 1 January 1994 and with the European Monetary Institute and European Central Bank thereafter. Sources of information include the Bank of England (25.3.1), Bank for International settlements, ONS inquiries to financial institutions (25.3.2) and rough estimates based on data from the International Passenger Survey (25.5.4).

Securities other than shares (F.3)

24.46   These are bearer instruments (i.e. bills, bonds, certificates of deposit, commercial paper, etc.) which are intended to circulate and whose nominal value is determined on issue. They are usually negotiable and traded on secondary markets or can be offset on the market, and do not grant the holder any ownership rights in the institutional unit issuing them. Information is obtained from Bank of England (25.3.1) and ONS inquiries (e.g. direct investment) (25.5.1). At present there are no derivatives data but new questions have been added to the inquiries to collect these in future.

Loans (F.4)

24.47 These are financial assets created when creditors lend funds to debtors, either directly or through brokers. Data are obtained from a variety of sources including the Bank of England (25.3.1), government departments, ONS inquiries to financial institutions (e.g. insurance companies, pension funds, etc.) (25.3.2) and the European Investment Bank. There is a problem in differentiating between bank loans and deposits in this sector. The assumption made is that assets of UK banks are loans and their liabilities are deposits.

Shares and other equity (F.5)

24.48 These are financial assets which represent property rights on corporations or quasi-corporations. They generally entitle the holder to a share in the profits and to a share in their net assets in the event of liquidation. Information comes from the Bank of England and ONS inquiries to financial institutions (e.g. insurance companies, unit trusts, etc. and the Financial Assets and Liabilities Inquiry) (25.3).

Insurance technical reserves (F.6)

24.49 Information is obtained from ONS inquiries (25.3.2) to the industry. The transactions of this sector in these reserves cover

Prepayment of insurance premiums & reserves for outstanding claims (F.62)

24.50 These are amounts already paid in insurance premiums for periods that extend beyond the current accounting period. Also included are the total estimated costs of settling all claims arising from events which have occurred up to the end of the accounting period (whether reported or not) less amounts already paid in respect of such claims. This item covers transactions of resident policyholders with non-resident corporations (liabilities) and *vice versa* (assets).

Other accounts receivable/payable (F.7)

24.51 These are financial claims which are created as a counterpart of a financial or non-financial transaction in cases where there is a timing difference between this transaction and the corresponding payment. It includes transactions in financial claims which stem from early or late payment for transactions in goods or services, distributive transactions or secondary trade in financial assets. They consist of the counterpart transactions in case payment is due and not yet paid, debts arising from income accruing over time and arrears. The two main components of this sector are trade credit and an accruals adjustment for re-insurance. Information is collected from the Bank of England (25.3) and the Export Credits Guarantee Department.

Net lending(+)/net borrowing(-) (B.9f)

24.52 This represents the balance for the financial account and is the difference between the net acquisition of financial assets and the net incurrence of liabilities. In principle it is the same as net lending/net borrowing derived from the capital account.

Statistical discrepancy

24.53 This is the adjustment between net lending/borrowing as measured from the capital account (B.9) and as measured from the financial account (B.9f). It represents the net sum of all errors and omissions in the accounts for the sector.

C. Balance sheets

Financial balance sheet (at end of period) (V.IV.3)

24.54 This table is derived from the same sources as the financial account, and contains the same headings, but shows financial assets and financial liabilities and net financial assets/liabilities, i.e. it relates to holdings whereas the financial account relates to flows. *See also* Chapters 17 and 18.

Balance of payments accounts and sources

A. Current account

Trade in goods (imports and exports)

24.55 Figures for trade in goods cover, broadly, goods which leave or enter the UK and whose ownership is transferred between UK and non-UK residents. They are based on the Overseas Trade Statistics (OTS) (25.5.3) which are compiled monthly by HM Customs and Excise from information provided by importers and exporters. Until 1993, the figures were based almost entirely on customs declarations. The system of recording used in the compilation of the OTS and which applied to all trade prior to 1993 and continues to apply to trade with non-EU countries is geared to the time at which declarations made by exporters and importers are received in the Tariff & Statistical Office of HM Customs and Excise. As different practices are acceptable in making declarations, timing differences occur.

24.56 For trade with EU countries, a new system was introduced on 1 January 1993 when, with the introduction of the single market, customs declarations were no longer required. The new system, Intrastat, makes use of regular VAT returns to identify businesses undertaking a large amount of overseas trade and to provide a check on its value. However, most of the information comes from 'supplementary declarations' – statistical forms which are submitted and processed on a monthly basis by large traders and provide some of the information formerly required on customs documents. These supplementary declarations cover about 97% of transactions (by value).

24.57 For consistency with the basis for estimating other transactions in the economic accounts, exports and imports of goods should be recorded at the time ownership of the goods changes. A reasonable assumption is that, on average, ownership changes at some time between the departure of the goods from the exporting country and their arrival in the importing country. In practice, the compilation of the figures is geared to the timing of declarations made by exporters and importers under both systems to HM Customs and Excise. Differences in timing between the recording of trade in goods and the recording of payment (or the granting of trade credit) will be reflected in the statistical discrepancy.

24.58 Certain adjustments are necessary to adjust the OTS data to be consistent with balance of payments (and economic accounts) definitions.

   a) *Valuation adjustments* are made to deduct some of the freight costs and insurance premiums from import figures. In the OTS, export values are recorded at the point of export from the country, i.e. free on board (f.o.b.), but import values include carriage, freight and insurance costs up to the point of entry into the importing country for both exports and imports. As the value required for economic accounts purposes is the f.o.b. value, estimates of the extra costs of freight services and insurance need to be deducted from import values. (These costs will be included in service transactions under transportation and insurance if produced by non-residents.) Freight estimates are derived from a variety of sources including freight rates, tonnages of goods arriving in the UK, estimated land distances covered, the earnings of UK airlines and tonnages landed at UK airports. The insurance premiums are estimated as a fixed percentage of the value of imports. The methodology for estimating freight and insurance for oil is under review as the previous source (a Department of Trade and Industry (DTI) inquiry to the industry) is no longer available.

b) *Coverage adjustments* are made to allow for items not included in the OTS and to exclude items where no change in ownership occurs. Estimates are also made to take account of estimated under-recording. In the past other adjustments have been made to take account of revised coverage and presentation of certain items such as gold, oil and aircraft. Adjustments in coverage required to bring OTS figures onto the required basis include:

i.) Second-hand ships sold or purchased abroad and new ships built and delivered abroad – these transactions are not recorded in the OTS; however, they should be included in *trade in goods* in the economic accounts. Supplementary information about these transactions is obtained from the DETR;

ii.) Additions and alterations to ships. When significant changes (other than basic repairs) are made to UK-owned and registered ships in yards located abroad this represents a component of trade in goods; estimates of such transactions are obtained from DETR.

iii) Goods exported by the Navy, Army and Air Force Institute (NAAFI) for the use of UK forces abroad. These goods are recorded in the OTS but need to be deducted as they are for sale to UK residents. Information on these exports is provided by the NAAFI.

iv) Goods not changing ownership; for example, contractors' plant temporarily leaving the country. These need to be deducted from the OTS. Estimates of value can be estimated from claims made to HM Customs and Excise for exemption from duty.

v) Exports by letter and parcel post. These are not included in the OTS. Information on the most important element of this trade, the export of books, is obtained from publishers and booksellers. Other estimates are based on information from Parcel Force and the Royal Mail.

vi) Transactions involving North Sea installations. There are various transactions that need to be adjusted for including goods imported directly to the installation from the rest of the world, the delivery of drilling rigs to or from the rest of the world, and adjustments to limit reporting for shared oil fields to the sector owned by the UK. Information comes from a variety of sources including inquiries to the petroleum and gas industries and DETR.

vii) Gold. The current UK treatment (under derogation until 2005) is described in the glossary entry. The OTS includes only finished manufactures of gold (e.g. jewellery), so an adjustment is made currently to include in exports the value added in refining gold and in the production of proof coins, and to include in imports the value of gold used for finished manufactures. The import adjustment is based on commercial statistics. ESA95 would require the inclusion of trade in semi-manufactures (such as rods, wire, etc), bullion and bullion coins.

c) Other adjustments include a net adjustment for exports taking into account under-recording to allow for declarations not sent to HM Customs and Excise, incorrect valuations and the declaration of value in foreign currency; and for imports from the EU for 1993 and subsequent years an adjustment for under-recording.

Trade in services (imports and exports)

24.59   This includes the provision of services by UK residents to non-residents and vice versa. Trade in services is disaggregated into the eleven broad categories of services specified in the table below.

Trade in services

1.   Transportation (Sea, Air and Other) - Passenger, freight and other
2.   Travel (Business and Personal)
3.   Communications services
4.   Construction services
5.   Insurance services
6.   Financial services
7.   Computer and information services
8.   Royalties and licence fees
9.   Other business services (Merchanting and other trade-related services; operational leasing services; miscellaneous business, professional and technical services)
10. Personal, cultural and recreational services (Audio-visual and related services; other cultural and recreational services)
11. Government services, not included elsewhere

Source: Fifth edition of IMF's Balance of Payments manual (BPM5)

24.60   *Transportation* covers sea, air and other (i.e. rail, land, internal waterways, space and pipeline) transport. It includes the movement of passengers and freight, and other related transport services, including chartering of ships or aircraft with crew, cargo handling, storage and warehousing, towing, pilotage and navigation, maintenance and cleaning, and commission and agents' fees associated with passenger/freight transportation. Information about passenger transport is principally obtained from the International Passenger Survey (IPS) (25.5.4); passenger fares for the channel tunnel are estimated from data provided by Le Shuttle and Eurostar. Other sources of information for freight and other services include the Chamber of Shipping, non-UK airlines, the Civil Aviation Authority, specialised returns from UK ports, Trinity House, Parcel Force, UK oil companies, HM Customs and Excise and other government departments.

24.61   In measuring freight costs, certain problems need to be addressed. Irrespective of payment basis (fob or cif), importers incur the following costs in addition to goods value - value of loss and damage incurred en route, freight services outside exporting country and insurance services (excess of premiums paid over claims made). The first represents part of price paid to the non-resident exporter and is always a debit entry; the other two only represent a debit entry when provided by non-residents. For exports, where UK residents provide freight and insurance services for journeys beyond the UK border, there is a credit entry in the services account. Where UK operators provide transport services for imports before they leave the exporting country, goods valuation (fob) is overstated and a offsetting credit entry is included under 'land transport'. Similarly an offsetting debit is included for foreign operators' carriage of UK exports within the UK.

24.62   *Travel* covers primarily goods and services provided to UK residents during trips of less than one year to the rest of the world (and vice versa) net of any purchases made with money earned or provided locally. The supply of sea or air transport from and to the UK is excluded here; it is included in transportation services. However, transport services used by overseas visitors within the UK and by UK residents travelling within or between overseas countries are included. The estimates derive principally from the IPS (25.5.4) which seeks information on expenditure from samples of visitors from the rest of the world leaving the UK and of UK residents returning from abroad. As the IPS does not cover travel between the UK and the Irish Republic, estimates are derived from statistics published by the Irish CSO.

24.63   Personal expenditure overseas by UK residents, other than expenditure by members of the armed forces etc. and expenditure on business accounts, is included in 'Household expenditure abroad' and therefore in final expenditure. An adjustment is made to exclude purchases by overseas visitors on both personal and business accounts, purchases by the Forces and other employees of overseas governments, and purchases by other non-residents. Such exports of services are thus excluded from domestic expenditure.

24.64   *Communications* covers two main categories of international transactions: telecommunications (telephone, telex, fax, e-mail, satellite, cable and business network services) and postal and courier services. Information is obtained through the ONS Overseas Trade in Services Inquiry (OTIS) (25.5.2) and direct from Parcel Force and the Royal Mail.

24.65   *Construction services* cover work done on construction projects and installations by employees of an enterprise in locations outside their resident economic territory. Where a permanent base is established which is intended to operate for a long period of time, the enterprise becomes part of the host economy and its services are excluded. The activity is included under direct investment. The source of information is the OTIS (25.5.2).

24.66   *Insurance services* cover the provision of various types of insurance to non-residents by resident insurance enterprises and vice versa. Types of insurance include freight insurance on goods being exported or imported, direct insurance (life, accident, health, fire, aviation, etc.) and re-insurance. Also included are agent commissions related to insurance transactions. The main sources of information are ONS inquiries to insurance companies and brokers (25.3), the OTIS (25.5.2) and administrative data from Lloyd's of London.

24.67   *Financial services* cover financial intermediary and auxiliary services other than those of insurance companies and pension funds. They include intermediary service fees associated with letters of credit, bankers' acceptances, lines of credit, financial leasing and foreign exchange transactions. Also included are commissions and other fees related to transactions in securities e.g. brokerage, underwritings, arrangements of swaps, options and other hedging instruments etc.; commissions of commodity futures traders; and services related to asset management, financial market operational and regulatory services, security custody services, etc. Estimates are based on returns from the Bank of England (for banks) (25.3.1), ONS inquiries (to securities dealers, fund managers, commodity traders, finance houses and credit guarantors etc.) (25.3.2) and directly from other sources including the Baltic Exchange. The commissions, fees and other trading revenue of securities dealers are included, but their dealing profits are not.

24.68   *Computer and information services* cover computer data and news related service transactions including data bases, such as development, storage and on-line time series; data processing; hardware consultancy; software implementation; maintenance and repair of computers and peripheral equipment; news agency services; and direct, non-bulk subscriptions to newspapers and periodicals. Information is obtained from the OTIS (25.5.2).

24.69   *Royalties and licence fees* cover the exchange of payments and receipts for the authorised use of intangible, non-produced, non-financial assets and proprietary rights (such as patents, copyrights, trademarks, industrial processes, franchises, etc.) and with the use, through licensing agreements, of produced originals or prototypes (such as manuscripts and films). Data are obtained through the OTIS (25.5.2).

24.70 Other business services cover a range of services including merchanting and other trade-related services, operational leasing (rental) without operators and miscellaneous business, professional and technical services.

a) Merchanting is defined as the purchase of a good by a resident from a non-resident and the subsequent resale of the good to another non-resident, without the good entering the compiling economy. The difference between the purchase and sale price is recorded as the value of merchanting services provided.
b) Operational leasing covers leasing (other than financial leasing) and charters of ships, aircraft and other transportation equipment without crews.
c) Miscellaneous services include legal, accounting, management consulting and public relations; advertising and market research; research and development; architectural, engineering and other technical services; agricultural, mining and on-site processing services associated with agricultural crops (protection against disease or insects), forestry, mining (analysis of ores) etc.; and other services such as placement of personnel, security and investigative services, translation, photographic services, etc.

Estimates are predominantly obtained from the OTIS (25.5.2). However, some specialised information comes from the Law Society and commercial bar for data on legal services, and from Lloyd's Register of Shipping for merchanting and other trade-related services.

24.71 *Personal, cultural and recreational services* are divided into:

- audiovisual and related services and
- other.

The first category covers services and associated fees relating to the production of motion pictures (on film or video tape), radio and television programmes (live or on tape), and musical recordings. It includes rentals, fees received by actors, directors, producers etc. and fees for distribution rights sold to the media for limited showings in specified areas. The second category covers all other personal, cultural and recreational services including those associated with museums, libraries, archives, provision of correspondence courses by teachers or doctors etc. Most of the information is obtained from OTIS, but there is a special ONS inquiry for the film and television industry (25.5.2).

24.72 *Government services* include all transactions by embassies, consulates, military units and defence agencies with residents of the economies in which they are located, i.e. the local purchase of goods and services and personal expenditure by diplomats, consular staff, military personnel etc. in the economies in which they are located. Other services included are transactions by other official entities such as aid missions and services, government tourist information and promotion offices, and the provision of joint military arrangements and peacekeeping forces (e.g. United Nations). Information comes directly from government departments (including the MOD and FCO), foreign embassies and United States airforce bases.

Income

24.73 This comprises compensation of employees, and investment income covering receipts and payments associated with holdings of external financial assets by residents and with liabilities to non-residents.

24.74 Compensation of employees covers wages, salaries and other benefits paid, in cash or in kind, to non-resident workers including, among others, cross-border workers, seasonal workers and local people employed in embassies, consulates, etc. The IPS (25.5.4) collects some limited information on earnings and expenditure abroad and the Irish Central Statistical Office provides limited data on cross-border working.

24.75   Investment income includes direct investment, portfolio investment and other investment income.

a) Direct investment income can be broken down into income on equity (dividends, branch profits and re-invested earnings) and on debt (interest). Estimates for all companies except banks are estimated from ONS inquiries to UK companies that either have non-resident affiliates or are affiliated to a non-resident parent (25.5.1). Information on earnings by UK registered banks from their branches, affiliates and associates in the rest of the world and by non-UK registered companies from their UK banking affiliates is collected by the Bank of England (25.3.1)

b) Portfolio investment income can be broken down into income on equity (dividends) and income on debt (interest). Estimates are based predominantly on inquiries by ONS and the Bank of England (25.3). For outward investment, earnings principally come from investment carried out by UK financial institutions (UK banks, securities dealers, insurance companies, pension funds, trusts, etc.). Earnings by members of Lloyd's of London are supplied directly. For inward investment, estimates for earnings on British government stocks, UK company securities and sterling Treasury bills are derived by applying the appropriate rate of interest/yield to the outstanding levels. There are no estimates available for earnings on banks' money market instruments at present.

c) Other investment income covers income earned on the United Kingdom's reserve assets (official reserves), interest on other capital (loans etc.), and imputed income to households from net equity in life insurance reserves and in pension funds. For monetary institutions, information is based on returns made by UK banks to the Bank of England (25.3.1). For the income of non-monetary financial institutions, there are a variety of sources including ONS inquiries to UK financial institutions (securities dealers, insurance companies, pension funds, etc.) (25.3), the Bank of England, government departments and international bodies such as the European Investment Bank and the Bank for International Settlements.

Current transfers

24.76   In addition to trade in goods and services and investment and employment income there is a further category of current transaction between residents and non-residents in which, unlike the others, there is no corresponding exchange of an actual good or service taking place. These transactions are termed current transfers. Current transfers are generally intended for immediate use.

24.77   These transfers can be sub-divided into those of general government and other:

a) General government transfers include some receipts, contributions and subscriptions from and to EU Institutions and other international bodies, bilateral aid and military grants. Information comes from government departments (HM Treasury, FCO, DID).

b) Other transfers cover cash gifts to dependants, legacies and gifts sent by post. Information on postal gifts is derived from an ONS inquiry, but other information is limited as there is no easy way to collect it directly. The data used are based on exchange controls records to 1979, updated from a variety of other sources, including the Family Expenditure Survey (25.4.5) and trends in UK personal disposable income for cash gifts, and the number of deaths in the UK and UK prices for legacies.

B. Capital and financial account

Capital account

24.78   This account comprises two components: capital transfers and the acquisition/disposal of non-produced, non-financial assets.

a) *Capital transfers* are those involving transfers of ownership of fixed assets, transfers of funds associated with acquisition or disposal of fixed assets, and agreed cancellation of liabilities by creditors without any counterparts being received in return. As with current transfers, they can be sub-divided into general government transfers (debt cancellation, other) and other (migrants' transfers, debt cancellation and other). The main sources of information are government departments (DID, HM Treasury) and the Bank of England (25.3.1).

b) *Acquisition/disposal of non-produced, non-financial assets* covers intangibles such as patents, copyrights, franchises, leases and other transferable contracts, goodwill, etc. and transactions involving tangible assets that may be used or needed for the production of goods and services but have not themselves been produced, e.g. land and sub-soil assets. The OTIS (25.5.2) collects information on the sale of copyrights and patents. Information about other items is minimal.

Financial account and balance sheets

24.79   These cover the following balance of payments categories:

a) *Direct investment* estimates include: the investor's share of the unremitted profits of subsidiary or associated companies (which is in proportion to the percentage holding of the investor in the associated company); the net acquisition of share or loan capital; changes in inter-company accounts; and changes in branch/head office indebtedness. They are based on ONS inquiries (25.5.1) and Bank of England inquiries (25.3.1).

b) *Portfolio investment* estimates include transactions in equity securities and debt securities (i.e. bonds and notes, money market instruments, financial derivatives such as options when the derivatives generate financial claims and liabilities. Information is obtained from ONS inquiries to financial institutions and Bank of England inquiries (25.3).

c) *Other investment* (which comprises trade credit, loans, currency and deposits and other assets and liabilities) is broken down into sectors:
    i)   Other government assets and liabilities include inter-government loans to and from the UK, subscriptions to international bodies and government liabilities on non-residents' holdings of sterling coinage. Also included is export credit taken over by the Export Credit Guarantee Department (ECGD) from UK banks and suppliers following settlement of insurance claims, together with long-term sterling assets acquired from UK banks under ECGD refinancing agreements.
    ii) Monetary authorities' liabilities cover estimates of sterling notes held by non-residents and is based on IPS data on tourists' expenditure (25.5.4).
    iii) Banks' assets and liabilities are estimates of foreign currency borrowing and lending and are based on information reported to the Bank of England.
    iv) Other sectors' assets and liabilities cover borrowing and lending by non-monetary financial intermediaries, securities dealers, non-financial corporations and any other UK residents. Estimates are derived from ONS inquiries to financial institutions and Bank of England inquiries.

d) Reserve assets consist of drawings on, and additions to, the monetary gold, convertible currencies and Special Drawing Rights held in the Exchange Equalisation Account which are recorded by the Bank of England.

Constant prices

24.80   In order to compile the expenditure measure of gross domestic product at constant prices it is necessary to prepare estimates of international trade in goods and services at constant prices. In general, current price information is deflated by appropriate and available price information.

24.81   For goods, the indicators of price movement ('deflators') for individual commodities used in the calculation of constant price estimates are based where possible on export and import price indices derived from the price information collected from UK manufacturers. Where export price indices (EPIs) are not available exports of manufactured goods are deflated by appropriate producer price indices (PPIs) adjusted by the relationship between EPIs and PPIs where both exist. Similarly, where import price indices (IPIs) are not available imports of manufactured goods are deflated by PPIs adjusted by the relationship between IPIs and PPIs where both exist. For commodities that are traded on world markets, price information on these markets is used to deflate both exports and imports. For other commodities, apart from manufactures, the indices are based on the value and quantity data provided by HM Customs and Excise.

24.82   For services, each series is converted using the most appropriate available data. For sea transport, freight rates and tonnages are used, and for air transport passenger mileage rates, combined with assorted price indices. For other services, extensive use is made of the appropriate components of the UK retail prices index, average earnings indices and price indices of various foreign countries, together with foreign currency exchange rates.

24.83   No attempt is made to deflate the other international flows – income, current and capital transfers and the financial account.

Chapter 25

Major sources for the accounts

Chapter 25 Major sources for the accounts

Summary

This chapter lists the main inquiries and administrative systems used as sources for the National Accounts.

It begins with a description of the Inter-departmental Business Register. This is followed by a table of the major inquiries, divided into five groups:

- Non-financial corporations – output and sales
- Non-financial corporations – other
- Financial corporations
- Households
- Rest of the world

Basic information is given on:

- population
- sample: frequency; whether voluntary or statutory; proportion of population represented
- information collected
- response rate
- calculations and adjustments
- further references.

Other sources are listed after the table.

Inter-Departmental Business Register

Background

25.1   The Inter-departmental Business Register (IDBR) was created by merging the Central Statistical Office business register with the Employment Department's Employment Statistics System. The Central Statistical Office register was based on traders registered for VAT with HM Customs and Excise. The Employment Statistics System was based on employers registered with the Inland Revenue Department for PAYE. Both VAT and PAYE information are used to update the IDBR. It has been designed in accordance with EU standards.

Coverage

The IDBR is a list of names and addresses of financial and non-financial corporations operating within the UK. The two sources cover all corporations with employees subject to income tax and those without employees but registered for VAT. (Exemptions from VAT registration include some health and education services; registration is voluntary for enterprises with turnover less than £49,000 a year.) The IDBR holds information on nearly 2m enterprises (see statistical units below), covering 98% of UK economic activity, excluding private households.

Statistical units

The administrative data on the IDBR are mapped on to three types of statistical unit, defined by EU regulation:

- Enterprise group – the group of all legal units under common control;

- Enterprise – the smallest autonomous group of legal units within an enterprise group.

- Local unit – the individual site or workplace where an activity takes place.

VAT registered traders are usually legal units (limited companies, sole proprietors, partnerships etc). PAYE employers may be parts of legal units. The enterprise is defined for purposes of the register as a grouping of VAT and PAYE units, operating one or more local units. For statistical purposes data are most usually collected at enterprise level. For some inquiries enterprises are split because they cover more than one industrial activity or more than one region.

Main data held on IDBR

The main data held are:

Administrative:
   registration number
   registration date
   deregistration date and reason

Enterprise group:
   group number
   country of ownership
   net assets (top 2000 groups only)
   turnover
   employment
   industrial classification

Enterprise:
   enterprise number
   turnover
   employment
   industrial classification
   legal status
   name and address

Local unit:
   local unit number
   employment
   industrial classification
   name and address.

Uses of IDBR

The IDBR is used as a sampling frame and source of general information for ONS surveys. It is also used by the Northern Ireland Department of Economic Development, the Department of the Environment, the Department of Trade and Industry, the Welsh Office and the Industry Department for Scotland. It permits improved co-ordination of government surveys, reducing the burden on respondents. The legal protection of data supplied by private corporations limits its use outside government departments.

References

Business monitor PA 1003: Size analyses of UK business

Business monitor PO1007: UK Directory of Manufacturing Businesses

Economic Trends, April 1992: The Inter-departmental Business Register

Economic Trends, November 1995: The Inter-departmental Business Register

Statistical News, Autumn 1995: Small and medium size enterprises - how many and how important?

EU Regulation no. 696/93 on the statistical units for the observation and analysis of the production system in the Union.

EU Regulation no. 2186/93 on Union co-ordination in drawing up business registers for statistical purposes.

25.1 Non-financial corporations – output and sales

25.1.1 Monthly production inquiries

Population	Sample	Information collected	Response rate	Adjustment for National Accounts	Further references
Non-financial corporations in production, except: gas and electricity; iron and steel; vehicles; some food and drink.	Monthly. Statutory. 199 industries, representing about 75% of manufacturing industry. All units with employment exceeding 50 or 150 (according to industry size). A stratified random sample of smaller units. Represents 65% of the population employment.	Turnover, exports and employment; engineering orders and export orders; merchanted goods.	80-83%	Imputations for non-responding units; grossing for non-sampled units. Industry totals estimated using the combined ratio estimator method. Adjustments for returns not in terms of calendar months. Data from the PPI inquiries used to deflate MPI data for index at constant prices. Prices collected at order stage. The home element of the deflator for engineering is lagged to reflect delay between orders and deliveries.	**Documentation** CSO Bulletin December 1992: Methodological Issues: Monthly Sales Inquiry.

25.1.2 Motor vehicle and engine production inquiries

Population	Sample	Information collected	Response rate	Adjustment for National Accounts	Further references
Vehicle and engine production corporations.	Monthly. Voluntary. 30 motor vehicle production corporations and 9 engine production corporations. Represents 95% of the population employment.	Numbers of cars produced, analysed by engine size; numbers of commercial vehicles produced, analysed by type of vehicle and gross weight; export numbers; car assembly and commercial vehicle assembly; holidays and production lost from industrial disputes. Numbers of petrol engines produced (analysed by size) and diesel engines produced; holiday information.	100%	No adjustments.	First Release: Motor vehicle production Business Monitor PM34.10: Car and commercial vehicle production.

25.1.3   Annual business inquiry into production

Population	Sample	Information collected	Response rate	Adjustment for National Accounts	Further references
Non-financial corporations in production and extraction	Annual. Statutory. All units with 200 or more employees. Stratified random sample of units with 10-199 employees. Less detailed questionnaire to units with 10-19 employees. Sub-sample of units with 20 or more employees for purchases information. Selected industries each year to cover all industries over 5-year period. Represents 73% of the population employment.	Employment, sales, work done, costs, duties and special leview, stock building and capital expenditure. Purchases information from sub-sample	88-90%	Estimates for non-responding units and for missing questions in shorter questionnaire. Sample results grossed.	News Releases: Annual Business Inquiry into Production Business Monitor PA 1002: Annual Business Inquiry into Production Summary Volume.

25.1.4   Prodcom

Prodcom is a harmonised Eurostat system of collecting product statistics required by EU regulation. It started in 1993. Eurostat maintains a list of products and associated codes. Member states are required to use this list as the basis for statistical inquiries into products sold, covering at least 90% of production in each NACE Rev1 Class

Population	Sample	Information collected	Response rate	Adjustment for National Accounts	Further references
Non-financial corporations in production.	Annual and quarterly. Statutory. All units with employment above a cut-off (varying with industry but never less than 20). Sample of smaller units. Represents 90% of employment in each NACE class.	Sales information; non-production activity (work done, merchanted goods, sales of work products etc.); total turnover.	80%	Imputation for non-response. Grossing for population estimates.	**Data** 1993-95: UK Markets annual and quarterly reports (published by Taylor Nelson AGB). 1996: Product sales and trade reports. **Documentation** Economic Trends, March 1993: Prodcom Economic Trends, February 1995: Measuring the pulse of the market: the Prodcom initiative.

25.1.5   Annual business inquiry into construction

Population	Sample	Information collected	Response rate	Adjustment for National Accounts	Further references
Non-financial corporations in construction.	Annual. Statutory. All units with 200 employees. Stratified random sample of units with less than 200 employees. All industries each year. Represents 39% of population employment.	Employment, sales, work done, costs, duties and special levies, stock building and capital expenditure.	85-88%	Estimates for non-responding units. Sample results grossed.	News Release: Annual Business Inquiry into Construction. Business Monitor PA500: Annual Business Inquiry into Construction

25.1.6   Direct labour organisations employment and output
Department of the Environment

Population	Sample	Information collected	Response rate	Adjustment for National Accounts	Further references
Direct labour organisations (DLOs) in local government and transport	Quarterly. Voluntary. All DLOs are included.	The value of housing and non-housing work, distinguishing new work from repair and maintenance; manpower, distinguishing professional, technical, clerical and operatives.	95%	Grossing for non-responding units.	Housing and Construction Statistics

25.1.7   Motor trades inquiry

Population	Sample	Information collected	Response rate	Adjustment for National Accounts	Further references
Non-financial corporations in Great Britain whose main activity is in the motor trades.	Annual. Statutory. All units with more than £15m turnover. Stratified random sample of smaller units. Represents 71% of population turnover.	Kind of business; turnover; stocks; employment costs; purchases; taxes and levies; capital expenditure.	83%	Grossing for non-selected and non-responding units.	Business Monitor SDA27: Motor trades. Business Monitor SDQ11: UK Service Sector **Documentation** Report of the Review of the Motor Trades Inquiry, 1993.

25.1.8 Annual retailing

Population	Sample	Information collected	Response rate	Adjustment for National Accounts	Further references
Non-financial corporations in retailing as primary or secondary activity.	Annual with 5-yearly benchmark surveys. Statutory. All units with turnover above £10m. Stratified random sample of smaller units. Represents 75% of population turnover.	Kind of business; turnover by commodity sold; number of outlets, employment costs; stocks, purchases; taxes and levies; capital expenditure.	80%	Grossing for non-selected and non-responding units.	Business monitor SDA25: Retailing. Business Monitor SDQ11: UK Service Sector **Documentation** Report of the Review of the Annual Retailing Inquiry, 1993

25.1.9 Annual wholesaling

Population	Sample	Information collected	Response rate	Adjustment for National Accounts	Further references
Non-financial corporations in wholesaling as primary activity (including commodity brokering, factoring, importing and exporting).	Annual. Statutory. All units with turnover above £30m. Stratified random sample of smaller units. Represents 73% of population turnover.	Kind of business, turnover, employment costs; stocks, purchases; taxes and levies; capital expenditure.	86%	Grossing for non-selected and non responding units.	Business Monitor SDA26: Wholesaling Business Monitor SDQ11 UK Service Sector. **Documentation** Report of the Review of the Annual Wholesaling Inquiry 1993.

25.1.10 Retail sales and commodity inquiries

Population	Sample	Information collected	Response rate	Adjustment for National Accounts	Further references
Non-financial corporations in retail trades.	Monthly. Statutory. All retailers with employment over 100. Stratified random sample of smaller retailers. Represents 60% of population in terms of employment. 66% in terms of turnover.	Turnover and employment. (Detailed data on commodity and credit provided collected from subsample.)	60%	Grossing for non-response. Population estimates from successive months.	First Release: Retail Sales Business monitor SDM28: Retail sales.
Non-financial corporations in retail trades.	Quarterly. Statutory. 80 retailers in relevant industry categories with turnover above £15m. Proportion of population not known.	Turnover by commodity.	70%	Grossing for non-response. Population estimates from successive quarters.	As above

25.1.11 Service trades, property and catering

Population	Sample	Information collected	Response rate	Adjustment for National Accounts	Further references
Non-financial corporations in road transport and ancillary transport services; communication services; renting; business services; professional and scientific services; health and education; personal miscellaneous services	Annual. Statutory. All units with turnover above £5m. Stratified random sample of smaller units. Represents 71% of population turnover.	Kind of business, turnover; employment costs; stocks, purchases; taxes and levies; capital expenditure.	85%	Grossing for non-selected and non-responding units.	Business Monitor SDA29: Service Trades Business Monitor SDQ11: UK Service Sector. **Documentation** Report of the Review of the Annual Service Trades Inquiry, 1995.
Non-financial corporations in catering and allied trades.	Annual. Statutory. All units with turnover above £3m. Stratified random sample of smaller units. Represents 56% of population turnover.	Kind of business; turnover; employment costs; purchases; capital expenditure; taxes and levies; stocks; whether accommodation offered.	80%	Grossing for non-selected and non-responding units.	Business Monitor SDA28: Catering and allied trades Business Monitor SDQ11: UK Service Sector. **Documentation** Report of the Review of the Annual Catering Inquiry, 1993
Non-financial corporations in property: investment, dealing, management.	Annual. Statutory. All units with turnover above £2m. Stratified random sample of smaller units. Represents 87% of population turnover.	Kind of business, turnover; purchases; employment costs; stocks; taxes and levies; capital expenditure.	83%	Grossing for non-selected and non-responding units.	Business Monitor SDQ11: UK Service Sector. **Documentation** Report of the Review of the Annual Property Inquiry, 1995.

25.1.12 Services turnover

Population	Sample	Information collected	Response rate	Adjustment for National Accounts	Further references
Non-financial service corporations in: wholesale distribution and dealing; transport and communication; business services; professional and scientific services; catering; motor trades; private health and education; personal services; cleaning; miscellaneous	Monthly and quarterly. Statutory. All units with employment over 100. Stratified random sample of smaller retailers. Represents 50% of population in terms of employment, 66% in terms of turnover.	Turnover and employment; value of commissions (for agents); purchases of media space and production services (for advertising agencies).	65%	Grossing for non-selected and non-responding units.	Quarterly News Release: Distribution and service trades. Business Monitor SDQ11: UK Service Sector.

25.2 Non-financial corporations - other

25.2.1 Airlines capital expenditure
Civil Aviation Authority

Population	Sample	Information collected	Response rate	Adjustment for National Accounts	Further references
Airlines	Quarterly. Statutory. 14 major airlines. Represents 97% of the population tonne kilometres available.	Capital transactions (mainly with aircraft and parts); income; expenditure.	100%	Estimates for non-aircraft capital expenditure.	Not published separately.

25.2.2 Capital expenditure

Population	Sample	Information collected	Response rate	Adjustment for National Accounts	Further references
Non-financial corporations in manufacturing; construction; other production; distribution; other services, excluding air-transport and shipping.	Quarterly. Statutory. All units with 300 or more employees. Sample of units with 20-299 employees. Units with less than 20 employees not yet in production. Represents 34% of the population employment.	Value of new building work; values of acquisitions and disposals of land, existing buildings, plant, machinery and other capital equipment; including finance leased assets.	75-80%	Net capital expenditure is calculated as acquisitions less disposals for land and existing buildings, vehicles and other capital equipment. Imputation for non-responders with 300 or more employees. Grossing for other strata and non-selected units. The value obtained for capital expenditure from companies not yet in production is added to the total. The current price values obtained are deflated to constant prices.	News releases: Capital expenditure. **Documentation** CSO Bulletin, Nov 1992; Methodological Issues - Capital Expenditure Inquiry.

25.2.3 Public corporations

Population	Sample	Information collected	Response rate	Adjustment for National Accounts	Further references
Public corporations.	Quarterly. Voluntary. Large public corporations. Percentage of population represented not known.	Sources and use of funds; borrowing; lending; other major transactions.	100%	Estimates for non-sampled smaller corporations.	Financial Statistics.

25.2.4 Stocks (inventories)

Population	Sample	Information collected	Response rate	Adjustment for National Accounts	Further references
Non-financial corporations in: Mining and Quarrying Manufacturing Electricity, Gas & Water Construction Motor Trades Wholesaling Retailing.	Quarterly. Statutory. All units in highest size-band. Stratified random sample of other size-bands with 20 or more employees. Represents 61% of the population employment. Voluntary panel provides monthly estimates for stockholding in computers, shipbuilding and aerospace. Represents 68% of the quarterly sample employment.	Values of stocks at beginning and end of period. Asset breakdowns appropriate to the industry, eg: Manufacturing: materials, stores, fuel, work in progress and finished goods. Construction: land, materials, fuel. Motor trades: new and used motor vehicles.	83-95%	Change in stock is value at end of period less value at start. Sample returns grossed. Imputation for non-responders in the 1:1 stratum. Non-responders in the other strata and non-selected are grossed. Book values deflated to constant prices and reflated to current price estimates.	News release: Stocks **Documentation** CSO Bulletin 84/91. November 1991: Methodological Issues - Production Stocks Inquiry.

25.2.5 Profits and financial assets

Population	Sample	Information collected	Response rate	Adjustment for National Accounts	Further references
Non-financial corporations apart from those in exploration for oil or natural gas production.	Quarterly. Statutory. Stratified sample of company groups with more than 20 employees. Represents 80% of population profits.	Pre-tax trading profit or loss; depreciation; rent income; exceptional and other large items.	83%	Imputation for non-responders and grossing for non-sampled units.	Financial Statistics.
Corporations holding financial instruments.	Quarterly. Statutory. Sample of company groups with over 2000 employees and smaller companies known to be major holders of financial assets. Proportion of population not known.	Levels of holdings of financial instruments.	86%	Imputation for non-responders and grossing for non-sampled units.	Financial Statistics.
Corporations holding financial instruments.	Annual. Statutory. Companies with over 2000 employees not included in quarterly sample.	Levels of holdings of financial instruments.	72%	Imputation for non-responders and grossing for non-sampled units.	Financial Statistics.

25.2.6   Trading profits
Inland Revenue

Population	Sample	Information collected	Response rate	Adjustment for National Accounts	Further references
Non-fincnaial corporations except those exploring for or producing oil and natural gas (the North Sea oil companies).	All corporations provide information by law. An annual sample is taken of about 13% of the population: all companies with trading profit or loss above £0.5m; 10% of the rest.	Trading profit or loss.	95% of sampled returns.	Grossing to provide UK estimate.	Inland Revenue Statistics.

25.2.7   Producer prices
The producer price indices measure the price movements of goods and services bought and sold by UK industries.

Population	Sample	Information collected	Response rate	Adjustment for National Accounts	Further references
Non-financial corporations in manufacturing.	Monthly. Statutory. Around 10k items from around 3k contributors. Represents nearly all manufacturing classes. Sample represents 44% of population.	Date and amount of price change.	95%	Weighting, determined by base year, changed every 5 years. Imputations for non-respondents.	First release: Producer prices Business Monitor MM22: Producer Price Indices. **Documentation** Economic Trends, No. 465, July 1992: Producer Price Indices - present practice, future developments and international comparisons.
Non-financial corporations exporting from GB.	Monthly. Statutory. All exporters above £1m a year in selected headings. Sample represents 30% of population.	Date and amount of price change.	95%	Weighting, determined by base year, changed every 5 years. Imputations for non-respondents.	**Documentation** Economic Trends April 1997: Deflation of trade in goods statistics..
Non-financial service corporations.	Quarterly. Statutory. About 1500 items are collected from 500 contributors covering 9 service industry headings. Represents over 30% of turnover in industries covered. To be expanded to cover most of corporate service sector by 1998.	Date and amount of price change.	85%	Weighting, determined by base year, changed every 5 years. Imputations for non-respondents.	**Documentation** Economic Trends, July 1996: Producer prices for services: development of a new price index.

25.2.8   Energy

Population	Sample	Information collected	Response rate	Adjustment for National Accounts	Further references
Electricity producers.	Quarterly. Voluntary. All companies generating 1 gigawatt hour or more per year. Represents 5% of population.	Capacity generated.	96%	No adjustment.	**Data** Energy Trends Annual Digest of Energy Statistics.
Electricity producers.	Annual. Voluntary. All companies generating 1 gigawatt hour or more per year. Represents 5% of population.	Capacity generated.	81%	No adjustment.	**Data** Energy Trends Annual Digest of Energy Statistics.
Corporations in oil and natural gas production and exploration.	Quarterly. Statutory. All licensees, contractors and agents in the industry.	Sales: purchases; expenditure.	92%	Estimates for non-respondents.	**Data** Energy Trends Annual Digest of Energy Statistics.
Fuels Inquiry					
Corporations in production.	Quarterly. Statutory. Stratified sample of 1200 from Annual Business Inquiry into Production. Represents 1% of population.	Quantities and cost of fuels purchased (oil, coke, gas, electricity).	90%	Estimates for non-respondents.	**Data** Energy Trends Annual Digest of Energy Statistics

25.2.9   Share register survey

Population	Sample	Information collected	Response rate	Adjustment for National Accounts	Further references
Registered companies with a full listing on the London Stock Exchange.	Annual. Voluntary. 200 companies selected with probability proportionate to market capitalisation. Represents 75% of the total value of quoted companies.	Value of quoted shares owned analysed by sector. Geographical breakdown of country of residence of Rest of the World shareholders. Analysis of shareholders in privatised companies and in companies in the Financial Times-Stock Exchange list of 100: industry of issuing company; size of shareholding.	99-100%	A sample of each company's shareholdings are analysed by National Accounts sector of the beneficial owner. Grossing is applied to provide estimates for the whole population.	Share ownership reports in Economic Trends

25.2.10 Capital issues

Bank of England database of security issues; London Stock Exchange data on listed securities.

Population	Sample	Information collected	Response rate	Adjustment for National Accounts	Further references
Market capital issues by UK resident corporations.	Continuous record. Stock Exchange database is complete record of listed securities. Bank of England capital issues database (CIDB) derived from press reports. Coverage not known exactly but considered comprehensive.	Name and sector of issuing company; issue data; redemption date; type of security; coupon, nominal value and value at issue. CIDB also records market on which security issued and value of further money raising issues. Stock Exchange database also records market values and prices of securities.	Not applicable.	None.	Bank of England: Capital Issues Statistical Release London Stock Exchange: Quality of Markets Monthly Fact Sheet; Stock Exchange Quarterly.

25.3 Financial corporations

25.3.1 Banks

Bank of England. Under the Banking Act 1987 authorised institutions are obliged to provide any statistical information the Bank requires in order to fulfil its responsibilities under the Act.

Population	Sample	Information collected	Response rate	Adjustment for National Accounts	Further references
Banks	Monthly and quarterly. Statutory. Quarterly: all banks. Monthly: banks with total footings of £300m+ or eligible liabilities of £30m+. Proportion of population not known.	Balance sheet.	100%	Flows derived from balance sheet levels. Adjusted for exchange rate changes and reclassifications.	Bank of England Monetary and Financial Statistics Bank of England Statistical Abstract Financial Statistics
Banks	Monthly. Statutory. Footings of £600m+ of eligible liabilities of £60m+. Proportion of population not known.	Further sectoral analysis of balance sheet.	100%	As above.	As above.
Banks	Quarterly. Statutory. Size criteria vary for sections of form. Proportion of population not known.	Further sectoral analysis of balance sheet; transactions in capital issues and investments; capital expenditure; direct investment.	100%	As above, as applicable.	As above.
Banks	Quarterly and annual. Statutory. (i) sample to be dertermined (ii) cut-off as for balance sheet survey; less for banks with large relevant business. Proportion of population not known.	Income and expenditure (i) total (ii) rest of world only	100%	Grossed to represent all banks.	Not published.

25.3.1   Banks - *cont...*

Population	Sample	Information collected	Response rate	Adjustment for National Accounts	Further references
Banks	Quarterly. Statutory. Footings of £600m+ or eligible liabilities of £60m+. Proportion of population not known.	Industrial analysis of deposits and lending.	100%	Flows derived from balance sheet levels. Adjusted for exchange rate changes and reclassifications.	Bank of England Statistical Abstract Financial Statistics
Banks	Quarterly. Statutory. Cut-off as for balance sheet survey; less for banks with large relevant business. Proportion of population not known.	Portfolio investment transactions for balance of payments.	100%	Grossed to represent all banks.	Not published.
Banks	Quarterly. Statutory. External business of £100m+ in foreign currency/ £20m+ in sterling. Proportion of population not known.	Country analysis of UK external liabilities and claims.	100%	Grossed to represent all banks.	Bank of England Statistical Abstract.
Banks	Quarterly. Statutory. All custodian banks.	Country analysis of custody holdings for non-residents.	100%	No adjustment.	Not published.
Banks	Annual. Statutory. All banks.	Geographical analysis of interest, dividend and other income flows from non-residents.	100%	No adjustment.	Not published.
Banks	Annual. Statutory. All banks.	Direct investment levels.	100%	No adjustment.	Bank of England Statistical Abstract.
Banks	Quarterly. Statutory. Sample of 59. Represents 10% of banks active in derivatives trade.	Analysis of derivatives issued/held.	Not known.	Grossed to represent all banks.	Not published.

25.3.2 Building societies

Building Societies Commission. Under the Building Societies Act 1986 building societies are obliged to submit returns to the Building Societies Commission.

Population	Sample	Information collected	Response rate	Adjustment for National Accounts	Further references
Building societies	Monthly. Statutory. All societies.	Balance sheet; analysis of changes in retail funding and commercial assets.	100%	Flows are derived from balance sheet levels. Adjusted for population changes.	Financial Statistics.
Building societies	Quarterly. Statutory. 40 largest societies. Represents 50% of population number.	Further sector analysis of funding and commercial assets.	100%	As above.	As above.
Building societies	Annual. Statutory. All societies.	Income and expenditure.	100%	No adjustment.	Not published.

25.3.3 Other financial corporations

Population	Sample	Information collected	Response rate	Adjustment for National Accounts	Further references
Insurance companies	Quarterly. Statutory. All companies over £100m premiums; some smaller companies. Represents 90% of population premiums.	Income and expenditure. Transactions in financial assets.	Income and expenditure: 95% Transactions in financial assets: 85%	Grossing for non-sampled units and non-response.	First Release: Institutional Investment Business Monitor MQ5: Insurance companies' and pension funds' investment. Financial Statistics.
Insurance companies	Annual. Statutory. All companies over £100m premiums; some smaller companies. Represents 90% of population premiums.	Income and expenditure. Balance sheet.	85%	As above.	As above.
Insurance companies with overseas direct investment or overseas subsidiaries	Annual. Statutory. All companies in population.	Transactions with overseas.	90%	Estimates of non-responders.	First Release: Overseas Direct Investment Business Monitor MA4: Overseas direct investment.

25.3.3　Other financial corporations - *cont...*

Population	Sample	Information collected	Response rate	Adjustment for National Accounts	Further references
Pension funds	Quarterly. Statutory. All largest pension funds and sample of smaller ones. Represents 50% of population market value.	Income and expenditure. Transactions in financial assets.	78%	Grossing for non-response and non-sampled units.	First Release: Institutional Investment Business Monitor MQ5: Insurance companies' and pension funds' investment. Financial Statistics.
Pension funds	Annual. Statutory. As above.	Income; expenditure. Balance sheet.	85%	As above.	As above.
Unit trusts	Quarterly. Voluntary. All unit trusts with over £2m assets. Represents 69% of population market value.	Income and expenditure; transactions in assets.	70%	As above.	Institutional Investment First Release. Financial Statistics UK Balance of Trade.
Unit trusts	Annual. Voluntary. All unit trusts with over £2m assets. Represents 50% of population market value.	Assets and liabilities; transactions in financial instruments.	69%	As above.	As above.
Investment trusts	Quarterly. Voluntary. All investment trusts with over £2m assets. Represents 54% of population market value.	Income and expenditure; transactions in assets.	55%	As above.	As above.
Investment trusts	Annual. Voluntary. All investment trusts with over £2m assets. Represents 49% of population market value.	Holdings, assets, liabilities.	59%	As above.	As above.

25.3.3 Other financial corporations - *cont...*

Population	Sample	Information collected	Response rate	Adjustment for National Accounts	Further references
Property unit trusts	Quarterly. Voluntary. All property unit trusts with over £2m assets. Represents 53% of population market value.	Income and expenditure; transactions in assets.	80%	As above.	As above.
Property unit trusts	Annual. Voluntary. All property unit trusts with over £2m assets. Represents 30% of population market value.	Holdings, assets, liabilities.	80%	As above.	As above.
Security dealers.	Quarterly. Statutory. 50 largest contributors to Securities and Futures Association statistics. Represents 80% of population turnover.	Income and expenditure; assets, liabilities; transactions in financial instruments.	90%	Grossed to Securities and Futures Authority return totals.	As above.
Credit Grantors.	Monthly. Statutory. 30 largest businesses. Represented 80% of consumer credit market at time of selection.	New credit advanced and credit outstanding.	96%	Grossing for non-response and non-sampled units.	Financial Statistics Economic Trends.
Non-bank credit grantors.	Quarterly. Voluntary. 7 largest businesses. Covered 50% of credit business at time of selection.	Holdings; assets; liabilities.	100%	Grossing for non-sampled units.	Business Monitor SDQ7 Financial Statistics
Finance lessors	Quarterly and annual. Fixed panel of 270 companies. Represents 85% of population number.	Capital expenditure and balance sheet data (quarterly). Income and expenditure (annual).	80%	Estimation for non-responders; grossing for non-responding and non-sampled units.	Not published elsewhere.

25.3.3   Other financial corporations - *cont...*

Population	Sample	Information collected	Response rate	Adjustment for National Accounts	Further references
Commodities traders	Annual. Statutory. All commodities traders.	Receipts from commodities trading.	Not known.	Estimates for non-response.	UK Balance of Payments.
Fund managers not covered by other surveys.	Annual. Voluntary. All population.	Income and expenditure.	75%	Estimates for non-response.	UK Balance of Payments.

25.3.4   Miscellaneous financial institutions

Bank of England

Population	Sample	Information collected	Response rate	Adjustment for National Accounts	Further references
Mortgage finance vehicles; specialist finance agencies and export credit vehicles; non-bank institutions mainly extending credit abroad; non-bank gold dealers.	Quarterly. Voluntary. Whole population.	Liabilities and assets at book value; transactions in capital issues and investments.	100%	No adjustments.	Financial Statistics. UK Balance of Payments.

25.4 Households

25.4.1 Earnings and employment

New Earnings Survey

Population	Sample	Information collected	Response rate	Adjustment for National Accounts	Further references
Employees	Annual. Statutory. Sent to employers. Represents 1% of employees.	Gross pay and hours worked, analysed by occupation, industry, geographical area and age.	94%	Grossing to estimate for whole population.	New Earnings Survey. Labour Market Trends.

Wages and Salaries Survey

Population	Sample	Information collected	Response rate	Adjustment for National Accounts	Further references
Employees	Monthly. Statutory. 8000 employers. Represents 40% of employees.	Gross pay and its composition; employee numbers for weekly and monthly paid.	97%	Weighted average wage calculated for each industrial group. Indices formed by calculating change in average from base year (currently 1990). Averages and indices aggregated for whole economy (represent 90% of population).	Labour Market Trends.

Annual Employment Survey

Population	Sample	Information collected	Response rate	Adjustment for National Accounts	Further references
Employees.	Annual. Statutory. 130k businesses covering 1m workplaces. All single-workplace businesses with 50+ employees and all multi-workplace businesses with 25+ are surveyed each year; others less regularly. Proportion of population represented not known.	Number of employees by sex, full-time or part-time status and location.	95%	Grossing is applied to cover for non-sampled units. Additional information from Ministry of Agriculture Fisheries and Foods and the Department of Agriculture and Fisheries for Scotland is used for agricultural sector estimates. Northern Ireland employment census data are added to provide estimate for the UK.	Booklets on Annual Employment Survey: 1. Results for GB 2. Results for counties and local authority districts. 3. Other analyses. Labour Market Trends (Annual article).

25.4.1 Earnings and employment - *cont...*

Short-term Turnover and Employment Surveys

Employment questions are attached to the Monthly Production Inquiries (25.2.1) and the quarterly Services Turnover inquiries (25.2.11).

Population	Sample	Information collected	Response rate	Adjustment for National Accounts	Further references
Employees in production.	Monthly. Statutory. All units above an employment threshold (normally 150 but lower in some industries). Stratified random sample of smaller units. Whole sample represents 65% of population employment.	Number of employees (analysed by sex quarterly). Industry.	95%	Estimates of movement are derived by region and industry. These are applied to the benchmark results of the Annual Employment Survey.	Labour Market Trends.
Employees in distribution and services.	Quarterly. Statutory. All units above 100 employment threshold. Stratified random sample of other units. Whole sample represents 50% of population employment.	Number of employees. Industry.	95%	As above.	Labour Market Trends.

25.4.2 Survey of personal incomes

Inland Revenue

Population	Sample	Information collected	Response rate	Adjustment for National Accounts	Further references
Individuals	Annual. Statutory. Information held by Inland Revenue tax offices. Stratified sample of 0.3% of population.	Income from employment, self-employment, pensions and benefits; deductions and reliefs.	96%	Grossed to represent population.	Inland Revenue Statistics.

25.4.3   Expenses and benefits
Inland Revenue

Population	Sample	Information collected	Response rate	Adjustment for National Accounts	Further references
Employees	Annual 9000 cases from Survey of Personal Incomes, biased towards those with expenses and benefits. Represents 0.05% of population.	Taxable benefits in kind and expenses (gross and net values), expense dispensations.	92% of sample matched with forms from Survey of Personal Incomes.	Grossed to represent population.	Inland Revenue Statistics.

25.4.4   National insurance certificates
Inland Revenue

Population	Sample	Information collected	Response rate	Adjustment for National Accounts	Further references
Employees	Annual. Sample of PAYE end of year documents. Represents 1% of population.	Pay, tax and National Insurance contributions.	95%	Grossed to national totals.	Not published separately.

25.4.5   Family Expenditure Survey

Population	Sample	Information collected	Response rate	Adjustment for National Accounts	Further references
Private households.	Continuous. Voluntary. 11,000 households a year. 0.05% of population.	For each individual over 16: age, sex, social status, economic status, occupation, details of expenditure on goods and services (recorded in a two-week diary), income.	66% for Great Britain; 60% for Northern Ireland.	Quarterly estimates of average household expenditure grossed for whole population. Results smoothed because sampling error leads to exaggerated period-to-period changes in average expenditure.	Family Spending, a Report on the Family Expenditure Survey. **Documentation** Statistical News no.106. Autumn 1994: The Family Expenditure Survey - some recent developments. The Family Expenditure Survey: Report on the first stage of the 1991 Census-linked study of survey non-respondents, June 1994.

25.4.6 National Food Survey

Ministry of Agriculture, Fisheries and Food

Population	Sample	Information collected	Response rate	Adjustment for National Accounts	Further references
Households in Great Britain.	Continuous. Voluntary. 12000 households a year. 0.05% of the population.	Description, quantity and cost of food entering the home during one week; the number of people present at each meal and type of meal served; characteristics of the household and its members; information on food and drink consumed outside the home (from sub-sample of half the main sample).	66% for main sample; 65% for sub-sample.	Grossing to provide estimate for consumption and expenditure per calendar quarter by UK civilian population.	National Food Survey.

25.4.7 Retail prices index

The general index of retail prices (RPI) measures the average change from month to month in the prices of goods and services in the UK.

Transactions by households.	Monthly. Voluntary. Sample of goods and services typical of most families. 120k price quotations for 600 goods and services. Shops visited in 180 areas of UK. 100 of 600 items collected centrally.	Market price including VAT.	100%	Components weighted to reflect different parts of the country and different types of shop. The index is chain-linked by linking average price changes for current year with previous year.	First Release: Retail Prices Index Business Monitor MM23: Retail Prices Index.

25.5 Rest of the World

25.5.1 Overseas direct investment

Population	Sample	Information collected	Response rate	Adjustment for National Accounts	Further references
Insurance brokers.	Quarterly. Voluntary. Sample of 40 insurance brokers. Represents 90% of the population investment overseas.	Total earnings in various currencies.	90%	Grossing for non-selected and non-responding units.	First Release: Balance of Payments UK Balance of Payments Financial Statistics.
Non-financial and non-monetary financial institutions with direct investment overseas.	Quarterly. Statutory. All oil companies with overseas direct investment. Stratified random sample of others (a subset of the annual sample). Represents 80% of population investment overseas.	Profit; interest; dividends; transactions in share and loan stock; inter-company account balances.	90%	Estimation for non-responding and non-sampled units known to have direct investment links. Estimates of exchange rate gains and losses.	First Release: Balance of Payments UK Balance of Payments Financial Statistics.
Non-financial and non-monetary financial institutions with direct investment overseas.	Annual. Statutory. All oil and insurance companies with overseas direct investment. Stratified random sample of others. Represents 95% of population investment overseas.	Profit; interest; dividends; transactions in share and loan stock; inter-company account balances; exchange rate gains/losses.	92%	Estimation for non-responding and non-sampled units known to have direct investment links.	First Release: Overseas Direct Investment. Business Monitor MA4: Overseas Direct Investment UK Balance of Payments Financial Statistics.

25.5.2 Overseas trade in services

Population	Sample	Information collected	Response rate	Adjustment for National Accounts	Further references
Non-financial corporations.	Quarterly. Statutory. Large traders. (The cut-off varies for different industries. For the larger industries it is £5m or more in overseas transactions). The proportion of the population represented is not known.	Receipts and payments, analysed by type of service involved and a geographical breakdown of the overseas transactions.	77%	Estimates for non-responders. Grossed to annual level.	First Release: Overseas earnings from royalties and services. First Release: Overseas transactions of UK consultancy firms. UK Balance of Payments Financial Statistics.
Non-financial corporations.	Annual. Statutory. All known traders and sample of others stratified by employment, turnover or overseas transactions. The proportion of the population represented is not known.	Receipts and payments, analysed by type of service involved and a geographical breakdown of the overseas concerns.	80%	Estimates for non-responders. Grossed to population.	As above.
Non-financial corporations.	Annual. Statutory. Known traders in films, television and other business services. The sampling method is under review.	Receipts and payments analysed by type of service involved and geographical breakdown.	70%	Estimates for non-responders.	News release: Overseas transactions of the film and TV industry. UK Balance of Payments Financial Statistics.

25.5.3   Overseas trade statistics

Intrastat
EU system for compiling statistics of trade in goods between member states.

Population	Sample	Information collected	Response rate	Adjustment for National Accounts	Further references
VAT registered companies trading in goods between member states of European Union.	Monthly and quarterly. Statutory. Returns (Global Declarations) from VAT registered businesses, usually quarterly. Detailed monthly declarations from traders whose arrivals and/or dispatches exceed threshold reviewed annually. Monthly sample accounts for 97% of UK trade with other EU states.	Global Declarations show total value of arrivals and dispatches in trade with other EU states. Detailed monthly declarations show commodity codes, statistical value, quantity; delivery terms, nature of transaction, member state traded with, country of import origin; mode of transport.	90%	For traders below the threshold for monthly returns estimates are made by commodity and country traded with.

Adjustments are made to overseas trade statistics in accordance with the IMF Balance of Payments Manual:

Additions are made to both despatches and arrivals in respect of recording errors, second-hand ships sold or purchased whilst abroad, and gold not treated as a financial asset; and to arrivals in respect of ships built and delivered abroad. Deductions are made from both despatches and arrivals in respect of revisions to the non-response estimates; and from arrivals in respect of freight and insurance. | Business Monitor MM20A: Overseas Trade Statistics of the United Kingdom with the World (including data for countries within the European Union: Intrastat).

Business Monitor MM24: Monthly Review of External Trade Statistics.

Business Monitor MQ20: Overseas Trade Statistics of the United Kingdom with countries within the European Union (Intra-EU) Trade: Intrastat.

UK Balance of Payments.

Documentation
European Regulations 3330/91, 2256/92, 3046/92, 3590/92.

Economic Trends, no. 471, January 1993: Intrastat.

Economic Trends, no. 490, August 1994: UK visible trade statistics - the Intrastat system. |

25.5.3 Overseas trade statistics - *cont...*

Extra-EU trade

Compiled from HM Customs and Excise Import and Export entries.

Population	Sample	Information collected	Response rate	Adjustment for National Accounts	Further references
Customs import and export declarations.	Collated monthly. All importers and exporters or the authorised agents registering completing customs entries (Single Administrative Document).	Commodity code, value, net mass, supplementary units, country of destination, dispatch and origin, port, flag and mode of transport.	100% of customs declarations.	Adjustments are made to overseas trade statistics in accordance with the IMF Balance of Payments Manual: Additions are made to both imports and exports in respect of second-hand ships sold or purchased whilst abroad, and gold not treated as a financial asset; to exports in respect of recording errors; and to imports in respect of ships built and delivered abroad. Deductions are made from both imports and exports in respect of goods not changing ownership; from exports in respect of foreign currency declarations; and from imports in respect of freight and insurance.	Business Monitor MA20: Overseas Trade Statistics of the United Kingdom. Business Monitor MM20: Overseas trade statistics of the UK with countries outside the EU. Business Monitor MM24: Monthly Review of External Trade Statistics. UK Balance of Payments. Documentation European Regulations 1172/95, 840/96, 2454/93 annexes 37 and 38.

25.5.4 International Passenger Survey

Population	Sample	Information collected	Response rate	Adjustment for National Accounts	Further references
Passengers entering or leaving the UK.	Continuous. Voluntary. At major airports sample of half days; fixed proportion of passengers. Minor airports sampled occasionally. Sea-routes covered by samples of sailings or by samples of days. 0.2% of population sampled.	Nationality; country of residence; country of origin/destination; reason for the visit; amount of expenditure during the visit; fare paid; length of visit.	84%	Weighting based on traffic flows and non-response at different periods and for different airports and sea-routes. Data on travel to and from the Irish Republic from the Irish Central Statistics Office. Estimates for total travel to and from the UK.	Business Monitor MQ6: Overseas travel and tourism International Migration Travel Trends.

Notes to table

- **Inquiries** are carried out by ONS unless otherwise specified.

- The **population** is that enumerated in the National Accounts. It relates to the UK unless otherwise specified.

- **Sample** details reflect the current situation. The proportion of the population represented by the sample is given approximately, based on recent data, in any convenient terms (e.g. number of units, turnover, employment).

- The **response rate** given is the approximate average over the last year or so. It relates to final results.

- **References** are labelled as "documentation" or "data". Where there is no indication the reference may be used for both. *Financial Statistics* should be taken to include the *Financial Statistics Explanatory Handbook*.

Other Inquiries

a. **The Baltic Exchange:** trade in financial services.

b. **The Chamber of Shipping** inquiry: the international earnings of the UK shipping industry.

c. **The Department of the Environment's** Construction Market Intelligence and Housing Data and Statistics: building stock and capital expenditure; surveys of construction output, prices and house condition.

d. **The Department of Social Security** Newcastle survey of administrative records: income data.

e. **The Department of Trade and Industry** product reporting system for oil stocks: reservoir performance and oil and natural gas stocks.

f. **The Department of Transport:** inquiry into shipowners' capital expenditure.

g. **Education funding councils and Independent Schools Information Survey:** capital spending in education.

h. **Forestry Commission:** administrative data on expenditure.

i. **The Government Actuary's Department:** pensions and social security contributions.

j. **HM Customs and Excise:** VAT data.

k. **HM Land Registry:** capital expenditure on land.

l. **HM Treasury:** Government Expenditure Monitoring System (GEMS) - expenditure by government departments.

m. **HM Treasury:** Public Expenditure Survey: Financial year expenditure by central and local government.

n. **The Law Society: overseas earnings of the legal professions.**

o. Lloyds: **sea transport and insurance data.**

p. London Metal Exchange: **metals stocks.**

q. The Ministry of Agriculture Fisheries and Food **Farm Business Survey and other sources: capital expenditure, stocks and trade in the agricultural sector.**

r. The Ministry of Defence: **armed services pay.**

s. The Office for National Statistics: **ad hoc inquiries into charities and residential nursing homes.**

Chapter 26

Calendar of economic events

Chapter 26 Calendar of economic events

1944

Jul   Bretton Woods agreement
Aug   White Paper on Employment Policy

1945

May   End of war in Europe
Jul   General Election won by Labour Party
Aug   Japan surrenders
Dec   Anglo-American financial agreement

1946

Jan   Bank of England nationalised
Jul   US Congress ratifies loan to UK

1947

Feb   Fuel crisis
Jul   Sterling made convertible
Aug   Convertibility suspended

1948

Feb   Wage freeze and dividend restraint
Apr   European Co-operation Act takes effect
Apr   Organisation for Economic Co-operation
      in Europe (OEEC) founded
Jul   Berlin airlift
Jul   Marshall Aid initiated

1949

Sep   Pound devalued from $4 to $2.80

1950

Feb   Labour returned to power at General
      Election
May   Announcement of Schumann Plan
Jun   Outbreak of Korean war
Jul   Agreement on European Payments Union
Dec   Marshall Aid suspended

1951

Oct   Conservative Party wins General Election
Nov   Monetary policy reinstituted
Nov   Record drain on UK's dollar reserves
Dec   London foreign exchange market
      re-opened

1952

Jan   Commonwealth finance ministers'
      conference

1953

Mar   Steel industry denationalised

1954

Jul   Food rationing ends

1955

Feb   Sterling made transferable
Apr   Conservatives increase majority at Election
Jul   Beginnning of credit squeeze
Oct   Balance of Payments crisis and Budget

1956

Oct   UK announces plan for European Free
      Trade Area (EFTA)
Nov   Suez: Anglo-French invasion of Egypt
Dec   Balance of Bayments crisis

1957

Aug   'Three Wise Men' appointed
Sep   Sterling crisis: bank rate raised to 7%
Dec   Sterling made convertible for non-residents

1958

Jan   Inauguration of European Economic
      Community (EEC)
Dec   Convertibility for European currencies

1959

Aug	Radcliffe Report on monetary system
Oct	Conservatives increase majority at Election
Nov	EFTA convention signed in Stockholm

1960

Apr	Special deposits and HP controls re-imposed
Sep	Organisation of Petroleum Exporting Countries (OPEC) founded
Dec	OEEC becomes OECD (Organisation for Economic Cooperation and Development) with the accession of US and Canada

1961

Apr	Graduated state pension scheme introduced
Jul	Plowden Report on control of public expenditure
Jul	Balance of Payments crisis; Mini-Budget applies 'regulator' for the first time; 'pay pause' introduced in public sector
Aug	UK applies for membership of EEC

1962

Mar	End of 'pay pause'
Apr	Budget ends tax on owner-occupation
Jul	UK discharges all debts to IMF
Oct	Release of Post-War Credits and removal of ceiling on bank advances
Nov	Purchase Tax on cars cut from 45% to 25%
Dec	Beginning of three-month cold spell

1963

Jan	Purchase Tax on TVs cut from 45% to 30%
Jan	UK application to join EEC rejected
Apr	Expansionary Budget: Income Tax cut
Oct	Harold Macmillan resigns as Prime Minister
Nov	US President John F. Kennedy assassinated

1964

Mar	'Beeching' cuts in railway system agreed
Jul	Resale price maintenance abolished
Oct	Labour Party wins General Election

Oct	Import surcharge and export rebates introduced to cut Balance of Payments deficit
Nov	Budget raises taxes and introduces new Corporation Tax

1965

Mar	Prices and Incomes Board established
Apr	Tax-raising Budget
Jun	Public spending cuts; tighter HP restrictions
Sep	National Economic Plan published
Nov	Rhodesia declares unilateral independence

1966

Mar	Move to decimal currency announced
Mar	Labour increases majority at Election
Jul	Sterling crisis: tax rises and credit restraint
Oct	Statutory pay policy initiated
Nov	Import surcharge abolished
Dec	EFTA internal tariffs abolished

1967

Mar	First landing of North Sea gas
May	UK applies to join EEC for the second time
Jun	Yom Kippur War cuts oil supplies
Jun	World's first cash dispenser installed
Nov	Pound devalued from $2.80 to $2.40

1968

Jan	£700m cuts in public spending
Mar	Two-tier gold market established
Mar	Budget raises £923m in taxes
Jul	Kennedy Round of tariff cuts implemented
Jul	Barclays Bank takes over Martin's
Aug	Soviet Union invades Czechoslovakia

1969

Apr	Budget raises £345 m in new taxes
Jul	First man on the Moon
Oct	Department of Economic Affairs disbanded

1970

Jun	Conservatives win General Election
Oct	Expenditure cuts of £330m linked to tax cuts

1971

Feb	Changeover to decimal currency
Mar	Budget announces introduction of VAT in 1973
Aug	US ends dollar-gold convertibility and thereby the era of fixed exchange rates initiated at Bretton Woods; sterling floated
Dec	General currency realigment following Smithsonian Agreement

1972

Feb	Power crisis due to strikes by coalminers and electricity supply workers
Sep	Anti-inflation programme announced including 90-day freeze on prices and incomes and establishment of Prices Commission and Pay Board

1973

Jan	UK joins EEC
Mar	European currencies floated against sterling
Apr	Value Added Tax (VAT) introduced
Oct	Outbreak of Middle East War followed by cut in oil supplies and quadrupling of oil prices
Oct	Further anti-inflation measures ('Phase 3') announced
Dec	Miners' overtime ban leads to announcement of three-day week for industry

1974

Feb	Beginning of full-scale miners' strike
Feb	General Election results in hung Parliament
Mar	Labour Government formed
Mar	Miners resume work and three-day week ends
Oct	General Election gives Labour a working majority
Dec	OPEC raises oil prices again and announces new unitary price system

1975

Jan	First index-linked National Savings schemes announced, for pensioners and small savers
Apr	Budget designed to cut Balance of Payments deficit and reduce public sector borrowing requirement

1976

Apr	Budget announces tax cuts contingent upon TUC agreement to low pay norm for Stage 2 of incomes policy
Jul	Public expenditure cuts of £1,000m announced
Sep	$4 billion standby credit sought from IMF
Oct	Pound falls to $1.57 – lowest ever
Dec	Further £1,000m of cuts announced as part of deal with IMF

1977

Jan	Chancellor announces intention to reduce £'s role as international reserve currency
Feb	Measures to encourage use of foreign currency for financing exports
Mar	Lib-Lab pact agreed
Jun	Additional bank holiday to celebrate Queen's silver jubilee
Aug	Unemployment peaks at 1.6 million
Sep	Government seeks voluntary 10% limit on earnings increases

1978

Jan	Official reserves reach $20.6 bn - highest ever
Feb	12-month increase in RPI is in single figures for the first time since 1973
May	Bank of England abandons market-related formula for determining MLR

1979

Jan	Cash limits on public spending introduced
Feb	Concordat between Government and TUC aimed at reducing inflation to 5% in 3 years
Mar	European Monetary System begins to operate (2 months late) without the UK
May	Conservative Party wins General Election; Margaret Thatcher becomes Prime Minister
Jun	Budget cuts Income Tax and raises VAT
Oct	Exchange controls are abolished

1980

Jan	Steel strike begins
Mar	Medium Term Financial Strategy announced
Jun	Britain becomes a net exporter of oil
May	Mount St Helens volcano erupts
Jun	Sixpence (2¹/₂p) piece discontinued
Jun	Agreement to reduce UK's budget contribution to EEC
Oct	Dollar exchange rate peaks at $2.39 per £
Nov	Ronald Reagan elected US President

1981

Jan	Bottom of worst post-War slump in Britain
Feb	*The Times* sold to Rupert Murdoch
Mar	Budget announces windfall tax on banks
Apr	Rioting in Brixton
Apr	Social Democratic Party founded
Jul	Cuts in university spending announced
Jul	Rioting in Toxteth
Jul	Prince of Wales marries Lady Diana Spencer
Aug	Minimum Lending Rate (MLR) suspended
Dec	Heavy snow causes chaos

1982

Feb	Laker Airlines collapses
Mar	British naval task force sent to Falklands
Jun	Ceasefire in Falklands
Jul	Hire purchase controls abolished
Aug	Barclays Bank starts opening on Saturdays
Sep	Unemployment reaches 3 million
Nov	Channel 4 Television begins transmission

1983

Apr	Pound coin issued for the first time
Jun	£450m EC budget rebate granted to UK
Jul	£500m public spending cuts announced
Sep	3% target set for public sector pay
Oct	European Parliament freezes budget rebate

1984

Mar	Miners' strike begins
Jun	Robert Maxwell buys *Daily Mirror*
Jun	Fontainebleau Summit agrees permanent settlement of UK's contribution to EEC

Oct	Bank of England rescues Johnson Matthey
Nov	British Telecom plc privatised
Dec	Agreement to hand over Hong Kong to China in 1997

1985

Jan	FT Index reaches 1,000 for the first time
Mar	End of year-long miners' strike
Mar	Dollar exchange rate bottoms out at $1.05/£
Dec	NatWest, Barclays and Lloyds Banks announce 'free banking'

1986

Jan	Michael Heseltine resigns from Government over Westland Helicopters affair
Feb	Single European Act signed
Mar	Budget cuts basic rate of income tax to 29% and introduces Personal Equity Plans (PEPs)
Mar	Greater London Council abolished
Apr	Chernobyl nuclear reactor disaster
Oct	Bus services deregulated
Oct	*The Independent* newspaper founded
Nov	'Big Bang' deregulates dealing in the City
Dec	British Gas privatisation

1987

Jan	Prosecutions for insider dealing in Guinness case
Jan	British Airways privatisation
Mar	Budget reduces basic rate of tax to 27%
Oct	Hurricane strikes Britain
Oct	'Black Monday': collapse of stock market

1988

Mar	Budget reduces basic rate of tax to 25%; top rate to 40%
Mar	BL sold to BMW
Jun	Barlow-Clowes collapses
Jul	*Piper Alpha* oil rig disaster
Sep	Worst ever UK trade deficit announced
Nov	George Bush elected US President
Dec	Salmonella outbreak in Britain

1989

Mar	*Exon Valdez* oil spillage disaster in Alaska
Apr	Chinese authorities quell dissidents in Tiananmen Square
Jul	*Blue Arrow* report from DTI
Oct	Nigel Lawson resigns as Chancellor
Nov	Ford takes over Jaguar
Nov	Fall of Berlin Wall

1990

Mar	Budget introduces tax-exempt savings accounts (TESSAs)
Apr	BSE ('mad cow disease') identified
Apr	New Education Act brings in student loans
Apr	Community Charge ('poll tax') introduced
Aug	Kuwait invaded by Iraq
Oct	Official reunification of Germany
Oct	UK enters Exchange Rate Mechanism
Nov	John Major replaces Mrs Thatcher as PM
Nov	Privatisation of electricity boards

1991

Jan	NHS internal market created
Jan	Gulf War begins
Jan	Central Statistical Office (forerunner of ONS) celebrates its 50th anniversary
Feb	Gulf War ends
Mar	Air Europe collapses
Mar	Budget restricts mortgage interest relief to basic rate; Corporation Tax reduced and VAT increased
Jul	BCCI closed by Bank of England
Sep	Rioting in Cardiff, Oxford and Birmingham
Nov	Robert Maxwell drowned
Nov	Maastricht agreement signed with UK opt-outs
Dec	Mikhial Gorbachev replaced by Boris Yeltsin as President of the Soviet Union

1992

Jan	Russia agrees to join the IMF
Jan	Bill McLennan appointed as Director of the Central Statistical Office and Head of the Government Statistical Service
Feb	'Delors Package' raises EC's spending limits to 1.37% of GDP to aid poorer member states
Mar	Budget reduces lower rate of income tax to 20% and announces that from next year Budgets will be in the autumn
Mar	Midland Bank agrees merger with Hong Kong and Shanghai Bank
Apr	Conservatives win General Election
May	Swiss vote in a referendum to join the IMF and IBRD
May	Reform of EC Common Agricultural Policy agreed, switching from farm price support to income support
Jul	Prince signs $108m recording deal, making him the highest paid artist in history
Sep	'Black Wednesday': UK leaves Exchange Rate Mechanism
Oct	North America Free Trade Agreement (NAFTA) signed
Nov	Fire at Windsor Castle
Dec	Plans for National Lottery announced

1993

Jan	Council Tax announced as replacement for Community Charge
Jan	University status given to polytechnics
Feb	Announcement that the Queen is to pay tax
Mar	Budget imposes VAT on domestic fuel
Nov	Parliament votes to relax Sunday trading rules
Nov	First autumn Budget cuts public expenditure and increases taxes
Dec	Uruguay Round of tariff reductions approved

1994

Jan	European Economic Area formed linking EU and EFTA
Apr	Eurotunnel opens
Aug	First IRA ceasefire begins
Oct	Brent Walker leisure group collapses
Nov	First draw of National Lottery
Nov	Broadly neutral Budget
Dec	Coal industry privatised

1995

Jan	EU expanded to include Sweden, Finland and Austria
Jan	World Trade Organisation succeeds GATT
Feb	Barings bank collapses
Aug	Hottest ever
Sep	Net Book Agreement suspended

1996

Jan	Gilt 'repo' market established
Apr	Office for National Statistics created
Mar	Rebates worth $1 billion paid to electricity consumers after break-up of National Grid
May	Railtrack privatised, reducing public service borrowing requirement by £1.1 billion
Sep	Privatisation of National Power and PowerGen reduces PSBR by a further £1.0 billion
Aug	CREST clearing system initiated

1997

Apr	Alliance & Leicester Building Society converts to a bank
May	Labour Party wins General Election
May	Chancellor announces operational independence for the Bank of England, decisions on interest rates to be taken by a new Monetary Policy Committee
Jun	Halifax Building Society converts to a bank
Jun	Norwich Union floated on the stock market
Jun	Economic and Fiscal Strategy Report announces new format for public finances, distinguishing between current and capital spending
Jul	Gordon Brown presents his first Budget, setting inflation target of $2^1/_2\%$
Jul	Woolwich Building Society converts to a bank
Jul	Bristol & West Building Society joins Bank of Ireland
Aug	Northern Rock Building Society converts to a bank
Aug	Stock market falls in Far East, Hang Seng Index ending 20 per cent lower than a year earlier
Aug	Economic and financial crisis in Russia

Glossary

Glossary

Above the line

Transactions in the production, current and capital accounts which are above the *Net lending* (+) / *Net borrowing (financial surplus or deficit)* line in the presentation used in the economic accounts. The financial transactions account is *below the line* in this presentation.

Acceptances

See *Bills and acceptances*.

Accruals basis

A method of recording transactions to relate them to the period when the exchange of ownership of the goods, services or financial asset applies. (See also *cash basis*). For example, value added tax accrues when the expenditure to which it relates takes place, but Customs and Excise receive the cash some time later. The difference between accruals and cash results in the creation of an asset and liability in the financial accounts, shown as *amounts receivable or payable* (F7).

Actual final consumption

The value of goods consumed by a sector but not necessarily purchased by that sector. See also *Final consumption expenditure, Intermediate consumption*.

Advance and progress payments

Payments made for goods in advance of completion and delivery of the goods. Also referred to as *stage payments*.

Affiliates

Branches, subsidiaries or associate companies.

American Depositary Receipts (ADR)

An ADR is a negotiable certificate that represents ownership of the securities of a non-US resident company. Although the securities underlying ADRs can be debt or money market instruments the large majority are equities. An ADR allows a non US resident company to introduce its equity into the US market in a form more readily acceptable to US investors without needing to disclose all the information normally required by the US Securities and Exchange Commission. A US depository bank will purchase the underlying foreign security and then issue receipts in dollars for those securities to the US investor. The receipts are registered. Investors can exchange their ADRs for the underlying security at any time. (See also *Depository receipts*.)

Assets

Entities over which ownership rights are enforced by institutional units, individually or collectively; and from which economic benefits may be derived by their owners by holding them over a period of time.

Associated companies

Companies in which the investing company has a substantial equity interest (which usually means that it holds between 20 per cent and 50 per cent of the equity share capital) and is in a position to exercise significant influence over the company.

Assurance

An equivalent term to insurance, commonly used in the life insurance business.

Auxiliary financial activities

Activities related to financial intermediation but which are not financial intermediation themselves. See *Financial auxiliaries*.

Balancing item

A balancing item is an accounting construct obtained by subtracting the total value of the entries on one side of an account from the total value for the other side. In the sector accounts in the former system of UK economic accounts the term referred to the difference between the *Financial Surplus or Deficit* for a sector and the sum of the financial transactions for that sector, currently designated the *statistical discrepancy*.

Balance of payments

A summary of the financial transactions between residents of a country and residents overseas in a given time period.

Balance of trade

The balance on trade in goods. The balance of trade is a summary of the imports and exports of goods across an economic boundary in a given period.

Balance sheet

A statement, drawn up at a particular point in time, of the value of *assets* owned and of the financial claims (*liabilities*) against the owner of these assets.

Banks (UK)

Strictly, all financial institutions located in the United Kingdom and recognised by the Bank of England as banks for statistical purposes up to late 1981 or as UK banks from then onwards. This category includes the UK offices of institutions authorised under the Banking Act (1987), the Bank of England, the National Girobank and the TSB Group plc. It may include branches of foreign banks where these are recognised as banks by the Bank of England, but not offices abroad of these or of any British-owned banks. An updated list of banks appears in each February's issue of the *Bank of England Quarterly Bulletin*. Institutions in the Channel Islands and the Isle of Man which have opted to adhere to the monetary control arrangements introduced in August 1981 were formerly included in the sector but are not considered to be residents of the United Kingdom under the ESA.

Bank of England

This comprises S.121, the central bank sub-sector of the *financial corporations* sector.

Bank of England - Issue Department

This part of the Bank of England deals with the issue of bank notes on behalf of central government and was formerly classified to central government though it is now part of the central bank sector. Its activities include, *inter alia*, market purchases of commercial bills from UK banks.

Basic prices

These prices are the preferred method of valuing output. They reflect the amount received by the producer for a unit of goods or services *minus* any taxes payable *plus* any subsidy receivable on that unit as a consequence of production or sale (i.e. the cost of production including subsidies). As a result the only taxes included in the basic price are taxes on the production process – such as business rates and any vehicle excise duty paid by businesses – which are not specifically levied on the production of a unit of output. Basic prices exclude any transport charges invoiced separately by the producer.

Below the line

The financial transactions account which shows the financing of *Net lending(+) / Net borrowing* (-) (formerly *financial surplus or deficit*).

Bills and acceptances

A bill is an unconditional undertaking by the drawer to pay the drawee a sum of money at a given date, usually three months ahead. A bill of exchange must be 'accepted' by the drawee before it becomes negotiable. This function is usually performed by an accepting house but bills can also be accepted by a bank or a trader. When a bill has been accepted the drawee can sell the bill on the money market, at a small discount, before it matures. Acceptances are recorded as financial instruments (F.34) issued by the drawee (usually a monetary institution) to the market, and as a loan (F.4) by the drawee to the drawer.

Blue Book

The informal name given to the Office for National Statistics' annual publication of economic accounts (the National Accounts).

Bond

A financial instrument that usually pays interest to the holder, issued by governments as well as companies and other institutions, e.g. local authorities. Most bonds have a fixed date on which the borrower will repay the holder. Bonds are attractive to investors since they can be bought and sold easily in a *secondary market*. Special forms of bonds include *deep discount bonds, equity warrant bonds, Eurobonds*, and *zero coupon bonds*.

Bonus issues

An issue of new shares to shareholders in proportion to the shares they already hold. Since the shares are free the issue raises no new capital for the company, and does not result in any transactions in the economic accounts. Also known as bonus shares.

Branch

An unincorporated office which is wholly owned by the parent company, and which is a permanent establishment as defined for UK income tax and double taxation relief purposes. A branch of a foreign company which is active in the United Kingdom may be a *notional resident* of the United Kingdom and be classified as a *quasi-corporation*.

British government securities

See *Gilts*.

Building society

Those institutions as defined in the Building Society Acts (1962 and 1986). They offer housing finance largely to the household sector and fund this largely by taking short term deposits from the household sector. They are part of the *monetary financial institutions* sub-sector.

Capital

Capital assets are those which contribute to the productive process so as to produce an economic return. In other contexts the word can be taken to include *tangible assets* (e.g. buildings, plant and machinery), *intangible assets* and financial capital. In the economic accounts, however, a more precise terminology is used, described in Chapter 6. See also *fixed assets, inventories*.

Capital formation

Acquisition *less* disposals of *fixed assets*, improvement of land, change in *inventories* and acquisition *less* disposals of *valuables*.

Capital transfers

Transfers which are related to the acquisition or disposal of assets by the recipient or payer. They may be in cash or kind, and may be imputed to reflect the assumption or forgiveness of debt.

Cash basis

The recording of transactions when cash or other assets are actually transferred, rather than on an *accruals* basis.

Central monetary institutions (CMIs)

Institutions (usually central banks) which control the centralised monetary reserves and the supply of currency in accordance with government policies, and which act as their governments' bankers and agents. In the UK this is equivalent to the Bank of England. In many other countries maintenance of the exchange rate is undertaken in this sector. In the United Kingdom this function is undertaken by central government (part of the Treasury) by use of the *Exchange Equalisation Account*.

Certificate of deposit

A short term interest-paying instrument issued by deposit-taking institutions in return for money deposited for a fixed period. Interest is earned at a given rate. The instrument can be used as security for a loan if the depositor requires money before the repayment date.

Chained index

An index number series which measures changes in consecutive years using weights updated periodically. These periodic changes are linked or chained together to produce comparisons over longer periods.

c.i.f.

The basis of valuation of imports for Customs purposes, it includes the cost of insurance premiums and freight services. These need to be deducted to obtain the f.o.b. valuation consistent with the valuation of exports which is used in the economic accounts.

Classification tables

See the *Financial Statistics Explanatory Handbook*.

COICOP (Classification of Individual Consumption by Purpose)

An international classification which groups consumption according to its function or purpose. Thus the heading *clothing*, for example, includes expenditure on garments, clothing materials, laundry and repairs.

Commercial paper

This is an unsecured *promissory note* for a specific amount and maturing on a specific date. The commercial paper market allows companies to issue short term debt direct to financial institutions who then market this paper to investors or use it for their own investment purposes.

Commonwealth Development Corporation

A public non-financial corporation which finances development projects overseas.

Consolidated Fund

An account of central government into which most government revenue (excluding borrowing and certain payments to government departments) is paid, and from which most government expenditure (excluding loans and National Insurance benefits) is paid.

Consumption

See *Final consumption, Intermediate consumption*.

Consumption of fixed capital

The amount of capital resources used up in the process of production in any period. It is not an identifiable set of transactions but an imputed transaction which can only be measured by a system of conventions.

Corporations

All bodies recognised as independent legal entities which are producers of market *output* and whose principal activity is the production of goods and services.

Counterpart

In a double-entry system of accounting each transaction gives rise to two corresponding entries. These entries are the counterparts to each other. Thus the counterpart of a payment by one sector is the receipt by another.

Coupon

The interest payable on the face value of a bond.

Central product classification

An internationally agreed detailed classification of products, applied to industries' purchases and sales. It should be distinguished from the industrial classification (see *NACE, SIC*) though at a broad level they are similar.

Currency swap

In a currency swap two parties exchange specified amounts of principal denominated in two different currencies. Principal amounts will be re-exchanged on an agreed timetable at exchange rates specified in the contract and not at the prevailing market rate. The parties will also make periodic exchanges of interest, again at exchange rates specified in the contract. Parties enter into a currency swap because it allows them to exploit their comparative advantage in different markets and hence both can reduce their borrowing costs.

Custody holdings

Securities held in trust by a third party.

Debenture

A long-term bond issued by a UK or foreign company and secured on *fixed assets*. A debenture entitles the holder to a fixed interest payment or a series of such payments.

Deep discount bond

See *zero-coupon bond*.

Depositary receipts

A vehicle by which regulatory or institutional barriers can be circumvented in order that an issuer or investor can access a particular market. Usually this circumvention is achieved by transforming the characteristics of a security into a form more acceptable or accessible to the end investor. Examples are *American depositary receipts*, bearer depositary receipts and global depositary receipts.

Depreciation

See *Consumption of fixed capital*.

Derivatives (F.34)

Financial instruments whose value is linked to changes in the value of another financial instrument, an indicator or a commodity. In contrast to the holder of a primary financial instrument (e.g. a government bond or a bank deposit), who has an unqualified right to receive cash (or some other economic benefit) in the future, the holder of a derivative has only a qualified right to receive such a benefit. Examples of derivatives are options and *swaps*.

DIM (Dividend and Interest Matrix)

The ONS Dividend and Interest Matrix uses holdings of financial instruments by each sector and certain interest flows to estimate flows of interest and dividends between the sectors of the economy.

Direct investment

Net investment by UK/overseas companies in their overseas/UK branches, subsidiaries or associated companies. A direct investment in a company means that the investor has a significant influence on the operations of the company. Investment includes not only acquisition of fixed assets, stock building and stock appreciation but also all other financial transactions such as additions to, or payments of, working capital, other loans and trade credit and acquisitions of securities. Estimates of investment exclude depreciation.

Discount market

That part of the market dealing with short-term borrowing. It is called the discount market because the interest on loans is expressed as a percentage reduction (discount) on the amount paid to the borrower. For example, for a loan of £100 face value when the discount rate is 5% the borrower will receive £95 but will repay £100 at the end of the term.

Dividend

A payment made to company shareholders from current or previously retained profits. See *DIM*.

ECGD

See *Export Credit Guarantee Department*.

Economically significant prices

These are prices whose level significantly affects the supply of the good or service concerned. Market output consists mainly of goods and services sold at 'economically significant' prices while non-market output comprises those provided free or at prices that are not economically significant.

ECU bill

A bill denominated in *European Currency Units*.

Enterprise

An *institutional unit* producing *market output*. Enterprises are found mainly in the *non-financial* and *financial corporations* sectors but exist in all sectors. Each enterprise consists of one or more *kind-of-activity units*.

Entrepreneur

One who organises business at his or her own risk; the owner of an *enterprise*. Entrepreneurial income is the income of that enterprise, *operating surplus* or *mixed income* as appropriate, *plus* receipts *less* payments of *property income*.

Equity

Equity is ownership or potential ownership of a company. An entity's equity in a company will be evidenced by ordinary shares. They differ from other financial instruments in that they confer ownership of something more than a financial claim. Shareholders are owners of the company whereas bond holders are merely outside creditors.

Equity warrant bond

Equity warrant bonds are debt securities which incorporate warrants which give the holder the option to purchase equity in the issuer, its parent company or another company during a pre-determined period or on one particular date. The warrants are detachable and may be traded separately from the debt security. The exercise of the equity warrant will normally increase the total capital funds of the issuer because the debt is not replaced by equity but remains outstanding until the date of its redemption. The warrant on the bond has a fixed strike price. The issue of equity warrant bonds reduces the funding costs for borrowers because the investor will generally accept a lower yield in anticipation of the future profit to be gained from exercising the warrant.

ESA

European System of National and Regional Accounts. An integrated system of economic accounts which is the European version of the System of National Accounts (SNA).

Eurobond

These securities are sold internationally by an international syndicate of banks and securities houses to international investors outside the domestic market of the country of issue. This feature gives borrowers access to a greater number and diversity of investors than would be possible in individual domestic markets, hence increasing liquidity and reducing costs. *Eurobonds* can be for any specified maturity but are usually medium- to long-term instruments (3–20 years). They are issued in bearer form and interest is paid without deduction of tax. *Eurobonds* are generally listed in either Luxembourg or London though this listing is not a requirement.

Euroclear

One of the major clearing systems for *Eurobonds*.

Eurocurrency

Currency traded in the Euromarket (e.g. Eurodollars, Eurosterling).

Euromarket

Euromarket is the name given to the international capital raising market. 'Euro' historically referred to the eurodollar market, which was defined as US dollars held in Europe. The key feature of the market is that securities are sold internationally rather than in the country of residence of the borrower (i.e. the domestic market). The international placement of securities means that national regulations regarding issuance may be avoided, reducing the costs of an issue for the borrower. The sale of securities internationally gives borrowers access to a greater number and diversity of investors than would be possible in the domestic market, hence enhancing liquidity and lowering cost.

European Currency Unit (ECU)

The ECU was officially introduced in 1979 in connection with the start of the European Monetary System (EMS). In the EMS the ECU serves as the basis for determining exchange rate parities and as a reserve asset and means of settlement. It is a composite currency which contains specified amounts of the currencies of the fifteen member states of the European Community weighted according to their economic importance, and is used in short-term finance.

European Investment Bank

This was set up to assist economic development within the European Union. Its members are the member states of the EU.

European Monetary Cooperation Fund

Central banks of member states of the European Monetary System deposit 20 per cent of their gold and foreign exchange reserves on a short-term basis with the European Monetary Cooperation Fund in exchange for ECUs. The Fund is the clearing house for central banks in the EMS.

European Monetary System

This was established in March 1979. Its most important element is the Exchange Rate Mechanism (ERM) whereby the exchange rates between the currencies of the participating member states (all EU countries except Greece and Portugal) are kept within set ranges. The UK joined the ERM on 8th October 1990. On 16th September 1992 sterling left the ERM and the EMS was suspended.

Exchange Cover Scheme (ECS)

A scheme first introduced in 1969 whereby UK public bodies raise foreign currency from overseas residents, either directly or through UK banks, and surrender it to the *Exchange Equalisation Account* in exchange for sterling for use to finance expenditure in the United Kingdom. HM Treasury sells the borrower foreign currency to service and repay the loan at the exchange rate that applied when the loan was taken out.

Exchange Equalisation Account (EEA)

An account of central government held by the Bank of England in which transactions in the official reserves are recorded. It is the means by which the government, through the Bank of England, influences exchange rates.

Export credit

Credit extended overseas by UK institutions primarily in connection with UK exports but also including some credit in respect of third-country trade.

Export Credits Guarantee Department (ECGD)

A government department whose main function is to provide insurance cover for export credit transactions.

Factor cost

In the former system of national accounts this was the basis of valuation which excluded the effects of taxes on expenditure and subsidies.

Final consumption expenditure

The expenditure on goods and services that are used for the direct satisfaction of individual needs or the collective needs of members of the community as distinct from their purchase for use in the productive process. It may be contrasted with *Actual final consumption*, which is the value of goods consumed but not necessarily purchased by that sector. See also *Intermediate consumption*.

Finance houses

Financial corporations that specialize in the financing of hire purchase arrangements.

Financial auxiliaries

Auxiliary financial activities are ones closely related to financial intermediation but which are not financial intermediation themselves, such as the repackaging of funds. Financial auxiliaries include such activities as insurance broking and fund management.

Financial corporations

All bodies recognised as independent legal entities whose principal activity is financial intermediation and/or the production of auxiliary financial services. However, the United Kingdom currently treats financial auxiliaries as non-financial corporations.

Financial intermediation

Financial intermediation is the activity by which an *institutional unit* acquires financial assets and incurs liabilities on its own account by engaging in financial transactions on the market. The assets and liabilities of financial intermediaries have different characteristics so that the funds are transformed or repackaged with respect to maturity, scale, risk, etc, in the financial intermediation process.

Financial leasing

A form of leasing in which the lessee contracts to assume the rights and responsibilities of ownership of leased goods from the lessor (the legal owner) for the whole (or virtually the whole) of the economic life of the asset. In the economic accounts this is recorded as the sale of the assets to the lessee, financed by an imputed loan (F.42). The leasing payments are split into interest payments and repayments of principal.

Financial Services Adjustment

Now renamed *FISIM* (see below) this is a feature temporarily carried over from the previous system. The output of many financial intermediation services is paid for not by charges, but by an interest rate differential. The value added of these industries is shown including their interest receipts *less* payments, in effect imputing charges for their services. However, GDP in total takes no account of this, and an adjustment is necessary to reconcile the two. For the treatment in the new SNA (to be implemented fully in the EU at a later date) see *FISIM*. Since most output of these industries is intermediate consumption of other industries the difference between the two methods in their effect on total GDP is relatively small.

Financial surplus or deficit (FSD)

The former term for *Net lending(+)/Net borrowing (-)*, the balance of all current and capital account transactions for an institutional sector or the economy as a whole.

FISIM

Financial Intermediation Services Indirectly Measured. The output of many financial intermediation services is paid for not by charges but by an interest rate differential. *FISIM* imputes charges for these services and corresponding offsets in property income, as described in the annex to Chapter 20. *FISIM*, an innovation of the 1993 SNA, has not yet been fully implemented in the UK economic accounts; the earnings are not yet allocated to the users of the services.

Fixed assets

Produced assets that are themselves used repeatedly or continuously in the production process for more than one year. They comprise buildings and other structures, vehicles and other plant and machinery and also plants and livestock which are used repeatedly or continuously in production, e.g. fruit trees or dairy cattle. They also include intangible assets such as computer software and artistic originals.

Flows

Economic flows reflect the creation, transformation, exchange, transfer or extinction of economic value. They involve changes in the volume, composition or value of an *institutional unit's* assets and liabilities. They are recorded in the *production, distribution and use of income* and *accumulation accounts*.

F.o.b.

Free on board, the valuation of imports and exports used in the economic accounts, including all costs invoiced by the exporter up to the point of loading on to the ship or aircraft but excluding the cost of insurance and freight from the country of consignment.

Futures

Instruments which give the holder the right to purchase a commodity or a financial asset at a future date.

GFCF

See *Gross fixed capital formation*.

Gilts

Bonds issued or guaranteed by the UK government. Also known as gilt-edged securities or British government securities.

Gold

The SNA and the IMF (in the 5th Edition of its Balance of Payments Manual) recognise three types of gold:

* monetary gold, treated as a financial asset;
* gold held as a store of value, to be included in valuables;
* gold as an industrial material, to be included in intermediate consumption or inventories.

This is a significant change from previous UK practice and presents problems such that the United Kingdom has received from the European Union a derogation from applying this fully until the year 2005.

The present treatment is as follows:

- In the accounts a distinction is drawn between gold held as a financial asset (financial gold) and gold held like any other commodity (commodity gold). Commodity gold in the form of finished manufactures together with net domestic and overseas transactions in gold moving into or out of finished manufactured form (i.e. for jewellery, dentistry, electronic goods, medals and proof - but not bullion - coins) is recorded in exports and imports of goods.

- All other transactions in gold (i.e. those involving semi-manufactures such as rods, wire, etc, or bullion, bullion coins or banking-type assets and liabilities denominated in gold, including official reserve assets) are treated as financial gold transactions and included in the financial account of the Balance of Payments.

The United Kingdom has adopted different treatment to avoid distortion of its visible trade account by the substantial transactions of the London bullion market.

Grants

Voluntary transfer payments. They may be current or capital in nature. Grants from government or the European Union to producers are *subsidies*.

Gross

Key economic series can be shown as gross (i.e. *before* deduction of the consumption of fixed capita) or net (i.e. *after* deduction). Gross has this meaning throughout this book unless otherwise stated.

Gross domestic product (GDP)

The total value of output in the economic territory. It is the balancing item on the production account for the whole economy. Domestic product can be measured *gross* or *net*. It is presented in the new accounts at market (or *purchasers'*) prices. A further distinction is that it can be at current or constant prices.

Gross fixed capital formation (GFCF)

Acquisition *less* disposals of *fixed assets* and the improvement of land.

Gross national disposable income

The income available to the residents arising from GDP, and receipts from, *less* payments to, the Rest of the World of employment income, property income and current transfers.

Gross value added (GVA)

The value generated by any unit engaged in production, and the contributions of individual sectors or industries to gross domestic product. It is measured at basic prices, excluding taxes *less* subsidies on products.

Hidden economy

Certain activities may be productive and also legal but are concealed from the authorities for various reasons – for example to evade taxes or regulation. In principle these, as well as economic production that is illegal, are to be included in the accounts but they are by their nature difficult to measure.

Holding gains or losses

Profit or loss obtained by virtue of the changing price of assets being held. Holding gains or losses may arise from either physical and financial assets.

Households (S.14)

Individuals or small groups of individuals as consumers and in some cases as entrepreneurs producing goods and market services (where such activities cannot be hived off and treated as those of a *quasi corporation*). See Chapter 10 for a fuller definition.

Imputation

The process of inventing a transaction where, although no money has changed hands, there has been a flow of goods or services. It is confined to a very small number of cases where a reasonably satisfactory basis for the assumed valuation is available.

Index-linked gilts

Gilts whose coupon and redemption value are linked to movements in the retail prices index.

Institutional unit

Institutional units are the individual bodies whose data is amalgamated to form the *sectors* of the economy. A body is regarded as an institutional unit if it has decision-making autonomy in respect of its principal function and either keeps a complete set of accounts or is in a position to compile, if required, a complete set of accounts which would be meaningful from both an economic and a legal viewpoint.

Institutional sector

See *Sector*.

Input-output

A detailed analytical framework based on supply and use tables. These are matrices showing the composition of output of individual industries by types of product and how the domestic and imported supply of goods is allocated between various intermediate and final uses, including exports. See Chapter 13 for a detailed account.

Intangible assets

Intangible fixed assets include mineral exploration, computer software and entertainment, literary or artistic originals. Expenditure on them is part of *gross fixed capital formation*. They exclude non-produced intangible assets such as patented entities, leases, transferable contracts and purchased goodwill, expenditure on which would be *intermediate consumption*.

Inter-company accounts

Accounts recording transactions between parent and subsidiary or associated companies and balances owed by one to the other.

Interest rate swaps

In an interest rate swap two parties swap interest payments/receipts on a specified nominal sum denominated in the same currency, e.g. one party exchanges payments of fixed rates of interest for floating rates of interest. Obligations under interest rate swaps are often met in the form of a net payment from one party to the other. There is no exchange of principal. Interest rate swaps developed so that parties to the swap could exploit their comparative advantage in accessing different markets. Banks, for example, are often able to raise fixed rate funds more cheaply than corporations although they prefer floating rate liabilities in order to match their floating rate assets. Corporations may be able to raise floating rate funds almost as cheaply as banks but they may prefer to pay fixed rates.

Intermediate consumption

The use of consumption goods and services in the production process. It may be contrasted with *final consumption* and *capital formation*.

International Monetary Fund (IMF)

A fund set up as a result of the Bretton Woods Conference in 1944 which began operations in 1947. It currently has about 180 member countries including most of the major countries of the world. The fund was set up to supervise the fixed exchange rate system agreed at Bretton Woods and to make available to its members a pool of foreign exchange resources to assist them when they have balance of payments difficulties. It is funded by member countries' subscriptions according to agreed quotas.

Inventories

Inventories (known as stocks in the former system) consist of finished goods (held by the producer prior to sale, further processing or other use) and products (materials and fuel) acquired from other producers to be used for *intermediate consumption* or resold without further processing.

Investment trust

An institution that invests its capital in a wide range of other companies' shares. Investment trusts issue shares which are listed on the London Stock Exchange and use this capital to invest in the shares of other companies. See also *Unit trusts*.

Kind-of-activity unit (KAU)

An *enterprise*, or part of an enterprise, which engages in only one kind of non-ancillary productive activity, or in which the principal productive activity accounts for most of the *value added*. Each *enterprise* consists of one or more kind-of-activity units.

Liability

A claim on an institutional unit by another body which gives rise to a payment or other transaction transferring assets to the other body. Conditional liabilities, i.e. where the transfer of assets only takes place under certain defined circumstances, are known as contingent liabilities.

Liquidity

The ease with which a financial instrument can be exchanged for goods and services. Cash is very liquid whereas a life assurance policy is less so.

Lloyd's of London

The international insurance and reinsurance market in London.

Marketable securities

Securities which can be sold on the open market.

Market output

Output of goods and services sold at *economically significant prices*.

Merchant banks

These are *monetary financial institutions* whose main business is primarily concerned with corporate finance and acquisitions.

Merger

Fusion of two or more companies into one. In a merger both the parties wish to join and do so on equal terms.

Mixed income

The balancing item on the generation of income account for unincorporated businesses owned by households. The owner or members of the same household often provide unpaid labour inputs to the business. The surplus is therefore a mixture of remuneration for such labour and return to the owner as *entrepreneur*.

Money market

The market in which short-term loans are made and short-term securities traded. 'Short term' usually applies to periods under one year but can be longer in some instances.

NACE

The industrial classification used in the European Union. Revision 1 is the 'Statistical classification of economic activities in the European Community in accordance with Council Regulation No. 3037/90 of 9th October 1990'.

National accounts

The economic accounts of the nation.

National income

See *Gross national disposable income* and *Real national disposable income*.

National Loans Fund

An account of HM Government set up under the National Loans Fund Act (1968) which handles all government borrowing and most domestic lending transactions.

Net

After deduction of the consumption of fixed capital. Also used in the context of financial accounts and balance sheets to denote, for example, assets *less* liabilities

Non-market output

Output of own account production or goods and services provided free or at prices that are not economically significant. Non-market output is produced mainly by the general government and NPISH sectors.

NPISH

Non-profit institutions serving households (S.15).

Operating surplus

The balance on the generation of income account. Households also have a mixed income balance. It may be seen as the surplus arising from the production of goods and services before taking into account flows of *property income*.

Operating leasing

The conventional form of leasing, in which the lessee makes use of the leased asset for a period in return for a rental while the asset remains on the balance sheet of the lessor. The leasing payments are part of the *output* of the lessor, and the *intermediate consumption* of the lessee. See also *Financial leasing*.

Ordinary share

The most common type of share in the ownership of a corporation. Holders of ordinary shares receive dividends. See also *Equity*.

Output for own final use

Production of output for *final consumption* or *gross fixed capital formation* by the producer. Also known as *own-account production*.

Own-account production

Production of output for *final consumption* or *gross fixed capital formation* by the producer. Also known as *output for own final use*.

Par value

A security's face or nominal value. Securities can be issued at a premium or discount to par.

Parent company

A company owning a majority share holding in a subsidiary company. In a Balance of Payments context this means a company with direct investments in other countries.

Pension funds

The institutions that administer pension schemes. Pension schemes are significant investors in securities. Self-administered funds are classified in the financial accounts as pension funds. Those managed by insurance companies are treated as long-term business of insurance companies. They are part of S.125, the *Insurance corporations and pension funds* sub-sector.

Perpetual Inventory Model (or Method) (PIM)

A method for estimating the level of assets held at a particular point of time by accumulating the acquisitions of such assets over a period and subtracting the disposals of assets over that period. Adjustments are made for price changes over the period. The PIM is used in the UK accounts to estimate the stock of fixed capital, and hence the value of the consumption of fixed capital, as described in Chapter 16.

Portfolio

A list of the securities owned by a single investor. In the Balance of Payments statistics, portfolio investment is investment in securities that does not qualify as *direct investment*.

Preference share

This type of share guarantees its holder a prior claim on dividends. The dividend paid to preference share holders is normally more than that paid to holders of ordinary shares. Preference shares may give the holder a right to a share in the ownership of the company (participating preference shares). However in the UK they usually do not, and are therefore classified as *bonds* (F.3).

Prices

See *economically significant prices, basic prices, producers' prices*.

Principal

The lump sum that is lent under a loan or a bond.

Private sector

Private non-financial corporations, financial corporations other than the Bank of England (and Girobank when it was publicly owned), households and the NPISH sector.

Promissory note

A security which entitles the bearer to receive cash. These may be issued by companies or other institutions. (See *commercial paper*).

Property income

Incomes that accrue from lending or renting financial or tangible non-produced assets, including land, to other units. See also *Tangible assets*.

Public corporations

These are public trading bodies which have a substantial degree of financial independence from the public authority which created them. A public corporation is publicly controlled to the extent that the public authority, i.e. central or local government, usually appoints the whole or a majority of the board of management. Such bodies comprise much the greater part of sub-sector S.11001, public non-financial corporations.

Public sector

Comprises general government *plus* public non-financial corporations *plus* the Bank of England. The concept is not part of the new SNA which includes public corporations with private corporations in the two corporate sectors.

Purchasers' prices

These are the prices paid by purchasers. They include transport costs, trade margins and taxes (unless the taxes are deductible by the purchasers from their own tax liabilities).

Quasi-corporations

Unincorporated *enterprises* that function as if they were corporations. For the purposes of allocation to sectors and sub-sectors they are treated as if they were corporations, i.e. separate units from those to which they legally belong. Three main types of quasi-corporation are recognised in the accounts: unincorporated enterprises owned by government which are engaged in market production, unincorporated enterprises (including partnerships) owned by households and unincorporated enterprises owned by foreign residents. The last group consists of permanent branches or offices of foreign enterprises and production units of foreign enterprises which engage in significant amounts of production in the territory over long or indefinite periods of time.

Real national disposable income (RNDI)

Gross national disposable income adjusted for changes in prices and in the terms of trade.

Refinanced export credit

Identified long-term credit extended for UK exports initially by banks, and refinanced with the *Export Credits Guarantee Department*.

Related companies

Branches, subsidiaries, associates or parents. See Chapter 19.

Related import or export credit

Trade credit between related companies, included in direct investment.

Rental

The amount payable by the user of a *fixed asset* to its owner for the right to use that asset in production for a specified period of time. It is included in the *output* of the owner and the *intermediate consumption* of the user.

Rents (D.45)

The property income derived from land and sub-soil assets. It should be distinguished in the current system from *rental* income derived from buildings and other fixed assets, which is included in *output* (P.1).

Repurchase agreement (Repo)

A deal in which an institution lends or 'sells' another institution a security and agrees to buy it back at a future date. Legal ownership does not change under a 'repo' agreement. It was previously treated as a change of ownership in the UK financial account but under the SNA is treated as a collateralised deposit (F.22).

Reserve assets

The UK official holdings of gold, convertible currencies, Special Drawing Rights, changes in the UK reserve position with the IMF and European currency. They include units acquired from swaps with the *European Monetary Co-operation Fund (EMCF)*.

Residents

These comprise general government, individuals, private non-profit-making bodies serving households and enterprises within the territory of a given economy. See Chapters 11 and 24.

Residual error

The term used in the former accounts for the difference between the measures of *gross domestic product* from the expenditure and income approaches.

Rest of the World

This *sector* records the counterpart of transactions of the whole economy with non-residents.

Saving

The balance on the *use of income account*. It is that part of disposable income which is not spent on *final consumption*, and may be positive or negative.

Sector

In the economic accounts the economy is split into different institutional sectors, i.e. groupings of units according broadly to their role in the economy. The main sectors are *non-financial corporations, financial corporations, general government, households and non-profit institutions serving households (NPISH)*. The *Rest of the World* is also treated as a sector for many purposes within the accounts. See Chapter 10 for a full account.

Secondary market

A market in which holders of financial instruments can re-sell all or part of their holding. The larger and more effective the secondary market for any particular financial instrument the more liquid that instrument is to the holder.

Securities

Tradeable or potentially tradeable financial instruments.

SIC

Standard Industrial Classification. The industrial classification applied to the collection and publication of a wide range of economic statistics. The current version, SIC 92, is consistent with *NACE, Rev.1.*

SNA

System of National Accounts, the internationally-agreed standard system for macroeconomic accounts. The latest version is described in *System of National Accounts 1993* (see Bibliography).

Special Drawing Rights (SDRs)

These are reserve assets created and distributed by decision of the members of the IMF. Participants accept an obligation, when designated by the IMF to do so, to provide convertible currency to another participant in exchange for SDRs equivalent to three times their own allocation. Only countries with a sufficiently strong Balance of Payments are so designated. SDRs may also be used in certain direct payments between participants in the scheme and for payments of various kinds to the IMF.

Stage payments

See *Advance and progress payments*.

Stocks, stockbuilding

The terms used in the former system corresponding to *inventories* and changes in inventories.

Subsidiaries

Companies owned or controlled by another company. Under Section 736 of the Companies Act (1985) this means, broadly speaking, that another company either holds more than half the equity share capital or controls the composition of the board of directors. The category also includes subsidiaries of subsidiaries.

Subsidies (D.3)

Current unrequited payments made by general government or the European Union to enterprises. Those made on the basis of a quantity or value of goods or services are classified as 'subsidies on products' (D.31). Other subsidies based on levels of productive activity (e.g. numbers employed) are designated *Other subsidies on production* (D.39).

Suppliers' credit

Export credit extended overseas directly by UK firms other than to related concerns.

Swaps

Swaps are a form of financial derivative which enables each party to the arrangement to exploit its comparative advantage in accessing particular markets, thereby lowering the cost of funds. Each party deals in the market where it can best exploit its comparative advantage, before exchanging cash flows with the other party. An intermediary brings the counterparties together. See also *currency swaps* and *interest rate swaps*.

Takeover

The acquisition of one company by another. The term implies that the acquisition is made without the full agreement of the acquired company.

Tangible assets

These comprise produced fixed assets and non-produced assets. Tangible *fixed* assets, the acquisition and disposal of which are recorded in *gross fixed capital formation* (P.51), comprise buildings and other structures (including historic monuments), vehicles, other machinery and equipment and cultivated assets in the form of livestock and trees yielding repeat products (e.g. dairy cattle, orchards). Tangible *non-produced* assets are assets such as land and sub-soil resources that occur in nature over which ownership rights have been established. Similar assets to which ownership rights have *not* been established are excluded as they do not qualify as economic assets. The acquisition and disposal of non-produced assets in principle is recorded separately in the capital account (K.2). The distinction between produced and non-produced assets is not yet fully possible for the United Kingdom.

Taxes

Compulsory unrequited transfers to central or local government or the European Union. Taxation is classified in the following main groups: taxes on production and imports (D.2), current taxes on income wealth, etc (D.5) and capital taxes (D.91).

Technical reserves (of insurance companies)

These reserves consist of pre-paid premiums, reserves against outstanding claims, actuarial reserves for life insurance and reserves for with-profit insurance. They are treated in the economic accounts as the property of policy-holders (see Annex 2 to Chapter 20).

Terms of trade

Ratio of the change in export prices to the change in import prices. An increase in the terms of trade implies that the receipts from the same quantity of exports will finance an increased volume of imports. Thus measurement of *real national disposable income* needs to take account of this factor.

Transfers

Unrequited payments made by one unit to another. They may be current transfers (D.5–7) or capital transfers (D.9). The most important forms of transfer are taxes, social contributions and benefits.

Treasury bills

Short-term securities or promissory notes which are issued by government in return for funding from the money market. In the United Kingdom every week the Bank of England invites tenders for sterling Treasury bills from the financial institutions operating in the market. ECU-denominated bills are issued by tender each month. Treasury bills are an important form of short-term borrowing for the government, generally being issued for periods of 3 or 6 months.

Unit trusts

Institutions within sub-sector S.123 through which investors pool their funds to invest in a diversified portfolio of securities. Individual investors purchase units in the fund representing an ownership interest in the large pool of underlying assets, i.e. they have an equity stake. The selection of assets is made by professional fund managers. Unit trusts therefore give individual investors the opportunity to invest in a diversified and professionally-managed portfolio of securities without the need for detailed knowledge of the individual companies issuing the stocks and bonds. They differ from *investment trusts* in that the latter are companies in which investors trade shares on the Stock Exchange, whereas unit trust units are issued and bought back on demand by the managers of the trust. The prices of unit trust units thus reflect the value of the underlying pool of securities, whereas the price of shares in investment trusts are affected by the usual market forces.

United Kingdom

Broadly, in the accounts, the United Kingdom comprises Great Britain plus Northern Ireland and that part of the continental shelf deemed by International convention to belong to the UK. See Chapter 11 for a fuller description.

Valuables

Goods of considerable value that are not used primarily for production or consumption but are held as stores of value over time. They consist of precious metals, precious stones, jewellery, works of art, etc. As a new category in the accounts the estimates for them are currently fairly rudimentary, though transactions are likely to have been recorded elsewhere in the accounts.

Valuation

See *Basic prices, Purchasers' prices, Factor cost.*

Value added

The balance on the production account: output *less* intermediate consumption. Value added may be measured *net* or *gross*.

Value added tax (VAT)

A tax paid by *enterprises*. In broad terms an enterprise is liable for VAT on the total of its taxable sales but may deduct tax already paid by suppliers on its inputs (*intermediate consumption*). Thus the tax is effectively on the value added by the enterprise. Where the enterprise cannot deduct tax on its inputs the tax is referred to as non-deductible. VAT is the main UK tax on products (D.31). See Annex 2 to Chapter 14.

Zero coupon bond / Deep discount bond

A debt security issued with no coupon or a substantially lower coupon than current interest rates. The bonds are issued at a discount to their nominal value, the discount reflecting the prevailing market interest rate. In the case of a zero coupon bond investors receive at maturity the difference between the purchase price and the nominal value of the bond (so-called 'uplift'). The longer the maturity of the bond the greater the discount against par value. In the economic accounts the sale of the security is recorded at the discounted price and interest flows equal to the discount are imputed

Bibliography

Bibliography

Economic Trends
Monthly (The Stationery Office)

Articles referred to

UK Economic Accounts	quarterly
Regional economic indicators	quarterly
Regional accounts published in two parts	annually
A glimpse of the hidden economy in the national accounts, K Macafee	1980
The Use of Supply side estimates in the national accounts, R Lynch and D Caplan	December 1991
Trade credit	September 1992
Rebasing the national accounts: the reasons and the likely effects	February 1993
Improvements to economic statistics	January 1994
Rich or poor? Purchasing power parities and international comparisons	July 1994
Quarterly integrated economic accounts: the United Kingdom approach	March 1997
Quarterly alignment adjustments in the UK national accounts	November 1997

Other relevant articles

National and sector balance sheets	October 1991
Improving economic statistics	February 1992
Sectoral analysis of banking statistics: a joint Bank of England/CSO study	March 1992
The inter-departmental business register (IDBR)	April 1992
The 1989 Share Register Survey	January 1991
The 1991 Share Register Survey	October 1991
The 1992 Share Register Survey	August 1992
The 1993 Share Register Survey	October 1993

Testing for bias in initial estimates	May 1992
	February 1993
	May 1993
	February 1994
	May 1994
	July 1994
	April 1995
	July 1995
	August 1996
Investigating the domestic interbank difference	June 1992
Producer price indices - present practice, future developments, and international comparisons	July 1992
Sector allocation of dividend and interest flows- a new framework	October 1992
The new UK Standard Industrial Classification of economic activities - SIC (92)	October 1992
The production of fully reconciled UK national and sector accounts for 1988–1991	November 1992
Environmental issues and the national accounts	November 1992
Developments in balance of payments statistics: problems and some solutions	December 1992
Intrastat	January 1993
Improving macro-economic statistics	January 1993
Transition to the new SIC (92)	February 1993
Measuring the contribution of financial institutions to the gross domestic product	May 1993
CSO's success in meeting national accounts targets in 1992–93	June 1993
Integrating the builders address file with the CSO business register	July 1993
The UK sector accounts	September 1993
The definition of PSBR	September 1993
Input output balance for the United Kingdom 1990	October 1993
Handling revisions in the national accounts	October 1993
The retail sales index and its use in consumers' expenditure	November 1993

A survey of expenditure in residential and nursing homes	November 1993
Charities' contribution to gross domestic product	December 1993
Developments in sources and methods for measuring overseas trade in non-financial services	March 1994
National accounts chain weighted price indicators	June 1994
UK visible trade statistics - the intrastat system	August 1994
Quarterly national accounts in the United Kingdom: an overview of the UK approach	April 1995
Changing the Blue Book	May 1995
Quarterly GDP: process and issues	October 1995
The inter-departmental business register	November 1995
A monthly indicator of GDP	March 1996
Environmental accounts - valuing the depletion of oil and gas reserves	April 1996
Measuring real growth: index numbers and chain linking	June 1996
Producer prices for services: development of a new price index	July 1996
Time use from a national accounts perspective	July 1996
The pilot UK environmental accounts	August 1996
A framework for social accounting matrices	September 1996
The use of quarterly current price output data in the national accounts	October 1996
Overseas trade in services: development of monthly estimates	November 1996
Charities' contribution to GDP: the results of the 1996 ONS survey of charities	November 1996
Developments in UK company securities statistics	December 1996
How far should economic theory and economic policy affect the design of national accounts?	December 1996
Service sector statistics: ONS plans and the President's task force	February 1997
Methodology series for UK national accounts	April 1997
Development of final expenditure price index	September 1997
Environmental input-output tables for the UK	October 1997
A household satellite account for the UK	October 1997
Improvements to business inquiries through the new IDBR	February 1998
Measuring public sector output	February 1998
Rebasing the national accounts	June 1998
Developing a methodology for measuring illegal activity in the UK national accounts	July 1998
New format for public finances	July 1998
Final expenditure prices index	August 1998
Forthcoming changes to the national accounts	August 1998
New measures of public sector output in the national accounts	October 1998

Other publications

System of National Accounts 1993, UN, OECD, IMF, EU	1993
European System of Accounts 1995, EU	1996
United Kingdom National Accounts: Sources and Methods:	
2nd Edition, HMSO	1968
3rd edition, HMSO	1985
Input-output methodological guide	Annual*
Financial Statistics	Monthly
Financial Statistics Explanatory Handbook	Annual

SMQ papers

Regional accounts methodology **
Measuring the accuracy of the national accounts **

UK National Accounts Papers

No. 1: The measurement of output in the estimation of GDP **

No. 2: A compilers' guide to the 1993 SNA **

No. 3: Data sources for the quarterly accounts **

* Published by ONS and available from ONS Sales Office

** Available from EAS Division, ONS

Printed in the United Kingdom for The Stationery Office J59991 9/98 C12 9385 9006

© National Accounts Concepts, Sources and Methods

Index

Index

Note: References preceded by 'T' are
to tables; others to paragraph numbers